Theory of Multipliers in Spaces of Differentiable Functions

Theory of Multipliers in Spaces of Differentiable Functions

V. G. Maz'ya &
T. O. Shaposhnikova

Pitman Advanced Publishing Program
Boston . London . Melbourne

PITMAN PUBLISHING INC.
1020 Plain Street, Marshfield, Massachusetts 02050

PITMAN PUBLISHING LIMITED
128 Long Acre, London WC2E 9AN

Associated Companies
Pitman Publishing Pty Ltd, Melbourne
Pitman Publishing New Zealand Ltd, Wellington
Copp Clark Pitman, Toronto

First published 1985

AMS Subject Classifications: (primary) 46E35, 46E15, 35J
(secondary) 45E10, 35B45, 31B15

ISSN 0743-0329

Library of Congress Cataloging in Publication Data

Maz'ya, V. G.
Theory of multipliers in spaces of differentiable functions.
"Pitman advanced publishing program."
Bibliography: p.
1. Function spaces. 2. Differentiable functions.
3. Multipliers (Mathematical analysis)
I. Shaposhnikova, T. O. II. Title.
QA323.M39 1985 515.7'3 84–14882
ISBN 0-273-08638-3

British Library Cataloguing in Publication Data

Maz'ya, V. G.
Theory of multipliers in spaces of differentiable functions.—(Monographs and studies in mathematics, ISSN 0743-0329; 23)
1. Function spaces. 2. Multipliers (Mathematical analysis)
I. Title. II. Shaposhnikova, T. O. III. Series
515.7'3 QA323

ISBN 0-273-08638-3

Filmset and printed in Northern Ireland by The Universities Press (Belfast) Ltd., and bound at the Pitman Press, Bath, Avon.

Contents

Preface

The subject of this book is the theory of multipliers in certain classes of differentiable functions which often occur in analysis. It is intended for students and researchers who are interested in function spaces and their relation to partial differential equations and operator theory. We discuss the description of multipliers, their properties and applications. The theory of Fourier multipliers is not dealt with in this book. A knowledge of basic Sobolev space theory is assumed: we confine ourselves here to the necessary statements, but the reader who is interested in proofs can find them in monographs by, for instance, Stein [1], S. Nikol'skii [1], Peetre [1] and Triebel [3], [4].

The monograph contains seven chapters, of which the first three concern multipliers in pairs of integer and fractional Sobolev spaces, Bessel potential spaces etc. Here we present conditions for a function to belong to different classes of multipliers. Topics related to the spaces of multipliers, which are also treated in Chapters 1–3, include imbedding and composite function theorems and the spectral properties of multipliers. The applications considered include the calculus of singular integral operators whose symbols are multipliers, as well as coercive estimates for solutions of elliptic boundary value problems in spaces of multipliers.

In Chapter 4 we study the essential norm of a multiplier. The trace and extension theorems for multipliers in Sobolev spaces are proved in Chapter 5. The next chapter deals with multipliers in a domain. In particular, we discuss the change of variables in norms of Sobolev spaces and present some implicit function theorems for multipliers. Chapter 7 presents applications of these results to L_p theory of elliptic boundary value problems in domains with non-smooth boundaries.

A more detailed description of the contents is given in the Introduction.

Throughout the book there is constant cross-reference between chapters and sections. Where there is more than one theorem, lemma, proposition etc. within a subsection, these are numbered in independent sequence and in cross-reference we simply refer to the individual numbered item: for example, Theorem 3.6.2/1 is Theorem 1 in Subsection 3.6.2, Lemma 3.4.2 is the single lemma in Subsection 3.4.2 and (3.9.2/17) is formula (17) of Subsection 3.9.2.

We conclude the book with a list of symbols and author and subject indexes. The bibliography contains only papers referred to in the text and published up to 1980.

Acknowledgements

We take pleasure in thanking Dr I. E. Verbitskii who, when the manuscript had already been sent to the publishers, kindly supplied us with the proof—just obtained—of our hypothesis on a simplified normalization of the space $M(H_p^m \to H_p^l)$ for $m > l$. This material forms Section 2.6.

Our cordial gratitude is due to Professor Gaetano Fichera, whose invaluable help and encouragement promoted the publication of this book.

We would like to thank Pitman Publishing for accepting this book for publication and we are indebted to our editor, Bridget Buckley, and to The Universities Press, Belfast, for a superb job on our manuscript.

Leningrad
September 1984

Vladimir Maz'ya
Tatyana Shaposhnikova

Introduction

By a multiplier acting from one functional space, S_1, into another, S_2, we mean a function which defines a bounded linear mapping of S_1 into S_2 by pointwise multiplication. Thus, with any pair of spaces S_1, S_2 we associate a third—the space of multipliers $M(S_1 \to S_2)$.

Multipliers arise in various problems of analysis and theory of differential and integral equations. Coefficients of differential operators can be naturally considered as multipliers. The same is true for symbols of more general pseudo-differential operators. Multipliers also appear in the theory of differentiable mappings preserving the Sobolev spaces. Solutions of boundary value problems can be sought in classes of multipliers. Because of their algebraic properties, multipliers are suitable objects for a generalization of the basic facts of calculus (theorems on superposition, implicit function theorems etc.).

The present book is devoted to the theory of multipliers in pairs of integer and fractional Sobolev spaces, Bessel potential spaces etc. Regardless of the substantiality and the usefulness of this theory, until recently it has attracted relatively little attention. The earliest papers devoted to the subject include that due to Devinatz and Hirschman [1], 1959, on the spectrum of the operator of multiplication in the space W_2^l, $2|l|<1$, on the unit circumference; two papers by Hirschman, [2], 1961 and [3], 1962, which also treat multipliers in W_2^l; and, finally, a study of multipliers in Bessel potential space due to Strichartz [1], 1967.

This monograph is based principally on the authors' results obtained in 1979–1980 and partly published in journals. The outline of the basic results given below is not exhaustive and does not follow the strict sequence of the subsequent exposition.

Description and Properties of Multipliers

According to a theorem due to one of the authors (Maz'ya [1], 1962), the inequality

$$\int_\Omega |\gamma(x)u(x)|^2 \, dx \leq \int_\Omega |\nabla u(x)|^2 \, dx \tag{1}$$

is valid for all $u \in C_0^\infty(\Omega)$ if, for any compact set e with $e \subset \Omega \subset R^n$,

$$\int_e |\gamma(x)|^2 \, dx \leqslant \frac{n-2}{4} \omega_n \operatorname{cap}_\Omega (e).$$

Moreover, the inequality

$$\int_e |\gamma(x)|^2 \, dx \leqslant (n-2)\omega_n \operatorname{cap}_\Omega (e)$$

is necessary for the validity of (1). Here $n>2$, ω_n is the surface area of a unit ball in R^n and $\operatorname{cap}_\Omega (e)$ is the Green capacity of the compact set e relative to Ω. Thus the condition

$$\sup_{e \subset \Omega} \frac{\|\gamma; e\|_{L_2}}{[\operatorname{cap}_\Omega (e)]^{1/2}} < \infty$$

is equivalent to the inclusion $\gamma \in M(\mathring{L}_2^1(\Omega) \to L_2(\Omega))$ where $\mathring{L}_2^1(\Omega)$ is the completion of C_0^∞ in the metrics of the Dirichlet integral.

This result serves as a model for the characterization of different spaces of multipliers which are presented in the book. For example, multipliers $\gamma : W_p^m(R^n) \to W_p^l(R^n)$ satisfy the relation

$$\|\gamma; R^n\|_{M(W_p^m \to W_p^l)} \sim \sup_{e \subset R^n} \frac{\|\nabla_l \gamma; e\|_{L_p}}{[\operatorname{cap}(e, W_p^m)]^{1/p}} + \sup_{e \subset R^n} \frac{\|\gamma; e\|_{L_p}}{[\operatorname{cap}(e, W_p^{m-l})]^{1/p}} \quad (2)$$

where $p \in (1, \infty)$, $m \geqslant l \geqslant 0$, ∇_l is the gradient of order l and $\operatorname{cap}(e, W_p^k)$ is the capacity generated by the norm in $W_p^k(R^n)$. For the case $m=l$ this implies the equivalence

$$\|\gamma; R^n\|_{MW_p^l} \sim \sup_{e \subset R^n} \frac{\|\nabla_l \gamma; e\|_{L_p}}{[\operatorname{cap}(e, W_p^l)]^{1/p}} + \|\gamma; R^n\|_{L_\infty}. \quad (3)$$

(Here and in what follows, $MS = M(S \to S)$.) The finiteness of the right-hand side of (2) is the necessary and sufficient condition for a function γ to belong to the space $M(W_p^m(R^n) \to W_p^l(R^n))$.

Recently I. E. Verbitskii showed that, for $m>l$, the second summand on the right in (2) can be replaced by $\sup_x \|\gamma; B_1(x)\|_{L_1}$. His result is presented in the last section of Chapter 2 within the more general context of Bessel potential spaces.

In two cases, $p=1$ and $pm>n$, we can describe the space $M(W_p^m(R^n) \to W_p^l(R^n))$ using no capacity. For $p=1$, the ball $B_\rho(x) = \{y \in R^n : |x-y| < \rho\}$, $\rho < 1$, plays the role of an arbitrary compact set e in (2). In other words,

$$\|\gamma; R^n\|_{M(W_1^m \to W_1^l)} \sim \sup_{x \in R^n, 0<\rho<1} (\rho^{m-n} \|\nabla_l \gamma; B_\rho(x)\|_{L_1} + \rho^{m-l-n} \|\gamma; B_\rho(x)\|_{L_1}). \quad (4)$$

In the case $pm>n$, one can change e by $B_1(x)$ in (2), i.e.

$$\|\gamma;R^n\|_{M(W_p^m\to W_p^l)}\sim\|\gamma;R^n\|_{W^l_{p,\mathrm{unif}}}. \tag{5}$$

Here and henceforth

$$\|\gamma;R^n\|_{S_{\mathrm{unif}}}=\sup_{z\in R^n}\|\gamma\eta_z;R^n\|_S,$$

where $\eta_z(x)=\eta(z-x)$, η is an arbitrary function in $C_0^\infty(R^n)$, $\eta=1$ on the ball $B_1=\{x:|x|<1\}$.

The greater part of Chapter 1 is devoted to the proof of the above-stated assertions.

Similar results for the pair $H_p^m(R^n)\to H_p^l(R^n)$ of Bessel potential spaces and the pair $W_p^m(R^n)\to W_p^l(R^n)$ of fractional order Sobolev spaces are obtained in the following two chapters. Relations analogous to (2)–(4), in which the corresponding 'fractional derivative' plays the role of the gradient, are valid for the spaces $M(H_p^m(R^n)\to H_p^l(R^n))$, $p\in(1,\infty)$, and $M(W_p^m(R^n)\to W_p^l(R^n))$, $p\in[1,\infty)$. The relation (5) can also be extended to the above-mentioned spaces of multipliers which, in particular, contains the result due to Strichartz [1], 1967: $MH_p^l(R^n)=H^l_{p,\mathrm{unif}}(R^n)$ for $pl>n$.

Relations similar to (2), together with upper and lower estimates for capacity, enable one to obtain necessary or sufficient conditions for being a multiplier which are formulated without the use of capacity. For example, in Chapter 2 we show that the space $M(H_p^m(R^n)\to H_p^l(R^n))$ with $m\geqslant l$, $mp\leqslant n$, contains functions which are 'uniformly locally' in the space $B_{q,\infty}^{n/q-m+l}$, where $q\geqslant p$, $\{n/q-m+l\}>0$, $n/q>m$. In the case $m=l$ we additionally require the boundedness of functions (for the space $B_{p,r}^s$ see, for instance, Nikol'skii [1], Peetre [1]). In particular, for the one-dimensional case the last result implies a sufficient condition in terms of q-variation which generalizes one due to Hirschman [2], 1962.

In Chapter 2 we also show that $H^l_{n/m,\mathrm{unif}}(R^n)\subset M(H_p^m(R^n)\to H_p^l(R^n))$ for $mp<n$, $l<m$ and $(H^l_{n/l,\mathrm{unif}}\cap L_\infty)(R^n)\subset MH_p^l(R^n)$ for $lp<n$. The latter imbedding was proved earlier, by a direct method, by Polking [1]. In Chapter 3 we prove that H can be replaced by W in the right-hand side only for $p\geqslant 2$.

In different parts of the book we study certain properties of multiplier spaces. We prove imbedding theorems of the form $M(H_p^m(R^n)\to H_p^l(R^n))\subset M(H_q^{m-j}(R^n)\to H_q^{l-j}(R^n))$, theorems on the composition $\varphi(\gamma)$ where γ is a multiplier, on the spectrum of a multiplier etc. For instance, in 2.5 a description of the point, residual and continuous spectra of multipliers in H_p^l and $H_{p'}^{-l}$ is presented. In Chapter 3 the following assertion, generalizing the theorem of Hirschman [1], is given: if a function φ satisfies the Hölder condition, $|\varphi(t+\tau)-\varphi(t)|\leqslant A\,|\tau|^\rho$, $|\tau|<1$, with $\rho \in (0, 1)$ and $\gamma \in M(W_p^m(R^n) \to W_p^l(R^n))$, $m \geq l$, $0 < l < 1$, $p > 1$, then $\varphi(\gamma) \in M(W_p^{m-l+r}(R^n) \to W_p^r(R^n))$ for any $r \in (0, l\rho)$.

In Chapter 6 we obtain equivalent expressions for the norm in the space $M(W_p^m(\Omega) \to W_p^l(\Omega))$, where m and l are integers, $m \geq l \geq 0$ and Ω is a Lipschitz domain. We prove that for such domains there exists a linear continuous extension operator $M(W_p^m(\Omega) \to W_p^l(\Omega)) \to M(W_p^m(R^n) \to W_p^l(R^n))$, $p \geq 1$, and that $M(W_p^m(\Omega) \to W_p^l(\Omega))$ is the space of restrictions to Ω of functions in $M(W_p^m(R^n) \to W_p^l(R^n))$. For an arbitrary domain the last property is not valid.

In Chapter 6 we also define and study the so-called (p, l)-diffeomorphisms, i.e. the bi-Lipschitz mappings $\varkappa : R^n \supset U \to V \subset R^n$ such that the elements of the Jacobi matrix $\varkappa'$ belong to the space of multipliers $MW_p^{l-1}(U)$. These mappings have a number of useful properties. They map $W_p^l(V)$ onto $W_p^l(U)$ and $\varkappa^{-1}$ is the (p, l)-diffeomorphism together with $\varkappa$. Moreover, the class of (p, l)-diffeomorphisms is closed with respect to a composition of mappings.

Using these properties of (p, l)-diffeomorphisms, we introduce the class of n-dimensional (p, l)-manifolds, on which the space W_p^l is properly defined.

One more subject considered in Chapter 6 is the class $T_p^{m,l}$ of mappings $\varkappa : U \to V$ which satisfy the inequality $\|u \circ \varkappa; U\|_{W_p^l} \leq c\, \|u; V\|_{W_p^m}$.

One of implicit function theorems proved in Chapter 6 is the following.

Let $G = \{(x, y) : x \in R^{n-1}, y > \varphi(x)\}$ *and let* u *be a function on* G *such that*

(i) grad $u \in MW_p^{l-1}(G)$, *where* l *is an integer*, $l \geq 2$,

(ii) $u(x, \varphi(x)) = 0$,

(iii) $\inf\, (\partial u/\partial y)(x, \varphi(x)) > 0$.

Then grad $\varphi \in MW_p^{l-1-1/p}(R^{n-1})$.

Essential Norm of a Multiplier

The norm in the quotient space of the space of multipliers modulo compact operators will be called the essential norm of a multiplier. In Chapter 4 we give two-sided estimates for the essential norm

$$\text{ess}\, \|\gamma; R^n\|_{M(W_p^m \to W_p^l)} = \inf_{\{T\}} \|\gamma - T\|_{W_p^m \to W_p^l},$$

where $\{T\}$ is the set of all compact operators: $W_p^m \to W_p^l$, $p \geq 1$, and m, l are simultaneously integer or fractional numbers. For example, if $p > 1$,

$mp \leq n$, m and l are integers, then

$$\operatorname{ess} \|\gamma; R^n\|_{M(W_p^m \to W_p^l)}$$
$$\sim \lim_{\delta \to 0} \sup_{\{e: \operatorname{diam} e \leq \delta\}} \left(\frac{\|\nabla_l \gamma; e\|_{L_p}}{[\operatorname{cap}(e, W_p^m)]^{1/p}} + \frac{\|\gamma; e\|_{L_p}}{[\operatorname{cap}(e, W_p^{m-l})]^{1/p}} \right)$$
$$+ \lim_{r \to \infty} \sup_{\{e \subset R^n \setminus B_r : \operatorname{diam} e \leq 1\}} \left(\frac{\|\nabla_l \gamma ff\ e\|}{[\operatorname{cap}(e, W_p^m)]^{1/p}} + \frac{\|\gamma; e\|_L}{[\operatorname{cap}(e, W_p^{m-l})]^{1/p}} \right). \tag{6}$$

For $mp > n$ the right-hand side of this relation has a simpler form, namely,

$$\operatorname{ess} \|\gamma; R^n\|_{M(W_p^m \to W_p^l)} \sim \overline{\lim_{|x| \to \infty}} \|\gamma; B_1(x)\|_{W_p^l} \quad \text{for } m > l,$$

$$\operatorname{ess} \|\gamma; R^n\|_{MW_p^l} \sim \|\gamma; R^n\|_{L_\infty} + \overline{\lim_{|x| \to \infty}} \|\gamma; B_1(x)\|_{W_p^l} \quad \text{for } m = l.$$

Making the right-hand sides of these and similar relations equal to zero, we obtain the characteristics of the space $\mathring{M}(W_p^m(R^n) \to W_p^l(R^n))$, $m > l$, of compact multipliers. In 4.2 we show that this space coincides with the completion of $C_0^\infty(R^n)$ in the norm of the space $M(W_p^m(R^n) \to W_p^l(R^n))$. In accordance with the latter assertion, by $\mathring{M}W_p^l(R^n)$ we denote the completion of $C_0^\infty(R^n)$ with respect to the norm of the space $MW_p^l(R^n)$. By virtue of a theorem proved in 4.3, the essential norm in $\mathring{M}W_p^l(R^n)$ is equivalent to the norm in $L_\infty(R^n)$. In addition, we note that for any multiplier in $W_p^l(R^n)$ the inequality

$$\|\gamma; R^n\|_{L_\infty} \leq \operatorname{ess} \|\gamma; R^n\|_{MW_p^l}$$

is valid (see 4.3).

Traces and Extensions of Multipliers

It is well known that the space $W_p^{l-1/p}(R^n)$with integer l is the space of traces on R^n of functions contained in the Sobolev space $W_p^l(R_+^{n+1})$, where $R_+^{n+1} = \{(x, y) : x \in R^n, y > 0\}$. In Chapter 5 we show that, similarly, for the space of multipliers $MW_p^{l-1/p}(R^n)$, the traces on R^n of functions in $MW_p^l(R_+^{n+1})$ belong to the space $MW_p^{l-1/p}(R^n)$ and there exists a linear continuous extension operator $MW_p^{l-1/p}(R^n) \to MW_p^l(R_+^{n+1})$. This result is contained in a more general assertion relating weighted Sobolev spaces. In the same Chapter 5 we show that, for $\{l - m/p\} > 0$, $1 \leq p < \infty$, the space $MW_p^{l-m/p}(R^n)$ coincides with the space of traces on R^n of functions in $MW_p^l(R^{n+m})$.

Applications of the Theory of Multipliers

In the course of the book we often dwell upon various applications of multipliers to the theory of differential and integral operators. In terms of multipliers we establish the bounds (in some cases two-sided) for a norm and for an essential norm of a differential operator acting in a pair of Sobolev spaces. The properties of a function in $M(W_2^m(R^n) \to W_2^l(R^n))$ are equivalent to the properties of the integral convolution operator considered as a mapping of $L_2(R^n; (1+|x|^2)^{m/2})$ into $L_2(R^n; (1+|x|^2)^{l/2})$. The basic tenets of the theory of singular integral operators acting in $W_p^l(R^n)$ can be generalized to operators with symbols belonging to spaces of multipliers. One can solve elliptic boundary value problems in spaces of multipliers. For multiplier norms of solutions, estimates similar to the known coercive L_p-estimates are valid. Here is the simplest example considered in 6.6: the Dirichlet problem

$$\Delta u = 0 \quad \text{in} \quad \Omega, \qquad u|_{\partial\Omega} = \varphi,$$

where Ω is a bounded Lipschitz domain in R^n, is uniquely solvable in the space $MW_2^1(\Omega)$ for any $\varphi \in MW_2^{1/2}(\partial\Omega)$. The solution satisfies the inequality

$$\|u; \Omega\|_{MW_2^1} \leq c\, \|\varphi; \partial\Omega\|_{MW_2^{1/2}}.$$

In the last chapter, multipliers are applied to the study of conditions on $\partial\Omega$ ensuring $W_p^l(\Omega)$-coercivity of the operator of the general elliptic boundary value problem. For $p(l-1) < n$, the condition we have found is the following: for any point of the boundary there exists a neighbourhood U and a function φ such that $U \cap \Omega = \{(x, y) \in U : x \in R^{n-1}, y > \varphi(x)\}$ and

$$\|\nabla\varphi; R^{n-1}\|_{MW_p^{l-1-1/p}} \leq \delta, \tag{7}$$

where δ is a small constant. We consider equivalent formulations of this requirement and obtain different sufficient conditions for its validity. For example, Ω satisfies (7) provided that φ has a small Lipschitz constant and belongs to the space $B_{q,p}^{l-1/p}(R^{n-1})$ for some q. In case $p(l-1) > n$, condition (7) should be replaced by $\varphi \in W_p^{l-1/p}(R^{n-1})$.

Special attention is paid to the Dirichlet boundary value problem. In particular, we show here that, for $p(l-1) > n$, the condition $\varphi \in W_p^{l-1/p}(R^{n-1})$ is not only sufficient but also necessary for solvability in $W_p^l(\Omega)$ of this problem in one of the two formulations we consider.

This is a summary of the book. The theory of multipliers in spaces of differentiable functions is still at the very beginning of its development and, undoubtedly, the discussion presented here by no means exhausts the possible scope of the theory. It suffices to remark that, simply by varying the pairs of function spaces S_1 and S_2, often encountered in analysis, we immediately arrive at new unsolved problems in the description and properties of $M(S_1 \to S_2)$.

1

Multipliers in Pairs of Sobolev Spaces

In this chapter we present the characteristics of multipliers acting in pairs of Sobolev spaces, which are the simplest functional spaces considered in the book. The concepts of this chapter prove to be prototypes for the subsequent study of multipliers in other pairs of spaces. In 1.1, 1.2 we give two-sided estimates for norms of operators which imbed the spaces W_p^m and w_p^m into the space of functions summable with order p, $1 \leq p < \infty$, with respect to a certain measure. Here W_p^k and w_p^k are completions of the space C_0^∞ with respect to the norms $\|\nabla_k u\|_{L_p} + \|u\|_{L_p}$ and $\|\nabla_k u\|_{L_p}$, $\nabla_k = \{\partial^k/\partial x_1^{\alpha_1} \cdots \partial x_n^{\alpha_n}\}$. When the domain is not indicated in the notation of a space or a norm then it is assumed to be R^n. Here and henceforth $C_0^\infty(\Omega) = D(\Omega)$ is the space of infinitely differentiable functions with compact supports in Ω.

Using the results of 1.1 and 1.2, we derive in 1.3 the necessary and sufficient conditions for a function to belong to the space of multipliers $M(W_p^m \to W_p^l)$ and $M(w_p^m \to w_p^l)$. In 1.4 we present a brief description of the space $M(W_p^m \to W_q^l)$. Subsection 1.5 gives the conditions sufficient for belonging to the space $M(W_p^m \to W_p^{-k})$, $k > 0$, and, finally, 1.6 deals with certain properties of multipliers.

1.1 Summability of Functions in W_p^m and w_p^m, $p > 1$, with Respect to an Arbitrary Measure

1.1.1 *Preliminaries*

In this and subsequent chapters we often use operators of the form $k(D) = F^{-1}k(\xi)F$, where F is the Fourier transform in R^n and k is a function or a vector-function which is called the symbol. In particular, $D = -i\nabla$, $D^\alpha = (-i)^{|\alpha|}/\partial x_1^{\alpha_1} \cdots \partial x_n^{\alpha_n}$. If k is a positively homogeneous function of degree 0 then $k(D)$ is called a singular integral operator.

The following assertion is a variant of the Mikhlin theorem on Fourier multipliers (see Lizorkin [2]).

Lemma 1. *Let the function k and its derivatives $\partial^m k/\partial\xi_{j_1},\ldots,\partial\xi_{j_m}$ $(0\leq j_1+\cdots+j_m=m\leq n,\ j_1,\ldots,j_m$ are distinct) be continuous on the set $\{\xi\in R^n:\xi_1\cdots\xi_n\neq 0\}$ and let*

$$\left|\xi_{j_1}\cdots\xi_{j_m}\frac{\partial^m k}{\partial\xi_{j_1}\cdots\partial\xi_{j_m}}\right|\leq \text{const}. \tag{1}$$

Then the operator $k(D)$ is continuous in L_p.

In particular, the singular integral operator with a symbol $k\in C^n(R^n\backslash 0)$ is continuous in L_p.

In what follows, the operators $(-\Delta)^{r/2}$ and $(1-\Delta)^{s/2}$ with symbols $|\xi|^r$ and $(1+|\xi|^2)^{s/2}$, $r>-n$, $s\in R^1$, play an important role.

Lemma 2. *Let $l=1,2,\ldots,1<p<\infty$. There exists a number $c>1$, depending only on n, p, l, such that*

$$c^{-1}\,\|(-\Delta)^{l/2}u\|_{L_p}\leq\|\nabla_l u\|_{L_p}\leq c\,\|(-\Delta)^{l/2}u\|_{L_p} \tag{2}$$

for all $u\in D$.

Proof. Let α be a multi-index of order l. Then $F^{-1}\xi^\alpha Fu=F^{-1}\xi^\alpha\,|\xi|^{-l}\,|\xi|^l\,Fu$. The function $\xi^\alpha\,|\xi|^{-l}$ satisfies the condition of Lemma 1 which implies the right-hand side of inequality (2). On the other hand,

$$|\xi|^l=|\xi|^{2l}\,|\xi|^{-l}=\left(\sum_{|\alpha|=l}c_\alpha\xi^\alpha\xi^\alpha\right)|\xi|^{-l},$$

where $c_\alpha=l!/\alpha!$, so

$$F^{-1}\,|\xi|^l\,Fu=\sum_{|\alpha|=l}c_\alpha F^{-1}\frac{\xi^\alpha}{|\xi|^l}\,\xi^\alpha Fu.$$

Applying Lemma 1 once more, we obtain the left-hand part of (2). □

An analogous proof has the following:

Lemma 3. *Let $l=1,2,\ldots,p\in(1,\infty)$. There exists a number $c>1$ depending only on n, p, l such that*

$$c^{-1}\,\|(1-\Delta)^{l/2}u\|_{L_p}\leq\|u\|_{W_p^l}\leq c\,\|(1-\Delta)^{l/2}u\|_{L_p} \tag{3}$$

for all $u\in D$.

The operator $(-\Delta)^{-l/2}$ is the integral convolution operator with the kernel $c\,|x|^{l-n}$, $c=\text{const}$, for $l\in(0,n)$. The operator $f\overset{I_l}{\to}|x|^{l-n}*f$ is usually called the Riesz potential of order l.

For $l>0$ the operator $(1-\Delta)^{-l/2}$ admits the representation $(1-\Delta)^{-l/2}f = G_l * f$, where G_l is the function with the Fourier transform $(1+|\xi|^2)^{-l/2}$. The function G_l can be written in the form

$$G_l(x) = c\int_0^\infty e^{-t-x^2/4t}t^{-n/2+l/2-1}\,dt$$

or in the form

$$G_l(x) = cK_{(n-l)/2}(|x|)\,|x|^{(l-n)/2},$$

where K_γ is the modified Bessel function of the third kind.

The function G_l is positive and decreases with the growth of $|x|$. It satisfies the following asymptotic estimates.

For $|x|\to 0$,

$$G_l(x) \sim \begin{cases} |x|^{l-n}, & 0<l<n, \\ \log|x|^{-1}, & l=n, \\ 1, & l>n. \end{cases} \tag{4}$$

For $|x|\to\infty$ the following relation holds:

$$G_l(x) \sim |x|^{(l-n-1)/2}e^{-|x|}. \tag{5}$$

Here and henceforth the values a and b are called equivalent ($a\sim b$) if their ratio is bounded and separated from zero by positive constants.

The integral operator $f \overset{J_l}{\to} G_l * f$ is called the Bessel potential of order l.

For properties of Riesz and Bessel potentials see Aronszajn, Mulla, Szeptycky [1], Stein [1], Strichartz [1].

We introduce the maximal Hardy–Littlewood operator M defined by $(Mf)(x) = \sup_{r>0}(1/\text{mes}_n B_r)\int_{B_r(x)}|f(y)|\,dy$. Here and henceforth $B_r(x) = \{y\in R^n : |y-x|<r\}$, $B_r = B_r(0)$. It is known (see Stein [1]) that the operator M is bounded in L_p.

1.1.2 Capacities Generated by Norms of Spaces w_p^m and W_p^m

We define the capacity $\text{cap}(e, w_p^m)$ of a compact set $e\subset R^n$ by the equality

$$\text{cap}(e, w_p^m) = \inf\{\|u\|^p_{w_p^m} : u\in D,\ u\geqslant 1 \text{ on } e\}.$$

Replacing here w_p^m by W_p^m, we obtain the definition of the capacity $\text{cap}(e, W_p^m)$.

These and analogous set functions used in the book have been the subject of active study during recent years (see Maz'ya [5], Maz'ya, Havin [1], [2], Meyers [1], Hedberg [1], Adams, Meyers [1], Sjödin [1], Adams [4], Adams, Hedberg [1]).

We describe certain simple properties of capacities $\text{cap}(e, w_p^m)$ and $\text{cap}(e, W_p^m)$ which will be used in this chapter.

Proposition 1. *The capacities* $\text{cap}(e, w_p^m)$ *and* $\text{cap}(e, W_p^m)$ *are non-decreasing functions of the set e.*

The proof is obvious.

Proposition 2. *If* $mp > n$, *then for all compact sets* $e \neq \varnothing$ *with diameter* $d(e) \leq 1$ *the relation* $\text{cap}(e, W_p^m) \sim 1$ *holds.*

Proof. Obviously, $\text{cap}(e, W_p^m) \leq \text{cap}(B_1, W_p^m)$. On the other hand, by Sobolev's theorem on the imbedding $W_p^m \subset L_\infty$, we have $c\,\|u\|_{W^m} \geq \|u\|_{L_\infty} \geq 1$ for any function $u \in D$ that exceeds unity on e. Consequently, $\text{cap}(e, W_p^m) \geq c^{-p}$. □

Proposition 3. *If* $mp < n$, $p \in (1, \infty)$, *then*

$$\text{cap}(e, w_p^m) \geq c(\text{mes}_n\, e)^{(n-mp)/n}. \tag{1}$$

The proof follows from the definition of the capacity and from Sobolev's inequality $\|u\|_{L_q} \leq c\,\|u\|_{w_p^m}$, where $q = pn/(n-mp)$, $u \in D$.

To prove an estimate similar to (1) in the case $mp = n$ we need the following known assertion (see Yudovič [1], Pohožaev [1], Trudinger [1]) which is given here with the proof for the reader's convenience.

Lemma. *If* $mp = n$, $p \in (1, \infty)$, *then*

$$\int \Phi\left(c \frac{|u|^{p'}}{\|u\|_{W_p^m}^{p'}}\right) dx \leq 1 \tag{2}$$

for all $u \in D$, *where c is a constant independent of u and* $\Phi(t) = e^t - \sum_{j=0}^{[p]} t^j/j!$.

Proof. Let $u = J_m f = G_m * f$. Taking into account Lemma 1.1.1/3, it is sufficient to give the proof under the assumption $\|f\|_{L_p} = 1$. Obviously,

$$\int \Phi(c\,|u|^{p'})\, dx = \sum_{j=[p]+1}^{\infty} \frac{c^j}{j!} \|u\|_{L_{p'j}}^{p'j}. \tag{3}$$

By Young's inequality for $q \geq p$,

$$\|u\|_{L_q} \leq \|G_m\|_{L_s} \|f\|_{L_p}, \qquad s = \frac{qp'}{q+p'}, \qquad p + p' = pp'. \tag{4}$$

Using estimates (1.1.1/4), (1.1.1/5) for the function G_m, one can easily show that

$$\|G_m\|^s_{L_s} \leqslant c_0 q, \tag{5}$$

where $c_0 = c_0(p, n)$. From (4), (5), where $q = p'j$, $s = p'j/(j+1)$, it follows that the right-hand side of (3) does not exceed

$$\sum_{j=[p]+1}^{\infty} c^j (c_0 p' j)^{j+1}/j!. \tag{6}$$

This series converges if $cc_0 p' e < 1$. Diminishing c, one can make the sum (6) arbitrary small. □

Proposition 4. *If $mp = n$, $p \in (1, \infty)$ and $d(e) \leqslant 1$, then*

$$\text{cap}(e, W^m_p) \geqslant c\left(\log \frac{2^n}{\text{mes}_n\, e}\right)^{1-p}. \tag{7}$$

Proof. Let $u \in D$, $u \geqslant 1$ on e. From (2) follows $\Phi(c\,\|u\|^{-p'}_{W^m_p})\, \text{mes}_n\, e \leqslant 1$. Hence

$$\Phi(c[\text{cap}\,(e, W^m_p)]^{1/(1-p)}) \leqslant (\text{mes}_n\, e)^{-1}. \tag{8}$$

Since the argument of the function Φ in (8) is separated from zero,

$$\exp\,(c[\text{cap}\,(e, W^m_p)]^{1/(1-p)}) \leqslant c_0(\text{mes}_n\, e)^{-1}. \quad \square$$

Proposition 5. *If $mp < n$, then* $\text{cap}\,(B_r, w^m_p) = cr^{n-mp}$.

Proof. Using the similarity transformation, we get

$$\text{cap}\,(B_r, w^m_p) = r^{n-mp}\, \text{cap}\,(B_1, w^m_p). \quad \square$$

Proposition 6. *If $mp < n$ and $0 < r \leqslant 1$, then* $\text{cap}\,(B_r, W^m_p) \sim r^{n-mp}$.

Proof. The lower bound for the capacity follows from (1). The upper one is obtained after the substitution of the function $x \to \eta(x/r)$, where $\eta \in D$, $\eta = 1$ on the ball B_1, into the norm $\|u\|_{W^m_p}$. □

Proposition 7. *If $mp = n$, $p \in (1, \infty)$ and $0 < r \leqslant 1$, then* $\text{cap}\,(B_r, W^m_p) \sim (\log 2/r)^{1-p}$.

Proof. The lower bound for the capacity follows from (7). Let us give the upper bound. We put $v(x) = (\log 2/r)^{-1} \log 2/|x|$. By α we denote a function in the space $C^\infty(R^1)$ such that $\alpha(t) = 0$ for $t < 0$, $\alpha(t) = 1$ for $t > 1$. Further, let $u(x) = \alpha[v(x)]$. Clearly, $u \in D(B_2)$, $u = 1$ on B_r. Moreover, one can easily check that $|\nabla_m u(x)| \leqslant c(\log 2/r)^{-1}\, |x|^{-m}$ on $B_2 \backslash B_r$. This implies

$$\operatorname{cap}(B_r, W_p^m) \leqslant c\,\|\nabla_m u; B_2\|_{L_p}^p$$
$$\leqslant c(\log 2/r)^{-p} \int_{B_2\setminus B_r} |x|^{-mp}\,dx = c(\log 2/r)^{1-p}. \quad \square$$

1.1.3 Estimate for the Integral of Capacity of a Set Bounded by a Level Surface

The following assertion is proved in Hedberg [3].

Lemma. *Let $0<\theta<1$, $0<r<n$ and let $I_r f$ be the Riesz potential of order r with a non-negative density f. Then*

$$(I_{r\theta}f)(x) \leqslant c[(I_r f)(x)]^{\theta}[(Mf)(x)]^{1-\theta}, \tag{1}$$

where M is the maximal operator.

Proof. For any $\delta>0$ we have

$$\int_{|y-x|>\delta} f(y)\,|x-y|^{r\theta-n}\,dy \leqslant \delta^{r(\theta-1)} \int_{|y-x|>\delta} f(y)\,|x-y|^{r-n}\,dy$$
$$\leqslant \delta^{r(\theta-1)}(I_r f)(x).$$

On the other hand,

$$\int_{|y-x|\leqslant\delta} f(y)\,|x-y|^{r\theta-n}\,dy = \sum_{k=0}^{\infty} \int_{\delta 2^{-k-1}<|y-x|<\delta 2^{-k}} f(y)\,|x-y|^{r\theta-n}\,dy$$
$$\leqslant c \sum_{k=0}^{\infty} (\delta 2^{-k})^{r\theta}(\delta 2^{-k})^{-n} \int_{|y-x|\leqslant\delta 2^{-k}} f(y)\,dy$$
$$\leqslant c\delta^{r\theta}(Mf)(x) \sum_{k=0}^{\infty} 2^{-kr\theta}. \tag{2}$$

Hence $(I_{r\theta}f)(x) \leqslant c(\delta^{r\theta}(Mf)(x)+\delta^{r(\theta-1)}(I_r f)(x))$. Putting here $\delta^r = (I_r f)(x)/(Mf)(x)$, we arrive at (1). $\square$

Corollary. *Let m be an integer, $0<m<n$, $I_m f=|x|^{m-n} * f$ with $f\geqslant 0$ and let F be a function in $C^m(0,\infty)$ such that $t^{k-1}|F^{(k)}(t)|\leqslant Q$, $k=0,1,\ldots,m$, $Q=$ const. Then*

$$|\nabla_m F(I_m f)| \leqslant cQ(Mf+|\nabla_m I_m f|) \quad \textit{almost everywhere in } R^n. \tag{3}$$

Proof. Let $u = I_m f$. One can verify by induction that

$$|\nabla_m F(u)| \leq c \sum_{k=1}^{m} |F^{(k)}(u)| \sum_{j_1+\cdots+j_k=m} |\nabla_{j_1} u| \cdots |\nabla_{j_k} u|. \tag{4}$$

Consequently

$$|\nabla_m F(u)| \leq cQ \sum_{k=1}^{m} \sum_{j_1+\cdots+j_k=m} \frac{|\nabla_{j_1} u|}{u^{1-j_1/m}} \cdots \frac{|\nabla_{j_k} u|}{u^{1-j_k/m}}. \tag{5}$$

Since $|\nabla_s u| \leq cI_{m-s}f$, then from (5) follows

$$|\nabla_m F(u)| \leq cQ\Big(|\nabla_m I_m f| + \sum_{k=1}^{m} \sum_{j_1+\cdots+j_k=m}{}' \frac{I_{m-j_1}f \cdots I_{m-j_k}f}{(I_m f)^{1-j_1/m} \cdots (I_m f)^{1-j_k/m}}\Big),$$

where the sum $\sum'$ is taken over the collections of numbers $j_1, \ldots, j_k$ less than m. Applying the lemma, we complete the proof. □

Our aim is the following:

Theorem. *Let $p \in (1, \infty)$, $m = 1, 2, \ldots$ and $mp < n$. Then, for any function $u \in D$,*

$$\int_0^\infty \operatorname{cap}(N_t, w_p^m) t^{p-1}\, dt \leq c \|u\|_{w_p^m}^p, \tag{6}$$

where $N_t = \{x : |u(x)| \geq t\}$ and c is a constant depending only on n, p, m.

Proof. Let $u = I_m f$ and $v = I_m |f|$. It is easily seen that $v \in C^m(R^n)$ and $v(x) = O(|x|^{m-n})$ as $|x| \to \infty$. Thus the set $\{x : v(x) \geq t\}$ is compact for any $t > 0$. Putting $t_j = 2^j$, $j = 0, \pm 1, \ldots$, and using $v(x) \geq |u(x)|$, we obtain

$$\int_0^\infty \operatorname{cap}(N_t, w_p^m) t^{p-1}\, dt \leq c \sum_{j=-\infty}^{+\infty} (t_{j+1} - t_j)^p \operatorname{cap}(\{x : v(x) \geq t_j\}, w_p^m). \tag{7}$$

Let $\gamma \in C^\infty(R^1)$, $\gamma(\tau) = 0$ for $\tau < \varepsilon$, $\gamma(\tau) = 1$ for $\tau > 1$, where $\varepsilon > 0$. We introduce the function $v \to F \in C^\infty(0, \infty)$ equal to

$$F_j(v) = t_j + (t_{j+1} - t_j)\gamma((v - t_j)(t_{j+1} - t_j)^{-1})$$

on the segment $[t_j, t_{j+1}]$.

According to the definition of capacity, the sum in the right-hand side of (7) does not exceed $\sum_{j=-\infty}^{\infty} \|F_j(v)\|_{w_p^m}^p = \|F(v)\|_{w_p^m}^p$. By virtue of the corollary, the last norm is majorized by

$$c(\|M|f|\|_{L_p} + \|\nabla_m I_m |f|\|_{L_p}). \tag{8}$$

Since the operator M and the singular integral operator $\nabla_m I_m$ are continuous in L_p, the sum (8) does not exceed $c\|f\|_{L_p}$. By Lemma 1.1.1/2, $\|f\|_{L_p} \sim \|u\|_{w_p^m}$. □

Remark 1. The existence of inequalities of the type (6) was demonstrated in Maz'ya [4], where (6) (and even a stronger inequality, in which the capacity of the condenser $N_t \backslash N_{2t}$ plays the role of the capacity of the set N_t) was obtained only for $m=1, 2$. In the more difficult case $m=2$ the proof was based on the 'smooth truncation' of a potential near equipotential surfaces. Unifying this device with Hedberg's inequality (1), Adams [3] obtained the above proof for all integers m.

Inequalities similar to (6) were obtained later for some other spaces. They will be formulated and used in subsequent chapters.

Remark 2. Together with (6) we shall use in this chapter the inequality

$$\int_0^\infty \operatorname{cap}(N_t, W_p^m) t^{p-1}\, dt \leq c\, \|u\|^p_{W_p^m} \tag{9}$$

where $p \in (1, \infty)$, $m=1, 2, \ldots$. The proof of (9) is similar to that of (6), the role of (1) being played by

$$(J_{r\theta} f)(x) \leq c[(J_r f)(x)]^\theta [(\mathcal{M} f)(x)]^{1-\theta}, \tag{10}$$

where $0<\theta<1$, $r>0$ and $J_r f$ is the Bessel potential of order r with non-negative density f. We do not dwell on a similar though more cumbersome proof of (10), since a more general inequality will be proved in Lemma 3.1.2/1.

The corollary and its proof remain valid for the change of I_m to J_m. To obtain (9), it is sufficient to use the following chain of inequalities:

$$\begin{aligned}\int_0^\infty \operatorname{cap}(N_t, W_p^m) t^{p-1}\, dt &\leq c \sum_{j=-\infty}^{\infty} (t_{j+1}-t_j)^p \operatorname{cap}(\{x : |v(x)| \geq t_j\}, W_p^m)\\ &\leq c \sum_{j=-\infty}^{\infty} \|F_j(v) - t_j\|^p_{W_p^m}\\ &\leq c(\|\nabla_m [F(v)]\|^p_{L_p} + \|v\|^p_{L_p})\end{aligned}$$

and to duplicate the end of the proof of the theorem.

1.1.4 Imbedding Theorems

A simple though important corollary of inequalities (1.1.3/6) and (1.1.3/9) is:

Theorem. (i) *Let $p \in (1, \infty)$, $m=1, 2, \ldots$ and let μ be a measure in R^n. Then the best constant C in*

$$\int |u|^p\, d\mu \leq C\, \|u\|^p_{W_p^m}, \qquad u \in D, \tag{1}$$

is equivalent to

$$\sup_e \frac{\mu(e)}{\mathrm{cap}\,(e, W_p^m)}, \tag{2}$$

where e is an arbitrary compact set in R^n *of positive capacity* $\mathrm{cap}\,(e, W_p^m)$. (Expressions similar to (2) often occur in this book. In what follows we do not mention the positivity of capacities in denominators.)

(ii) *For* $mp<n$ *an analogous assertion holds, with W replaced by w.*

Proof. (i) From the definition of Lebesgue integral we obtain $\int |u|^p \,d\mu = \int_0^\infty \mu(N_t)\,d(t^p)$, where $N_t=\{x:|u(x)|\geq t\}$. Therefore

$$\int |u|^p \,d\mu \leq \sup_e \frac{\mu(e)}{\mathrm{cap}\,(e, W_p^m)} \int_0^\infty \mathrm{cap}\,(N_t, W_p^m)\,d(t^p).$$

Now (1) follows from (1.1.3/9).

Minimizing the right-hand side of (1) over the set $\{u\in D: u\geq 1 \text{ on } e\}$, we get $C\geq \sup_e (\mu(e)/\mathrm{cap}\,(e, W_p^m))$. □

Case (ii) is considered in the same way.

Lemma 1. The best constants C_0 *and C in the inequalities*

$$\int |\nabla_l u|^p \,d\mu \leq C_0 \|u\|^p_{w_p^m},$$

$$\int |u|^p \,d\mu \leq C \|u\|^p_{w_p^{m-l}}, \tag{3}$$

where $m>l$ *and* $u\in D$, *are equivalent.*

Proof. The estimate $C_0\leq cC$ is obvious. We show that $C_0\geq cC$. It is clear that $u=\sum_{|\alpha|=l}(l!/\alpha!)D^{2\alpha}(-\Delta)^{-l}u$. From (3) and Lemma 1.1.1/2 we get

$$\int |D^{2\alpha}(-\Delta)^{-l}u|^p \,d\mu \leq C_0 \|D^\alpha(-\Delta)^{-l}u\|^p_{w_p^m} \leq cC_0 \|u\|^p_{w_p^{m-l}}.$$

So $\int |u|^p \,d\mu \leq cC_0 \|u\|^p_{w_p^{m-l}}$. □

Lemma 2. The best constants C_0 *and C in the inequalities*

$$\int (|\nabla_l u|^p + |u|^p)\,d\mu \leq C_0 \|u\|^p_{W_p^m},$$

$$\int |u|^p \,d\mu \leq C \|u\|^p_{W_p^{m-l}}, \tag{4}$$

where $m>l$ *and* $u\in D$, *are equivalent.*

Proof. The estimate $C_0 \leqslant cC$ is obvious. We prove the converse. Let $x \to \sigma$ be a smooth positive function on $[0, \infty)$, equal to x for $x > 1$. For any $u \in D$ we have the representation $u = (-\Delta)^l[\sigma(-\Delta)]^{-l}u + T(-\Delta)$, where T is a function from $D([0,\infty))$. Since $(-\Delta)^l = (-1)^l \sum_{|\alpha|=l} (l!/\alpha!)D^{2\alpha}$, then, by applying (4) and Lemmas 1.1.1/4, 1.1.1/3, we obtain

$$\int |u|^p \, d\mu \leqslant cC_0(\|\nabla_l[\sigma(-\Delta)]^{-l}u\|^p_{W_p^m} + \|Tu\|^p_{W_p^m})$$
$$\leqslant c_1 C_0 \|u\|^p_{W_p^{m-l}}. \quad \square$$

We give one more expression equivalent to the best constant C in (1).

Corollary 1. The exact constant C in (1) is equivalent to $\sup_{\{e:d(e)\leqslant 1\}} (\mu(e)/\text{cap}\,(e, W_p^m))$, *where $d(e)$ is the diameter of e.*

Proof. The lower bound for C immediately follows from the theorem. We verify the upper bound. Let $\varkappa$ be an arbitrary compact set in R^n. Further, let closed cubes Q_j form the coordinate grid with step $n^{-1/2}$ and let $2Q_j$ be homothetic open cubes with double edge length. We denote by u a function in D, $u \geqslant 1$ on $\varkappa$. Let η_j be a function in $D(2Q_j)$ equal to unity on Q_j. Since the multiplicity of intersection of $2Q_j$ is finite and depends only on n, then

$$\sum_j \text{cap}\,(\varkappa \cap Q_j,\, W_p^m) \leqslant c_1 \sum_j \|\eta_j u; 2Q_j\|^p_{W_p^m}$$
$$\leqslant c_2 \sum_j \|u; 2Q_j\|^p_{W_p^m} \leqslant c_3 \|u\|^p_{W_p^m}.$$

Minimizing the last norm, we get

$$\text{cap}\,(\varkappa, W_p^m) \geqslant c \sum_j \text{cap}\,(\varkappa \cap Q_j,\, W_p^m). \tag{5}$$

Clearly,

$$\mu(\varkappa \cap Q_j) \leqslant \sup_{\{e:d(e)\leqslant 1\}} \frac{\mu(e)}{\text{cap}\,(e, W_p^m)}\, \text{cap}(\varkappa \cap Q_j,\, W_p^m).$$

Summing this over j and using (5), we arrive at the inequality

$$\mu(\varkappa) \leqslant c \sup_{\{e:d(e)\leqslant 1\}} \frac{\mu(e)}{\text{cap}\,(e, W_p^m)}\, \text{cap}\,(\varkappa, W_p^m). \quad \square$$

Corollary 2. If $mp > n$, then the best constant in (1) is equivalent to $\sup_{x \in R^n} \mu(B_1(x))$.

Proof. By Proposition 1.1.2/2, $\operatorname{cap}(e, W_p^m) \sim 1$ for any non-empty compact set e with $d(e) \leq 1$. It remains to use Corollary 1. □

Using the estimates for capacity by Lebsegue measure obtained in Propositions 1.1.2/3 and 1.1.2/4, we immediately get:

Proposition 1. The following inequalities hold:

$$\sup_{\{e:d(e)\leq 1\}} \frac{\mu(e)}{\operatorname{cap}(e, W_p^m)} \leq \begin{cases} c \sup\limits_{\{e:d(e)\leq 1\}} \dfrac{\mu(e)}{(\operatorname{mes}_n e)^{(n-pm)/n}} & \text{for } mp<n, \\ c \sup\limits_{\{e:d(e)\leq 1\}} \left(\log \dfrac{4^n}{\operatorname{mes}_n e}\right)^{p-1} \mu(e) & \text{for } mp=n; \end{cases}$$

$$\sup_e \frac{\mu(e)}{\operatorname{cap}(e, w_p^m)} \leq c \sup_e \frac{\mu(e)}{(\operatorname{mes}_n e)^{(n-pm)/n}} \qquad \text{for } mp<n.$$

A direct corollary of Propositions 1.1.2/5–1.1.2/7 is

Proposition 2. The following inequalities hold:

$$\sup_e \frac{\mu(e)}{\operatorname{cap}(e, W_p^m)} \geq \begin{cases} c \sup\limits_{x\in R^n, \rho\in(0,1)} \rho^{mp-n}\mu(B_\rho(x)) & \text{for } mp<n, \\ c \sup\limits_{x\in R^n, \rho\in(0,1)} \left(\log \dfrac{2}{\rho}\right)^{p-1} \mu(B_\rho(x)) & \text{for } mp=n; \end{cases}$$

$$\sup_e \frac{\mu(e)}{\operatorname{cap}(e, w_p^m)} \geq c \sup_{x\in R^n, \rho>0} \rho^{mp-n}\mu(B_\rho(x)) \qquad \text{for } mp<n.$$

Clearly, these two propositions, together with the theorem, Lemmas 1, 2 and Corollary 1, give upper and lower bounds for exact constants in (1), (3), (4) which do not contain a capacity. □

According to the paper of D. R. Adams [3], inequality (1.1.4/1) is valid if and only if

$$\int \left(\int_e G_m(x-y)\, d\mu(y)\right)^{p'} dy \leq \text{const}\, \mu(e)$$

for any compactum $e \subset R^n$. This is a corollary of Theorem 1.1.4 and the next assertion.

Proposition 3. The following relation holds:

$$\sup_e \frac{\mu(e)}{\operatorname{cap}(e, W_p^m)} \sim \sup_e \left[\frac{\int (J_m\mu_e(x))^{p'}\, dx}{\mu(e)}\right]^{p-1} \tag{6}$$

where the suprema are taken either over arbitrary compacta $e \subset R^n$ or over compacta whose diameters do not exceed unity and μ_e designates the restriction of measure μ to e.

Proof. Let the left- and right-hand sides of the relation to be proved be denoted by A and B, respectively. Further, let u be an arbitrary function in C_0^∞ with $u \geq 1$ on e. We have

$$\mu(e) \leq \int u(x)\, d\mu_e(x) \leq \|(1-\Delta)^{-m/2}\mu_e\|_{L_{p'}} \|(1-\Delta)^{m/2}u\|_{L_p}$$

which can be rewritten as $\mu(e) \leq c\,\|J_m\mu_e\|_{L_{p'}}\|u\|_{W_p^m}$. Minimizing the right-hand side over all functions u, we get

$$\mu(e) \leq cB^{1/p}\mu(e)^{1/p'}[\text{cap}\,(e, W_p^m)]^{1/p},$$

i.e. $A \leq cB$. Now we verify the converse estimate. According to part (i) of Theorem 1.1.4,

$$\int |u|^p\, d\mu \leq cA\,\|u\|^p_{W_p^m}$$

for all $u \in C_0^\infty$. Consequently,

$$\left|\int u\, d\mu_e\right|^p \leq cA\mu(e)^{p-1}\,\|(1-\Delta)^{m/2}u\|^p_{L_p}$$

and so $\|J_m\mu_e\|_{L_{p'}} \leq cA^{1/p}\mu(e)^{1/p'}$. Thus $B \leq cA$. □

The relation

$$\sup_e \frac{\mu(e)}{\text{cap}\,(e, w_p^m)} \sim \sup_e \left[\frac{\int (I_m\mu_e(x))^{p'}\, dx}{\mu(e)}\right]^{p-1}, \tag{7}$$

where e is an arbitrary compactum in R^n, $mp<n$, $p>1$, can be established in precisely the same manner.

1.2 Summability of Functions in w_1^m and W_1^m with Respect to an Arbitrary Measure

In this section we present two-sided estimates for best constants in inequalities of the type (1.1.4/1) with $p=1$, obtained by Maz'ya [10].

1.2.1 The Case $m=1$

The following lemma gives a representation of the n-dimensional variation of a function as an integral of the area of a level surface.

We denote by Ω an arbitrary open set in R^n.

Lemma 1. *Let $u \in C^\infty(\Omega)$ and $N_t = \{x \in \Omega : |u(x)| \geq t\}$. Then*

$$\int_\Omega |\nabla u(x)|\, dx = \int_0^\infty s(\Omega \cap \partial N_t)\, dt, \tag{1}$$

where s is the $(n-1)$-dimensional area (it is well known that ∂N_t is a smooth $(n-1)$-dimensional manifold for almost all $t > 0$).

For more general classes of functions, formula (1) is proved in Kronrod [1] for $n = 2$ and in Federer [1].

We give a simple proof of (1) for $u \in C^\infty(\Omega)$.

Proof. Let $w = (w_1, \ldots, w_n)$, $w_j \in D(\Omega)$. Integrating by parts, we obtain

$$\int_\Omega w \nabla u\, dx = -\int_\Omega u \operatorname{div} w\, dx = -\int_{u>0} u \operatorname{div} w\, dx - \int_{u<0} u \operatorname{div} w\, dx.$$

From the definition of the Lebesgue integral we get $\int_{u>0} u \operatorname{div} w\, dx = \int_0^\infty dt \int_{u>t} \operatorname{div} w\, dx$.

For almost all $t > 0$

$$\int_{u>t} \operatorname{div} w\, dx = -\int_{u=t} w\nu\, ds = -\int_{u=t} \frac{w\, \nabla u}{|\nabla u|}\, ds,$$

where ν is an inward normal to $\{x : u(x) \geq t\}$. Therefore

$$\int_{u>0} u \operatorname{div} w\, dx = -\int_0^\infty \int_{u=t} \frac{w\, \nabla u}{|\nabla u|}\, ds\, dt.$$

The transformation of the integral $\int_{u<0} u \operatorname{div} w\, dx$ is quite similar. So

$$\int_\Omega w\, \nabla u\, dx = \int_0^\infty dt \int_{\Omega \cap \partial N_t} \frac{w\, \nabla u}{|\nabla u|}\, ds.$$

We set in this identity

$$w = \phi_\nu \frac{\nabla u}{(|\nabla u|^2 + \nu^{-1})^{1/2}},$$

where $\nu = 1, 2, \ldots$ and $\{\phi_\nu\}$ is a non-decreasing sequence of non-negative functions in $D(\Omega)$, convergent to unity in Ω. Then

$$\int_\Omega \phi_\nu \frac{(\nabla u)^2\, dx}{(|\nabla u|^2 + \nu^{-1})^{1/2}} = \int_0^\infty dt \int_{\Omega \cap \partial N_t} \frac{\phi_\nu\, |\nabla u|\, ds}{(|\nabla u|^2 + \nu^{-1})^{1/2}}.$$

Passing to the limit as $\nu \to \infty$ with the use of the monotone convergence theorem, we obtain (1). □

Corollary 1. *Let $u \in C^\infty(\Omega)$ and let Φ be a non-negative lower semi-continuous function on Ω. Then $\int_\Omega \Phi(x) |\nabla u(x)|\, dx = \int_0^\infty dt \int_{\Omega \cap \partial N_t} \Phi(x)\, ds_x$.*

The proof is a consequence of the following chain of identities:

$$\int_\Omega \Phi(x) |\nabla u(x)|\, dx = \int_0^\infty \int_{\Phi(x)>\tau} |\nabla u(x)|\, dx\, d\tau$$

$$= \int_0^\infty \int_0^\infty s(\{x \in \partial N_t : \Phi(x) > \tau\})\, d\tau\, dt$$

$$= \int_0^\infty dt \int_{\Omega \cap \partial N_t} \Phi(x)\, ds_x.$$

Formula (1) leads to the relation between the estimate

$$\int |u|\, d\mu \leq C \|\nabla u\|_{L_1}, \qquad u \in D, \tag{2}$$

and some isoperimetric inequality. Namely, we have the following assertion:

Lemma 2. *The exact constant C in (2) is equal to*

$$\sup_g \frac{\mu(g)}{s(\partial g)}, \tag{3}$$

where g is any open set in R^n with compact closure and smooth boundary.

Proof. We have $\int |u|\, d\mu = \int_0^\infty \mu(N_t)\, dt \leq \sup_g (\mu(g)/s(\partial g)) \int_0^\infty s(\partial N_t)\, dt$, which, together with Lemma 1, gives the upper bound for C.

Let $\delta(x) = \text{dist}\,(x, g)$, $g_t = \{x : \delta(x) < t\}$. It is well known that there exists a small $\varepsilon > 0$ such that the surface ∂g_t is smooth for $t \leq \varepsilon$. We substitute the function $u_\varepsilon(x) = \alpha[\delta(x)]$, where $\alpha \in C^\infty([0, \infty))$, $\alpha(0) = 1$, $\alpha(t) = 0$ for $t > \varepsilon$, into (2). According to Corollary 1,

$$\int |\nabla u_\varepsilon|\, dx = \int_0^\varepsilon \alpha'(t) s(\partial g_t)\, dt.$$

Since $s(\partial g_t) \to s(\partial g)$ as $t \to +0$, then

$$\int |\nabla u_\varepsilon|\, dx \to s(\partial g). \tag{4}$$

On the other hand,

$$\int |u_\varepsilon|\, d\mu \geq \mu(g). \tag{5}$$

Combining (4), (5) and (2), we obtain $\mu(g) \leq Cs(\partial g)$. □

The following more general assertion, which will be used in 3.4.1, is proved in the same way.

Proposition. *The best constant* C *in* $\int |u|\, d\mu \leq C\, \|\Phi \nabla u\|_{L_1}$, *where* $\Phi \in C(R^n)$ *and* u *is an arbitrary function in* $D(R^n)$, *is equal to* $\sup_g (\mu(g)/\int_{\partial g} \Phi(x)\, ds_x)$. *Here* g *is any open set in* R^n *with a compact closure, bounded by a smooth surface, as in Lemma 2.*

Further, we prove that

$$\sup_g \frac{\mu(g)}{s(\partial g)} \sim \sup_{x \in R^n, r>0} r^{1-n} \mu(B_r(x)). \tag{6}$$

With this aim in view we present certain known auxiliary assertions. We start with the formulation of the classical Besicovitch covering theorem (see Guzman [1]).

Lemma 3. *Let* E *be a bounded set in* R^n *and let* $B_{r(x)}(x)$ *be a ball with* $r(x) > 0$ *and* $x \in E$. *By* L *we denote the totality of these balls. Then one can choose a sequence of balls* $\{B^{(m)}\}$ *from* L *such that*

(i) $E \subset \bigcup_m B^{(m)}$;

(ii) *there exists a number* N, *depending only on the dimension of the space, such that every point of the space belongs to not more than* N *balls from* $\{B^{(m)}\}$;

(iii) *the balls* $(1/3)B^{(m)}$ *are disjoint;*

(iv) $\bigcup_{B \in L} B \subset \bigcup_m 4B^{(m)}$.

(Here αB *and* B *are concentric balls with ratio of radii being equal to* α.*)*

We present one more well-known geometric lemma.

Lemma 4. *Let* g *be an open subset of* R^n *with a smooth boundary such that* $2\,\mathrm{mes}_n (B_r \cap g) = \mathrm{mes}_n (B_r)$. *Then* $s(B_r \cap \partial g) \geq c_n r^{n-1}$, *where* c_n *is a positive constant depending only on* n.

Proof. Let χ and ψ be characteristic functions of sets $g \cap B_r$ and $B_r \backslash g$. For any vector $z \neq 0$ we introduce the projection mapping p_z onto a $(n-1)$-dimensional subspace orthogonal to z. By Fubini's theorem,

$$\begin{aligned}(1/4)(\mathrm{mes}_n (B_1) r^n)^2 &= \mathrm{mes}_n (g \cap B_r)\, \mathrm{mes}_n (B_r \backslash g) \\ &= \iint \chi(x)\psi(y)\, dx\, dy = \iint \chi(x)\psi(x+z)\, dz\, dx \\ &= \int_{|z| \leq 2r} \mathrm{mes}_n (\{x : x \in B_r \cap g, (x+z) \in B_r \backslash g\})\, dz.\end{aligned}$$

Since any segment which joins $x \in g \cap B_r$ with $(x+z) \in B_r \backslash g$, intersects $B_r \cap \partial g$, the last integral does not exceed

$$2r \int_{|z|\leq 2r} \text{mes}_{n-1} [p_z(B_r \cap \partial g)] \leq (2r)^{n+1} \text{mes}_n (B_1) s(B_r \cap \partial g). \quad \square$$

Remark. This simple proof is due to Federer [2]. It does not present the best constant c_n which is equal to the volume of a $(n-1)$-dimensional unit ball (see Burago, Maz'ya [1]).

The following covering lemma is due to Gustin [1]. We give here the proof found by Federer [2].

Lemma 5. *Let g be a bounded open subset of R^n with smooth boundary. There exists a covering of g by a sequence of balls with radii ρ_i $(i=1, 2, \dots)$ such that*

$$\sum_j \rho_j^{n-1} \leq cs(\partial g), \tag{7}$$

where c is a constant which depends only on n.

Proof. Each point $x \in g$ is the centre of the ball $B_r(x)$ for which

$$\frac{\text{mes}_n (B_r(x) \cap g)}{\text{mes}_n (B_r(x))} = \frac{1}{2}. \tag{8}$$

(This ratio is a continuous function of r equal to 1 for small r and tending to zero as $r \to \infty$.) By Lemma 3, there exists a sequence of disjoint balls $B_{r_j}(x_j)$ for which $g \subset \bigcup_{j=1}^{\infty} B_{3r_j}(x_j)$. From Lemma 4 and from (8) we get $s(B_{r_j}(x_j) \cap \partial g) \geq c_n r_j^{n-1}$. Therefore

$$s(\partial g) \geq \sum_j s(B_{r_j}(x_j) \cap \partial g) \geq 3^{1-n} c_n \sum_j (3r_j)^{n-1}.$$

Thus, $\{B_{3r_j}(x_j)\}$ is the required covering. $\square$

Corollary 2. *The best constant in (2) is equivalent to $K = \sup_{x \in R^n, r>0} r^{1-n} \mu(B_r(x))$.*

Proof. By Lemma 2 it is sufficient to show that for any admissible set g we have $\mu(g) \leq cKs(\partial g)$. Let $\{B_{\rho_j}(x_j)\}$ be a covering of g constructed in Lemma 5. It is clear that

$$\mu(g) \leq \sum_j \mu(B_{\rho_j}(x_j)) \leq K \sum_j \rho_j^{n-1} \leq cKs(\partial g). \quad \square$$

1.2.2 The Case $m \geqslant 1$

Theorem 1. *Let m and l be integers, $m \geqslant l \geqslant 0$. Then the best constant in*

$$\int |\nabla_l u| \, d\mu \leqslant C \|u\|_{w_1^m}, \qquad u \in D, \tag{1}$$

is equivalent to

$$K = \sup_{x \in R^n, r>0} r^{m-l-n} \mu(B_r(x)). \tag{2}$$

Proof. (i) The estimate $C \geqslant cK$: we set in (1) $u(\xi) = (x_1 - \xi_1)^l \varphi(r^{-1}(x-\xi))$, where $\varphi \in D(B_2)$, $\varphi = 1$ on B_1. Since

$$\int |\nabla_l u| \, d\mu \geqslant l! \, \mu(B_r(x)), \qquad \|\nabla_m u\|_{L_1} = c r^{n-m+l},$$

then $C \geqslant cK$.

(ii) The estimate $C \leqslant cK$: it suffices to give the proof for $l = 0$. We have

$$\int |u| \, d\mu(x) = c \int \left| \int \frac{(\xi - x) \nabla_\xi u(\xi)}{|\xi - x|^n} \, d\xi \right| d\mu(x) \leqslant c \int |\nabla u| \, P_\mu \, dx, \tag{3}$$

where $P_\mu = |x|^{1-n} * \mu$. For $m = 1$ the result is contained in Corollary 1.2.2. The last integral in (3) does not exceed

$$c \sup_{x \in R^n, r>0} \left(r^{m-n-1} \int_{B_r(x)} P_\mu(\xi) \, d\xi \right) \|\nabla u\|_{w_1^{m-1}}.$$

Clearly,

$$\int_{B_r(x)} P_\mu(\xi) \, d\xi = \int_{B_r(x)} d\xi \int_{B_{2r}(x)} |\xi - \sigma|^{1-n} \, d\mu(\sigma)$$
$$+ \int_{B_r(x)} d\xi \int_{R^n \setminus B_{2r}(x)} |\xi - \sigma|^{1-n} \, d\mu(\sigma).$$

The first integral in the right-hand side is majorized by $cr\mu(B_{2r}(x))$ and the second one is not more than

$$c r^n \int_{R^n \setminus B_{2r}(x)} |x - \sigma|^{1-n} \, d\mu(\sigma)$$
$$= c(n-1) r^n \int_{2r}^{\infty} \mu\{\sigma : 2r \leqslant |x - \sigma| < t\} t^{-n} \, dt.$$

So $r^{m-n-1} \int_{B_r(x)} P_\mu(\xi) \, d\xi \leqslant c \sup_{x \in R^n, r>0} r^{m-n} \mu(B_r(x))$. □

Remark 1. It is clear that for $m-l>n$ the finiteness of (2) means $\mu=0$. In the case $m-l=n$, the value (2) is equal to $\mu(R^n)$.

We give an analogue of Theorem 1 for the space W_1^m.

Theorem 2. *Let m and l be integers, $m \geq l \geq 0$. Then the best constant in*

$$\int |\nabla_l u| \, d\mu \leq C \, \|u\|_{W_1^m}, \qquad u \in D, \tag{4}$$

is equivalent to

$$K = \sup_{x \in R^n, r \in (0,1]} r^{m-l-n} \mu(B_r(x)). \tag{5}$$

Proof. The estimate $C \geq cK$ is obtained in precisely the same way as the analogous one in Theorem 1. To prove the converse inequality we introduce a partition of unity $\{\varphi_j\}_{j\geq 1}$ subjected to the covering of R^n by unit balls with centres in nodes of sufficiently small coordinate grid. We apply Theorem 1 to the integral $\int |\nabla_l(\varphi_j u)| \, d\mu_j$, where μ_j is the restriction of μ to the support of the function φ_j. Then

$$\int |\nabla_l u| \, d\mu \leq \sum_j \int |\nabla_l(\varphi_j u)| \, d\mu_j \leq cK \sum_j \|\varphi_j u\|_{W_1^m}$$

$$\leq cK \, \|u\|_{W_1^m}. \qquad \square$$

Remark 2. Obviously, in the case $m-l \geq n$ the value K defined in (5) is equal to $\sup_{x \in R^n} \mu(B_1(x))$.

1.3 Description of Spaces $M(W_p^m \to W_p^l)$ and $M(w_p^m \to w_p^l)$

In this section we study multipliers acting from one Sobolev space into another. Subsection 1.3.1 contains basic definitions and lemmas. In 1.3.2 we give a description of spaces $M(W_p^m \to W_p^l)$, $M(w_p^m \to w_p^l)$, $p>1$, and derive two-sided estimates for norms in these spaces. The results presented here are obtained by the authors in [2]. In 1.3.3 we consider specifically the simplest case $mp>n$. From necessary and sufficient conditions established in 1.3.2 we derive in 1.3.7 simpler and distinct necessary or sufficient conditions for belonging to classes $M(W_p^m \to W_p^l)$ and $M(w_p^m \to w_p^l)$. The description of spaces $M(W_1^m \to W_1^l)$ and $M(w_1^m \to w_1^l)$ is given in 1.3.5 (see Maz'ya, Shaposhnikova [10]).

1.3.1 Definitions and Auxiliary Assertions

We say that a function γ, defined on R^n, belongs to the space $M(W_p^m \to W_p^l)$, if $\gamma u \in W_p^l$ for all $u \in W_p^m$.

By analogy, $\gamma \in M(w_p^m \to w_p^l)$ if $\gamma u \in w_p^l$ for all $u \in w_p^m$.

When referring to the space $M(w_p^m \to w_p^l)$ we shall always assume that $mp < n$, $p > 1$ or $m \leq n$, $p = 1$.

Let $\gamma \in M(W_p^m \to W_p^l)$, $u_n \to u$ in W_p^m, and let $\gamma u_n \to v$ in W_p^l. Then there exists a sequence of numbers $\{n_k\}_{k \geq 1}$ such that $u_{n_k}(x) \to u(x)$, $\gamma(x) u_{n_k}(x) \to v(x)$, for almost all x with respect to mes_n. Hence $v = \gamma u$ almost everywhere in R^n. Thus, the operator $W_p^m \ni u \to \gamma u \in W_p^l$ is closed. Since it is defined on the whole of W_p^m, it is bounded by Banach's theorem.

In the same way we obtain the boundedness of the operator $w_p^m \ni u \to \gamma u \in w_p^l$, where $\gamma \in M(w_p^m \to w_p^l)$.

As the norms in spaces $M(W_p^m \to W_p^l)$ and $M(w_p^m \to w_p^l)$ we introduce the norms of operators of multiplication

$$\|\gamma\|_{M(W_p^m \to W_p^l)} = \sup \{\|\gamma u\|_{W_p^l} : \|u\|_{W_p^m} \leq 1\},$$
$$\|\gamma\|_{M(w_p^m \to w_p^l)} = \sup \{\|\gamma u\|_{w_p^l} : \|u\|_{w_p^m} \leq 1\}.$$

We shall write MW_p^l and Mw_p^l instead of $M(W_p^l \to W_p^l)$ and $M(w_p^l \to w_p^l)$.

By $W_{p,\text{loc}}^l$ we denote the space $\{u : \eta u \in W_p^l$ for all $\eta \in D\}$. Obviously, $M(W_p^m \to W_p^l) \subset W_{p,\text{loc}}^l$ and $M(w_p^m \to w_p^l) \subset W_{p,\text{loc}}^l$.

We introduce the space $W_{p,\text{unif}}^l = \{u : \sup_{z \in R^n} \|\eta_z u\|_{W_p^l} < \infty\}$, where $\eta_z(x) = \eta(x - z)$, $\eta \in D$, $\eta = 1$ on B_1. Let $\|u\|_{W_{p,\text{unif}}^l} = \sup_{z \in R^n} \|\eta_z u\|_{W_p^l}$.

The definitions presented above can obviously be extended to other pairs S_1, S_2 of Banach spaces of functions on R^n which will be considered in the book. In what follows, $M(S_1 \to S_2) = \{\gamma : \gamma u \in S_2$ for all $u \in S_1\}$ and

$$\|\gamma\|_{M(S_1 \to S_2)} = \sup \{\|\gamma u\|_{S_2} : \|u\|_{S_1} \leq 1\}.$$

We shall write MS instead of $M(S \to S)$. We associate with S the space $S_{\text{loc}} = \{u : \eta u \in S$ for all $\eta \in D\}$ and we associate the space S_{unif} with the norm $\|u\|_{S_{\text{unif}}} = \sup_{z \in R^n} \|\eta_z u\|_S$.

We now return to the spaces $M(W_p^m \to W_p^l)$ and $M(w_p^m \to w_p^l)$.

Multipliers satisfy the inequalities

$$\|\gamma\|_{M(W_p^{m-j} \to W_p^{l-j})} \leq c\, \|\gamma\|_{M(W_p^m \to W_p^l)}^{(l-j)/l}\, \|\gamma\|_{M(W_p^{m-l} \to L_p)}^{j/l}, \tag{1}$$
$$\|\gamma\|_{M(w_p^{m-j} \to w_p^{l-j})} \leq c\, \|\gamma\|_{M(w_p^m \to w_p^l)}^{(l-j)/l}\, \|\gamma\|_{M(w_p^{m-l} \to L_p)}^{j/l}, \qquad j = 0, 1, \ldots, l, \tag{2}$$

which follow from the interpolation property of Sobolev spaces (see Calderon [1] and also Triebel [4]).

In the following lemma, γ_h denotes the mollification of γ with radius h, i.e.

$$\gamma_h(x) = h^{-n} \int K\left(\frac{x-\xi}{h}\right)\gamma(\xi)\, d\xi,$$

where $K \in D$, $K \geqslant 0$ and $\|K\|_{L_1} = 1$.

Lemma 1. The inequalities

$$\|\gamma_h\|_{M(W_p^m \to W_p^l)} \leqslant \|\gamma\|_{M(W_p^m \to W_p^l)} \leqslant \varliminf_{h\to 0} \|\gamma_h\|_{M(W_p^m \to W_p^l)} \tag{3}$$

are valid. The same is true if the space $M(W_p^m \to W_p^l)$ is replaced by $M(w_p^m \to w_p^l)$.

Proof. Let $u \in D$. By Minkowski's inequality,

$$\left(\int \left|\nabla_{j,x} \int h^{-n}K(\xi/h)\gamma(x-\xi)u(x)\, d\xi\right|^p dx\right)^{1/p}$$
$$\leqslant \int h^{-n}K(\xi/h)\left(\int |\nabla_{j,y}[\gamma(y)u(y-\xi)]|^p\, dy\right)^{1/p} d\xi,$$

where $j = 0, 1$. So

$$\|\gamma_h u\|_{W_p^l} \leqslant \|\gamma\|_{M(W_p^m \to W_p^l)} \int h^{-n}K(\xi/h)\left[\left(\int |\nabla_{m,y}(u(y-\xi))|^p\, dy\right)^{1/p}\right.$$
$$\left. + \left(\int |u(y-\xi)|^p\, dy\right)^{1/p}\right] d\xi \leqslant \|\gamma\|_{M(W_p^m \to W_p^l)} \|u\|_{W_p^m}.$$

This implies the left-hand side of (3). The right-hand side of (3) follows from

$$\|\gamma u\|_{W_p^l} = \varliminf_{h\to 0} \|\gamma_h u\|_{W_p^l} \leqslant \varliminf_{h\to 0} \|\gamma\|_{M(W_p^m \to W_p^l)} \|u\|_{W_p^m}.$$

For the space $M(w_p^m \to w_p^l)$ the proof is precisely the same. □

Lemma 2. Let $\psi \in D(B_2)$, $\psi = 1$ on B_1 and $\psi_r(x) = \psi(x/r)$, where $r > 0$. Then, for $lp < n$, $\|\psi_r\|_{Mw_p^l} \leqslant c(l, n, p)$.

Proof. For any $u \in D$ we have

$$\|\psi_r u\|_{w_p^l} \leqslant c \sum_{j=0}^{l} \| \, |\nabla_j \psi_r| \, |\nabla_{l-j} u| \, \|_{L_p} \leqslant c \sum_{j=0}^{l} \| \, |x|^{-j} \, |\nabla_{l-j} u| \, \|_{L_p}.$$

The last sum is estimated by Hardy's inequality,

$$\| \, |x|^{-j} v\|_{L_p} \leqslant c \, \|\nabla_j v\|_{L_p}, \qquad jp < n, \quad v \in D. \; \square$$

Corollary. *The following inequalities are valid*

$$c\,\|\psi_r\gamma\|_{M(w_p^m\to w_p^l)}\leqslant\|\gamma\|_{M(w_p^m\to w_p^l)}\leqslant\varliminf_{r\to\infty}\|\psi_r\gamma\|_{M(w_p^m\to w_p^l)}. \tag{4}$$

Proof. The left-hand side follows from Lemma 2 and the obvious inequality

$$\|\psi_r\gamma\|_{M(w_p^m\to w_p^l)}\leqslant\|\psi_r\|_{Mw_p^l}\,\|\gamma\|_{M(w_p^m\to w_p^l)}.$$

Noticing that

$$\|\gamma u\|_{w_p^l}=\lim_{r\to\infty}\|\psi_r\gamma u\|_{w_p^l}\leqslant\varliminf_{r\to\infty}\|\psi_r\gamma\|_{M(w_p^m\to w_p^l)}\,\|u\|_{w_p^m},$$

we obtain the right-hand side of (4). □

The following assertion is a particular case of Lemma 1.1.4/2.

Lemma 3. *Let $\gamma\in L_{p,\text{loc}}$, $p\in(1,\infty)$ and let u be an arbitrary function in D. The best constant in $\|\gamma\nabla_l u\|_{L_p}+\|\gamma u\|_{L_p}\leqslant C\,\|u\|_{W_p^m}$ is equivalent to $\|\gamma\|_{M(W_p^{m-l}\to L_p)}$.*

An obvious corollary of Lemma 1.1.4/1 is

Lemma 4. *Let $\gamma\in L_{p,\text{loc}}$, $p\in(1,\infty)$ and let u be an arbitrary function in D. The best constant C in $\|\gamma\nabla_l u\|_{L_p}\leqslant C\,\|u\|_{w_p^m}$ is equivalent to the norm $\|\gamma\|_{M(W_p^{m-l}\to L_p)}$.*

1.3.2 Characterization of Multipliers (The Case $p>1$)

In this subsection we present necessary and sufficient conditions for a function to belong to spaces $M(W_p^m\to W_p^l)$ and $M(w_p^m\to w_p^l)$.

Lemma 1. (i) *If $\gamma\in M(W_p^m\to W_p^l)\cap M(W_p^{m-l}\to L_p)$, $p\in(1,\infty)$, then $D^\alpha\gamma\in M(W_p^m\to W_p^{l-|\alpha|})$ for any multi-index α of positive order $|\alpha|\leqslant l$. The estimate*

$$\|D^\alpha\gamma\|_{M(W_p^m\to W_p^{l-|\alpha|})}\leqslant\varepsilon\,\|\gamma\|_{M(W_p^{m-l}\to L_p)}+c(\varepsilon)\,\|\gamma\|_{M(W_p^m\to W_p^l)} \tag{1}$$

is valid, where ε is an arbitrary positive number.

(ii) *Assertion (i) remains true for the change of W to w.*

Proof. (i) Let $u\in W_p^l$ and let φ be an arbitrary function in D. By the Leibniz formula

$$D^\alpha(\varphi u)=\sum_{\{\beta:\alpha\geqslant\beta\geqslant 0\}}\frac{\alpha!}{\beta!\,(\alpha-\beta)!}D^\beta\varphi D^{\alpha-\beta}u$$

we have

$$\int \varphi u D^\alpha \gamma \, dx = \int \gamma D^\alpha(\varphi u)\, dx$$
$$= \sum_{\{\beta:\alpha \geq \beta \geq 0\}} \frac{\alpha!}{\beta!\,(\alpha-\beta)!} \int \gamma D^\beta \varphi D^{\alpha-\beta} u \, dx$$
$$= \int \varphi \sum_{\{\beta:\alpha \geq \beta \geq 0\}} \frac{\alpha!}{\beta!\,(\alpha-\beta)!} D^\beta(\gamma D^{\alpha-\beta} u)\, dx.$$

Therefore

$$uD^\alpha \gamma = \sum_{\{\beta:\alpha \geq \beta \geq 0\}} \frac{\alpha!}{\beta!\,(\alpha-\beta)!} D^\beta(\gamma D^{\alpha-\beta} u)$$

which implies the estimate

$$\|uD^\alpha \gamma\|_{W_p^{l-|\alpha|}} \leq c \sum_{\{\beta:\alpha \geq \beta \geq 0\}} \|\gamma D^{\alpha-\beta} u\|_{W_p^{l-|\alpha|+|\beta|}}.$$

Hence it suffices to prove (1) for $|\alpha| = 1$, $l \geq 1$. We have

$$\|u\, \nabla \gamma\|_{W_p^{l-1}} \leq \|u\gamma\|_{W_p^l} + \|\gamma\, \nabla u\|_{W_p^{l-1}}$$
$$\leq (\|\gamma\|_{M(W_p^m \to W_p^l)} + \|\gamma\|_{M(W_p^{m-1} \to W_p^{l-1})})\, \|u\|_{W_p^m}.$$

Estimating the norm $\|\gamma\|_{M(W_p^{m-1} \to W_p^{l-1})}$ by means of (1.3.1/1), we arrive at (1).

The proof in case (ii) is precisely the same. □

Clearly, the problem of description of the space $M(W_p^m \to L_p)$ was solved in part (i) of Theorem 1.1.4. Namely,

$$\|\gamma\|_{M(W_p^m \to L_p)} \sim \sup_e \frac{\|\gamma; e\|_{L_p}}{[\text{cap}\,(e, W_p^m)]^{1/p}}. \tag{2}$$

According to Corollary 1.1.4/1, relation (2) can be rewritten in the form

$$\|\gamma\|_{M(W_p^m \to L_p)} \sim \sup_{\{e: d(e) \leq 1\}} \frac{\|\gamma; e\|_{L_p}}{[\text{cap}\,(e, W_p^m)]^{1/p}}. \tag{3}$$

Moreover, by Theorem 1.1.4,

$$\|\gamma\|_{M(w_p^m \to L_p)} \sim \sup_e \frac{\|\gamma; e\|_{L_p}}{[\text{cap}\,(e, w_p^m)]^{1/p}}. \tag{4}$$

We now obtain two-sided estimates for norms in spaces $M(W_p^m \to W_p^l)$ and $M(w_p^m \to w_p^l)$, $p \in (1, \infty)$, formulated in terms of spaces $M(W_p^k \to L_p)$ and $M(w_p^k \to L_p)$.

We begin with lower bounds.

Lemma 2. (i) *Let* $\gamma \in M(W_p^m \to W_p^l)$. *Then*

$$\|\nabla_l \gamma\|_{M(W_p^m \to L_p)} + \|\gamma\|_{M(W_p^{m-l} \to L_p)} \leq c\, \|\gamma\|_{M(W_p^m \to W_p^l)}. \tag{5}$$

(ii) *The same is true if W is changed to w.*

Proof. (i) First we assume that $\gamma \in M(W_p^{m-l} \to L_p)$. It is clear that

$$\|\gamma \nabla_l u\|_{L_p} \leq \|\gamma\|_{M(W_p^m \to W_p^l)} \|u\|_{W_p^m} + c \sum_{\substack{|\alpha|+|\beta|=l,\\ |\beta|\neq 0}} \|D^\alpha u D^\beta \gamma\|_{L_p}$$

$$\leq \Big(\|\gamma\|_{M(W_p^m \to W_p^l)} + c \sum_{j=1}^{l} \|\nabla_j \gamma\|_{M(W_p^{m-l-j} \to L_p)}\Big) \|u\|_{W_p^m}. \tag{6}$$

By Lemma 1,

$$\|\nabla_j \gamma\|_{M(W_p^{m-l+j} \to L_p)} \leq \varepsilon\, \|\gamma\|_{M(W_p^{m-l} \to L_p)} + c(\varepsilon)\, \|\gamma\|_{M(W_p^{m-l+j} \to W_p^j)}.$$

Using (1.3.1/1) to estimate the last norm, we get

$$\|\nabla_j \gamma\|_{M(W_p^{m-l+j} \to L_p)} \leq \varepsilon\, \|\gamma\|_{M(W_p^{m-l} \to L_p)} + c(\varepsilon)\, \|\gamma\|_{M(W_p^m \to W_p^l)}.$$

We substitute this inequality into (6). Then

$$\|\gamma \nabla_l u\|_{L_p} \leq (\varepsilon\, \|\gamma\|_{M(W_p^{m-l} \to L_p)} + c(\varepsilon)\, \|\gamma\|_{M(W_p^m \to W_p^l)})\, \|u\|_{W_p^m}. \tag{7}$$

At the same time,

$$\|\gamma u\|_{L_p} \leq \|\gamma\|_{M(W_p^m \to W_p^l)} \|u\|_{W_p^m}. \tag{8}$$

Summing (7) with (8) and using Lemma 1.3.1/3, we arrive at

$$\|\gamma\|_{M(W_p^{m-l} \to L_p)} \leq \varepsilon\, \|\gamma\|_{M(W_p^{m-l} \to L_p)} + c(\varepsilon)\, \|\gamma\|_{M(W_p^m \to W_p^l)}.$$

Consequently,

$$\|\gamma\|_{M(W_p^{m-l} \to L_p)} \leq c\, \|\gamma\|_{M(W_p^m \to W_p^l)}. \tag{9}$$

It remains to get rid of the assumption $\gamma \in M(W_p^{m-l} \to L_p)$. Since $\gamma \in M(W_p^m \to W_p^l)$, then $\|\gamma\eta\|_{L_p} \leq C\, \|\eta\|_{W_p^m}$, where $\eta \in D(B_2(x))$, $\eta = 1$ on $B_1(x)$, x being an arbitrary point in R^n. Hence $\sup_x \|\gamma; B_1(x)\|_{L_p} < \infty$ and, for any $k = 0, 1, \ldots$, there exists a constant C_h such that $|\nabla_k \gamma_h| \leq C_h$. Since the function γ_h is bounded together with all its derivatives, then γ_h is a multiplier in W_p^k for any $k = 1, 2, \ldots$ and, *a fortiori*, $\gamma_h \in M(W_p^{m-1} \to L_p)$. So

$$\|\gamma_h\|_{M(W_p^{m-l} \to L_p)} \leq c\, \|\gamma_h\|_{M(W_p^m \to W_p^l)}.$$

Lemma 1.3.1/1 makes it possible to pass here to the limit as $h \to 0$. Thus we obtain (9) for all $\gamma \in M(W_p^m \to W_p^l)$.

Let us estimate the first summand in the left-hand side of (5). We have

$$\|u\nabla_l\gamma\|_{L_p} \leq \|\gamma\|_{M(W_p^m \to W_p^l)} \|u\|_{W_p^m} + c \sum_{\substack{|\alpha|+|\beta|=l,\\ \alpha\neq 0}} \|D^\alpha u D^\beta \gamma\|_{L_p}$$

$$\leq \Big(\|\gamma\|_{M(W_p^m \to W_p^l)} + c \sum_{j=0}^{l-1} \|\nabla_j\gamma\|_{M(W_p^{m-l+j} \to L_p)}\Big) \|u\|_{W_p^m}$$

which, together with Lemma 1 and inequality (9), yields

$$\|u\nabla_l\gamma\|_{L_p} \leq c(\|\gamma\|_{M(W_p^m \to W_p^l)} + \|\gamma\|_{M(W_p^{m-l} \to L_p)}) \|u\|_{W_p^m}$$

$$\leq c\, \|\gamma\|_{M(W_p^m \to W_p^l)} \|u\|_{W_p^m}.$$

This immediately implies the estimate

$$\|\nabla_l\gamma\|_{M(W_p^m \to L_p)} \leq c\, \|\gamma\|_{M(W_p^m \to W_p^l)}.$$

(ii) The proof is the same as in case (i), but using the function $\psi_r\gamma_h$ instead of γ_h. Obviously, $\psi_r\gamma_h \in M(w_p^{m-l} \to L_p)$, if $\gamma \in M(w_p^m \to w_p^l)$. The pass to the limit as $r \to \infty$ and $h \to 0$ is justified by Lemma 1.3.1/1 and the corollary that follows from Lemma 1.3.1/2. □

The following lemma presents the converse result.

Lemma 3. (i) *The following inequality holds:*

$$\|\gamma\|_{M(W_p^m \to W_p^l)} \leq c(\|\nabla_l\gamma\|_{M(W_p^m \to L_p)} + \|\gamma\|_{M(W_p^{m-l} \to L_p)}). \tag{10}$$

(ii) *The same is true for the change of W to w.*

Proof. (i) It it is sufficient to assume that the right-side of (10) is finite. Inequality (1.3.1/1) together with Lemma 2 implies

$$\|\nabla_j\gamma\|_{M(W_p^{m-l+j} \to L_p)} \leq c\, \|\gamma\|_{M(W_p^m \to W_p^l)}^{j/l} \|\gamma\|_{M(W_p^{m-l} \to L_p)}^{1-j/l}, \quad j = 1, \ldots, l-1. \tag{11}$$

Let $u \in D$. We have

$$\|\nabla_l(\gamma u)\|_{L_p} \leq c \sum_{j=0}^{l} \|\, |\nabla_j\gamma|\, |\nabla_{l-j}u|\, \|_{L_p}$$

$$\leq c\Big(\|\nabla_l\gamma\|_{M(W_p^m \to L_p)} + \|\gamma\|_{M(W_p^{m-l} \to L_p)} + \sum_{j=1}^{l-1} \|\nabla_j\gamma\|_{M(W_p^{m-l+j} \to L_p)}\Big) \|u\|_{W_p^m}.$$

From this and (11) we obtain

$$\|\nabla_l(\gamma u)\|_{L_p} \leq c(\|\nabla_l\gamma\|_{M(W_p^m \to L_p)} + \|\gamma\|_{M(W_p^{m-l} \to L_p)}) \|u\|_{W_p^m}.$$

It remains to note that $\|\gamma u\|_{L_p} \leq \|\gamma\|_{M(W_p^{m-l} \to L_p)} \|u\|_{W_p^{m-l}}$.

The proof of part (ii) is precisely the same. □

Combining the formulations of Lemmas 2 and 3, we arrive at the following result.

Theorem 1. (i) *Let m, l be integers, $p \in (1, \infty)$. A function γ belongs to the space $M(W_p^m \to W_p^l)$ if and only if $\gamma \in W_{p,\mathrm{loc}}^l$, $\nabla_l \gamma \in M(W_p^m \to L_p)$ and $\gamma \in M(W_p^{m-l} \to L_p)$.*

The following relation holds:

$$\|\gamma\|_{M(W_p^m \to W_p^l)} \sim \|\nabla_l \gamma\|_{M(W_p^m \to L_p)} + \|\gamma\|_{M(W_p^{m-l} \to L_p)}.$$

(ii) *The same is true if everywhere W_p^k is replaced by w_p^k.*

Relations (2) and (4) enable one to transcribe this theorem into another form.

Theorem 2. (i) *A function γ belongs to the space $M(W_p^m \to W_p^l)$, $p \in (1, \infty)$, if and only if $\gamma \in W_{p,\mathrm{loc}}^l$ and*

$$\|\nabla_l \gamma; e\|_{L_p}^p \leqslant c \operatorname{cap}(e, W_p^m),$$
$$\|\gamma; e\|_{L_p}^p \leqslant c \operatorname{cap}(e, W_p^{m-l})$$

for any compact $e \subset R^n$.

The following relation holds:

$$\|\gamma\|_{M(W_p^m \to W_p^l)} \sim \sup_e \left(\frac{\|\nabla_l \gamma; e\|_{L_p}}{[\operatorname{cap}(e, W_p^m)]^{1/p}} + \frac{\|\gamma; e\|_{L_p}}{[\operatorname{cap}(e, W_p^{m-l})]^{1/p}} \right). \tag{12}$$

(ii) *The same is true if everywhere W_p^k is replaced by w_p^k.*

We distinguish an important particular case of Theorem 2 for $m = l$.

Corollary. (i) *A function γ belongs to the space MW_p^l, $p \in (1, \infty)$, if and only if $\gamma \in W_{p,\mathrm{loc}}^l \cap L_\infty$ and $\|\nabla_l \gamma; e\|_{L_p}^p \leqslant c \operatorname{cap}(e, W_p^l)$ for any compact $e \subset R^n$.*

The following relation holds:

$$\|\gamma\|_{MW_p^l} \sim \sup_e \frac{\|\nabla_l \gamma; e\|_{L_p}}{[\operatorname{cap}(e, W_p^l)]^{1/p}} + \|\gamma\|_{L_\infty}. \tag{13}$$

(ii) *The same is true for the change of W_p^k to w_p^k.*

Remark 1. Taking into account (5), we may restrict ourselves to compact sets e with $d(e) \leqslant 1$, when formulating part (i) of Theorem 2 and its corollary. Then relations (12), (13) can be rewritten as

$$\|\gamma\|_{M(W_p^m \to W_p^l)} \sim \sup_{\{e:d(e)\leq 1\}} \left(\frac{\|\nabla_l \gamma; e\|_{L_p}}{[\text{cap}\,(e, W_p^m)]^{1/p}} + \frac{\|\gamma, e\|_{L_p}}{[\text{cap}\,(e, W_p^{m-l})]^{1/p}} \right), \tag{14}$$

$$\|\gamma\|_{MW_p^l} \sim \sup_{\{e:d(e)\leq 1\}} \frac{\|\nabla_l \gamma; e\|_{L_p}}{[\text{cap}\,(e, W_p^l)]^{1/p}} + \|\gamma\|_{L_\infty}. \tag{15}$$

Remark 2. Obviously, (12) is equivalent to

$$\|\gamma\|_{M(W_p^m \to W_p^l)} \sim \sup_{e \subset \text{supp}\,\gamma} \left(\frac{\|\nabla_l \gamma; e\|_{L_p}}{[\text{cap}\,(e, W_p^m)]^{1/p}} + \frac{\|\gamma; e\|_{L_p}}{[\text{cap}\,(e, W_p^{m-l})]^{1/p}} \right). \tag{16}$$

In the same way we may add the condition $e \subset \text{supp}\,\gamma$ into (13)–(15).

Remark 3. According to a recent result by I. E. Verbitskii, the second summand on the right in (12) can be replaced by the norm $\|\gamma\|_{L_{1,\text{unif}}}$ for $m > l$. The proof of this assertion is not elementary and is postponed to the end of Chapter 2.

Remark 4. The relations (1.1.4/6) and (1.1.4/7) together with Theorem 2 imply representations for the norms in $M(W_p^m \to W_p^l)$ and in $M(w_p^m \to w_p^l)$ which do not involve a capacity. For instance,

$$\|\gamma\|_{MW_p^l} \sim \sup_{\{e:d(e)\leq 1\}} \left[\frac{\int (J_l \chi_e\, |\nabla_l \gamma|^p\,(x))^{p'}\,dx}{\int_e |\nabla_l \gamma(x)|^p\,dx} \right]^{(p-1)/p} + \|\gamma\|_{L_\infty},$$

where χ_e is the characteristic function of the set e.

For $p=2$ this relation takes the form

$$\|\gamma\|_{MW_2^l} \sim \sup_{\{e:d(e)\leq 1\}} \left[\frac{\int_e \int_e G_{2l}(x-y)\, |\nabla_l \gamma(x)|^2\, |\nabla_l \gamma(y)|^2\,dx\,dy}{\int_e |\nabla_l \gamma(x)|^2\,dx} \right]^{1/2} + \|\gamma\|_{L_\infty},$$

which immediately implies the estimate

$$\|\gamma\|_{MW_2^l} \leq c \left[\sup_{\{x,\xi:|x-\xi|<1\}} \left(\int_{B_1(\xi)} G_{2l}(x-y)\, |\nabla_l \gamma(y)|^2\,dy \right)^{1/2} + \|\gamma\|_{L_\infty} \right].$$

1.3.3 *The Space* $M(W_p^m \to W_p^l)$ *for* $mp > n$, $p > 1$

In the case $mp > n$ the space $M(W_p^m \to W_p^l)$ admits a simpler description. First we notice that $\|\gamma\|_{W^l_{p,\text{unif}}} \sim \sup_{x \in R^n} \|\gamma; B_1(x)\|_{W_p^l}$. Since $\text{cap}\,(e, W_p^k) \leq c$ under the condition $d(e) \leq 1$, then from (1.3.2/4) follows

$$\|\gamma\|_{M(W_p^m \to W_p^l)} \geq c\, \|\gamma\|_{W^l_{p,\text{unif}}}.$$

Thus, $W^l_{p,\text{unif}} \subset M(W^m_p \to W^l_p)$. We shall show that the converse inclusion holds as well, provided that $mp > n$.

Theorem. *If $mp > n$, $p \in (1, \infty)$, then*

$$\|\gamma\|_{M(W^m_p \to W^l_p)} \sim \sup_{x \in R^n} \|\gamma; B_1(x)\|_{W^l_p}. \tag{1}$$

Proof. It suffices to prove that the last supremum majorizes the right-hand side of (1.3.2/14). Since cap $(e, W^m_p) \sim 1$ for all $e \neq \varnothing$, $d(e) \leq 1$,

$$\sup_{\{e: d(e) \leq 1\}} \frac{\|\nabla_l \gamma; e\|_{L_p}}{[\text{cap}(e, W^m_p)]^{1/p}} \leq c \sup_{x \in R^n} \|\nabla_l \gamma; B_1(x)\|_{L_p}.$$

By analogy, for $(m-l)p > n$,

$$\sup_{\{e: d(e) \leq 1\}} \frac{\|\gamma; e\|_{L_p}}{[\text{cap}\,(e, W^{m-l}_p)]^{1/p}} \leq c \sup_{x \in R^n} \|\gamma; B_1(x)\|_{L_p}.$$

Proposition 1.1.4/1 implies that for $(m-l)p \leq n$ the left-hand side of the last inequality does not exceed $c \sup_{x \in R^n} \|\gamma; B_1(x)\|_{L_q}$, where

$$\begin{cases} q = n/(m-l) & \text{for} \quad (m-l)p < n, \\ q > p & \text{for} \quad (m-l)p = n. \end{cases}$$

It remains to note that $W^l_p(B_1) \subset L_q(B_1)$, provided that $mp > n$. □

Remark. We have deduced (1) using the results of the previous subsection. However, it can be established directly. For $m = l$, the direct proof is due to Strichartz [1].

1.3.4 One-sided Estimates for Norms of Multipliers (The case $mp \leq n$)

For $mp \leq n$ we can give distinct upper and lower bounds for norms in $M(W^m_p \to W^l_p)$, $M(w^m_p \to w^l_p)$ which involve no capacity. Thus we obtain separate necessary and sufficient conditions for a function to belong to these spaces of multipliers.

The following two propositions are immediate corollaries of Theorem 1.3.2/2 and of Proposition 1.1.4/2.

Proposition 1. *The following estimates hold:*

$$\|\gamma\|_{M(W_p^m \to W_p^l)} \geq \begin{cases} c \sup\limits_{x\in R^n, r\in(0,1)} r^{m-n/p}(\|\nabla_l \gamma; B_r(x)\|_{L_p} + r^{-l}\|\gamma; B_r(x)\|_{L_p}) \\ \qquad \text{for } mp<n \text{ and } p>1; \\ c \sup\limits_{x\in R^n, r\in(0,1)} ((\log 2/r)^{(p-1)/p}\|\nabla_l \gamma; B_r(x)\|_{L_p} \\ \qquad + r^{-l}\|\gamma; B_r(x)\|_{L_p}) \\ \qquad \text{for } mp=n, \quad p>1. \end{cases}$$

Proposition 2. *The inequality*

$$\|\gamma\|_{M(w_p^m \to w_p^l)} \geq c \sup_{x\in R^n, r>0} r^{m-n/p}(\|\nabla_l \gamma; B_r(x)\|_{L_p} + r^{-l}\|\gamma; B_r(x)\|_{L_p})$$

is true for $mp<n$, $p\in(1,\infty)$.

The next assertion contains upper bounds for the norm in $M(W_p^m \to W_p^l)$.

Proposition 3. *The following inequalities hold:*

$$\|\gamma\|_{M(W_p^m \to W_p^l)} \leq \begin{cases} c\left(\sup\limits_{\{e:d(e)\leq 1\}} \dfrac{\|\nabla_l \gamma; e\|_{L_p}}{(\mathrm{mes}_n\, e)^{1/p-m/n}} + \sup\limits_{x\in R^n}\|\gamma; B_1(x)\|_{L_p}\right) \\ \qquad \text{for } mp<n, \quad p<1, \quad l<m, \\ c\left(\sup\limits_{\{e:d(e)\leq 1\}} (\log(2^n/\mathrm{mes}_n\, e))^{(p-1)/p}\|\nabla_l \gamma; e\|_{L_p} \right. \\ \qquad \left. + \sup\limits_{x\in R^n}\|\gamma; B_1(x)\|_{L_p}\right) \\ \qquad \text{for } mp=n, \quad p>1, \quad l<m. \end{cases}$$

In the case $m=l$ one should replace $\sup_{x\in R^n}\|\gamma; B_1(x)\|_{L_p}$ by $\|\gamma\|_{L_\infty}$.

Proof. First we note that (1.3.2/14) and Proposition 1.1.4/1 imply

$$\|\gamma\|_{M(W_p^m \to W_p^l)} \leq \begin{cases} c \sup\limits_{\{e:d(e)\leq 1\}} (\mathrm{mes}_n\, e)^{m/n-1/p}(\|\nabla_l \gamma; e\|_{L_p} \\ \qquad + (\mathrm{mes}_n\, e)^{-l/n}\|\gamma; e\|_{L_p}) \\ \qquad \text{for } mp<n, \quad p>1, \\ c \sup\limits_{\{e:d(e)\leq 1\}} ((\log(2^n/\mathrm{mes}_n\, e))^{(p-1)/p}\|\nabla_l \gamma; e\|_{L_p} \\ \qquad + (\mathrm{mes}_n\, e)^{-l/n}\|\gamma; e\|_{L_p}) \\ \qquad \text{for } mp=n, \quad p>1. \end{cases}$$

The result follows for $m=l$.

Suppose $m > l$. Let e be a compact in R^n with $d(e) \leqslant 1$ and let B be an open unit ball, $\bar{B} \supset e$. The function γ satisfies the Sobolev integral representation (see Sobolev [1])

$$\gamma(x) = \sum_{|\beta|<l} x^\beta \int_B \varphi_\beta(y)\gamma(y)\,dy + \sum_{|\alpha|=l} \int_B \frac{f_\alpha(x, r, \theta)}{r^{n-l}} \nabla^\alpha \gamma(y)\,dy,$$

where $x \in B$, $r = |y-x|$, $\theta = (y-x)r^{-1}$, $\varphi_\beta \in D(B)$ and $f_\alpha(x, r, \theta)$ are infinitely differentiable functions of their arguments. So

$$|\gamma(x)| \leqslant c\left(\|\gamma; B\|_{L_1} + \int_B \frac{|\nabla_l \gamma(y)|}{r^{n-l}}\,dy\right) \tag{1}$$

and, consequently, for $mp \leqslant n$

$$(\text{mes}_n\, e)^{(m-l)/n-1/p} \|\gamma; e\|_{L_p}$$
$$\leqslant c\left[\|\gamma; B\|_{L_p} + (\text{mes}_n\, e)^{(m-l)/n-1/p}\left(\int_e \left(\int_{|z|<2} \frac{|\nabla_l \gamma(x+z)|\,dz}{|z|^{n-l}}\right)^p dx\right)^{1/p}\right].$$

We estimate the integral over the ball $|z|<2$ by the sum

$$\int_{B^{(0)}} + \sum_{j=0}^{N} \int_{B^{(j+1)} \setminus B^{(j)}},$$

where $B^{(j)} = \{z : |z| \leqslant 2^j (\text{mes}_n\, e)^{1/n}\}$ and $2^N (\text{mes}_n\, e)^{1/n} \leqslant 1$. By Minkowski's inequality,

$$\left(\int_e \left(\int_{B^{(0)}} \cdots dz\right)^p dx\right)^{1/p} \leqslant \int_{B^{(0)}} \frac{dz}{|z|^{n-l}} \left(\int_e |\nabla_l \gamma(x+z)|^p\,dx\right)^{1/p}.$$

We introduce the notation

$$s(\gamma) = \sup_{\{e : d(e) \leqslant 1\}} \frac{\|\nabla_l \gamma; e\|_{L_p}}{(\text{mes}_n\, e)^{1/p - m/n}}.$$

Clearly, the right-hand side of the last inequality does not exceed

$$s(\gamma)(\text{mes}_n\, e)^{1/p-m/n} \int_{B^{(0)}} \frac{dz}{|z|^{n-l}} \leqslant cs(\gamma)(\text{mes}_n\, e)^{1/p-(m-l)/n}.$$

We estimate the integral over the spherical layer $B^{(j+1)} \setminus B^{(j)}$ as follows:

$$\int_{B^{(j+1)} \setminus B^{(j)}} \cdots dz \leqslant c(2^j (\text{mes}_n\, e)^{1/n})^{l-n/p} \left(\int_{B^{(j+1)} \setminus B^{(j)}} |\nabla_l \gamma(x+z)|^p\,dz\right)^{1/p}$$
$$\leqslant cs(\gamma)(2^j (\text{mes}_n\, e)^{1/n})^{l-m}.$$

Consequently,

$$\Big(\int_e \Big(\sum_{j=0}^{N} \int_{B^{(j+1)}\setminus B^{(j)}} \cdots dz\Big)^p dx\Big)^{1/p}$$

$$\leqslant cs(\gamma) \sum_{j=0}^{N} (2^j (\text{mes}_n\, e)^{1/n})^{l-m} (\text{mes}_n\, e)^{1/p}$$

$$\leqslant cs(\gamma)(\text{mes}_n\, e)^{1/p-(m-l)/n}.$$

Thus,

$$(\text{mes}_n\, e)^{(m-l)/n-1/p} \, \|\gamma; e\|_{L_p}$$

$$\leqslant c\Big[\|\gamma; B\|_{L_p} + \sup_{e\subset \bar{B}} (\text{mes}_n\, e)^{m/n-1/p} \|\nabla_l \gamma; e\|_{L_p}\Big]. \quad \square \tag{2}$$

Sometimes Propositions 1–3 enable one to verify conditions for inclusion of individual functions into spaces of multipliers. We give two examples of this kind.

Example 1. Let $\mu > 0$ and let $\gamma(x) = \eta(x) \exp(i\,|x|^{-\mu})$, where $\eta \in C_0^\infty$, $\eta(0) = 1$. Clearly, $|\nabla_l \gamma(x)| \sim |x|^{-l(\mu+1)}$ as $x \to 0$. So $\gamma \in W_p^l$ is equivalent to $n > pl(\mu+1)$.

By virtue of Theorem 1.3.3, the same inequality is necessary and sufficient for γ to belong to $M(W_p^m \to W_p^l)$ for $mp > n$.

Suppose $mp < n$. We have $\|\nabla_l \gamma; B_r\|_{L_p} \sim \|\,|x|^{-l(\mu+1)}; B_r\|_{L_p}$ and $\lim_{r\to 0} r^{m-n/p} \|\nabla_l \gamma; B_r\|_{L_p} = \infty$ for $m < l(\mu+1)$. According to Proposition 1, this means that $\gamma \bar{\in} M(W_p^m \to W_p^l)$ for $m < l(\mu+1)$. If $m \geqslant l(\mu+1)$ then, for any compact set e with $d(e) \leqslant 1$,

$$\|\nabla_l \gamma; e\|_{L_p} \leqslant c\,\|\,|x|^{-l(\mu+1)}; e\|_{L_p} \leqslant (\text{mes}_n\, e)^{-l(\mu+1)/n+1/p}.$$

This, together with Proposition 3, implies $\gamma \in M(W_p^m \to W_p^l)$. Thus, for $mp < n$,

$$\gamma \in M(W_p^m \to W_p^l) \Leftrightarrow m \geqslant l(\mu+1).$$

In the same way we verify that, for $mp = n$, $\gamma \in M(W_p^m \to W_p^l) \Leftrightarrow m > l(\mu+1)$.

Example 2. Let $\mu, \nu > 0$, $\eta \in C_0^\infty(B_1(0))$, $\eta(0) = 1$ and $\gamma(x) = \eta(x) \times (\log |x|^{-1})^{-\nu} \exp(i(\log |x|^{-1})^\mu)$. Clearly, $|\nabla_l \gamma(x)| \sim c\,|x|^{-l} (\log |x|^{-1})^{l(\mu-1)-\nu}$. Using the same arguments as in Example 1, from this and Propositions

1 and 3 we obtain

$$\gamma \in W_p^l \Leftrightarrow l(\mu-1) < \nu - 1/p,$$
$$\gamma \in MW_p^l \Leftrightarrow l(\mu-1) \leq \nu - 1$$

for $lp = n$.

From Proposition 3 immediately follows:

Corollary 1. (i) *If* $mp < n$, $p > 1$, $l < m$, *then*

$$\|\gamma\|_{M(W_p^m \to W_p^l)} \leq c\, \|\gamma\|_{W^l_{n/m,\,\mathrm{unif}}}.$$

(ii) *If* $mp < n$, $p > 1$, *then*

$$\|\gamma\|_{MW_p^m} \leq c\left(\sup_{x \in R^n} \|\nabla_m \gamma; B_1(x)\|_{L_{n/m}} + \|\gamma\|_{L_\infty}\right).$$

We prove the assertion analogous to Proposition 3 for the space $M(w_p^m \to w_p^l)$.

Proposition 4. (i) *Let* $mp < n$, $p \in (1, \infty)$. *Then the following inequality holds:*

$$\|\gamma\|_{Mw_p^m} \leq c\left(\sup_e \frac{\|\nabla_m \gamma; e\|_{L_p}}{(\mathrm{mes}_n\, e)^{1/p - m/n}} + \|\gamma\|_{L_\infty}\right). \tag{3}$$

(ii) *If* $m > l$ *and*

$$\lim_{r \to \infty} r^{-n}\, \|\gamma; B_r\|_{L_1} = 0, \tag{4}$$

then the following inequality holds:

$$\|\gamma\|_{M(w_p^m \to w_p^l)} \leq c \sup_e \frac{\|\nabla_l \gamma; e\|_{L_p}}{(\mathrm{mes}_n\, e)^{1/p - m/n}}. \tag{5}$$

Proof. From part (ii) of Theorems 1.1.4 and 1.3.2/2 follows

$$\|\gamma\|_{M(w_p^m \to w_p^l)} \leq c \sup_e (\mathrm{mes}_n\, e)^{m/n - 1/p} (\|\nabla_l \gamma; e\|_{L_p} + (\mathrm{mes}_n\, e)^{-l/n}\, \|\gamma; e\|_{L_p}),$$

which is equivalent to (3) for $m = l$.

Using (4) and the fact that the right-hand side of (5) is finite, one can easily deduce the validity of Sobolev's integral representation

$$\gamma(x) = c \int \sum_{|\alpha| = l} \frac{\theta^\alpha}{\alpha!} \nabla^\alpha \gamma(y) \frac{dy}{r^{n-l}}, \qquad c = c(n, l). \tag{6}$$

Repeating the concluding part of the proof of Proposition 3 with obvious changes, we arrive at

$$\sup_e \frac{\|\gamma; e\|_{L_p}}{(\text{mes}_n\, e)^{1/p-(m-l)/n}} \leq c \sup_e \frac{\|\nabla_l \gamma; e\|_{L_p}}{(\text{mes}_n\, e)^{1/p-m/n}}. \quad \square$$

Remark. Condition (4) is necessary for a function γ to belong to the space $M(w_p^m \to w_p^l)$, $m > l$, since, according to Proposition 2,

$$\|\gamma\|_{M(w_p^m \to w_p^l)} \geq c r^{m-l-n/p} \|\gamma; B_r\|_{L_p}$$

for all $r > 0$.

Corollary 2. (i) *If* $mp < n$, $p \in (1, \infty)$, *then*

$$\|\gamma\|_{Mw_p^m} \leq c(\|\nabla_m \gamma\|_{L_{n/m}} + \|\gamma\|_{L_\infty}). \tag{7}$$

(ii) *If* $mp < n$, $p \in (1, \infty)$, $l < m$ *and* $\gamma \in w_{n/m}^l$, *then* $\gamma \in M(w_p^m \to w_p^l)$ *and the following inequality holds:*

$$\|\gamma\|_{M(w_p^m \to w_p^l)} \leq c \|\nabla_l \gamma\|_{L_{n/m}}. \tag{8}$$

Proof. (i) Estimate (7) follows from (3) by Hölder's inequality. (ii) By virtue of Sobolev's theorem, $w_{n/m}^l \subset L_{n/(m-l)}$. Hence $r^{-n} \|\gamma; B_r\|_{L_1} = O(r^{l-m})$ and so (5) holds. Applying Hölder's inequality to the right-hand side of (5), we arrive at (8). $\square$

1.3.5 Characterization of Multipliers (The Case $p = 1$)

Lemma 1. *The following relation holds:*

$$\|\gamma\|_{M(W_1^m \to W_1^l)} \sim \sup_{x \in R^n, r \in (0,1)} r^{m-n} (\|\nabla_l \gamma; B_r(x)\|_{L_1} + r^{-l} \|\gamma; B_r(x)\|_{L_1}). \tag{1}$$

Before we prove this lemma, we make the following:

Remark 1. Since

$$r^{j-l} \|\nabla_j \gamma; B_r(x)\|_{L_1} \leq c(\|\nabla_l \gamma; B_r(x)\|_{L_1} + r^{-l} \|\gamma; B_r(x)\|_{L_1}),$$

then (1) can be rewritten in the form

$$\|\gamma\|_{M(W_1^m \to W_1^l)} \sim \sup_{x \in R^n, r \in (0,1)} r^{m-n} \sum_{j=0}^{l} \|\nabla_j \gamma; B_r(x)\|_{L_1}. \tag{2}$$

Proof of Lemma 1. Let $u(y) = \varphi((y-x)/r)$, where $r \in (0, 1)$ for $m < n$ and $r = 1$ for $m \geq n$, $\varphi \in D(B_2)$, $\varphi = 1$ on B_1. We substitute the function u

into

$$\|\gamma u\|_{W_1^l} \leqslant \|\gamma\|_{M(W_1^m \to W_1^l)} \|u\|_{W_1^m}.$$

Since supp $\gamma u \subset B_{2r}(x)$, then, for $j = 0, 1, \ldots, l$,

$$\|\nabla_j(\gamma u); B_{2r}(x)\|_{L_1} \leqslant cr^{l-j} \|\gamma u\|_{W_1^l}.$$

Consequently, $r^{j-l} \|\nabla_j \gamma; B_r(x)\|_{L_1} \leqslant c \|\gamma\|_{M(W_1^m \to W_1^l)}$ and the required lower bound for the norm $\|\gamma\|_{M(W_1^m \to W_1^l)}$ is obtained.

Now we turn to the upper bound. We have

$$\|\nabla_l(\gamma u)\|_{L_1} \leqslant c \sum_{j=0}^{l} \| |\nabla_j \gamma| \, |\nabla_{l-j} u| \|_{L_1}.$$

By Theorem 1.2.2/2,

$$\| |\nabla_j \gamma| \, |\nabla_{l-j} u| \|_{L_1} \leqslant c \sup_{x \in R^n, r \in (0,1)} r^{m-l+j-n} \|\nabla_j \gamma; B_r(x)\|_{L_1} \|\nabla_{l-j} u\|_{W_1^{m-l+j}}.$$

Therefore

$$\|\nabla_l(\gamma u)\|_{L_1} \leqslant c \sup_{x \in R^n, r \in (0,1)} r^{m-n} \sum_{j=0}^{l} r^{j-l} \|\nabla_j \gamma; B_r(x)\|_{L_1} \|u\|_{W_1^m}.$$

According to the same Theorem 1.2.2/2,

$$\|\gamma u\|_{L_1} \leqslant c \sup_{x \in R^n, r \in (0,1)} r^{m-l-n} \|\gamma; B_r(x)\|_{L_1} \|u\|_{W_1^{m-l}}.$$

Summing the last two inequalities and taking into account Remark 1, we complete the proof. □

Theorem 1. (i) *If $m \geqslant n$, $m \geqslant l$, then*

$$\|\gamma\|_{M(W_1^m \to W_1^l)} \sim \sup_{x \in R^n} \|\gamma; B_1(x)\|_{W_1^l}. \tag{3}$$

(ii) *If $l < n$, then*

$$\|\gamma\|_{MW_1^l} \sim \sup_{x \in R^n; r \in (0,1)} r^{l-n} \|\nabla_l \gamma; B_r(x)\|_{L_1} + \|\gamma\|_{L_\infty}. \tag{4}$$

(iii) *If $l < m < n$, then*

$$\|\gamma\|_{M(W_1^m \to W_1^l)} \sim \sup_{x \in R^n, r > 0} r^{m-n} \|\nabla_l \gamma; B_r(x)\|_{L_1} + \sup_{x \in R^n} \|\gamma; B_1(x)\|_{L_1}. \tag{5}$$

Proof. Relations (3) and (4) are immediate corollaries of Lemma 1. It remains to prove (5). The lower bound for the norm in $M(W_1^m \to W_1^l)$

follows directly from (1). By (1.3.4/1),

$$\|\gamma; B_r(z)\|_{L_1} \leq c\left(r^n \|\gamma; B_1(z)\|_{L_1} + \int_{B_r(z)} \int_{B_1(z)} \frac{|\nabla_l \gamma(y)|}{|x-y|^{n-l}}\, dy\, dx\right)$$

$$\leq c\left(r^n \|\gamma; B_1(z)\|_{L_1} + r^l \|\nabla_l \gamma; B_{2r}(z)\|_{L_1}\right.$$

$$\left. + \int_{B_r(z)} \int_{B_1(z)\setminus B_{2r}(z)} |\nabla_l \gamma(y)| \frac{dy\, dx}{|y|^{n-l}}\right).$$

It is clear that

$$\int_{B_1(z)\setminus B_{2r}(z)} |\nabla_l \gamma(y)| \frac{dy}{|y|^{n-l}} \leq c r^{l-m+n} \sup_{\rho \in (0,1)} \rho^{m-n} \|\nabla_l \gamma; B_\rho(z)\|_{L_1}.$$

Consequently,

$$r^{m-n-l} \|\gamma; B_r(z)\|_{L_1} \leq c\left(\|\gamma; B_1(z)\|_{L_1} + \sup_{\rho \in (0,1)} \rho^{m-n} \|\nabla_l \gamma; B_\rho(z)\|_{L_1}\right).$$

By comparing this inequality with (1), we complete the proof. □

Next we obtain a similar result for the space $M(w_1^m \to w_1^l)$.

The following auxiliary assertion is proved in the same way as Lemma 1, but applying Theorem 1.2.2/1 instead of Theorem 1.2.2/2.

Lemma 2. *If $m \leq n$, then*

$$\|\gamma\|_{M(w_1^m \to w_1^l)} \sim \sup_{x \in R^n, r>0} r^{m-n} (\|\nabla_l \gamma; B_r(x)\|_{L_1} + r^{-l} \|\gamma; B_r(x)\|_{L_1}). \tag{6}$$

Theorem 2. (i) *If $l \leq n$, then*

$$\|\gamma\|_{Mw_1^l} \sim \sup_{x \in R^n, r>0} r^{l-n} \|\nabla_l \gamma; B_r(x)\|_{L_1} + \|\gamma\|_{L_\infty}. \tag{7}$$

(ii) *If $l < m \leq n$ and* $\lim_{r\to\infty} r^{-n} \|\gamma; B_r\|_{L_1} = 0$, *then*

$$\|\gamma\|_{M(w_1^m \to w_1^l)} \sim \sup_{x \in R^n, r>0} r^{m-n} \|\nabla_l \gamma; B_r(x)\|_{L_1}. \tag{8}$$

Proof. Relation (7) follows from (6). To prove (8) it suffices to use the integral representation (1.3.4/6) and to repeat the proof of part (iii) of Theorem 1 with obvious changes. □

Remark. From (6) it follows that the convergence of the average of $|\gamma|$ in the ball B_r to zero as $r \to \infty$ is necessary for a function γ to belong to $M(w_1^m \to w_1^l)$, $m > l$.

One can easily derive interpolation inequalities for elements of spaces $M(W_1^m \to W_1^l)$ and $M(w_1^m \to w_1^l)$ from Theorems 1 and 2.

Corollary 1. (i) *The following inequality holds:*

$$\|\gamma\|_{M(W_1^{m-j} \to W_1^{l-j})} \leq c\, \|\gamma\|_{M(W_1^m \to W_1^l)}^{1-j/l}\, \|\gamma\|_{M(W_1^{m-l} \to L_1)}^{j/l} \tag{9}$$

with $j = 0, 1, \ldots, l$.

(ii) *The same is true if* W_1^k *is replaced by* w_1^k.

Proof. (i) Applying the similarity transformation to the known and easily verified inequality

$$\|\nabla_{l-j}\gamma; B_1\|_{L_1} \leq c\, \|\gamma; B_1\|_{W_1^l}^{1-j/l}\, \|\gamma; B_1\|_{L_1}^{j/l},$$

we obtain

$$r^{j-l}\, \|\nabla_{l-j}\gamma; B_r(x)\|_{L_1} \leq c\left(\sum_{k=0}^{l} r^{k-l}\, \|\nabla_k\gamma; B_r(x)\|_{L_1}\right)^{1-j/l} \|\gamma; B_r(x)\|_{L_1}^{j/l}.$$

Consequently,

$$\begin{aligned} &r^{m-n+j-l}\, \|\nabla_{l-j}\gamma; B_r(x)\|_{L_1} + \|\gamma; B_r(x)\|_{L_1} \\ &\leq c\left(r^{m-n} \sum_{k=0}^{l} r^{k-l}\, \|\nabla_k\gamma; B_r(x)\|_{L_1}\right)^{1-j/l} (r^{m-n}\, \|\gamma; B_r(x)\|_{L_1})^{j/l}. \end{aligned}$$

Reference to (2) completes the proof.

(ii) The proof is the same. One should use relation (6). □

1.4 The Space $M(W_p^m \to W_q^l)$

In the preceding section we gave the description of classes $M(W_p^m \to W_p^l)$ and $M(w_p^m \to w_p^l)$ for non-negative integers m, l. Henceforth we present analogous results for multipliers in some other pairs of spaces.

In this book consideration is limited to classes of the type $M(\mathfrak{A}_p^m \to \mathfrak{A}_p^l)$, i.e. to multipliers acting in one scale of spaces and preserving the summability degree p. Generalization to pairs $(\mathfrak{A}_p^m, \mathfrak{B}_q^l)$ has been almost unknown until now. It is clear that sometimes such generalization needs no new technique. Certainly, this is true for the class $M(W_p^m \to W_q^l)$ (m and l are integers) which is briefly described in the present section.

Using the same arguments as in part (i) of Theorem 1.3.2/1, we obtain

$$\|\gamma\|_{M(W_p^m \to W_q^l)} \sim \|\nabla_l\gamma\|_{M(W_p^m \to L_q)} + \|\gamma\|_{M(W_p^{m-l} \to L_q)}, \tag{1}$$

where $p, q \in (1, \infty)$. Thus, as in Subsection 1.3.2, the problem is reduced to the description of the class $M(W_p^m \to L_q)$. This can be obtained immediately from the following assertions on the best constant C in the inequality $(\int |u|^q \, d\mu)^{1/q} \leq C\|u\|_{W_p^m}$, $u \in D$.

Lemma 1 (Adams [1], $p>1$; Maz'ya [10], $p=1$). *If* $1 \leq p < q$, $mp < n$, *then*

$$C \sim \sup_{x\in R^n, r\in(0,1]} r^{m-n/p}[\mu(B_r(x))]^{1/q}. \tag{2}$$

Lemma 2 (Maz'ya, Preobraženskii [1]). *If* $1<p<q$, $mp=n$, *then*

$$C \sim \sup_{x\in R^n, r\in(0,1]} (\log 2/r)^{1/p'}[\mu(B_r(x))]^{1/q}.$$

Lemma 3. *If* $1<p<q$, $mp>n$ *or* $1=p<q$, $m \geq n$, *then*

$$C \sim \sup_{x\in R^n} [\mu(B_1(x))]^{1/q}. \tag{3}$$

This is an obvious corollary of Sobolev's theorem ensuring the imbedding $W_p^m \subset L_\infty$ for $mp>n$.

Lemma 4 (Maz'ya [12]). (i) *If* $0<q<p$, $p>1$, *then*

$$C \sim \sup_{\{\sigma\}} \left(\sum_{j=-\infty}^{+\infty} \frac{[\mu(g_j \backslash g_{j+1})]^{p/(p-q)}}{[\text{cap}\,(g_j, W_p^m)]^{q/(p-q)}} \right)^{(p-q)/pq},$$

where σ *is an arbitrary sequence of open sets* $\{g_j\}_{j=-\infty}^{\infty}$ *such that* $\overline{g_{j+1}} \subset g_j$.

(ii) *If* $0<q<p$, $p>1$, $mp>n$ *or* $0<q<p=1$, $m \geq n$, *then*

$$C \sim \left(\sum_{j=0}^{\infty} [\mu(Q^{(j)})]^{p/(p-q)} \right)^{(p-q)/pq}, \tag{4}$$

where $\{Q^{(j)}\}$ *is a sequence of closed cubes which form a coordinate grid in* R^n.

Lemmas 1–4 together with (1) lead to the following assertions on equivalent norms in spaces $M(W_p^m \to W_q^l)$, $p>1$, $p \neq q$.

Theorem 1. *If* $1<p<q$, *then*

$$\|\gamma\|_{M(W_p^m \to W_q^l)} \sim \begin{cases} \sup\limits_{x\in R^n, r\in(0,1)} r^{m-n/p} \|\nabla_l \gamma; B_r(x)\|_{L_q} + \sup\limits_{x\in R^n} \|\gamma; B_1(x)\|_{L_q} \\ \quad \text{for} \quad mp<n; \\ \sup\limits_{x\in R^n, r\in(0,1)} (\log 2/r)^{1/p'} \|\nabla_l \gamma; B_r(x)\|_{L_q} + \sup\limits_{x\in R^n} \|\gamma; B_1(x)\|_{L_q} \\ \quad \text{for} \quad mp=n; \\ \sup\limits_{x\in R^n} \|\gamma; B_1(x)\|_{W_q^l} \quad \text{for} \quad mp>n. \end{cases}$$

Theorem 2. (i) *If* $1 \leq q < p$, *then*

$$\|\gamma\|_{M(W_p^m \to W_q^l)} \sim \sup_{\{\sigma\}} \left(\sum_{j=-\infty}^{\infty} \left(\frac{\|\nabla_l \gamma; g_j \backslash g_{j+1}\|_{L_q}}{[\operatorname{cap}(g_j, W_p^m)]^{1/p}} + \frac{\|\gamma; g_j \backslash g_{j+1}\|_{L_q}}{[\operatorname{cap}(g_j, W_p^{m-l})]^{1/p}} \right)^{pq/(p-q)} \right)^{(p-q)/pq}.$$

(ii) *If* $1 \leq q < p$, $mp > n$, *then*

$$\|\gamma\|_{M(W_p^m \to W_q^l)} \sim \left(\sum_{j=0}^{\infty} \|\gamma; Q^{(j)}\|_{W_q^l}^{pq/(p-q)} \right)^{(p-q)/pq}.$$

(The notations $\{\sigma\}$ and $\{Q^{(j)}\}$ have the same sense as in Lemma 4.)

Using the same arguments as in Subsection 1.3.5, one can derive the following result from relations (2), (3), (4) with $p = 1$.

Theorem 3. (i) *If* $q \geq 1$ *and* $m < n$, *then*

$$\|\gamma\|_{M(W_1^m \to W_q^l)} \sim \sup_{x \in R^n, r \in (0,1)} r^{m-n} \|\nabla_l \gamma; B_r(x)\|_{L_q} + \sup_{x \in R^n} \|\gamma; B_1(x)\|_{L_q}.$$

(ii) *If* $q \geq 1$ *and* $m \geq n$, *then*

$$\|\gamma\|_{M(W_1^m \to W_q^l)} \sim \sup_{x \in R^n} \|\gamma; B_1(x)\|_{W_q^l}.$$

1.5 The Space $M(W_p^m \to W_p^{-k})$

We consider multipliers acting into the space of generalized functions W_p^{-k}, $k \geq 0$, $p \in (1, \infty)$.

This space is defined as dual to $W_{p'}^k$, $p' + p = pp'$. It is well known that $u \in W_p^{-k}$ if and only if $u = \operatorname{div}_k \mathbf{U}$, where $\mathbf{U}$ is a vector $\{u_\alpha\}_{|\alpha| \leq k}$ with components in L_p and $\operatorname{div}_k \mathbf{U} = \sum_{|\alpha| \leq k} i^{|\alpha|} D^\alpha u_\alpha$. The norm in W_p^{-k} is defined by $\|u\|_{W_p^{-k}} = \inf_{\mathbf{U}} \sum_{|\alpha| \leq k} \|D^\alpha u_\alpha\|_{L_p}$.

Since

$$\|\gamma\|_{M(W_p^{-l} \to W_p^{-m})} = \|\bar{\gamma}\|_{M(W_{p'}^m \to W_{p'}^l)},$$

then, by Theorem 1.3.2/2,

$$\|\gamma\|_{M(W_p^{-l} \to W_p^{-m})} \sim \sup_{e \subset R^n} \left(\frac{\|\nabla_l \gamma; e\|_{L_{p'}}}{[\operatorname{cap}(e, W_{p'}^m)]^{1/p'}} + \frac{\|\gamma; e\|_{L_{p'}}}{[\operatorname{cap}(e, W_{p'}^{m-l})]^{1/p'}} \right). \tag{1}$$

The following theorem contains sufficient conditions for a function to belong to the space $M(W_p^m \to W_p^{-k})$, $m, k \geq 0$.

Theorem. (i) *Let* $p \in (1, \infty)$, $m \leq k$. *If* $\gamma = \operatorname{div}_k \mathbf{\Gamma}$, *where* $\mathbf{\Gamma} = \{\gamma_\alpha\}_{|\alpha| \leq k}$ *and* $\gamma_\alpha \in M(W_{p'}^k \to W_{p'}^{k-m}) \cap M(W_p^m \to L_p)$, *then* $\gamma \in M(W_p^m \to W_p^{-k})$. *The mapping* $\{\gamma_\alpha\}_{|\alpha| \leq k} \to \gamma$ *is continuous.*

(ii) *Let* $p \in (1, \infty)$, $m \geq k$. *If* $\gamma = \operatorname{div}_m \mathbf{\Gamma}$, *where* $\mathbf{\Gamma} = \{\gamma_\beta\}_{|\beta| \leq m}$ *and* $\gamma_\beta \in M(W_p^m \to W_p^{m-k}) \cap M(W_{p'}^k \to L_{p'})$, *then* $\gamma \in M(W_p^m \to W_p^{-k})$. *The mapping* $\{\gamma_\beta\}_{|\beta| \leq m} \to \gamma$ *is continuous.*

Proof. (i) Let $u \in W_p^m$, $m \leq k$. Since

$$uD^\alpha \gamma_\alpha = \sum_{\lambda \leq \alpha} c_{\lambda\alpha} D^\lambda (\gamma_\alpha D^{\alpha-\lambda} u), \qquad c_{\lambda\alpha} = \text{const},$$

(see page 29) then

$$\|\gamma u\|_{W_p^{-k}} \leq c \sum_{|\lambda| \leq |\alpha| \leq k} \|\gamma_\alpha D^{\alpha-\lambda} u\|_{W_p^{|\lambda|-k}}$$

$$\leq c \sum_{|\lambda| \leq |\alpha| \leq k} \|\gamma_\alpha\|_{M(W_p^{m-k+|\lambda|} \to W_p^{|\lambda|-k})} \|u\|_{W_p^{m+|\alpha|-k}}.$$

Applying interpolation inequality (1.3.1/1), we obtain

$$\|\gamma u\|_{W_p^{-k}} \leq c \sum_{|\lambda| \leq |\alpha| \leq l} \|\gamma_\alpha\|_{M(W_p^{m-k} \to W_p^{-k})}^{(k-|\lambda|)/k} \|\gamma_\alpha\|_{M(W_p^m \to L_p)}^{|\lambda|/k} \|u\|_{W_p^m}.$$

It remains to note that $\|\gamma_\alpha\|_{M(W_p^{m-k} \to W_p^{-k})} = \|\gamma_\alpha\|_{M(W_{p'}^k \to W^{k-m})}$.

(ii) Let $m \geq k$. Since $\|\gamma\|_{M(W_p^m \to W_p^{-k})} = \|\gamma\|_{M(W_{p'}^k \to W_{p'}^{-m})}$, the result follows from part (i). □

Since $M(W_2^m \to W_2^l) \subset M(W_2^{m-l} \to L_2)$, the formulation of the theorem is simpler for $p = 2$.

Corollary. *If* $\gamma = \operatorname{div}_s \mathbf{\Gamma}$, *where* $\mathbf{\Gamma} = \{\gamma_\alpha\}_{|\alpha| \leq s}$, $s = \max\{k, m\}$ *and* $\gamma_\alpha \in M(W_2^s \to W_2^{|k-m|})$, *then* $\gamma \in M(W_2^m \to W_2^{-k})$.

We conclude the section with a simple description of non-negative elements in the space $M(W_2^m \to W_2^{-m})$.

Theorem 2. *Let* $\gamma \geq 0$. *Then* $\gamma \in M(W_2^m \to W_2^{-m})$ *if and only if* $\gamma^{1/2} \in M(W_2^m \to L_2)$. *Moreover,*

$$\|\gamma\|_{M(W_2^m \to W_2^{-m})} = \|\gamma^{1/2}\|_{M(W_2^m \to L_2)}^2.$$

Proof. Let $u \in W_2^m$, $v \in W_2^m$. We have

$$\left| \int \gamma u \bar{v} \, dx \right| \leq \|\gamma^{1/2} u\|_{L_2} \|\gamma^{1/2} v\|_{L_2} \leq \|\gamma^{1/2}\|_{M(W_2^m \to L_2)}^2 \|u\|_{W_2^m} \|v\|_{W_2^m}.$$

Consequently, $\|\gamma\|_{M(W_2^m \to W_2^{-m})} \leq \|\gamma^{1/2}\|_{M(W_2^m \to L_2)}^2$.

Next we get the converse inequality. It is clear that

$$\left|\int \gamma u\bar{v}\,dx\right| \leq \|\gamma\|_{M(W_2^m\to W_2^{-m})}\|u\|_{W_2^m}\|v\|_{W_2^m}.$$

Putting here $u=v$, we find

$$\int |\gamma^{1/2}u|^2\,dx \leq \|\gamma\|_{M(W_2^m\to W_2^{-m})}\|u\|^2_{W_2^m}.$$

Thus

$$\|\gamma\|_{M(W_2^m\to W_2^{-m})} \geq \|\gamma^{1/2}\|^2_{M(W_2^m\to L_2)}. \quad \square$$

This theorem, together with Corollary 1.1.4 and Proposition 1.1.4/3, shows that for $\gamma\geq 0$ the following relations hold:

$$\|\gamma\|_{M(W_2^m\to W_2^{-m})} \sim \sup_{\{e:d(e)\leq 1\}} \frac{\int_e \gamma\,dx}{\operatorname{cap}(e, W_2^m)}$$

$$\sim \sup_{\{e:d(e)\leq 1\}} \frac{\int_e\int_e G_{2m}(x-y)\gamma(x)\gamma(y)\,dx\,dy}{\int_e \gamma\,dx}.$$

1.6 Certain Properties of Multipliers

In this section we study certain simple properties of elements of the space $M(W_p^m\to W_p^l)$, $p\geq 1$.

Proposition 1. *The space $M(W_p^m\to W_p^l)$ is contained in $M(W_p^{m-j}\to W_p^{l-j})$, $j=1,\ldots,l$, and the following estimate holds:*

$$\|\gamma\|_{M(W_p^{m-j}\to W_p^{l-j})} \leq c\,\|\gamma\|_{M(W_p^m\to W_p^l)}.$$

Proof. The inequality $\|\gamma\|_{M(W_p^{m-l}\to L_p)} \leq c\,\|\gamma\|_{M(W_p^m\to W_p^l)}$ is proved in Lemma 1.3.2/2 for $p>1$ and in Lemma 1.3.5/1 for $p=1$. It remains to use interpolation inequalities (1.3.1/1), (1.3.5/9). $\square$

Proposition 2. *If function γ depends only on variables $x_1,\ldots,x_s$, $s<n$, and $\gamma\in M(W_p^m(R^s)\to W_p^l(R^s))$, then $\gamma\in M(W_p^m(R^n)\to W_p^l(R^n))$ and the following estimate holds:*

$$\|\gamma; R^n\|_{M(W_p^m\to W_p^l)} \leq c\,\|\gamma; R^s\|_{M(W_p^m\to W_p^l)}.$$

The proof is obvious.

Proposition 3. *If $\gamma \in M(W_p^m \to W_p^l)$ and k is a positive integer satisfying the inequality $k \leq m/(m-l)$, then $\gamma^k \in M(W_p^m \to W_p^{m-k(m-l)})$. The estimate*

$$\|\gamma^k\|_{M(W_p^m \to W_p^{m-k(m-l)})} \leq c\, \|\gamma\|^k_{M(W_p^m \to W_p^l)}$$

is valid.

The proof is obvious.

Proposition 4. *The following estimate is true:*

$$\|\gamma\|_{L_\infty} \leq \|\gamma\|_{MW_p^l}. \tag{1}$$

Proof. For any $N = 1, 2, \ldots$ and for arbitrary $u \in D$ we have

$$\|\gamma^N u\|_{L_p}^{1/N} \leq \|\gamma^N u\|_{W_p^l}^{1/N} \leq \|\gamma\|_{MW_p^l} \|u\|_{W_p^l}^{1/N}.$$

Passing to the limit as $N \to \infty$, we obtain (1). □

Proposition 5. *Let $\gamma \in MW_p^l$ and let σ be a segment on the real axis such that $\gamma(x) \in \sigma$ for almost all $x \in R^n$. Further let $f \in C^{l-1,1}(\sigma)$. Then $f(\gamma) \in MW_p^l$ and the following estimate holds:*

$$\|f(\gamma)\|_{MW_p^l} \leq c \sum_{j=0}^{l} \|f^{(j)}; \sigma\|_{L_\infty} \|\gamma\|^j_{MW_p^l}.$$

Proof. The assertion is obvious for $l = 1$. Let it be proved for $l - 1$. For all $u \in D$ we have

$$\|uf(\gamma)\|_{W_p^l} \leq \|f(\gamma)\nabla u\|_{W_p^{l-1}} + \|uf'(\gamma)\nabla\gamma\|_{W_p^{l-1}} + \|uf(\gamma)\|_{L_p}. \tag{2}$$

By induction assumption the first summand in the right-hand side does not exceed

$$c\, \|\nabla u\|_{W_p^{l-1}} \sum_{j=0}^{l-1} \|f^{(j)}; \sigma\|_{L_\infty} \|\gamma\|^j_{MW_p^{l-1}}.$$

By the same reasoning the second summand in the right-hand side of (2) is not more than

$$c\, \|u\nabla\gamma\|_{W_p^{l-1}} \sum_{j=0}^{l-1} \|f^{(j+1)}; \sigma\|_{L_\infty} \|\gamma\|^j_{MW_p^{l-1}}.$$

From (1.3.1/1) and Proposition 1 it follows that

$$\|\nabla\gamma\|_{M(W_p^l \to W_p^{l-1})} \leq c\, \|\gamma\|_{MW_p^l}, \qquad \|\gamma\|_{MW_p^{l-1}} \leq c\, \|\gamma\|_{MW_p^l}.$$

So the right-hand side in (2) is dominated by

$$c\,\|u\|_{W_p^l}\sum_{j=0}^{l}\|f^{(j)};\sigma\|_{L_\infty}\|\gamma\|^j_{MW_p^l}.\quad\square$$

Corollary. *If $\gamma\in MW_p^l$ and $\|\gamma^{-1}\|_{L_\infty}<\infty$, then $\gamma^{-1}\in MW_p^l$ and $\|\gamma^{-1}\|_{MW_p^l}\leq c\,\|\gamma^{-1}\|^{l+1}_{L_\infty}\|\gamma\|^l_{MW_p^l}$.*

The proof immediately follows from Proposition 5 for $f(\gamma)=\gamma^{-1}$ and from the inequality

$$\|\gamma^{-1}\|_{L_\infty}\|\gamma\|_{MW_p^l}\geq 1$$

which is the consequence of Proposition 4.

Remark. All the assertions of the present section obviously can be reformulated for the space $M(w_p^m\to w_p^l)$.

1.7 The Maximal Algebra in W_p^l

Let A be a subset of some Banach function space. A is called a multiplication algebra if for any $u\in A$ and $v\in A$ their product uv also belongs to A and if there exists a positive number C such that $\|uv\|\leq C\,\|u\|\,\|v\|$.

For $lp\leq n$, $p>1$, or for $l<n$, $p=1$, the space W_p^l contains unbounded functions which obviously are not multipliers in W_p^l (see, for example, (1.6/1)). So the space W_p^l is not a multiplication algebra for the values of p and l given above. It is not difficult to describe the maximal algebra contained in W_p^l. If $u\in A$, then for any $N=1,2,\ldots,$

$$\|u^N\|^{1/N}_{L_p}\leq\|u^N\|^{1/N}_{W_p^l}\leq c\,\|u\|_{W_p^l}.$$

Consequently, $A\subset W_p^l\cap L_\infty$. On the other hand, it is well known that the intersection $W_p^l\cap L_\infty$ is a multiplication algebra. In fact, for any $u,v\in W_p^l\cap L_\infty$,

$$\begin{aligned}\|\nabla_l(uv)\|_{L_p}&\leq c\sum_{k=0}^{l}\|\,|\nabla_k u|\,|\nabla_{l-k}v|\,\|_{L_p}\\&\leq c\sum_{k=0}^{l}\|\nabla_k u\|_{L_{pl/k}}\|\nabla_{l-k}v\|_{L_{pl/(l-k)}}\\&\leq c\sum_{k=0}^{l}\|u\|^{(l-k)/l}_{L_\infty}\|u\|^{k/l}_{W^l}\|v\|^{k/l}_{L_\infty}\|v\|^{(l-k)/l}_{W_p^l}.\end{aligned}$$

Here we have used the Gagliardo [1]–Nirenberg [1] inequality

$$\|\nabla_j u\|_{L_{pl/j}} \leqslant c\, \|u\|_{L_\infty}^{(l-j)/l}\, \|u\|_{W_p^l}^{j/l}, \qquad j=1,\ldots,l-1.$$

Thus the space $W_p^l \cap L_\infty$ is the maximal algebra contained in W_p^l.

Since, by Sobolev's theorem, $W_p^l \subset L_\infty$ for $lp>n$, $p>1$, or for $l \geqslant n$, $p=1$, then the space W_p^l is a multiplication algebra for indicated values of p and l. This known assertion obviously also follows from (1.3.3/1), (1.3.5/3).

Remark. The conditions on orders of summability and smoothness for concrete function spaces to be multiplication algebras have been studied by many authors (see Strichartz [1], Peetre [1], Herz [1], Bennet, Gilbert [1], Johnson [1], [2], Triebel [1]–[3], Zolesio [1], Bliev [1], [2], Kalyabin [1]).

2

Multipliers in Pairs of Potential Spaces

In this chapter we study the space of multipliers $M(H_p^m \to H_p^l)$, $m \geq l \geq 0$, where H_p^s is the space of Bessel potentials of order s with densities in L_p. The introductory Section 2.1 gives information on Bessel linear and nonlinear potentials, on capacity and on imbedding theorems, which is used in subsequent sections. The characterization of the space $M(H_p^m \to H_p^l)$ is given in 2.2. In 2.3 we obtain sufficient conditions for a function to belong to $M(H_p^m \to H_p^l)$, the conditions being formulated in terms of classes $H_{n/m}^l$ and $B_{q,\infty}^\mu$. In 2.4 certain properties of elements of $M(H_p^m \to H_p^l)$ are studied. In particular, we consider the imbedding $M(H_p^m \to H_p^l) \subset M(H_q^{m-j} \to H_q^{l-j})$. A description of the spectrum of multipliers in H_p^l and $H_{p'}^{-l}$ is given in 2.5. Finally, 2.6 contains a theorem on a simplified equivalent normalization in $M(H_p^m \to H_p^l)$ for $m > l$.

2.1 The Bessel Potential Space and its Properties

Here we present some known facts on Bessel potential spaces without proofs.

2.1.1 The Space H_p^m

For any real μ we put

$$\Lambda^\mu = (-\Delta + 1)^{\mu/2} = F^{-1}(1 + |\xi|^2)^{\mu/2} F,$$

where F is the Fourier transform in R^n.

We introduce the space H_p^m $(1 < p < \infty,\ m \geq 0)$ obtained by completion of the space D with respect to the norm $\|u\|_{H_p^m} = \|\Lambda^m u\|_{L_p}$.

If m is an integer then, according to Lemma 1.1.1/3, the spaces W_p^m and H_p^m coincide.

From the definition of the space H_p^m, there easily follows:

Theorem 1. *The function u belongs to the space H_p^m if and only if $u = \Lambda^{-m} f$, where $f \in L_p$.*

In other words, Theorem 1 states that each element of H_p^m is the Bessel potential with density in L_p (cf. 1.1.1).

Definition. Let $(S_m u)(x)=|\nabla_m u(x)|$ for integer $m>0$ and

$$(S_m u)(x)=\left(\int_0^\infty \left[\int_{B_1} |\nabla_{[m]}u(x+\theta y)-\nabla_{[m]}u(x)|\, d\theta\right]^2 y^{-1-2\{m\}}\, dy\right)^{1/2} \quad (1)$$

for fractional $m>0$.

The following assertion is due to Strichartz [1].

Theorem 2. *The following relations hold:*

$$\|(-\Delta)^{m/2}u\|_{L_p}\sim\|S_m u\|_{L_p},$$
$$\|\Lambda^m u\|_{L_p}\sim\|S_m u\|_{L_p}+\|u\|_{L_p}.$$

This implies the theorem on uniform localization for the space H_p^m (see Strichartz [1]).

Theorem 3. *Let $\{B^{(j)}\}_{j\geqslant 0}$ be a covering of R^n by balls with unit diameter. Let the covering have a finite multiplicity, depending only on n. Further, let $O^{(j)}$ be the centre of the ball $B^{(j)}$, $O^{(0)}=0$ and $\eta_j(x)=\eta(x-O^{(j)})$, where $\eta\in D(2B^{(0)})$, $\eta=1$ on $B^{(0)}$. Then*

$$\|u\|_{H_p^m}\sim\left(\sum_{j\geqslant 0}\|u\eta_j\|_{H_p^m}^p\right)^{1/p}. \quad (2)$$

We formulate the Sobolev imbedding theorem for the space H_p^m.

Theorem 4. (i) *If $mp<n$, $p\leqslant q\leqslant np/(n-mp)$ or $mp=n$, $p\leqslant q<\infty$, then, for all $u\in H_p^m$, $\|u\|_{L_q}\leqslant c\,\|u\|_{H_p^m}$.*

(ii) *If $mp>n$, then, for all $u\in H_p^m$, $\|u\|_{L_\infty}\leqslant c\,\|u\|_{H_p^m}$.*

We also need the following refinement of part (i) in Theorem 4 essentially due to Yudovič [1] (see also Pohožaev [1] and Trudinger [1]).

Theorem 5. *If $mp=n$, $p\in(1,\infty)$, then, for all $u\in H_p^m$,*

$$\int\Phi\left(c\frac{|u|^{p'}}{\|u\|_{H_p^m}^{p'}}\right)dx\leqslant 1,$$

where $\Phi(t)=e^t-\sum_{j=0}^{[p]}t^j/j!$.

The proof of this inequality for the case of integer m, which was given in Subsection 1.1.2, is suitable also for $\{m\}>0$.

2.1.2 The Capacity Generated by the Norm in H_p^m

For any compact set $e \subset R^n$ we define the capacity by

$$\operatorname{cap}(e, H_p^m) = \inf \{\|u\|_{H_p^m}^p : u \in D,\ u \geq 1 \text{ on } e\}.$$

If E is any subset of R^n, then the numbers

$$\underline{\operatorname{cap}}(E, H_p^m) = \sup \{\operatorname{cap}(e, H_p^m) : e \subset E,\ e \text{ is a compact set}\},$$

$$\overline{\operatorname{cap}}(E, H_p^m) = \inf \{\underline{\operatorname{cap}}(G, H_p^m) : G \supset E,\ G \text{ is an open set}\}$$

are called the inner and outer capacities of the set E.

We formulate some known properties of the capacity (see Maz'ya, Havin [1], [2], Meyers [1]).

1. If the set $e \subset R^n$ is compact, then $\overline{\operatorname{cap}}(e, H_p^m) = \operatorname{cap}(e, H_p^m)$.
2. If $E_1 \subset E_2 \subset R^n$, then $\underline{\operatorname{cap}}(E_1, H_p^m) \leq \underline{\operatorname{cap}}(E_2, H_p^m)$, $\overline{\operatorname{cap}}(E_1, H_p^m) \leq \overline{\operatorname{cap}}(E_2, H_p^m)$.
3. Let $\{E_k\}_{k=1}^\infty$ be a sequence of sets in R^n, $E = \bigcup_k E_k$. Then $\overline{\operatorname{cap}}(E, H_p^m) \leq \sum_{k=1}^\infty \overline{\operatorname{cap}}(E_k, H_p^m)$.
4. Any analytic (in particular, any Borel) subset E of the space R^n is measurable with respect to the capacity $\operatorname{cap}(\cdot, H_p^m)$ (i.e. $\overline{\operatorname{cap}}(E, H_p^m) = \underline{\operatorname{cap}}(E, H_p^m)$).

If the inner and the outer capacities of a set E are equal, then their value is called the capacity of E and is denoted $\operatorname{cap}(E, H_p^m)$.

The capacity of a compact set may be defined as follows:

$$\operatorname{cap}(e, H_p^m) = \inf \{\|f\|_{L_p}^p : f \in L_p,\ f \geq 0,\ J_m f \geq 1 \text{ on } e\} \tag{1}$$

(see Meyers [1]).

We cite certain 'metric' properties of the capacity which will be used later.

Proposition 1. *If $mp > n$, then for all compact sets $e \neq \varnothing$ with $d(e) \leq 1$ the relation $\operatorname{cap}(e, H_p^m) \sim 1$ holds.*

The proof follows from part (ii) of Theorem 2.1.1/4 (see Proposition 1.1.1/2).

Proposition 2. *If $mp < n$, then*

$$\operatorname{cap}(e, H_p^m) \geq c(\operatorname{mes}_n e)^{(n-mp)/n}. \tag{2}$$

The proof follows from part (i) of Theorem 2.1.1/4.

Proposition 3. *If $mp=n$ and $d(e)\leqslant 1$ then*

$$\operatorname{cap}(e, H_p^m)\geqslant c\left(\log\frac{2^n}{\operatorname{mes}_n e}\right)^{1-p}. \tag{3}$$

Proof. See Theorem 2.1.1/5 and the proof of Proposition 1.1.2/4.

Proposition 4. (i) *If $mp<n$ and $0<r\leqslant 1$, then* $\operatorname{cap}(B_r, H_p^m)\sim r^{n-mp}$.
(ii) *If $mp=n$, $0<r\leqslant 1$, then* $\operatorname{cap}(B_r, H_p^m)\sim(\log 2/r)^{1-p}$.
(iii) *If $r>1$, then* $\operatorname{cap}(B_r, H_p^m)\sim r^n$.

For the proof of these relations see Meyers [1].
In the next proposition $\{B^{(j)}\}$ is the same covering as in Theorem 2.1.1/3.

Proposition 5. *For any compact e,*

$$\operatorname{cap}(e, H_p^m)\sim\sum_{j\geqslant 0}\operatorname{cap}(e\cap B^{(j)}, H_p^m).$$

Proof. Let $u\in D$, $u\geqslant 1$ on e. From the definition of the capacity follows

$$\sum_{j\geqslant 0}\operatorname{cap}(e\cap B^{(j)}, H_p^m)\leqslant\sum_{j\geqslant 0}\|u\eta_j\|_{H_p^m}^p,$$

where $\{\eta_j\}_{j\geqslant 0}$ is the sequence defined in Theorem 2.1.1/3. By (2.1.1/2) the right-hand side is dominated by $c\,\|u\|_{H_p^m}^p$. Minimizing this value, we obtain the lower bound for $\operatorname{cap}(e, H_p^m)$. The required upper bound is an immediate corollary of the semi-additivity of the capacity. □

2.1.3 Nonlinear Potentials

By $V_{p,m}\mu$ we denote the nonlinear Bessel potential of measure μ, i.e., $V_{p,m}\mu=J_m(J_m\mu)^{p'-1}$.

The potential $V_{p,m}\mu$ satisfies the following 'rough' maximum principle (cf. Maz'ya, Havin [1], [2], Adams, Meyers [1]).

Proposition 1. *There exists a constant $\mathfrak{M}$ depending only on n, p, m and such that*

$$(V_{p,m}\mu)(x)\leqslant\mathfrak{M}\sup\{(V_{p,m}\mu)(x): x\in\operatorname{supp}\mu\}. \tag{1}$$

The following assertion contains basic properties of the so-called capacitary measure (see Maz'ya, Havin [1], [2], Meyers [1]).

Proposition 2. *Let E be a subset of R^n. If $\overline{\text{cap}}\,(E, H_p^m)<\infty$, then there exists the unique measure μ_E with properties:*

1. $\|J_m\mu_E\|_{L_{p'}}^{p'} = \overline{\text{cap}}\,(E, H_p^m)$,
2. $(V_{p,m}\mu_E)(x)\geqslant 1$ *for (p, m)-quasi all* $x\in E$.

(Here '(p, m)-quasi everywhere' means 'everywhere except the set of zero outer capacity $\text{cap}\,(\cdot, H_p^m)$*'.)*

3. $\text{supp}\ \mu_E\subset \bar{E}$,
4. $\mu_E(\bar{E}) = \overline{\text{cap}}\,(E, H_p^m)$,
5. $(V_{p,m}\mu_E)(x)\leqslant 1$ *for all* $x\in \text{supp}\ \mu_E$.

The measure μ_E is called the capacitary measure of the set E and $V_{p,m}\mu_E$ is called the capacitary potential of the set E.

In addition we notice that the capacity $\text{cap}\,(e, H_p^m)$ may be defined by

$$\text{cap}\,(e, H_p^m) = \sup\{\mu(e) : \text{supp}\ \mu\subset e \text{ and } (V_{p,m}\mu)(x)\leqslant 1 \text{ on supp } \mu\}$$

(see Maz'ya, Havin [2]).

2.1.4 An Imbedding Theorem

Our aim is the generalization to H_p^m of the theorem in Subsection 1.1.4 on the summability with respect to a measure μ of functions in W_p^m.

We begin with the estimate of the integral of the capacity of a set bounded by a level surface. This estimate contains (1.1.3/9) as a particular case.

Theorem 1. *Let $u\in H_p^m$ and $N_t = \{x : |u(x)|\geqslant t\}$. Then*

$$\int_0^\infty \text{cap}\,(N_t, H_p^m)t^{p-1}\,dt \leqslant C\,\|u\|_{H_p^m}^p. \tag{1}$$

The best constant in this inequality admits the bounds

$$C\leqslant \begin{cases} (p')^{p-1}\mathfrak{M} & \text{for } p\geqslant 2,\\ (p')^p p^{-1}\mathfrak{M}^{p-1} & \text{for } p<2.\end{cases}$$

Here $\mathfrak{M}$ is the constant in (2.1.3/1).

Proof. Let, for the sake of brevity, $c(t) = \text{cap}\,(N_t, H_p^m)$. We may prove the theorem under the assumption $u = J_m f$, $f\geqslant 0$, $f\in L_p$. By μ_t we denote the capacitary measure of the set N_t (see Proposition 2.1.3/2).

The left-hand side of (1) does not exceed

$$\int_0^\infty \int J_m f\,d\mu_t t^{p-2}\,dt = \int f\,dx \int_0^\infty J_m\mu_t t^{p-2}\,dt$$

which is not more than $\|f\|_{L_p}\,\|\int_0^\infty t^{p-2}J_m\mu_t\,dt\|_{L_{p'}}$. Thus the theorem will be proved if we derive the estimate

$$\int\left(\int_0^\infty t^{p-2} J_m\mu_t \, dt\right)^{p'} dx \leq \int_0^\infty c(t)t^{p-1}\, dt. \tag{2}$$

First we notice that, by the maximum principle,

$$\int (J_m\mu_t)^{p'-1} J_m\mu_t \, dx \leq \mathfrak{M} c(t). \tag{3}$$

Then we separately consider the cases $p \geq 2$ and $p<2$. Let $p \geq 2$. We rewrite the left-hand side of (2) as

$$p' \int\int_0^\infty J_m\mu_\tau \left(\int_\tau^\infty J_m\mu_t t^{p-2}\, dt\right)^{p'-1} \tau^{p-2}\, d\tau\, dx.$$

By Hölder's inequality this expression is dominated by

$$p'\left(\int\int_0^\infty \tau^{p-1}(J_m\mu_\tau)^{p'}\, d\tau\, dx\right)^{2-p'}\left(\int\int_0^\infty (J_m\mu_\tau)^{p'-1}\int_\tau^\infty J_m\mu_t t^{p-2} dt\, d\tau\, dx\right)^{p'-1}$$

which, by virtue of (3), does not exceed

$$p'\mathfrak{M}^{p'-1}\left(\int_0^\infty \|J_m\mu_t\|^{p'}_{L_{p'}} t^{p-1}\, dt\right)^{2-p'}\left(\int_0^\infty c(t)t^{p-1}\, dt\right)^{p'-1}.$$

So (1) is proved for $p \geq 2$.

Let $p<2$. The left-hand side of (2) is equal to

$$p' \int\int_0^\infty J_m\mu_t t^{p-2}\, dt\left(\int_0^t J_m\mu_\tau \tau^{p-2}\, d\tau\right)^{p'-1} dx,$$

and therefore, by Minkowski's inequality, it is not more than

$$p'\int_0^\infty \left(\int_0^t\left(\int (J_m\mu_\tau)^{p'-1} J_m\mu_t\, dx\right)^{p-1} \tau^{p-2}\, d\tau\right)^{p'-1} t^{p-2}\, dt.$$

Estimating this value with the help of (3), we obtain that it is majorized by

$$p'\mathfrak{M}\int_0^\infty c(t)\left(\int_0^t \tau^{p-2}\, d\tau\right)^{p'-1} t^{p-2}\, dt$$

and so (1) follows for $p<2$. □

Remark 1. Estimate (1) is due to Maz'ya [8] for $p \geq 2$. The restriction $p \geq 2$ was removed by Dahlberg [1] whose proof is also based on the 'smooth truncation' of potential near equipotential surfaces and on subtle estimates of potentials with non-negative densities. Hansson [1] found a new proof of (1) without the use of 'smooth truncation'. His argument is suitable for a large class of potentials with general kernels. Here we have given a proof of (1), borrowed from Maz'ya [12], which is based on Hansson's idea but appears to be somewhat simpler.

Now, by using (1), we can prove the following theorem in the same way as Theorem 1.1.4.

Theorem 2. *The best constant C in*

$$\int |u|^p \, d\mu \leqslant C \, \|u\|^p_{H^m_p}, \qquad u \in D,$$

is equivalent to $\sup_e (\mu(e)/\text{cap}(e, H^m_p))$, *where e is an arbitrary compact set of positive capacity* $\text{cap}(e, H^m_p)$.

Remark 2. From Proposition 2.1.2/5 follows

$$\sup_e \frac{\mu(e)}{\text{cap}(e, H^m_p)} \sim \sup_{\{e : d(e) \leqslant 1\}} \frac{\mu(e)}{\text{cap}(e, H^m_p)}. \tag{4}$$

Remark 3. Obviously, the constant C in Theorem 2 satisfies the inequality

$$C \geqslant \sup_{x \in R^n, \rho \in (0, 1/2)} \frac{\mu(B_\rho(x))}{\text{cap}(B_\rho(x), H^m_p)}. \tag{5}$$

If the converse estimate (up to a factor $c = c(n, p, l)$) were valid we could manage without the notion of capacity in this book (see Proposition 2.1.2/4). According to Proposition 2.1.2/1, this is really the case for $mp > n$ when the right-hand and left-hand sides of (5) are equivalent to $\sup\{\mu(B_1(x)) : x \in R^n\}$. However, as was noted by Adams [3], the finiteness of the right-hand side of (5) for $n \geqslant mp$ does not imply $C < \infty$.

Before we prove the last assertion we recall the definition of the Hausdorff φ-measure of a set $E \subset R^n$, where φ is a non-decreasing positive function on [0, 1]: namely,

$$H(E, \varphi) = \lim_{\varepsilon \to +0} \inf_{\{B^{(i)}\}} \sum_i \varphi(r_i).$$

Here $\{B^{(i)}\}$ is any covering of E by open balls $B^{(i)}$ with radii $r_i < \varepsilon$. We put $\varphi(t) = t^{n-mp}$ for $n > mp$ and $\varphi(t) = |\log t|^{1-p}$ for $n = mp$. Let E be a Borel set in R^n such that $d(E) < 1$ and $0 < H(E, \varphi) < \infty$. We may assume E to be closed and bounded, since any Borel set of positive Hausdorff measure contains a closed subset with the same property.

By Frostman's theorem (see Carleson [1], Theorem 1, Ch. 2) there exists a non-zero measure μ with support in E such that $\mu(B_\rho(x)) \leqslant c\varphi(\rho)$

with a constant c independent of x and ρ. By virtue of Proposition 2.1.2/4, this means that the right-hand side of (5) is finite.

On the other hand, by the theorem of Meyers [1] and Maz'ya, Havin [1], [2], the finiteness of the measure $H(E, \varphi)$ implies cap $(E, H_p^m) = 0$. So $C = \infty$ (see Theorem 2), although the right-hand side of (5) is finite.

2.2 Description of the Space $M(H_p^m \to H_p^l)$

In this section we mainly follow Maz'ya and Shaposhnikova [11].

2.2.1 Auxiliary Assertions

We formulate the Calderon interpolation theorem [1] for spaces of Bessel potentials, of which a particular case is inequality (1.3.1), used in the study of the space $M(W_p^m \to W_p^l)$.

Proposition. *Let $p_0, p_1 \in (1, \infty)$, $\theta \in (0, 1)$, $\mu \in R^1$ and*

$$\frac{1}{p} = \frac{\theta}{p_0} + \frac{1-\theta}{p_1}, \qquad l = \theta l_0 + (1-\theta) l_1.$$

Further, let L be a linear operator, mapping $H_{p_0}^{l_0+\mu} \cap H_{p_1}^{l_1+\mu}$ into $H_{p_0}^{l_0} \cap H_{p_1}^{l_1}$ and admitting an extension to continuous operators: $H_{p_0}^{l_0+\mu} \to H_{p_0}^{l_0}$ and $H_{p_1}^{l_1+\mu} \to H_{p_1}^{l_1}$. Then L can be extended to a continuous operator: $H_p^{l+\mu} \to H_p^l$ and the following interpolation inequality

$$\|L\|_{H_p^{l+\mu} \to H_p^l} \leqslant c\, \|L\|^{\theta}_{H_{p_0}^{l_0+\mu} \to H_{p_0}^{l_0}} \|L\|^{1-\theta}_{H_{p_1}^{l_1+\mu} \to H_{p_1}^{l_1}}$$

holds.

Setting $L = \gamma$ in this proposition, we obtain

$$\|\gamma\|_{M(H_p^{l+\mu} \to H_p^l)} \leqslant c\, \|\gamma\|^{\theta}_{M(H_{p_0}^{l_0+\mu} \to H_{p_0}^{l_0})} \|\gamma\|^{1-\theta}_{M(H_{p_1}^{l_1+\mu} \to H_{p_1}^{l_1})}. \tag{1}$$

Lemma 1. *Let γ_ρ be a mollification of $\gamma \in H^l_{p,\mathrm{loc}}(R^n)$, $1 < p < \infty$, with kernel $K \geqslant 0$ and radius ρ. Then*

$$\|\gamma_\rho\|_{M(H_p^m \to H_p^l)} \leqslant \|\gamma\|_{M(H_p^m \to H_p^l)} \leqslant \varliminf_{\rho \to 0} \|\gamma_\rho\|_{M(H_p^m \to H_p^l)}. \tag{2}$$

Proof. Let $u \in D$ and $\{l\} > 0$. It is clear that

$$\|\gamma_\rho u\|_{H_p^l} = \Big\{ \int \Big(\int_0^\infty \Big[\int_{B_1} \Big| \int \rho^{-n} K(\xi/\rho)$$

$$\times \nabla_{[l],x} (Q(x+\theta y, \xi) - Q(x, \xi))\, d\xi \Big|\, d\theta \Big]^2 y^{-1-2\{l\}}\, dy \Big)^{p/2} dx \Big\}^{1/p}$$

$$+ \Big\{ \int \Big| \int \rho^{-n} K(\xi/\rho) Q(x, \xi)\, d\xi \Big|^p dx \Big\}^{1/p},$$

where $Q(x, \xi) = \gamma(x-\xi)u(x)$. By the Minkowski inequality

$$\|\gamma_\rho u\|_{H_p^l} \leq \int \rho^{-n} K(\xi/\rho)\, \|S_l Q(\cdot, \xi)\|_{L_p}\, d\xi + \int \rho^{-n} K(\xi/\rho)\, \|Q(\cdot, \xi)\|_{L_p}\, d\xi.$$

Since

$$\|Q(\cdot, \xi)\|_{H_p^l} \leq \|\gamma\|_{M(H_p^m \to H_p^l)}\, \|u\|_{H_p^m}$$

then the left estimate (2) follows. The right inequality (2) results from

$$\|\gamma u\|_{H_p^l} = \lim_{\rho \to 0} \|\gamma_\rho u\|_{H_p^l} \leq \underline{\lim}_{\rho \to 0} \|\gamma_\rho\|_{M(H_p^m \to H_p^l)}\, \|u\|_{H_p^m}.$$

In the case $\{l\} = 0$ the proof is like that of Lemma 1.3.1/1. □

Lemma 2. *Let* $1 < p < \infty$, $0 < \nu < \mu$ *and* $\varphi \in L_{p\mu,\text{loc}}$, $\varphi \geq 0$. *Then*

$$\sup_e \left(\frac{\int_e \varphi^{\nu p}\, dx}{\text{cap}(e, H_p^\nu)} \right)^{1/\nu} \leq c \sup_e \left(\frac{\int_e \varphi^{\mu p}\, dx}{\text{cap}(e, H_p^\mu)} \right)^{1/\mu},$$

where e is any compact set of positive measure.

Proof. Let $u \in D$ and $f = J_\nu u$. By Lemma 1.1.3

$$\int \varphi^{\nu p} |u|^p\, dx \leq c \int \varphi^{\nu p} (J_\mu |f|)^{\nu p/\mu} (Mf)^{(\mu-\nu)p/\mu}\, dx$$

$$\leq c \left(\int \varphi^{\mu p} (J_\mu |f|)^p\, dx \right)^{\nu/\mu} \left(\int (Mf)^p\, dx \right)^{(\mu-\nu)/\mu},$$

where M is the Hardy–Littlewood maximal operator. Using the continuity of M in L_p and Theorem 2.1.4/2, we get

$$\int \varphi^{\nu p} |u|^p\, dx \leq c\, \|f\|_{L_p}^{(\mu-\nu)p/\mu} \sup_e \left(\frac{\int_e \varphi^{\mu p}\, dx}{\text{cap}(e, H_p^\mu)} \right)^{\nu/\mu} \|J_\mu |f|\|_{H_p^\mu}^{\nu p/\mu}.$$

(Here we have used Theorem 2.1.1/1, according to which the function $J_\mu |f|$ can be approximated in H_p^μ by a sequence of functions in D.) Since the last norm does not exceed $c\, \|f\|_{L_p}$,

$$\int \varphi^{\nu p} |u|^p\, dx \leq c \sup_e \left(\frac{\int_e \varphi^{\mu p}\, dx}{\text{cap}(e, H_p^\mu)} \right)^{\nu/\mu} \|u\|_{H_p^\nu}^p.$$

It remains to apply Theorem 2.1.4/2. □

2.2.2 Imbedding of $M(H_p^m \to H_p^l)$ into $M(H_p^{m-l} \to L_p)$

In Theorem 2.1.4/2 and Remark 2.1.4/2 the following description of $M(H_p^k \to L_p)$ is contained:

Lemma 1. *The following relations hold:*

$$\|\gamma\|_{M(H_p^k \to L_p)} \sim \sup_e \frac{\|\gamma; e\|_{L_p}}{[\mathrm{cap}\,(e, H_p^k)]^{1/p}},$$

$$\|\gamma\|_{M(H_p^k \to L_p)} \sim \sup_{\{e: d(e) \leqslant 1\}} \frac{\|\gamma; e\|_{L_p}}{[\mathrm{cap}\,(e, H_p^k)]^{1/p}}.$$

Lemma 2. *The inequality*

$$\|\gamma\|_{M(H_p^{m-l} \to L_p)} \leqslant c\, \|\gamma\|_{M(H_p^m \to H_p^l)} \tag{1}$$

is valid.

Proof. Let $\gamma \in M(H_p^m \to H_p^l)$ and let γ_ρ be a mollification of γ with radius ρ. Since $M(H_p^m \to H_p^l) \subset M(H_p^m \to L_p)$, then $\gamma \in L_{p,\mathrm{unif}}$. Therefore $\gamma_\rho \in L_\infty$ and consequently $\gamma_\rho \in M(H_p^{m-l} \to L_p)$. This property of mollifications will be used in what follows.

1. The case $m \geqslant 2l$. Let $u = J_{m-l} f$, $f \in L_p$. By Lemma 1.1.3,

$$|u| \leqslant c(J_m\, |f|)^{1-l/m} (Mf)^{l/m}.$$

Hence

$$\|\gamma_\rho u\|_{L_p} \leqslant c\, \|f\|_{L_p}^{l/m}\, \|\gamma_\rho^{l/(l-m)} \gamma_\rho J_m\, |f|\|_{L_p}^{1-l/m}.$$

This and Lemma 1 imply

$$\|\gamma_\rho u\|_{L_p} \leqslant c\, \|f\|_{L_p}^{l/m}\, \|\gamma_\rho J_m\, |f|\|_{H_p^l}^{1-l/m} \sup_e \left(\frac{\int_e |\gamma_\rho|^{pl/(m-l)}\, dx}{\mathrm{cap}\,(e, H_p^l)} \right)^{(m-l)/pm}.$$

By Lemma 2.2.1/2 with $\varphi = |\gamma_\rho|^{1/(m-l)}$, $\nu = l$, $\mu = m - l$, the last supremum does not exceed

$$c \left(\sup_e \frac{\|\gamma_\rho; e\|_{L_p}}{[\mathrm{cap}\,(e, H_p^{m-l})]^{1/p}} \right)^{l/m}$$

which, together with Lemma 1, gives

$$\|\gamma_\rho u\|_{L_p} \leqslant c\, \|\gamma_\rho\|_{M(H_p^{m-l} \to L_p)}^{l/m}\, \|\gamma_\rho\|_{M(H_p^m \to H_p^l)}^{1-l/m}\, \|f\|_{L_p}^{l/m}\, \|J_m\, |f|\|_{H_p^m}^{1-l/m}$$

$$\leqslant c\, \|\gamma_\rho\|_{M(H_p^{m-l} \to L_p)}^{l/m}\, \|\gamma_\rho\|_{M(H_p^m \to H_p^l)}^{1-l/m}\, \|u\|_{H_p^{m-l}}.$$

The reference to Lemma 2.2.1/1 yields (1) for $m \geqslant 2l$.

2. Suppose $m=l$. For any positive integer N we have

$$\|\gamma^N u\|_{L_p}^{1/N} \leqslant \|\gamma^N u\|_{H_p^l}^{1/N} \leqslant \|\gamma\|_{MH_p^l} \|u\|_{H_p^l}^{1/N}.$$

Consequently, $\|\gamma\|_{ML_p} = \|\gamma\|_{L_\infty} \leqslant \|\gamma\|_{MH_p^l}$.

3. Now let $2l > m > l$. By ε we denote a positive number such that $\varepsilon < m-l$. Since $m-l+\varepsilon > 2\varepsilon$, then according to the first part of the proof

$$\|\gamma_\rho\|_{M(H_p^{m-l} \to L_p)} \leqslant c\, \|\gamma_\rho\|_{M(H_p^{m-l+\varepsilon} \to H_p^\varepsilon)}.$$

By (2.2.1/1) we have

$$\|\gamma_\rho\|_{M(H_p^{m-l+\varepsilon} \to H_p^\varepsilon)} \leqslant c\, \|\gamma_\rho\|_{M(H_p^{m-l} \to L_p)}^{1-\varepsilon/l} \|\gamma_\rho\|_{M(H_p^m \to H_p^l)}^{\varepsilon/l}$$

which, together with the preceding estimate and Lemma 2.2.1/1, implies (1). □

2.2.3 Estimates for Derivatives of a Multiplier

Lemma 1. *If $\gamma \in M(H_p^m \to H_p^l)$, then $D^\alpha \gamma \in M(H_p^m \to H_p^{l-|\alpha|})$ for any multi-index α, $|\alpha| \leqslant l$. The estimate*

$$\|D^\alpha \gamma\|_{M(H_p^m \to H_p^{l-|\alpha|})} \leqslant c\, \|\gamma\|_{M(H_p^m \to H_p^l)} \tag{1}$$

holds.

Proof. It suffices to consider the case $|\alpha|=1$, $l \geqslant 1$. Obviously

$$\|u\, \nabla\gamma\|_{H_p^{l-1}} \leqslant \|u\gamma\|_{H_p^l} + \|\gamma\, \nabla u\|_{H_p^{l-1}}$$
$$\leqslant (\|\gamma\|_{M(H_p^m \to H_p^l)} + \|\gamma\|_{M(H_p^{m-1} \to H_p^{l-1})})\, \|u\|_{H_p^m}.$$

Using (2.2.1/1) and (2.2.2/1), we obtain

$$\|\gamma\|_{M(H_p^{m-1} \to H_p^{l-1})} \leqslant c_1\, \|\gamma\|_{M(H_p^{m-l} \to L_p)}^{1-1/l} \|\gamma\|_{M(H_p^m \to H_p^l)}^{1/l}$$
$$\leqslant c_2\, \|\gamma\|_{M(H_p^m \to H_p^l)}.$$

Consequently,

$$\|u\, \nabla\gamma\|_{H_p^{l-1}} \leqslant c\, \|\gamma\|_{M(H_p^m \to H_p^l)} \|u\|_{H_p^m}. \quad \square$$

This and Lemma 2.2.2/2 give:

Corollary. *If $\gamma \in M(H_p^m \to H_p^l)$, then $D^\alpha \gamma \in M(H_p^{m-l+|\alpha|} \to L_p)$ for any multi-index α of order $|\alpha| \leqslant l$. The inequality*

$$\|D^\alpha \gamma\|_{M(H_p^{m-l+|\alpha|} \to L_p)} \leqslant c\, \|\gamma\|_{M(H_p^m \to H_p^l)}$$

is true.

2.2.4 Two More Auxiliary Assertions

Lemma 1. *Let* $m \geq l$, $0 < \delta < l < 1$. *Then*

$$S_{l-\delta}\gamma \leq c(S_l\gamma + \|\gamma\|_{M(H_p^{m-l} \to L_p)})^{1-\delta/m} \|\gamma\|^{\delta/m}_{M(H_p^{m-l} \to L_p)}. \tag{1}$$

Proof. For any $R > 0$

$$(S_{l-\delta}\gamma)(x) \leq c\left(R^\delta \left[\int_0^R \left(\int_{B_1} |\gamma(x+\theta y) - \gamma(x)|\, d\theta\right)^2 y^{-1-2l}\, dy\right]^{1/2} + \left[\int_R^\infty \left(\int_{B_1} |\gamma(x+\theta y)|\, d\theta\right)^2 y^{-1-2(l-\delta)}\, dy\right]^{1/2} + |\gamma(x)|\, R^{\delta - l}\right).$$

Since

$$c\,|\gamma(x)| \leq \int_{B_1} |\gamma(x+\theta y) - \gamma(x)|\, d\theta + \int_{B_1} |\gamma(x+\theta y)|\, d\theta,$$

then

$$cR^{-l}\,|\gamma(x)| \leq \left(\int_R^\infty \left(\int_{B_1} |\gamma(x+\theta y) - \gamma(x)|\, d\theta\right)^2 y^{-1-2l}\, dy\right)^{1/2} + \left(\int_R^\infty \left(\int_{B_1} |\gamma(x+\theta y)|\, d\theta\right)^2 y^{-1-2l}\, dy\right)^{1/2}.$$

Henceforth $R \leq 1$. We have

$$(S_{l-\delta}\gamma)(x) \leq c\left[R^\delta (S_l\gamma)(x) + \left(\int_R^1 \left(\int_{B_1} |\gamma(x+\theta y)|\, d\theta\right)^2 y^{-1-2(l-\delta)}\, dy\right)^{1/2} + \left(\int_1^\infty \left(y^{-n}\int_{B_y} |\gamma(x+s)|\, ds\right)^2 y^{-1-2(l-\delta)}\, dy\right)^{1/2}\right]$$

$$\leq c\left[R^\delta (S_l\gamma)(x) + R^{\delta-m} \sup_{x\in R^n, \rho\in(0,1)} \frac{\|\gamma; B_\rho(x)\|_{L_p}}{\rho^{n/p-m+l}} + \sup_{x\in R^n} \|\gamma; B_1(x)\|_{L_p}\right].$$

It is clear that the last summand can be thrown off by changing the constant c. By Lemma 2.2.2/1,

$$\sup_{x\in R^n, \rho\in(0,1)} \frac{\|\gamma; B_\rho(x)\|_{L_p}}{\rho^{n/p-m+l}} \leq c\, \|\gamma\|_{M(H_p^{m-l} \to L_p)}.$$

So for all $R \in (0, 1]$

$$(S_{l-\delta}\gamma)(x) \leq cR^\delta[(S_l\gamma)(x) + R^{-m}\, \|\gamma\|_{M(H_p^{m-l} \to L_p)}]. \tag{2}$$

If $(S_{l-\delta}\gamma)(x) \leqslant \|\gamma\|_{M(H_p^{m-l} \to L_p)}$, we arrive at (1) by putting $R=1$ in (2). In the opposite case, (1) follows from (2), where

$$R^m = \frac{\|\gamma\|_{M(H_p^{m-l} \to L_p)}}{(S_l\gamma)(x)}. \quad \square$$

We formulate some asymptotic properties of the kernel G_l of the Bessel potential which will be used later (see Azonszajn, Mulla, Szeptycky [1]):

(i) For $|x| \to 0$,

$$G_l(x) \sim |x|^{l-n} \qquad \text{if} \quad 0<l<n, \tag{3}$$

$$|G_l(x) - c_1 \log |x|^{-1}| \leqslant c_2 \qquad \text{if} \quad l=n, \tag{4}$$

$$|G_l(x) - c| \leqslant c\, |x|^{\min(l-n,1)} \qquad \text{if} \quad l>n, \tag{5}$$

$$|\nabla G_l(x)| \leqslant \begin{cases} c\, |x|^{l-n-1} & \text{if} \quad l \leqslant n+1, \\ c & \text{if} \quad l \geqslant n+1. \end{cases} \tag{6}$$

(ii) For $|x| \to \infty$,

$$G_l(x) \sim |x|^{(l-n-1)/2} e^{-|x|}, \tag{7}$$

$$|\nabla G_l(x)| \sim |x|^{(l-n-1)/2} e^{-|x|}. \tag{8}$$

By virtue of these relations we prove the following:

Lemma 2. *The estimate*

$$|G_l(x) - G_l(y)| \leqslant c\, |x-y|^{\delta}\, [G_{l-\delta}(x/4) + G_{l-\delta}(y/4)], \tag{9}$$

where $\delta \in (0, 1]$, is valid.

Proof. It suffices to consider the case $|y|>|x|$. If $2\,|x-y|<|x|$, then by (6), for $|x|<2$,

$$|G_l(x) - G_l(y)| \leqslant \begin{cases} c\, |x-y|\, |x|^{l-n-1}, & l \leqslant n+1, \\ c\, |x-y|, & l \geqslant n+1, \end{cases}$$

and for $|x| \geqslant 2$ we deduce from (7) that

$$|G_l(x) - G_l(y)| \leqslant c\, |x-y|\, |x|^{(l-n-1)/2}\, e^{-|x|/2}.$$

These estimates and (8) yield

$$|G_l(x) - G_l(y)| \leqslant c\, |x-y|^{\delta}\, G_{l-\delta}(x/4)$$

for $2\,|x-y|<|x|$.

Now let $2|x-y|>|x|$. If $l>n$ and $|y|<2$, then by virtue of (5) we have

$$|G_l(x)-G_l(y)| \leq c\,|y|^{\min(l-n,1)}$$

which, together with $|y| \leq 3\,|x-y|$, gives

$$|G_l(x)-G_l(y)| \leq c\,|x-y|^{\delta}\,|y|^{\min(l-n,1)-\delta}.$$

Combining this estimate with (3)–(5), we obtain (9) in the case $l>n$.

For the same values of x and y we have

$$|G_l(x)-G_l(y)| \leq c_1\,|\log(|x|/|y|)|+c_2 \leq c(|y|/|x|)^{\delta} \leq c\,|x-y|^{\delta}\,|x|^{-\delta}$$

if $l=n$ and

$$|G_l(x)-G_l(y)| \leq c(|x|^{l-n}+|y|^{l-n}) \leq c\,|x-y|^{\delta}\,(|x|^{l-n-\delta}+|y|^{l-n-\delta})$$

if $l<n$. Using (3) again, we arrive at (9).

It remains to deal with the case $2|x-y| \geq |x|$, $|y| \geq 2$. By (3)–(5) and (7) we obtain

$$|G_l(x)-G_l(y)| \leq cG_l(x) \leq c\,|x-y|^{\delta}\,|x|^{-\delta}\,G_l(x) \leq c\,|x-y|^{\delta}\,G_{l-\delta}(x/4)$$

for $|x|>1$. If $|x|<1$, then $|x-y| \geq 1$ and therefore

$$|G_l(x)-G_l(y)| \leq cG_l(x) \leq c\,|x-y|^{\delta}\,G_l(x) \leq c\,|x-y|^{\delta}\,G_{l-\delta}(x/4). \quad \square$$

2.2.5 Upper Bound for the Norm of a Multiplier

We obtain a sufficient condition for a function to belong to the space $M(H_p^m \to H_p^l)$.

Lemma. *Let $\gamma \in H^l_{p,\text{loc}}$, $S_l\gamma \in M(H_p^m \to L_p)$ and $\gamma \in M(H_p^{m-l} \to L_p)$. Then $\gamma \in M(H_p^m \to H_p^l)$ and*

$$\|\gamma\|_{M(H_p^m \to H_p^l)} \leq c(\|S_l\gamma\|_{M(H_p^m \to L_p)}+\|\gamma\|_{M(H_p^{m-l} \to L_p)}). \tag{1}$$

(The function S_l was introduced in Definition 2.1.1.)

Proof. By virtue of Theorem 2.1.1/2 one can easily verify that a function in L_∞ with uniformly bounded derivatives of any order belongs to the space $M(H_p^m \to H_p^l)$.

Let γ_ρ be a mollification of γ with non-negative kernel and radius ρ. Since $M(H_p^m \to H_p^l) \subset M(H_p^m \to L_p)$, then $\gamma \in L_{p,\text{unif}}$. Therefore $\nabla_j\gamma_\rho \in L_\infty$, $j=0,1,\ldots$ and $\gamma_\rho \in M(H_p^m \to H_p^l)$.

For the case $\{l\}=0$ the assertion was proved in Lemma 1.3.2/3.

Suppose $\{l\}>0$. For any $u\in D$,

$$\|\gamma_\rho u\|_{L_p}\leqslant\|\gamma_\rho\|_{M(H_p^{m-l}\to L_p)}\|u\|_{H_p^{m-l}}. \tag{2}$$

Clearly,

$$\|S_l(\gamma_\rho u)\|_{L_p}\leqslant c\sum_{\substack{0\leqslant\beta\leqslant\alpha\\|\alpha|=[l]}}\|S_{\{l\}}(D^\beta\gamma_\beta D^{\alpha-\beta}u)\|_{L_p}.$$

First consider the items corresponding to multi-indices of order $j<[l]$. By Lemma 2.2.3/1 and interpolation inequality (2.2.1/1), for any $k\in(0,l-j)$ we have

$$\begin{aligned}\|\nabla_j\gamma_\rho\|_{M(H_p^{m-k}\to H_p^{l-k-j})}&\leqslant c\,\|\gamma_\rho\|_{M(H_p^{m-k}\to H_p^{l-k})}\\&\leqslant c\,\|\gamma_\rho\|^{1-k/l}_{M(H_p^m\to H_p^l)}\|\gamma_\rho\|^{k/l}_{M(H_p^{m-l}\to L_p)}.\end{aligned} \tag{3}$$

Consequently, for $|\beta|=j<[l]$,

$$\|S_{\{l\}}(D^\beta\gamma_\rho D^{\alpha-\beta}u)\|_{L_p}\leqslant c\,\|\gamma_\rho\|^{(\{l\}+j)/l}_{M(H_p^m\to H_p^l)}\|\gamma_\rho\|^{([l]-j)/l}_{M(H_p^{m-l}\to L_p)}\|u\|_{H_p^m}. \tag{4}$$

This, together with (2), implies

$$\begin{aligned}\|\gamma_\rho u\|_{H_p^l}\leqslant&(\varepsilon\,\|\gamma_\rho\|_{M(H_p^m\to H_p^l)}+c(\varepsilon)\,\|\gamma_\rho\|_{M(H_p^{m-l}\to L_p)})\,\|u\|_{H_p^m}\\&+\|S_{\{l\}}(u\,\nabla_{[l]}\gamma_\rho)\|_{L_p}\end{aligned} \tag{5}$$

for any $\varepsilon>0$. It remains to estimate the last summand in the right-hand side. We have

$$\begin{aligned}\|S_{\{l\}}(u\,\nabla_{[l]}\gamma_\rho)\|_{L_p}&\leqslant\|uS_l\gamma_\rho\|_{L_p}+\||\nabla_{[l]}\gamma_\rho|\,S_{\{l\}}u\|_{L_p}\\&+\left\{\int\left(\int_0^\infty\left[\int_{B_1}|u(x+\theta y)-u(x)|\,|\nabla_{[l]}\gamma_\rho(x+\theta y)-\nabla_{[l]}\gamma_\rho(x)|\,d\theta\right]^2\right.\right.\\&\left.\left.\times y^{-1-2\{l\}}\,dy\right)^{p/2}dx\right\}^{1/p}.\end{aligned} \tag{6}$$

Obviously,

$$\|uS_l\gamma_\rho\|_{L_p}\leqslant c\,\|S_l\gamma_\rho\|_{M(H_p^m\to L_p)}\|u\|_{H_p^m}. \tag{7}$$

Consider the second norm in the right-hand side of (6). Applying Minkowski's inequality, we obtain

$$S_{\{l\}}u\leqslant\Lambda^{-[l]}S_{\{l\}}\Lambda^{[l]}u.$$

This and (3) yield

$$\begin{aligned}\||\nabla_{[l]}\gamma_\rho|\,S_{\{l\}}u\|_{L_p}&\leqslant\|\nabla_{[l]}\gamma_\rho\|_{M(H_p^{m-\{l\}}\to L_p)}\|\Lambda^{\{l\}-m}S_{\{l\}}\Lambda^{m-\{l\}}u\|_{H_p^{m-\{l\}}}\\&\leqslant c\,\|\gamma_\rho\|^{[l]/l}_{M(H_p^m\to H_p^l)}\|\gamma_\rho\|^{\{l\}/l}_{M(H_p^{m-l}\to L_p)}\|u\|_{H_p^m}.\end{aligned} \tag{8}$$

Now we turn to the estimate of the third summand in the right-hand side of (6). Let $u = \Lambda^{-m} f$ and let δ be a sufficiently small positive number. We notice that

$$|u(x+\theta y) - u(x)| \leq \int |G_m(x-\xi+\theta y) - G_m(x-\xi)|\, |f(\xi)|\, d\xi$$

and use Lemma 2.2.4/2. Then

$$|u(x+\theta y) - u(x)| \leq c y^{\delta} \left[(\Lambda^{\delta-m} |f|)\left(\frac{x+\theta y}{4}\right) + (\Lambda^{\delta-m} |f|)\left(\frac{x}{4}\right) \right].$$

So the third summand in (6) does not exceed

$$c\left\{ \int \left(\int_0^{\infty} \left[\int_{B_1} (\Lambda^{\delta-m} |f|)\left(\frac{x+\theta y}{4}\right) \right.\right.\right.$$
$$\left.\left.\left. \times |\nabla_{[l]}\gamma_\rho(x+\theta y) - \nabla_{[l]}\gamma_\rho(x)|\, d\theta \right]^2 y^{-1-2(\{l\}-\delta)}\, dy \right)^{p/2} dx \right\}^{1/p}$$
$$+ c\left\{ \int \left[(\Lambda^{\delta-m} |f|)\left(\frac{x}{4}\right) \right]^p (S_{l-\delta}\gamma_\rho)^p\, dx \right\}^{1/p}.$$

After simple calculations we obtain that this sum is majorized by

$$c\left(\left\| (\Lambda^{\delta-m} |f|)\left(\frac{\cdot}{4}\right) \nabla_{[l]}\gamma_\rho \right\|_{H_p^{\{l\}-\delta}} + \left\| |\nabla_{[l]}\gamma_\rho|\, S_{\{l\}-\delta}\left[(\Lambda^{\delta-m} |f|)\left(\frac{\cdot}{4}\right) \right] \right\|_{L_p} \right.$$
$$\left. + \left\| (\Lambda^{\delta-m} |f|)\left(\frac{\cdot}{4}\right) S_{\{l\}-\delta}(\nabla_{[l]}\gamma_\rho) \right\|_{L_p} \right).$$

Let the last norms be denoted by N_1, N_2 and N_3 respectively. Using (3), we get

$$N_1 \leq \|\nabla_{[l]}\gamma_\rho\|_{M(H_p^{m-\delta} \to H_p^{\{l\}-\delta})} \left\| (\Lambda^{\delta-m} |f|)\left(\frac{\cdot}{4}\right) \right\|_{H_p^{m-\delta}}$$
$$\leq c\, \|\gamma_\rho\|^{1-\delta/l}_{M(H_p^m \to H_p^l)}\, \|\gamma_\rho\|^{\delta/l}_{M(H_p^{m-l} \to L_p)}\, \|f\|_{L_p}. \tag{9}$$

Similarly,

$$N_2 \leq \|\nabla_{[l]}\gamma_\rho\|_{M(H_p^{m-\{l\}} \to L_p)} \left\| S_{\{l\}-\delta}\left[(\Lambda^{\delta-m} |f|)\left(\frac{\cdot}{4}\right) \right] \right\|_{H_p^{m-\{l\}}}$$
$$\leq c\, \|\nabla_{[l]}\gamma_\rho\|_{M(H_p^{m-\{l\}} \to L_p)} \left\| (\Lambda^{\delta-m} |f|)\left(\frac{\cdot}{4}\right) \right\|_{H_p^{m-\delta}}$$
$$\leq c\, \|\gamma_\rho\|^{[l]/l}_{M(H_p^m \to H_p^l)}\, \|\gamma_\rho\|^{\{l\}/l}_{M(H_p^{m-l} \to L_p)}\, \|f\|_{L_p}. \tag{10}$$

Now we turn to the estimate of the norm N_3. According to Lemma 2.2.4/1,

$$S_{\{l\}-\delta}(\nabla_{[l]}\gamma_\rho) \leq c(S_l\gamma_\rho)^{1-\delta/m}\, \|\nabla_{[l]}\gamma_\rho\|^{\delta/m}_{M(H_p^{m-\{l\}} \to L_p)} + c\, \|\nabla_{[l]}\gamma_\rho\|_{M(H_p^{m-\{l\}} \to L_p)}.$$

Further, we notice that

$$\Lambda^{\delta-m}|f| \leqslant c(\Lambda^{-m}|f|)^{1-\delta/m}(Mf)^{\delta/m}$$

(see Lemma 1.1.3). Thus we have proved the estimate

$$N_3 \leqslant c\,\|\nabla_{[l]}\gamma_\rho\|^{\delta/m}_{M(H_p^{m-\{l\}}\to L_p)} \left\| \left[(\Lambda^{-m}|f|)\left(\frac{\cdot}{4}\right)S_l\gamma_\rho\right]^{1-\delta/m} \left[(Mf)\left(\frac{\cdot}{4}\right)\right]^{\delta/m} \right\|_{L_p}$$
$$+ c\,\|\nabla_{[l]}\gamma_\rho\|_{M(H_p^{m-\{l\}}\to L_p)} \|\Lambda^{\delta-m}|f|\|_{L_p}.$$

Consequently,

$$N_3 \leqslant c\,\|\nabla_{[l]}\gamma_\rho\|^{\delta/m}_{M(H_p^{m-\{l\}}\to L_p)} \|Mf\|^{\delta/m}_{L_p} \left\|(\Lambda^{-m}|f|)\left(\frac{\cdot}{4}\right)S_l\gamma_\rho\right\|^{1-\delta/m}_{L_p}$$
$$+ c\,\|\nabla_{[l]}\gamma_\rho\|_{M(H_p^{m-\{l\}}\to L_p)} \|\Lambda^{\delta-m}|f|\|_{H_p^{m-\delta}}.$$

Applying (3), we finally obtain

$$N_3 \leqslant c\,\|\gamma_\rho\|^{\delta[l]/ml}_{M(H_p^m\to H_p^l)} \|\gamma_\rho\|^{\delta\{l\}/ml}_{M(H^m\to L_p)}$$
$$\times (\|S_l\gamma_\rho\|_{M(H_p^m\to L_p)} + \|\gamma_\rho\|^{[l]/l}_{M(H_p^m\to H_p^l)} \|\gamma_\rho\|^{\{l\}/l}_{M(H_p^{m-l}\to L_p)})^{1-\delta/m} \|f\|_{L_p}. \quad (11)$$

Summing estimates (9)–(11), we conclude that

$$\left\{\int\left(\int_0^\infty\left[\int_{B_1} |u(x+\theta y)-u(x)|\right.\right.\right.$$
$$\left.\left.\left.\times|\nabla_{[l]}\gamma_\rho(x+\theta y)-\nabla_{[l]}\gamma_\rho(x)|\,d\theta\right]^2 y^{-1-2\{l\}}\,dy\right)^{p/2} dx\right\}^{1/p}$$
$$\leqslant (\varepsilon\,\|\gamma_\rho\|_{M(H_p^m\to H_p^l)} + \varepsilon\,\|S_l\gamma_\rho\|_{M(H_p^m\to L_p)} + c(\varepsilon)\,\|\gamma_\rho\|_{M(H_p^{m-l}\to L_p)})\,\|u\|_{H_p^m}. \quad (12)$$

This, together with (7) and (8), makes it possible to deduce from (6) that

$$\|S_{\{l\}}(u\,\nabla_{[l]}\gamma_\rho)\|_{L_p} \leqslant (\varepsilon\,\|\gamma_\rho\|_{M(H_p^m\to H_p^l)} + c\,\|S_l\gamma_\rho\|_{M(H_p^m\to L_p)}$$
$$+ c(\varepsilon)\,\|\gamma_\rho\|_{M(H_p^{m-l}\to L_p)})\,\|u\|_{H_p^m}.$$

The substitution of this estimate into (5) leads to (1) for γ_ρ. Reference to Lemma 2.2.1/1 completes the proof. □

2.2.6 Lower Bound for the Norm of a Multiplier

We prove the assertion converse to Lemma 2.2.5.

Lemma. *If $\gamma \in M(H_p^m \to H_p^l)$, then $\gamma \in H^l_{p,\mathrm{loc}}$, $S_l\gamma \in M(H_p^m \to L_p)$ and $\gamma \in M(H_p^{m-l} \to L_p)$. The following inequality holds:*

$$\|S_l\gamma\|_{M(H_p^m\to L_p)} + \|\gamma\|_{M(H_p^{m-l}\to L_p)} \leqslant c\,\|\gamma\|_{M(H_p^m\to H_p^l)}. \quad (1)$$

Proof. It is clear that $\gamma \in H_{p,\text{loc}}^l$ and the estimate of $\|\gamma\|_{M(H_p^{m-l} \to L_p)}$ is contained in Lemma 2.2.2/2.

For integer l the assertion was proved in Lemma 1.3.2/2. Suppose $\{l\} > 0$. For any $u \in D$,

$$\|S_{\{l\}}(u\, \nabla_{[l]}\gamma_\rho)\|_{L_p} \le \|S_l(\gamma_\rho u)\|_{L_p} + \sum_{\substack{0 \le \beta < \alpha \\ |\alpha| = [l]}} \|S_{\{l\}}(D^\beta \gamma_\rho D^{\alpha-\beta} u)\|_{L_p}$$

and therefore

$$\|uS_l\gamma_\rho\|_{L_p} \le \|\gamma_\rho u\|_{H_p^l} + \sum_{\substack{0 \le \beta < \alpha \\ |\alpha| = [l]}} \|S_{\{l\}}(D^\beta \gamma_\rho D^{\alpha-\beta} u)\|_{L_p}$$

$$+ \| |\nabla_{[l]}\gamma_\rho|\, S_{\{l\}} u\|_{L_p} + \left\{ \int \left(\int_0^\infty \left[\int_{B_1} |u(x+\theta y) - u(x)| \right.\right.\right.$$

$$\left.\left.\left. \times |\nabla_{[l]}\gamma_\rho(x+\theta y) - \nabla_{[l]}\gamma_\rho(x)|\, d\theta \right]^2 y^{-1-2\{l\}}\, dy \right)^{p/2} dx \right\}^{1/p}.$$

This and estimates (2.2.5/4), (2.2.5/8), (2.2.5/12) give

$$\|uS_l\gamma_\rho\|_{L_p} \le (\varepsilon\, \|S_l\gamma_\rho\|_{M(H_p^m \to L_p)}$$

$$+ c_1 \|\gamma_\rho\|_{M(H_p^m \to H_p^l)} + c\, \|\gamma_\rho\|_{M(H_p^{m-l} \to L_p)})\, \|u\|_{H_p^m}$$

$$\le (\varepsilon\, \|S_l\gamma_\rho\|_{M(H_p^m \to L_p)} + c_2\, \|\gamma_\rho\|_{M(H_p^m \to H_p^l)})\, \|u\|_{H_p^m}.$$

Minimizing the last norm over the set $\{u \in D : u \ge 1 \text{ on } e\}$, we arrive at

$$\|S_l\gamma_\rho\|_{M(H_p^m \to L_p)} \le c\, \|\gamma_\rho\|_{M(H_p^m \to H_p^l)}.$$

It remains to use Lemma 2.2.1/1. □

2.2.7 Description of Spaces of Multipliers

Combining Lemmas 2.2.5 and 2.2.6, we obtain the following result.

Theorem 1. *Let $m \ge l \ge 0$, $p \in (1, \infty)$. A function γ belongs to $M(H_p^m \to H_p^l)$ if and only if $\gamma \in H_{p,\text{loc}}^l$, $S_l\gamma \in M(H_p^m \to L_p)$ and $\gamma \in M(H_p^{m-l} \to L_p)$. The relation*

$$\|\gamma\|_{M(H_p^m \to H_p^l)} \sim \|S_l\gamma\|_{M(H_p^m \to L_p)} + \|\gamma\|_{M(H_p^{m-l} \to L_p)}$$

holds.

This, together with Lemma 2.2.2/1, gives the following assertion:

Theorem 2. *A function γ belongs to $M(H_p^m \to H_p^l)$, $m \ge l \ge 0$, $p \in (1, \infty)$ if and only if $\gamma \in H_{p,\text{loc}}^l$ and, for any compactum $e \subset R^n$,*

$$\|S_l\gamma; e\|_{L_p}^p \le c \operatorname{cap}(e, H_p^m),$$

$$\|\gamma; e\|_{L_p}^p \le c \operatorname{cap}(e, H_p^{m-l}).$$

The following relations hold:

$$\|\gamma\|_{M(H_p^m\to H_p^l)} \sim \sup_e \left(\frac{\|S_l\gamma; e\|_{L_p}}{[\text{cap}\,(e, H_p^m)]^{1/p}} + \frac{\|\gamma; e\|_{L_p}}{[\text{cap}\,(e, H_p^{m-l})]^{1/p}}\right), \tag{1}$$

$$\|\gamma\|_{M(H_p^m\to H_p^l)} \sim \sup_{\{e:d(e)\leq 1\}} \left(\frac{\|S_l\gamma; e\|_{L_p}}{[\text{cap}\,(e, H_p^m)]^{1/p}} + \frac{\|\gamma; e\|_{L_p}}{[\text{cap}\,(e, H_p^{m-l})]^{1/p}}\right). \tag{2}$$

Remark 1. From (1) and (2) for $m=l$ follows

$$\|\gamma\|_{MH_p^l} \sim \sup_e \frac{\|S_l\gamma; e\|_{L_p}}{[\text{cap}\,(e, H_p^l)]^{1/p}} + \|\gamma\|_{L_\infty},$$

$$\|\gamma\|_{MH_p^l} \sim \sup_{\{e:d(e)\leq 1\}} \frac{\|S_l\gamma; e\|_{L_p}}{[\text{cap}\,(e, H_p^l)]^{1/p}} + \|\gamma\|_{L_\infty}.$$

Remark 2. Replacing Λ^m by $(-\Delta)^{m/2}$ in the definition of the space H_p^m, we obtain the space h_p^m.

Using the Strichartz norm $\|S_m u\|_{L_p}$ in h_p^m and Sobolev's theorem on the imbedding $h_p^m \subset L_q$ with $q = mp/(n-mp)$, $n > mp$, one can easily prove that any function $\psi \in D$ belongs to the space Mh_p^m of multipliers in h_p^m. This immediately implies that the norm in Mh_p^m of the function ψ_r with values $\psi_r(x) = \psi(x/r)$, $r > 0$, does not depend on r.

This enables one to proceed as in the proofs of Lemmas 2.2.5 and 2.2.6, with γ_ρ replaced by $\psi_r\gamma_\rho$, where $\psi \in D$, $\psi = 1$ on B_1, passing to the limit as $\rho \to 0$, $r \to \infty$ at the final step (see Subsection 1.3).

Thus we are led to the following relation for the norm in the space of multipliers $M(h_p^m \to h_p^l)$, $0 \leq l \leq m < n/p$:

$$\|\gamma\|_{M(h_p^m\to h_p^l)} \sim \sup_e \left(\frac{\|S_l\gamma; e\|_{L_p}}{[\text{cap}\,(e, h_p^m)]^{1/p}} + \frac{\|\gamma; e\|_{L_p}}{[\text{cap}\,(e, h_p^{m-l})]^{1/p}}\right). \tag{3}$$

The capacity cap (e, h_p^k) is defined in the same way as cap (e, H_p^k).

Remark 3. According to Stein's theorem [2], the norm in H_p^l for $p \geq 2$, $0 < l < 1$, is equivalent to $\|Tu\|_{L_p} + \|u\|_{L_p}$, where

$$(Tu)(x) = \left(\int |u(x+y) - u(x)|^2 \frac{dy}{|y|^{n+2l}}\right)^{1/2}.$$

This and the inequality $|T(\gamma u) - uT\gamma| \leq \|\gamma\|_{L_\infty} Tu$ imply

$$\|\gamma\|_{MH_p^l} \sim \sup_e \frac{\|T\gamma; e\|_{L_p}}{[\text{cap}\,(e, H_p^l)]^{1/p}} + \|\gamma\|_{L_\infty},$$

where $p \geq 2$, $0 < l < 1$.

2.2.8 Equivalent Norms in $M(H_p^m \to H_p^l)$

In this subsection we obtain one more norm for the space $M(H_p^m \to H_p^l)$.

Lemma 1. *If $\gamma \in M(H_p^m \to H_p^l)$, $m > l$, $p \in (1, \infty)$ and k is an integer, $1 \leqslant k \leqslant m/(m-l)$, then $\gamma^k \in M(H_p^m \to H_p^{m-k(m-l)})$. Moreover,*

$$\|\gamma^k\|_{M(H_p^m \to H_p^{m-k(m-l)})} \leqslant c\, \|\gamma\|^{k\theta(k)}_{M(H_p^{m-l} \to L_p)}\, \|\gamma\|^{k(1-\theta(k))}_{M(H^m \to H_p^l)}, \tag{1}$$

where $\theta(k) = (k-1)(m-l)/2l$.

Proof. We have

$$\|\gamma^k\|_{M(H_p^m \to H_p^{m-k(m-l)})} \leqslant c\, \|\gamma\|_{M(H_p^{m-(k-1)(m-l)} \to H_p^{m-k(m-l)})} \times \|\gamma^{k-1}\|_{M(H_p^m \to H_p^{m-(k-1)(m-l)})}. \tag{2}$$

Suppose that (1) is proved for $k-1$. Then from (2) and the interpolation inequality

$$\|\gamma\|_{M(H_p^{m-(k-1)(m-l)} \to H_p^{m-k(m-l)})} \leqslant c\, \|\gamma\|^{(k-1)(m-l)/l}_{M(H_p^{m-l} \to L_p)}\, \|\gamma\|^{(m-k(m-l))/l}_{M(H_p^m \to H_p^l)}$$

we obtain

$$\|\gamma^k\|_{M(H_p^m \to H_p^{m-k(m-l)})} \leqslant c\, \|\gamma\|^{(k-1)(m-l)/l + (k-1)\theta(k-1)}_{M(H_p^m \to L_p)} \times \|\gamma\|^{m-k(m-l)/l + (k-1)(1-\theta(k-1))}_{M(H_p^m \to H_p^l)}$$

which is equivalent to (1). □

Corollary 1. *The following inequality holds:*

$$\sup_e \frac{\|\gamma; e\|_{L_{pm/(m-l)}}}{[\operatorname{cap}(e, H_p^m)]^{(m-l)/mp}} \leqslant c\, \|\gamma\|^{1-\mu}_{M(H_p^{m-l} \to L_p)}\, \|\gamma\|^{\mu}_{M(H_p^m \to H_p^l)},$$

where $\mu = k(1-\theta(k))$, $k = [m/(m-l)]$, $\theta(k) = (k-1)(m-l)/2l$.

Proof. By Lemma 2.2.1,

$$\||\gamma|^{m/(m-l)} u\|_{L_p} \leqslant c \left(\sup_e \frac{\int_e |\gamma|^{p\alpha}\, dx}{\operatorname{cap}(e, H_p^{\alpha(m-l)})} \right)^{1/p} \|\gamma^k u\|_{H_p^{\alpha(m-l)}},$$

where α is a fractional part of $m/(m-l)$. Applying Lemmas 2.2.2/1 and 2.2.1/2, we get from the last estimate that

$$\||\gamma|^{m/(m-l)} u\|_{L_p} \leqslant c\, \|\gamma\|^{\alpha}_{M(H_p^{m-l} \to L_p)}\, \|\gamma^k u\|_{H_p^{\alpha(m-l)}}.$$

It remains to use Lemma 1. □

Theorem. *The following relation holds:*

$$\|\gamma\|_{M(H_p^m\to H_p^l)}\sim\sup_e\left(\frac{\|\gamma;e\|_{L_{mp/(m-l)}}}{[\operatorname{cap}(e,H_p^m)]^{(m-l)/mp}}+\frac{\|S_l\gamma;e\|_{L_p}}{[\operatorname{cap}(e,H_p^m)]^{1/p}}\right),\tag{3}$$

where $p>1$, $m\geqslant l\geqslant 0$.

Proof. The upper estimate for the norm in $M(H_p^m\to H_p^l)$ follows from Lemma 2.2.5 and from Lemma 2.2.1/2 which implies

$$\sup_e\frac{\|\gamma;e\|_{L_p}}{[\operatorname{cap}(e,H_p^{m-l})]^{1/p}}\leqslant c\sup_e\frac{\|\gamma;e\|_{L_{mp/(m-l)}}}{[\operatorname{cap}(e,H_p^m)]^{(m-l)/mp}}.$$

The lower bound is deduced from Corollary 1 and Lemma 2.2.6. □

From this and (2.1.4/4) we obtain

Corollary 2. *The relation*

$$\|\gamma\|_{M(H_p^m\to H_p^l)}\sim\sup_{\{e:d(e)\leqslant 1\}}\left(\frac{\|\gamma;e\|_{L_{mp/(m-l)}}}{[\operatorname{cap}(e,H_p^m)]^{(m-l)/mp}}+\frac{\|S_l\gamma;e\|_{L_p}}{[\operatorname{cap}(e,H_p^m)]^{1/p}}\right)$$

holds.

2.2.9 The Space $M(H_p^m\to H_p^l)$ for $mp>n$

Theorem. *If* $mp>n$, $p\in(1,\infty)$, *then* $\|\gamma\|_{M(H_p^m\to H_p^l)}\sim\|\gamma\|_{H^l_{p,\mathrm{unif}}}$.

Proof. The required lower bound for the norm in $M(H_p^m\to H_p^l)$ follows from the inequality

$$\|\gamma\eta_z\|_{H_p^l}\leqslant c\,\|\gamma\|_{M(H_p^m\to H_p^l)}\,\|\eta_z\|_{H_p^m}.$$

We obtain the upper bound. Since H_p^m is imbedded into L_∞, then $\operatorname{cap}(e,H_p^m)\sim 1$ for any compactum e with $d(e)\leqslant 1$. Therefore from (2.2.8/3) it follows that

$$\|\gamma\|_{M(H_p^m\to H_p^l)}\sim\sup_{z\in R^n}(\|\gamma;B_{1/2}(z)\|_{L_{mp/(m-l)}}+\|S_l\gamma;B_{1/2}(z)\|_{L_p}).$$

According to the Sobolev imbedding theorem, $\|\gamma\eta_z\|_{L_q}\leqslant c\,\|\gamma\eta_z\|_{H_p^l}$, where $q\leqslant\infty$ for $lp>n$, q is any positive number if $lp=n$ and $q\leqslant pn(n-lp)^{-1}$ if $lp<n$. Consequently,

$$\|\gamma;B_{1/2}(z)\|_{L_{mp/(m-l)}}\leqslant c\|\gamma\eta_z\|_{H_p^l}.$$

If l is an integer, then the assertion is proved.

Let l be fractional. We have

$$\|S_l\gamma;B_{1/2}(z)\|_{L_p}\leqslant\|S_l(\gamma\eta_z);B_{1/2}(z)\|_{L_p}+\|S_l[\gamma(1-\eta_z)];B_{1/2}(z)\|_{L_p}.$$

The first norm in the right-hand side does not exceed $\|\gamma\eta_z\|_{H_p^l}$ by Theorem 2.1.1/2 and the second one is not more than

$$c \sup_{x\in B_{1/2}(z)} \left(\int_{1/2}^{\infty} \left[y^{-n}\int_{|s|<y} |\nabla_{[l]}[\gamma(x+s)(1-\eta_z(x+s))]|\, ds\right]^2 y^{-1-2\{l\}}\, dy\right)^{1/2}$$

$$\leqslant c_1 \sup_{x\in R^n} \|\gamma; B_1(x)\|_{W_p^{[l]}} \leqslant c_2 \sup_{x\in R^n} \|\gamma\eta_x\|_{H_p^l}. \quad \square$$

Remark. The coincidence of MH_p^m and $H_{p,\text{unif}}^m$ for $mp>n$ is a result by Strichartz [1] obtained earlier by a direct method.

2.2.10 One-sided Estimates for the Norm in $M(H_p^m \to H_p^l)$

We present here some lower and, separately, upper bounds for the norm in $M(H_p^m \to H_p^l)$, $mp \leqslant n$, which contain no capacity and which follow easily from the results of this section.

From Proposition 2.1.2/4 and Corollary 2.2.8/2 there immediately follows:

Proposition 1. *If $mp<n$, then*

$$\|\gamma\|_{M(H_p^m\to H_p^l)} \geqslant c \sup_{x\in R^n, r\in(0,1)} (r^{-m+n/p}\|S_l\gamma; B_r(x)\|_{L_p} + r^{(-m+n/p)(1-l/m)}\|\gamma; B_r(x)\|_{L_{mp/(m-l)}})$$

and if $mp=n$, then

$$\|\gamma\|_{M(H_p^m\to H_p^l)} \geqslant c \sup_{x\in R^n, r\in(0,1)} [(\log 2r^{-1})^{(p-1)/p}\|S_l\gamma; B_r(x)\|_{L_p} + (\log 2r^{-1})^{(p-1)(m-l)/pn}\|\gamma; B_r(x)\|_{L_{n/(m-l)}}].$$

A corollary of Theorem 2.2.7/2 and estimates (2.1.2/2), (2.1.2/3) is the following:

Proposition 2. *If $mp<n$, then*

$$\|\gamma\|_{M(H_p^m\to H_p^l)} \leqslant c \sup_{\{e: d(e)\leqslant 1\}} (\text{mes}_n\, e)^{m/n-1/p}(\|S_l\gamma; e\|_{L_p} + (\text{mes}_n\, e)^{-l/n}\|\gamma; e\|_{L_p}) \tag{1}$$

and if $mp=n$, then

$$\|\gamma\|_{M(H_p^m\to H_p^l)} \leqslant c \sup_{\{e: d(e)\leqslant 1\}} ((\log(2^n/\text{mes}_n\, e))^{1/p'}\|S_l\gamma; e\|_{L_p} + (\text{mes}_n\, e)^{-l/n}\|\gamma; e\|_{L_p}). \tag{2}$$

In connection with (1) we make the following:

Remark. By the Marcinkiewicz space $\mathfrak{M}_\alpha$ we mean the linear set of functions in a domain $\Omega \subset R^n$ for which

$$\sup_{0<t<\infty} t[m(t)]^\alpha < \infty, \tag{3}$$

where $\alpha \in (0, 1)$ and $m(t) = \text{mes}_n \{x \in \Omega : |f(x)| \geq t\}$. We denote the left-hand side of (3) by $\|f; \Omega\|^*_{\mathfrak{M}_\alpha}$ and set

$$\|f; \Omega\|_{\mathfrak{M}_\alpha} = \sup_{e \subset \Omega} \frac{\int_e |f| \, dx}{(\text{mes}_n \, e)^{1-\alpha}}.$$

It is known that

$$(1-\alpha) \|f; \Omega\|_{\mathfrak{M}_\alpha} \leq \|f; \Omega\|_{\mathfrak{M}^*_\alpha} \leq \|f; \Omega\|_{\mathfrak{M}_\alpha} \tag{4}$$

(cf. Krasnosel'skii, Zabreyko, Pustyl'nik, Sobolevskii [1], 8.3, ch. 2). In fact, for all e,

$$\int_e |f| \, dx \leq \|f; \Omega\|_{\mathfrak{M}_\alpha} (\text{mes}_n \, e)^{1-\alpha}$$

and, in particular,

$$\int_{|f| \geq t} |f| \, dx \leq \|f; \Omega\|_{\mathfrak{M}_\alpha} [m(t)]^{1-\alpha}.$$

On the other hand,

$$\int_{|f| \geq t} |f| \, dx \geq t m(t).$$

Consequently,

$$t[m(t)]^\alpha \leq \|f; \Omega\|_{\mathfrak{M}_\alpha}$$

and the right inequality (4) is proved.

Now let $(0, \text{mes}_n \, e) \ni \mu \to F(\mu)$ be a non-increasing function which satisfies

$$m(t) = \text{mes}_1 (\{\mu : F(\mu) \geq t\}).$$

It is clear that

$$\int_e |f| \, dx \leq \int_0^{\text{mes}_n e} F(\mu) \, d\mu.$$

Moreover,

$$F(\mu) \leq \mu^{-\alpha} \|f; \Omega\|_{\mathfrak{M}^*_\alpha}.$$

So

$$\int_e |f| \, dx \leq \|f; \Omega\|_{\mathfrak{M}^*_\alpha} \int_0^{\operatorname{mes}_n e} \mu^{-\alpha} \, d\mu = \|f; \Omega\|_{\mathfrak{M}^*_\alpha} \frac{(\operatorname{mes}_n e)^{1-\alpha}}{1-\alpha}$$

and the left estimate (4) is proved. □

Now it is clear that (1) can be rewritten in terms of Marcinkiewicz spaces. For example, for $m = l < n/p$ this inequality is equivalent to

$$\|\gamma\|_{MH_p^l} \leq c\Big(\|\gamma\|_{L_\infty} + \sup_{x \in R^n} \|S_l\gamma; B_1(x)\|_{\mathfrak{M}^*_{l/n}}\Big). \tag{5}$$

An estimate of such a kind, in which $p = 2$, $0 < l < 1$ and the function

$$x \to \left(\int |\gamma(x) - \gamma(y)|^2 \frac{dy}{|x-y|^{n+2l}}\right)^{1/2}$$

plays the role of $S_l\gamma$, was given by Hirschman [1]. In the case $l \in (0, 1)$, $lp < n$, inequality (5) was proved by Strichartz [1] with the use of interpolation methods (see also Stegenga [2], where the case $m = l \in (0, 1)$, $p \geq 2$, $lp = n$ is considered as well).

2.3 Sufficient Conditions for Inclusion into $M(H_p^m \to H_p^l)$

In this section we derive upper bounds for the norm in $M(H_p^m \to H_p^l)$ in terms of Bessel potential spaces and Besov spaces on the basis of Proposition 2.2.10/2. These results were obtained by Maz'ya and Shaposhnikova [11].

2.3.1 Estimates for the Norm of a Multiplier in Terms of the Space $H^l_{n/m}$

Proposition 1. (i) *If $lp < n$ and $\gamma \in H^l_{n/l,\mathrm{unif}} \cap L_\infty$, then $\gamma \in MH_p^l$ and*

$$\|\gamma\|_{MH_p^l} \leq c(\|\gamma\|_{H^l_{n/l,\mathrm{unif}}} + \|\gamma\|_{L_\infty}). \tag{1}$$

(ii) *If $mp < n$, $l < m$ and $\gamma \in H^l_{n/m,\mathrm{unif}}$, then $\gamma \in M(H_p^m \to H_p^l)$ and*

$$\|\gamma\|_{M(H_p^m \to H_p^l)} \leq c \, \|\gamma\|_{H^l_{n/m,\mathrm{unif}}}.$$

Proof. Let $\eta \in D(B_1)$, $\eta = 1$ on $B_{1/2}$ and $\eta_z(x) = \eta(x - z)$. By Theorem 2.1.1/3,

$$\|\gamma\|_{M(H_p^m \to H_p^l)} \leq c \sup_{z \in R^n} \|\eta_z\gamma\|_{M(H_p^m \to H_p^l)}.$$

This and (2.2.10/1) imply that the norm of γ in $M(H_p^m \to H_p^l)$ has the majorant

$$c \sup_{z\in R^n} (\|S_l(\eta_z\gamma)\|_{L_{n/m}} + \|\eta_z\gamma\|_{L_{n/(m-l)}}).$$

Now, using Theorem 2.1.1/2, we obtain the estimate

$$\|\gamma\|_{M(H_p^m \to H_p^l)} \leqslant c(\|\gamma\|_{H^l_{n/m,\text{unif}}} + \|\gamma\|_{L_{n/(m-l),\text{unif}}})$$

which coincides with (1) for $m = l$.

Since by the Sobolev theorem $H^l_{n/m} \subset L_{n/(m-l)}$ for $m > l$, then

$$\|\gamma\|_{L_{n/(m-l),\text{unif}}} \leqslant c\, \|\gamma\|_{H^l_{n/m,\text{unif}}}. \quad \square$$

Estimate (1) was obtained in the paper of Polking [1] who used an analogue of the Leibniz formula for the nonlinear operator of fractional differentiation:

$$(D^s_{p,\theta}u)(x) = \left(\int_0^\infty \left(\int_{B_1} |u(x+\rho y) - u(x)|^p \,dy\right)^{\theta/p} \frac{d\rho}{\rho^{1+s\theta}}\right)^{1/\theta}.$$

For $\theta = \infty$, by definition,

$$(D^s_{p,\infty}u)(x) = \sup_{0<\rho<\infty} \rho^{-s}\left(\int_{B_1} |u(x+\rho y) - u(x)|^p \,dy\right)^{1/p}.$$

Here we assume that $0<s<1$, $1\leqslant p<\infty$, $1\leqslant\theta\leqslant\infty$ and $u\in L_{p,\text{loc}}$.

The proof of Polking is interesting in itself, and we reproduce it, omitting the proof of the next auxiliary assertion.

Lemma 1 (Polking [1]). *Let $0<s<1$, $1\leqslant p\leqslant\theta$, $2\leqslant\theta<\infty$. If $1<r<\infty$ and $r>np/(n+sp)$ then there exists a constant c depending only on s, p, θ, r, n and such that*

$$\|D^s_{p,\theta}u\|_{L_r} \leqslant c\, \|u\|_{H^s_r} \tag{2}$$

for all $u\in H^s_r$.

An analogue of the Leibniz formula for $D^s_{p,\theta}$ is contained in the following:

Lemma 2. (Polking [1]). *Let $1/p_1 + 1/p_2 = 1/p$, $1/\theta_1 + 1/\theta_2 = 1/\theta$, $s_1 + s_2 = s$. Then, for all $u\in L_{p_1,\text{loc}}$, $v\in L_{p_2,\text{loc}}$, the inequality*

$$(D^s_{p,\theta}(uv))(x) \leqslant |u(x)|\,(D^s_{p,\theta}v)(x) + |v(x)|\,(D^s_{p,\theta}u)(x) + (D^{s_1}_{p_1,\theta_1}u)(x)(D^{s_2}_{p_2,\theta_2}v)(x) \tag{3}$$

is valid.

If $u \in L_\infty$ *then the simpler estimate*

$$(D^s_{p,\theta}(uv))(x) \leqslant \|u\|_{L_\infty} (D^s_{p,\theta}v)(x) + |v(x)|\,(D^s_{p,\theta}u)(x) \tag{4}$$

holds.

Proof. First we note that

$$\begin{aligned} &u(x+\rho y)v(x+\rho y) - u(x)v(x) \\ &\quad = u(x)[v(x+\rho y) - v(x)] + v(x)[u(x+\rho y) - u(x)] \\ &\qquad + [u(x+\rho y) - u(x)][v(x+\rho y) - v(x)]. \end{aligned}$$

Consequently,

$$\begin{aligned} &(D^s_{p,\theta}(uv))(x) \\ &\quad \leqslant |u(x)|\,(D^s_{p,\theta}v)(x) + |v(x)|\,(D^s_{p,\theta}u)(x) \\ &\qquad + \left(\int_0^\infty \left(\int_{B_1} |u(x+\rho y) - u(x)|^p\,|v(x+\rho y) - v(x)|^p\,dy\right)^{\theta/p} \frac{d\rho}{\rho^{1+s\theta}}\right)^{1/\theta}. \end{aligned}$$

Inequality (3) follows by application of Hölder's inequality to the last summand.

Inequality (4) results from the identity

$$\begin{aligned} u(x+\rho y)v(x+\rho y) - u(x)v(x) &= u(x+\rho y)[v(x+\rho y) - v(x)] \\ &\quad + v(x)[u(x+\rho y) - u(x)]. \quad \square \end{aligned}$$

We pass to the above-mentioned result of Polking.

Proposition 2. *Let* $1 < p < \infty$, $lp < n$. *Then there exists a constant* c *depending only on* p *and* l *and such that*

$$\|\gamma u\|_{H_p^l} \leqslant c(\|\gamma\|_{L_\infty} + \|(-\Delta)^{l/2}\gamma\|_{L_{n/l}})\,\|u\|_{H_p^l}$$

for all $u \in H_p^l$, $\gamma \in H_{n/l}^l$.

Proof. It suffices to consider the case $\{l\} > 0$. By the usual Leibniz formula we have

$$\|\gamma u\|_{H_p^l} \leqslant c\left(\sum_{|\alpha|+|\beta| \leqslant [l]} \|D^\alpha \gamma D^\beta u\|_{L_p} + \sum_{|\alpha|+|\beta| \leqslant [l]} \|D^\alpha \gamma D^\beta u\|_{H_p^{\{l\}}}\right). \tag{5}$$

By the Strichartz theorem the norms $\|v\|_{H_p^l}$ and $\|D^{\{l\}}_{1,2} v\|_{L_p} + \|v\|_{L_p}$ are equivalent. Therefore the second sum on the right in (5) can be replaced by

$$\sum_{|\alpha|+|\beta| \leqslant [l]} \|D^{\{l\}}_{1,2}(D^\alpha \gamma D^\beta u)\|_{L_p}.$$

We restrict ourselves by the bound for this sum since other terms on the right in (5) can be estimated in a similar way and even more simply.

We choose positive numbers σ_1, σ_2 such that $\sigma_1+\sigma_2=\{l\}$ and put $p_1=n/(|\alpha|+\sigma_1)$, $p_2=n/(n-|\alpha|-\sigma_1)$. By Lemma 2

$$\begin{aligned}&\|D^{\{l\}}_{1,2}(D^\alpha\gamma D^\beta u)\|_{L_p}\\ &\quad\leq\|D^\alpha\gamma D^{\{l\}}_{1,2}(D^\beta u)\|_{L_p}+\|D^\beta u D^{\{l\}}_{1,2}(D^\alpha\gamma)\|_{L_p}\\ &\quad\quad+\|D^{\sigma_1}_{p_1,2p_1}(D^\alpha\gamma)D^{\sigma_2}_{p_2,2p_2}(D^\beta u)\|_{L_p}.\end{aligned}\tag{6}$$

For $\alpha=0$ the first summand on the right does not exceed the product $\|\gamma\|_{L_\infty}\|D^{\{l\}}_{1,2}(D^\beta u)\|_{L_p}$ which, obviously, is not more than $c\,\|\gamma\|_{L_\infty}\|u\|_{H^l_p}$. With the exception of this particular case, all the norms on the right in (6) are estimated in a similar way. We therefore consider only the third.

Let $q_1=n/(|\alpha|+\sigma_1)$, $q_2=np/(n-|\alpha|\,p-\sigma_1 p)$. Then $1/q_1+1/q_2=1/p$, $q_1=p_1$ and $q_2>p_2$. Using Hölder's inequality and then Lemma 1, we find

$$\begin{aligned}&\|D^{\sigma_1}_{p_1,2p_1}(D^\alpha\gamma)D^{\sigma_2}_{p_2,2p_2}(D^\beta u)\|_{L_p}\\ &\quad\leq\|D^{\sigma_1}_{p_1,2p_1}(D^\alpha\gamma)\|_{L_{q_1}}\|D^{\sigma_2}_{p_2,2p_2}(D^\beta u)\|_{L_{q_2}}\leq c\,\|D^\alpha\gamma\|_{H^{\sigma_1}_{q_1}}\|D^\beta u\|_{H^{\sigma_2}_{q_2}}.\end{aligned}$$

It remains to be noted that by the Sobolev theorem

$$\begin{aligned}&\|D^\alpha\gamma\|_{H^{\sigma_1}_{q_1}}\leq c\,\|\gamma\|_{H_{n/l}},\\ &\|D^\beta u\|_{H^{\sigma_2}_{q_2}}\leq c\,\|u\|_{H^l_p}.\end{aligned}$$

The result follows. □

2.3.2 Auxiliary Assertions

Lemma 1. *Let $q\geq 1$, $\{\mu\}>0$, $\mu q<n$. Further, let $v\in W^{[\mu]}_q(B_1)$. Then*

$$\sup_{e\subset B_1}\frac{\|v;e\|_{L_q}}{(\text{mes}_n\,e)^{\mu/n}}\leq c\left(\sup_{e\subset B_1}\frac{\|\nabla_{[\mu]}v;e\|_{L_q}}{(\text{mes}_n\,e)^{\{\mu\}/n}}+\|v;B_1\|_{L_q}\right).\tag{1}$$

Proof. It suffices to put $m=n/q-\{\mu\}$, $l=[\mu]$, $p=q$ in (1.3.4/2).

Lemma 2. *Let $q\geq 1$, $\mu>0$, $\{\mu\}>0$, $v\in W^{[\mu]}_q(B_2)$. Then*

$$\begin{aligned}&\sup_{e\subset B_1}(\text{mes}_n\,e)^{-\{\mu\}/n}\|\nabla_{[\mu]}v;e\|_{L_q}\\ &\quad\leq c\left[\sup_{h\in B_1}|h|^{-\{\mu\}}\|\Delta_h\,\nabla_{[\mu]}v;B_1\|_{L_q}+\sup_{x\in B_1,0<r<1}r^{-\mu}\,\|v;B_r(x)\|_{L_q}\right],\end{aligned}$$

where $(\Delta_h w)(x)=w(x+h)-w(x)$.

Proof. Let $Q \in D(B_1)$, $\int Q\,dx = 1$, $\rho \in (0, 1)$, and let e be an arbitrary compact set in B_1. For any multi-index α of order $[\mu]$ we have

$$|D^\alpha v(x)| \leqslant \left|\rho^{-n}\int Q(h/\rho)(D^\alpha v(x+h) - D^\alpha v(x))\,dh\right| + \rho^{-n-[\mu]}\left|\int (D^\alpha Q)(h/\rho)v(x+h)\,dh\right|.$$

Consequently,

$$\int_e |D^\alpha v|^q\,dx \leqslant c\Big[\rho^{-qn}\int_e \Big(|Q(h/\rho)|^{1/q'}\,|Q(h/\rho)|^{1/q} \times |D^\alpha v(x+h) - D^\alpha v(x)|\,dh\Big)^q dx + (\text{mes}_n\, e)\rho^{-q[\mu]-n}\sup_{x\in B_1}\int_{B_\rho(x)} |v(h)|^q\,dh\Big].$$

Applying Hölder's inequality to the first summand in the right-hand side, we obtain

$$\int_e |D^\alpha v|^q\,dx \leqslant c\rho^{q\{\mu\}}\Big[\sup_{h\in B_\rho} |h|^{-q\{\mu\}}\int_e |D^\alpha v(x+h) - D^\alpha v(x)|^q\,dx + (\text{mes}_n\, e)\rho^{-n}\sup_{x\in B_1, r\in(0,1)} r^{-\mu q}\,\|v; B_r(x)\|_{L_q}^q\Big].$$

It remains to put $\rho = 2^{-1}(\text{mes}_n\, e)^{1/n}$ and to multiply both sides in the last inequality by $\rho^{-q\{\mu\}}$. □

Let $\{\mu\} > 0$. We define the space $B_{q,\infty}^\mu$ of functions in R^n with the norm

$$\sup_{h\in R^n} |h|^{-\{\mu\}}\,\|\Delta_h \nabla_{[\mu]} v\|_{L_q} + \|v\|_{W_q^{[\mu]}}. \tag{2}$$

It is known (cf. Besov, Il'in, Nikol'skii [1]) that the norm (2) is equivalent to

$$\sum_{i=1}^n \sup_{t\in R^1} |t|^{-\mu}\,\|\Delta_{te_i}^s v\|_{L_q} + \|v\|_{L_q},$$

where $s > \mu$ and $\Delta_{te_i}^s$ is the difference of order s in the direction of the unit vector e_i. One can easily verify that the norm in $B_{q,\infty,\text{unif}}^\mu$ is equivalent to any of the following norms:

$$\sup_{x\in R^n, h\in B_1} |h|^{-\{\mu\}}\,\|\Delta_h \nabla_{[\mu]} v; B_1(x)\|_{L_q} + \sup_{x\in R^n} \|v; B_1(x)\|_{L_q},$$

$$\sum_{i=1}^n \sup_{x\in R^n, |t|<1} |t|^{-\mu}\,\|\Delta_{te_i}^s v; B_1(x)\|_{L_q} + \sup_{x\in R^n} \|v; B_1(x)\|_{L_q}.$$

Next we present an assertion which, together with its proof, is due to V. P. Il'in (personal communication):

Lemma 3. *Let v be a function in L_q with a compact support and let $q \geq 1$, $\mu > 0$, $\mu q < n$. Then*

$$\sup_{x \in R^n, r>0} r^{-\mu} \|v; B_r(x)\|_{L_q} \leq c \sum_{i=1}^{n} \sup_{t \in R^1} |t|^{-\mu} \|\Delta^s_{te_i} v\|_{L_q}. \tag{3}$$

Proof. From the Il'in integral representation (see formula (97) p. 107 of Besov, Il'in, Nikol'skii [1]) follows

$$|v(z)| \leq \sum_{i=1}^{n} \int_0^\infty \frac{d\sigma}{\sigma^{n+2}} \int_{R^1} |M(t/\sigma)|\, dt \int |\Delta^s_{te_i} v(y+te_i)| \left|\Omega\left(\frac{y-z}{\sigma}\right)\right| dy,$$

where $M \in D(0, 1)$, $\Omega \in D((0, 1)^n)$.

By U_i we denote the i-th summand in the right-hand side. Represent U_i as the sum $V_i + W_i$ of two integrals over σ so that the integration in V_i is taken over $\sigma \in [0, r]$. By Minkowski's inequality,

$$\|V_i; B_r(x)\|_{L_q} \leq \int_0^r \frac{d\sigma}{\sigma^{n+2}} \int_{R^1} |M(t/\sigma)|\, dt$$
$$\times \left\| \int |\Delta^s_{te_i} v(y+te_i)| \left|\Omega\left(\frac{y-z}{\sigma}\right)\right| dy; B_r(x) \right\|_{L_q}.$$

Applying Minkowski's inequality once more, we obtain

$$\left\| \int |\Delta^s_{te_i} v(y+te_i)| \left|\Omega\left(\frac{y-z}{\sigma}\right)\right| dy; B_r(x) \right\|_{L_q} \leq c\sigma^n \|\Delta^s_{te_i} v\|_{L_q}.$$

So

$$\|V_i; B_r(x)\| \leq c \int_0^r \frac{d\sigma}{\sigma^{n+2}} \sigma^n \int_0^\sigma t^\mu\, dt \sup_{t \in R^1} |t|^{-\mu} \|\Delta^s_{te_i} v\|_{L_q}$$
$$\leq cr^\mu \sup_{t \in R^1} |t|^{-\mu} \|\Delta^s_{te_i} v\|_{L_q}. \tag{4}$$

By Hölder's inequality,

$$\sup_{z \in B_r(x)} |W_i(z)| \leq c_1 \int_r^\infty \frac{d\sigma}{\sigma^{n+2}} \int_0^\sigma \sigma^{n/q'} \left[\sup_{t \in R^1} |t|^{-\mu} \|\Delta^s_{te_i} v\|_{L_q}\right] \tau^\mu\, d\tau$$
$$= cr^{\mu - n/q} \sup_{t \in R^1} |t|^{-\mu} \|\Delta^s_{te_i} v\|_{L_q}.$$

Consequently,

$$\|W_i; B_r(x)\|_{L_q} \leqslant cr^{n/q} \sup_{z \in B_r(x)} |W_i(z)| \leqslant cr^{\mu} \sup_{t \in R^1} |t|^{-\mu} \|\Delta^s_{te_i} v\|_{L_q}.$$

This, together with (4), implies (3). □

From Lemmas 1–3 there immediately follows:

Corollary. *Let $\{\mu\}>0$, $q \geqslant 1$, $\mu q \leqslant n$ and let $v \in W^{[\mu]}_{q,\mathrm{loc}}$. Then*

$$\sup_{\{e: d(e) \leqslant 1\}} (\mathrm{mes}_n\, e)^{-\mu/n}((\mathrm{mes}_n\, e)^{[\mu]/n} \|\nabla_{[\mu]} v; e\|_{L_q} + \|v; e\|_{L_q})$$
$$\leqslant \begin{cases} c\, \|v\|_{B^{\mu}_{q,\infty,\mathrm{unif}}} & \text{for} \quad \mu q < n, \\ c(\|v\|_{B^{\mu}_{q,\infty,\mathrm{unif}}} + \|v\|_{L_\infty}) & \text{for} \quad \mu q = n. \end{cases}$$

2.3.3 Estimates for the Norm of a Multiplier in Terms of the Space $B^{\mu}_{q,\infty}$

Theorem. *Let $q \geqslant p$, $\mu = n/q - m + l$, $\mu > l$, $\{\mu\} > 0$.*
(i) *If $\gamma \in B^{\mu}_{q,\infty,\mathrm{unif}} \cap L_\infty$, then $\gamma \in MH^l_p$ and*

$$\|\gamma\|_{MH^l_p} \leqslant c\Big(\sup_{x \in R^n, h \in B_1} |h|^{-\{\mu\}} \|\Delta_h \nabla_{[\mu]} \gamma; B_1(x)\|_{L_q} + \|\gamma\|_{L_\infty}\Big). \tag{1}$$

(ii) *If $\gamma \in B^{\mu}_{q,\infty,\mathrm{unif}}$, then $\gamma \in M(H^m_p \to H^l_p)$ and*

$$\|\gamma\|_{M(H^m_p \to H^l_p)}$$
$$\leqslant c\Big(\sup_{x \in R^n, h \in B_1} |h|^{-\{\mu\}} \|\Delta_h \nabla_{[\mu]} \gamma; B_1(x)\|_{L_q} + \sup_{x \in R^n} \|\gamma; B_1(x)\|_{L_q}\Big). \tag{2}$$

Proof. Let $m \geqslant l$. It is sufficient to assume that the difference $\varepsilon = n/q - m$ is small since the general case follows by interpolation between pairs $\{H^{m-l}_p, L_p\}$ and $\{H^{n/q-\varepsilon}_p, H^{\mu-\varepsilon}_p\}$ (cf. (2.2.1/1)). So we assume that $1 - \{l\} > n/q - m > 0$.

We introduce the function Φ_x on $(0, \infty)$ with values

$$\Phi_x(y) = \int_{B_y} |\nabla_{[l]} \gamma(x+z) - \nabla_{[l]} \gamma(x)|\, dz.$$

Clearly,

$$\|S_l \gamma; e\|^p_{L_p} = \int_e \Big(\int_0^\infty [\Phi_x(y)]^2 y^{-1-2(\{l\}+n)}\, dy \Big)^{p/2} dx,$$

where e is a compact set with $d(e) \leqslant 1$. Since Φ_x is an increasing function, the internal integral in the right-hand side is not more than

$$(\{l\}+n)\int_0^\infty \frac{\Phi_x(y)\,dy}{y^{1+\{l\}+n}}\int_y^\infty \Phi_x(t)\frac{dt}{t^{1+\{l\}+n}}=\frac{\{l\}+n}{2}\left(\int_0^\infty \Phi_x(y)\frac{dy}{y^{1+\{l\}+n}}\right)^2.$$

We express the last integral as the sum of two integrals $i_1(x)+i_2(x)$, of which the first is over semi-axis $y>|e|^{1/n}$, where $|e|=\text{mes}_n\, e$. We have

$$\int_e i_1(x)^p\,dx\leqslant\int_e\left\{\int_{|e|^{1/n}}^\infty\left(y^{-n}\int_{B_y}|\nabla_{[l]}\gamma(x+z)|\,dz+|\nabla_{[l]}\gamma(x)|\right)\frac{dy}{y^{1+\{l\}}}\right\}^p dx.$$

This and Minkowski's inequality imply

$$\int_e i_1(x)^p\,dx\leqslant\left\{\int_{|e|^{1/n}}^\infty\left(\int_e\left[y^{-n}\int_{B_y}|\nabla_{[l]}\gamma(x+z)|\,dz\right.\right.\right.$$
$$\left.\left.\left.+|\nabla_{[l]}\gamma(x)|\right]^p dx\right)^{1/p}\frac{dy}{y^{1+\{l\}}}\right\}^p.$$

So

$$\int_e i_1(x)^p\,dx\leqslant c\left\{\int_{|e|^{1/n}}^\infty\left(\int_e\left[y^{-n}\int_{B_y}|\nabla_{[l]}\gamma(x+z)|^p\,dz\right.\right.\right.$$
$$\left.\left.\left.+|\nabla_{[l]}\gamma(x)|^p\right]dx\right)^{1/p}\frac{dy}{y^{1+\{l\}}}\right\}^p$$
$$\leqslant c\,|e|^{1-p/q}\left\{\int_{|e|^{1/n}}^\infty\left[y^{-n}\int_{B_y}\left(\int_e|\nabla_{[l]}\gamma(x+z)|^q\,dx\right)^{1/q}dz\right.\right.$$
$$\left.\left.+\left(\int_e|\nabla_{[l]}\gamma(x)|^q\,dx\right)^{1/q}\right]\frac{dy}{y^{1+\{l\}}}\right\}^p.$$

By $N(\gamma)$ we denote the right-hand sides of (1) and (2). According to Corollary 2.3.2, where $\mu=n/q-m+l$,

$$\left(\int_e|\nabla_{[l]}\gamma(x+z)|^q\,dx\right)^{1/q}\leqslant c\,|e|^{1/q-m/n}N(\gamma).$$

Consequently,

$$\int_e i_1(x)^p\,dx\leqslant c\,|e|^{1-mp/n}N(\gamma)^p. \tag{3}$$

Applying Minkowski's inequality, we obtain

$$\int_e i_2(x)^p\,dx=\int_e\left(\int_0^{|e|^{1/n}}\int_{B_y}|\nabla_{[l]}\gamma(x+z)-\nabla_{[l]}\gamma(x)|\,dz\frac{dy}{y^{1+\{l\}+n}}\right)^p dx$$
$$\leqslant\left\{\int_0^{|e|^{1/n}}\int_{B_y}\left(\int_e|\nabla_{[l]}\gamma(x+z)-\nabla_{[l]}\gamma(x)|^p\,dx\right)^{1/p}dz\frac{dy}{y^{1+\{l\}+n}}\right\}^p.$$

By Hölder's inequality the last expression does not exceed

$$|e|^{1-p/q}\left\{\int_0^{|e|^{1/n}}\int_{B_y}\left(\int_e |\nabla_{[l]}\gamma(x+z)-\nabla_{[l]}\gamma(x)|^q\,dx\right)^{1/q} dz\frac{dy}{y^{1+\{l\}+n}}\right\}^p.$$

Since

$$|z|^{q(m-\{l\})-n}\int_e |\nabla_{[l]}\gamma(x+z)-\nabla_{[l]}\gamma(x)|^q\,dx\le cN(\gamma)^q,$$

then

$$\int_e i_2(x)^p\,dx\le c\,|e|^{1-p/q}N(\gamma)^p\left(\int_0^{|e|^{1/n}}\int_{B_y}|z|^{-m+\{l\}+n/q}\,dz\frac{dy}{y^{1+\{l\}+n}}\right)^p$$
$$=c\,|e|^{1-pm/n}N(\gamma)^p. \tag{4}$$

Summing (3) with (4), we arrive at

$$\|S_l\gamma;e\|_{L_p}^p\le c\,|e|^{1-pm/n}N(\gamma)^p.$$

From Hölder's inequality and Corollary 2.3.2 for $\mu=n/q-m+l$ we obtain

$$\|\gamma;e\|_{L_p}\le|e|^{1/p-1/q}\|\gamma;e\|_{L_q}\le c\,|e|^{1/p-1/q}|e|^{1/q-(m-l)/n}N(\gamma). \tag{5}$$

Reference to Proposition 2.2.10/2 completes the proof. □

2.3.4 Estimate of the Norm of a Multiplier in $MH_p^l(R^1)$ by q-Variation

Hirschman [2] obtained the following sufficient condition for a function γ to belong to the class MW_2^l on a unit circumference C: γ is bounded and has a finite q-variation $\mathrm{Var}_q(\gamma)$ for some q, $2<q<1/l$.

Here q-variation is defined by

$$\mathrm{Var}_q(\gamma)=\sup\left(\sum_{j=0}^{m-1}|\gamma(t_{j+1})-\gamma(t_j)|^q\right)^{1/q}, \tag{1}$$

the supremum being taken over all partitions of the circumference C by points t_j.

Using Theorem 2.3.3, one may easily derive a sufficient condition for a function to belong to $MH_p^l(R^1)$ which, for $p=2$, coincides with Hirschman's condition up to the change of R^1 to C.

We define the local q-variation of a function γ defined on R^1 by (1), the supremum being taken over all choices of a finite number of points $t_0<t_1<\cdots<t_m$ on any interval σ of unit length.

Since evidently $\int_\sigma|\gamma(t+h)-\gamma(t)|^q\,dt\le c\,|h|\,[\mathrm{Var}_q(\gamma)]^q$, we arrive at the following:

Corollary. *Let* $n=1$, $q\geqslant p$, $lq<1$. *If* $\gamma\in L_\infty$ *and* $\operatorname{Var}_q(\gamma)<\infty$, *then* $\gamma\in MH^l_p$ *and*

$$\|\gamma\|_{MH^l_p}\leqslant c(\|\gamma\|_{L_\infty}+\operatorname{Var}_q(\gamma)).$$

2.4 Some Properties of Elements of $M(H^m_p\to H^l_p)$

The following assertion is a generalization of Proposition 1.6/1.

Proposition 1. *If* $k\in[0,l]$, *then* $M(H^m_p\to H^l_p)\subset M(H^{m-l+k}_p\to H^k_p)$ *and*

$$\|\gamma\|_{M(H^{m-l+k}_p\to H^k_p)}\leqslant c\,\|\gamma\|_{M(H^m_p\to H^l_p)}.$$

Proof. The imbedding $M(H^m_p\to H^l_p)\subset M(H^{m-l}_p\to L_p)$ and the corresponding inequality for norms was proved in Lemma 2.2.2/2. It remains to use the interpolation inequality

$$\|\gamma\|_{M(H^{m-l+k}_p\to H^k_p)}\leqslant c\,\|\gamma\|^{k/l}_{M(H^m_p\to H^l_p)}\,\|\gamma\|^{(l-k)/l}_{M(H^{m-l}_p\to L_p)}$$

which is a particular case of the estimate (2.2.1/1). □

Proposition 2. (i) *If* $mp>n$, $1<q<\infty$, $0\leqslant k<l$ *and* $k\leqslant l+n(1/q-1/p)$, *then* $M(H^m_p\to H^l_p)\subset M(H^{m-l+k}_q\to H^k_q)$ *and*

$$\|\gamma\|_{M(H^{m-l+k}_q\to H^k_q)}\leqslant c\,\|\gamma\|_{M(H^m_p\to H^l_p)}. \tag{1}$$

(ii) *If* $mp=n$, $0\leqslant k\leqslant l$, $q>1$ *and* $k<l-m+n/q$, *then* $M(H^m_p\to H^l_p)\subset M(H^{m-l+k}_q\to H^k_q)$ *and inequality* (1) *is valid.*

Proof. (i) According to Theorem 2.2.9, $M(H^m_p\to H^l_p)=H^l_{p,\,\mathrm{unif}}$. Since $mp>n$, then $M(H^m_p\to H^l_p)\subset H^l_{n/m,\,\mathrm{unif}}$ for $m>l$ and $MH^m_p\subset H^m_{n/m,\,\mathrm{unif}}\cap L_\infty$ for $m=l$. This, together with Proposition 2.3.1/1, implies $M(H^m_p\to H^l_p)\subset M(H^m_q\to H^l_q)$ for any $q\in(1,n/m)$. Applying Proposition 1 and interpolating with respect to q between $1+\varepsilon$ and p (cf. (2.2.1/1)), we obtain the inclusion $M(H^m_p\to H^l_p)\subset M(H^{m-l+k}_q\to H^k_q)$ for all $k\in[0,l]$, $q\in(1,p]$.

For $q\in(p,\infty)$ we put $s=l+n(1/q-1/p)$. It is clear that $s<l$ and $q(s+m-l)>n$. By Theorem 2.2.9 and by the Sobolev imbedding theorem, $M(H^m_p\to H^l_p)=H^l_{p,\,\mathrm{unif}}\subset H^s_{q,\,\mathrm{unif}}$. Since $q(s+m-l)>n$ then, applying Theorem 2.2.9 once more, we obtain $M(H^m_p\to H^l_p)\subset M(H^{m-l+s}_q\to H^s_q)$. Further, by Proposition 1, $M(H^m_p\to H^l_p)\subset M(H^{m-l+k}_q\to H^k_q)$ for all $k\leqslant s$, $q\in[p,\infty)$.

(ii) By virtue of Theorem 2.2.9, $M(H_p^m \to H_p^l) \subset H^l_{n/m,\text{unif}}$ for $m > l$ and $MH_p^m \subset H^m_{n/m,\text{ unif}} \cap L_\infty$. According to the Sobolev imbedding theorem, $H^l_{n/m,\text{ unif}} \subset H^k_{n/r,\text{ unif}}$, where $r = m - l + k$, $k < l$. This and Proposition 2.3.1/1 imply $M(H_p^m \to H_p^l) \subset M(H_q^{m-l+k} \to H_q^k)$ for any $q \in (1, n/r)$. □

Next we present an imbedding theorem for the space MH_p^m.

Proposition 3. *The space MH_p^m is imbedded into MH_q^k, $k \leqslant m$, $1 < q < \infty$, provided that*

(i) $mp > n$, $k \leqslant m + n(1/q - 1/p)$,

(ii) $mp \leqslant n$, $k < mp/q$.

Proof. For $mp \geqslant n$ the assertion is proved in Proposition 2. Let $mp < n$. We start with the case $q \geqslant p$. If $\gamma \in MH_p^m$ then $\gamma \in L_\infty$ and hence $\gamma \in ML_r$ for any $r \in (1, \infty)$. The interpolation between H_p^m and L_r with $r \in [q, \infty)$ (cf. inequality (2.2.1/1)) implies the imbedding $MH_p^m \subset MH_q^k$ with $0 < k < mp/q$, $q \geqslant p$. Now suppose $q < p$, $k \leqslant m$ and $\gamma \in MH_p^m$. Then by Theorem 2.2.7/2

$$\sup_e \frac{\|S_m\gamma;\, e\|^p_{L_p}}{\text{cap}\,(e, H_p^m)} < \infty.$$

This and Lemma 2.2.1/2 yield

$$\sup_e \frac{\|S_m\gamma;\, e\|^q_{L_q}}{\text{cap}\,(e, H_p^{mq/p})} < \infty.$$

Applying the inequality $\text{cap}\,(e, H_p^{mq/p}) \leqslant c\,\text{cap}\,(e, H_q^m)$ with $q < p$ (see Adams, Meyers [1]), we find

$$\sup_e \frac{\|S_m\gamma;\, e\|^q_{L_q}}{\text{cap}\,(e, H_q^m)} < \infty.$$

Since $\gamma \in L_\infty$ then by Theorem 2.2.7/2 we get $\gamma \in MH_q^m$. Again, by interpolation between H_q^m and L_q we obtain $\gamma \in MH_q^k$ with $k \leqslant m$, $q < p$. The result follows. □

Remark 1. For $mp \geqslant n$ this proposition is proved by Strichartz [1]. In the case $mp < n$ the result proved in the paper mentioned is incomplete. Above we have presented a stronger statement communicated to us by I. E. Verbitskii who also showed that these imbeddings cannot be improved. For $q \leqslant p$, $k \leqslant m$ the latter can be easily verified by an example of lacunary series. The proof of the exactness is much more difficult in the case $q > p$, $k \leqslant mp/q$, in particular, for $k = mp/q$ it can be performed by using Blaschke products.

Next we give a simple sufficient condition for a function to belong to the space $M(H_p^m(R^n)\to H_p^l(R^n))$ formulated in terms of $M(H_p^m(R^1)\to H_p^l(R^1))$.

Proposition 4. *If for all* $j=1,\ldots,n$

$$\operatorname*{ess\,sup}_{(x_1,\ldots,x_{j-1},x_{j+1},\ldots,x_n)} \|\gamma(x_1,\ldots,x_{j-1},\cdot,x_{j+1},\ldots,x_n;R^1\|_{M(H_p^m\to H_p^l)}<\infty,$$

then $\gamma\in M(H_p^m(R^n)\to H_p^l(R^n))$.

Proof. It can easily be verified that the functions

$$\frac{(1+|\xi|^2)^{l/2}}{\sum_{j=1}^n(1+\xi_j^2)^{l/2}},\qquad \left(\frac{1+\xi_k^2}{1+|\xi|^2}\right)^{l/2},\qquad k=1,2,\ldots,n,$$

satisfy the condition of Lemma 1.1.1/1. Therefore

$$\|u\|_{H_p^l}\sim\sum_{j=1}^n\left\|\left(1-\frac{\partial^2}{\partial x_j^2}\right)^{l/2}u\right\|_{L_p}. \tag{2}$$

The conclusion is obvious. □

From (2) we immediately obtain

Proposition 5. *If function* γ *depends only on variables* $x_1,\ldots,x_s$, $s<n$, *then* $\gamma\in M(H_p^m(R^n)\to H_p^l(R^n))$ *if and only if*

$$\gamma\in M(H_p^m(R^s)\to H_p^l(R^s)).$$

Moreover,

$$\|\gamma;R^n\|_{M(H_p^m\to H_p^l)}\sim\|\gamma;R^s\|_{M(H_p^m\to H_p^l)}.$$

Proof. The upper bound for the norm in $M(H_p^m(R^n)\to H_p^l(R^n))$ immediately follows from Proposition 4. Let $\eta\in C_0^\infty(R^{n-s})$, $\eta\neq 0$. By virtue of (2) for $v\in H_p^m(R^s)$ we have

$$\|\gamma v;R^s\|_{H_p^l}\leqslant c\,\|\gamma\eta v;R^n\|_{H_p^l}\leqslant c\,\|\gamma;R^n\|_{M(H}$$
$$\leqslant c\,\|\gamma;R^n\|_{M(H_p^m\to H_p^l)}\|v;R^s\|_{H_p^m},$$

which yields the lower bound for the norm in $M(H_p^m(R^n)\to H_p^l(R^n))$. □

Corollary 1. *The characteristic function* χ *of the half-space* R_+^n *belongs to* $MH_p^l(R^n)$ *if and only if* $lp<1$.

Proof. By Proposition 5 we may limit consideration to the case $n=1$.

According to Theorem 2.3.3, for $lp<1$

$$\|\chi; R^1\|_{MH_p^l} \leqslant c\Big(\sup_{x\in R^1, |h|<1} |h|^{-1/p} \|\Delta_h \chi; B_1(x)\|_{L_p} + 1\Big).$$

Since the right-hand side is bounded, $\chi \in MH_p^l(R^1)$.

Now let $lp=1$. By η we denote a function in $C_0^\infty(R^1)$ which is equal to unity in a neighbourhood of the point O. Since $(F\chi)(\xi) = i\xi^{-1} + \pi\,\delta(\xi)$, then $[F(\eta\chi)](\xi) = i\xi^{-1} + O(|\xi|^{-2})$ for large $|\xi|$. Therefore

$$\Lambda^l(\eta\chi) = F^{-1}(1+\xi^2)^{l/2} F(\eta\chi) = c\,|x|^{-l}\operatorname{sgn} x + O(1)$$

for small $|x|$ and so $\chi \,\overline{\in}\, H_{p,\mathrm{loc}}^l(R^1)$. Consequently, $\chi \,\overline{\in}\, MH_p^l(R^1)$. □

Remark 2. Corollary 1 is well known. For $p=2$ it was proved by Hirschman [2]. The case $p\in(1,\infty)$ was considered by Shamir [1] and Strichartz [1]. An analogous assertion for the space W_p^l is given by Lions and Magenes [1]. The conditions for the function χ to belong to classes of multipliers in more general functional spaces have been studied by Triebel [1]–[3].

Remark 3. Corollary 1 for $n=1$ and Proposition 4 immediately imply that the characteristic function of any convex open subset of R^n is the multiplier in $H_p^l(R^n)$ for $lp<1$. It is even sufficient to assume that the multiplicity of the intersection of the set with almost any straight line parallel to one coordinate axis is bounded.

We give some more simple properties of the class MH_p^l.

Proposition 6. *The following relation holds:* $\|\gamma\|_{L_\infty} \leqslant \|\gamma\|_{MH_p^l}$.

This estimate can be deduced in precisely the same way as (1.6/1).

The following assertion complements Proposition 1.6/5, where the case $\{l\}=0$ is considered.

Proposition 7. *Let* $\{l\}>0$, $\gamma\in MH_p^l$ *and let* σ *be a segment of the real axis such that* $\gamma(x)\in\sigma$ *for almost all* $x\in R^n$. *Further, let* $f\in C^{[l],1}(\sigma)$. *Then* $f(\gamma)\in MH_p^l$ *and we have the estimate*

$$\|f(\gamma)\|_{MH_p^l} \leqslant c\sum_{j=0}^{[l]+1} \|f^{(j)}; \sigma\|_{L_\infty} \|\gamma\|^j_{MH_p^l}.$$

Proof. Let $l\in(0,1)$. By Theorem 2.1.1/2, for all $u\in D$

$$\|uf(\gamma)\|_{H_p^l} \leqslant c(\|S_l(uf(\gamma))\|_{L_p} + \|uf(\gamma)\|_{L_p}).$$

Since

$$S_l(uf(\gamma)) \leq |u|\, S_l f(\gamma) + \|f(\gamma)\|_{L_\infty} S_l u$$
$$\leq |u|\, \|f';\sigma\|_{L_\infty} S_l\gamma + \|f(\gamma)\|_{L_\infty} S_l u,$$

we have $\|uf(\gamma)\|_{H_p^l} \leq c(\|f'\|_{L_\infty} \|S_l\gamma\|_{M(H_p^l \to L_p)} + \|f(\gamma)\|_{L_\infty})\, \|u\|_{H_p^l}$. This, together with Lemma 2.2.6, implies the required estimate

$$\|f(\gamma)\|_{MH_p^l} \leq c(\|f'\|_{L_\infty} \|\gamma\|_{MH_p^l} + \|f(\gamma)\|_{L_\infty}).$$

It remains to proceed by induction on $[l]$ (cf. the proof of Proposition 1.6/5). □

This and Proposition 6 immediately imply

Corollary 2. *If* $\gamma \in MH_p^l$, $\{l\} > 0$, *and* $\|\gamma^{-1}\|_{L_\infty} < \infty$ *then* $\gamma^{-1} \in MH_p^l$ *and*
$$\|\gamma^{-1}\|_{MH_p^l} \leq c\, \|\gamma^{-1}\|_{L_\infty}^{[l]+2} \|\gamma\|_{MH^l}^{[l]+1}.$$

2.5 The Spectrum of Multipliers in H_p^l and $H_{p'}^{-l}$

2.5.1 *Preliminary Information*

We present certain definitions and the simplest facts of the spectral theory.

Let X be a complex Banach space and let A be a bounded linear operator in X.

Definition 1. The set of complex values λ, for which the operator $(\lambda I - A)^{-1}$ exists, is defined on the whole of X and is bounded, is called the resolvent set $\rho(A)$ of the operator A. The complement of $\rho(A)$ is called the spectrum $\sigma(A)$ of A.

It is known that the resolvent set $\rho(A)$ is open and that the function $(\lambda I - A)^{-1}$ is analytic on $\rho(A)$.

Definition 2. The value $r(A) = \sup |\sigma(A)|$ is called the spectral radius of the operator A.

The Gelfand formula holds:

$$r(A) = \lim_{m\to\infty} \sqrt[m]{(\|A^m\|)} \tag{1}$$

Definition 3. The operator A is called quasinilpotent if $\lim_{m\to\infty} \sqrt[m]{(\|A^m\|)} = 0$.

The next three definitions give a classification of points of the spectrum.

Definition 4. The set of points $\lambda \in \sigma(A)$ such that the mapping $\lambda I - A$ is not one-to-one is called the pointwise spectrum and is denoted by $\sigma_p(A)$. In other words, $\lambda \in \sigma_p(A)$ if and only if there exists a nontrivial solution $u \in X$ of the equation $(\lambda I - A)u = 0$. Elements of σ_p are called eigenvalues.

Definition 5. The set of numbers $\lambda \in \sigma(A)$ for which the mapping $\lambda I - A$ is one-to-one and the range of $\lambda I - A$ is not dense in X is called the residual spectrum and is denoted by $\sigma_r(A)$.

Definition 6. The set of numbers $\lambda \in \sigma(A)$ for which the mapping $\lambda I - A$ is one-to-one and the range of $\lambda I - A$ is dense in X but does not coincide with X is called the continuous spectrum of A and is denoted by $\sigma_c(A)$.

It is clear that the sets $\sigma_p(A)$, $\sigma_r(A)$ and $\sigma_c(A)$ are disjoint. By the Banach theorem on isomorphism, the condition $(\lambda I - A)X \neq X$ in Definition 5 is unnecessary and therefore

$$\sigma(A) = \sigma_p(A) \cup \sigma_r(A) \cup \sigma_c(A). \tag{2}$$

Let A^* be the operator adjoint of A. Definitions 4–6 imply

$$\sigma_r(A) \subset \overline{\sigma_p(A^*)} \subset \sigma_r(A) \cup \sigma_p(A), \tag{3}$$

where the overbar denotes the complex conjunction.

2.5.2 The Spectrum of a Multiplier. The Main Theorem

In the present and subsequent subsections the role of the space X is played by H_p^l and $H_{p'}^{-l}$; the multipliers are considered as an operator A.

Corollaries 1.6, 2.4/2 and the imbedding $MH_p^l \subset L_\infty$ immediately imply:

Corollary 1. The number λ belongs to the spectrum of a multiplier $\gamma \in MH_p^l$ if and only if $(\gamma - \lambda)^{-1} \bar{\in} L_\infty$ or, which is equivalent, for any $\varepsilon > 0$ the set $\{x : |\gamma(x) - \lambda| < \varepsilon\}$ has positive n-dimensional measure.

Since the adjoint operator of $\gamma \in MH_p^l$ is the multiplier $\bar{\gamma}$ in $H_{p'}^{-l}$, Corollary 1 implies:

Corollary 2. The number λ belongs to the spectrum of $\gamma \in MH_{p'}^{-l}$ if and only if $(\gamma - \lambda)^{-1} \bar{\in} L_\infty$.

From Corollaries 1 and 2 we immediately obtain that the spectral radius $r(\gamma)$ of a multiplier γ in H_p^l or in $H_{p'}^{-l}$ is equal to $\|\gamma\|_{L_\infty}$.

This and (2.5.1/1) imply $\lim_{m\to\infty} \sqrt[m]{(\|\gamma^m\|_{MH_p^l})} = \|\gamma\|_{L_\infty}$. So the only quasi-nilpotent multiplier is zero. In other words, the algebra MH_p^l is semisimple.

This is a generalization of results obtained in Devinatz, Hirschman [1] for $p=2$, $2l<1$.

The main theorem of the present section contains a description of the decomposition (2.5.1/2) for multipliers in H_p^l and $H_{p'}^{-l}$. Before we pass to its formulation we present certain auxiliary definitions and results.

Definition 1. The function u is called (p, l)-refined if for any $\varepsilon>0$ one can find an open set ω such that $\text{cap}(\omega, H_p^l)<\varepsilon$ and u is continuous on $R^n \backslash \omega$.

For proofs of the next assertions see Maz'ya, Havin [2].

Proposition 1. *For any $u \in H_{p,\text{loc}}^l$ there exists a (p, l)-refined Borel function which coincides with u almost everywhere.*

Proposition 2. *If two (p, l)-refined functions u_1 and u_2 are equal almost everywhere then they are equal (p, l)-quasi everywhere.*

Henceforth in this section all the functions are assumed to be (p, l)-refined and Borel.

The following assertion is proved for integer l (see Maz'ya [14]) and for fractional l in Adams and Polking [1] for compacta. The passage to arbitrary sets does not need new arguments if one uses Proposition 2.1.3/2.

Proposition 3. *Let $E \subset R^n$. The capacity $\overline{\text{cap}}\,(E, H_p^l)$ is equivalent to the set function $\inf\{\|v\|_{H_p^l}^p : v \in \mathfrak{M}(E)\}$, where $\mathfrak{M}(E)$ is the collection of (p, l)-refined functions equal to unity (p, l)-quasi everywhere on E and satisfying the inequalities $0 \leqslant v \leqslant 1$.*

Definition 2. The set $E \subset R^n$ is called the set of uniqueness for the space H_p^l if the conditions $u \in H_p^l$, $u(x)=0$ for (p, l)-quasi all $x \in R^n \setminus E$ imply $u=0$.

The description of sets of uniqueness for H_p^l is given by Hedberg [2] and Polking [2]. The first result of such a kind for $H_2^{1/2}$ on a circumference is due to Ahlfors and Beurling [1].

Proposition 4 (Hedberg [2]). *Let E be a Borel subset of R^n. The following conditions are equivalent:*

(i) *E is the set of uniqueness for H_p^l;*

(ii) $\operatorname{cap}(G\backslash E, H_p^l) = \operatorname{cap}(G, H_p^l)$ *for any open set G;*

(iii) $\overline{\lim}_{\rho\to 0} \rho^{-n} \operatorname{cap}(B_\rho(x)\backslash E, H_p^l) > 0$ *for almost all x.*

If $lp > n$ then E is the set of uniqueness if and only if it has no interior points.

Now we state a theorem which gives a characteristic of the sets $\sigma_p(\gamma)$, $\sigma_r(\gamma)$ and $\sigma_c(\gamma)$ for a multiplier γ in H_p^l or $H_{p'}^{-l}$.

Theorem. (i) *Let $\gamma \in MH_p^l$ and $\lambda \in \sigma(\gamma)$.*

1. *$\lambda \in \sigma_p(\gamma)$ if and only if the set $Z_\lambda = \{x : \gamma(x) = \lambda\}$ does not satisfy any of conditions* (i)–(iii) *of Proposition* 4.

2. *$\lambda \in \sigma_r(\gamma)$ if and only if the set Z_λ satisfies any of the conditions of Proposition* 4 *and* $\operatorname{cap}(Z_\lambda, H_p^l) > 0$.

3. *$\lambda \in \sigma_c(\gamma)$ if and only if* $\operatorname{cap}(Z_\lambda, H_p^l) = 0$.

(ii) *Let $\gamma \in MH_{p'}^{-l}$ and $\lambda \in \sigma(\gamma)$.*

1. *$\lambda \in \sigma_p(\gamma)$ if and only if* $\operatorname{cap}(Z_\lambda, H_p^l) > 0$.

2. *$\lambda \in \sigma_c(\gamma)$ if and only if* $\operatorname{cap}(Z_\lambda, H_p^l) = 0$ *(hence the set $\sigma_r(\gamma)$ is empty).*

An obvious corollary of Propositions 1 and 2 is the following assertion.

Lemma. *Let γ be a (p, l)-refined function in MH_p^l. The equation $(\gamma - \lambda)u = 0$ has a nontrivial solution in H_p^l if and only if there exists a (p, l)-refined non-zero function in H_p^l vanishing (p, l)-quasi everywhere outside Z_λ.*

This lemma shows that item (i)1 of the theorem immediately follows from Proposition 4. The proof of other items of the theorem is contained in the subsequent subsection.

2.5.3 Proof of Theorem 2.5.2

Below we shall use the following assertion.

Lemma. *Let γ be a (p, l)-refined function in MH_p^l and let $Z_0 = \{x : \gamma(x) = 0\}$. If* $\operatorname{cap}(Z_0, H_p^l) = 0$, *then the set γH_p^l is dense in H_p^l.*

Proof. Let $f \in D$ and let $N_\tau = \{x \in \operatorname{supp} f : |\gamma(x)| \leq \tau\}$. By ε we denote a small positive number and by ω we mean an open set with $\operatorname{cap}(\omega, H_p^l) < \varepsilon$ and such that γ is continuous on $R^n\backslash\omega$. Let G designate a neighbourhood of the set $N_0\backslash\omega$ with $\operatorname{cap}(G, H_p^l) < \varepsilon$.

We note that $N_\tau \backslash \omega \subset G$ for small enough $\tau > 0$. In fact, if for any $\tau > 0$ there exists a point $x_\tau \in N_\tau \backslash \omega$ which is not contained in G then, by continuity of γ outside ω, the limit point x_0 of the family $\{x_\tau\}$ is in $N_0 \backslash \omega$, contrary to the definition of G.

Consequently, $\text{cap}\,(N_\tau \backslash \omega, H_p^l) < \varepsilon$ for small values of τ and

$$\text{cap}\,(N_\tau, H_p^l) \leq \text{cap}\,(N_\tau \backslash \omega, H_p^l) + \text{cap}\,(\omega, H_p^l) < 2\varepsilon.$$

Thus, $\text{cap}\,(N_\tau, H_p^l) \to 0$ as $\tau \to 0$.

By $\{w_\tau\}_{\tau>0}$ we denote a family of functions in $\mathfrak{M}(N_\tau)$ such that $\lim_{\tau\to 0} \|w_\tau\|_{H_p^l} = 0$ (see Proposition 2.5.2/3). Further, we put $u_{\tau,\delta} = (1 - w_\tau)\bar{\gamma} f/(\gamma\bar{\gamma} + \delta)$ where $\delta > 0$. Since $(1 - w_\tau) f \in H_p^l$, $\bar{\gamma} \in MH_p^l$, $\gamma\bar{\gamma} \in MH_p^l$ and $\gamma\bar{\gamma} + \delta \geq \delta$, then $u_{\tau,\delta} \in H_p^l$. We have

$$f - \gamma u_{\tau,\delta} = w_\tau f + \delta(1 - w_\tau) f/(\gamma\bar{\gamma} + \delta).$$

Let φ be a smooth increasing function on $[0, +\infty)$, $\varphi(0) = \tau^2/4$, $\varphi(t) = t$ for $t > \tau^2/2$. Since $1 - w_\tau = 0$ quasi everywhere on N_τ,

$$f - \gamma u_{\tau,\delta} = w_\tau f + \delta(1 - w_\tau) f/[\varphi(\gamma\bar{\gamma}) + \delta].$$

Using the inequality $\varphi(\gamma\bar{\gamma}) + \delta > \tau^2/4$, we obtain from Propositions 1.6/5, 2.4/7 and Corollary 2.4/2 that the norm $\|[\varphi(\gamma\bar{\gamma}) + \delta]^{-1}\|_{MH_p^l}$ is uniformly bounded with respect to δ. Therefore

$$\|f - \gamma u_{\tau,\delta}\|_{H_p^l} \leq \|w_\tau f\|_{H_p^l} + \delta k(\tau)$$

where $k(\tau)$ does not depend on δ. We put $\delta(\tau) = \tau/k(\tau)$. Then

$$\|f - \gamma u_{\tau,\delta(\tau)}\|_{H_p^l} \leq c\, \|w_\tau\|_{H_p^l} + \tau$$

and so $\gamma u_{\tau,\delta(\tau)} \to f$ as $\tau \to 0$ in H_p^l. $\square$

In the next three propositions γ is a (p, l)-refined function from MH_p^l.

Proposition 1. *The number λ is contained in the pointwise spectrum of a multiplier γ in $H_{p'}^{-l}$ if and only if* $\text{cap}\,(Z_\lambda, H_p^l) > 0$.

Proof. Sufficiency: let R be so large that $\text{cap}\,(Z_\lambda \cap B_R, H_p^l) > 0$ and let μ be the capacitary measure of $Z_\lambda \cap B_R$. Note that whatever be a (p, l)-refined function $u \in H_p^l$ for (p, l)-quasi all $x \in Z_\lambda \cap B_R$ we have $u(x)(\gamma(x) - \lambda) = 0$. By Proposition 2.1.3/2 the last equality holds μ-almost everywhere. Therefore $\int u(\gamma - \lambda)\, d\mu = 0$. In other words, $(\gamma - \lambda)\mu = 0$. Since

$$\|\mu\|_{H_{p'}^{-l}}^{p'} = \|J_l \mu\|_{L_{p'}}^{p'} = \text{cap}\,(Z_\lambda \cap B_R, H_p^l) < \infty,$$

then $\lambda \in \sigma_p(\gamma)$.

Necessity: let $\lambda \in \sigma_p(\gamma)$. Then there exists a distribution $T \in H_{p'}^{-l}$, $T \neq 0$, such that $(\gamma-\lambda)T=0$. Therefore $(T, (\gamma-\lambda)u)=0$ for all $u \in H_p^l$ and the set $(\gamma-\lambda)H_p^l$ is not dense in H_p^l. The result follows by application of the lemma. □

Proposition 2. *The number λ is contained in the residual spectrum of a multiplier in H_p^l if and only if $\lambda \bar{\in} \sigma_p(\gamma)$ and $\text{cap}\,(Z_\lambda, H_p^l)>0$.*

Proof. Sufficiency: since $\text{cap}\,(Z_\lambda, H_p^l)>0$ then, by Proposition 1, $\bar{\lambda}$ is an eigenvalue of the multiplier $\bar{\gamma}$ in $H_{p'}^{-l}$. This and (2.5.1/3) imply $\lambda \in \sigma_r(\gamma) \cup \sigma_p(\gamma) = \sigma_r(\gamma)$.

Necessity: let $\lambda \in \sigma_r(\gamma)$. By (2.5.1/3) $\bar{\lambda}$ is an eigenvalue of the multiplier $\bar{\gamma}$ in $H_{p'}^{-l}$. So, according to Proposition 1, $\text{cap}\,(Z_\lambda, H_p^l)>0$. □

Proposition 3. *The multiplier γ in $H_{p'}^{-l}$ has no residual spectrum.*

Proof. Let $\lambda \in \sigma_r(\gamma)$. By virtue of (2.5.1/3), $\bar{\lambda}$ is an eigenvalue of $\bar{\gamma}$ in H_p^l. This and item (i)1 of Theorem 2.5.2 imply $\text{cap}\,(G\backslash Z_\lambda, H_p^l) < \text{cap}\,(G, H_p^l)$ for some open set $G \subset R^n$. Since $\text{cap}\,(Z_\lambda, H_p^l) > \text{cap}\,(G, H_p^l) - \text{cap}\,(G\backslash Z_\lambda, H_p^l)$, then $\text{cap}\,(Z_\lambda, H_p^l)>0$; according to Proposition 1, this means that $\lambda \in \sigma_p(\gamma)$. So we arrive at a contradiction. □

Thus the statements of Theorem 2.5.2 concerning the pointwise and residual spectrum are proved. The characteristic of the continuous spectrum obviously follows from these criteria and the relation (2.5.1/2).

2.6 The Verbitskii Theorem on Equivalent Normalization of the Space $M(H_p^m \to H_p^l)$

In this section the proof of the following assertion due to Verbitskii is presented.

Theorem. *Let $0<l<m\leq n/p$, $p \in (1, \infty)$. The relation*

$$\|\gamma\|_{M(H_p^m\to H_p^l)} \sim \sup_e \frac{\|S_l\gamma; e\|_{L_p}}{[\text{cap}\,(e, H_p^m)]^{1/p}} + \sup_{x\in R^n} \|\gamma; B_1(x)\|_{L_1}$$

is valid.

2.6.1 *The Case of Fractional l*

By $f(x, y)$ we denote the Poisson integral for a function $f \in L_{1,\text{unif}}$, i.e.

$$f(x, y) = \int \frac{yf(t)\,dt}{(|x-t|^2+y^2)^{(n+1)/2}}, \qquad (x, y) \in R_+^{n+1}.$$

Proposition 1. *The following identity is valid*

$$\gamma(x) = \gamma(x, 1) - \frac{\partial \gamma(x, 1)}{\partial y} + \frac{1}{2}\frac{\partial^2 \gamma(x, 1)}{\partial y^2} - \cdots + \frac{(-1)^k}{k!}\frac{\partial^k \gamma(x, 1)}{\partial y^k}$$
$$+ \frac{(-1)^{k+1}}{k!}\int_0^1 \frac{\partial^{k+1}\gamma(x, y)}{\partial y^{k+1}} y^k \, dy.$$

This proposition follows from a direct integration by parts.

Proposition 2. *For any $k = 0, 1, \ldots$ we have*

$$\left\| \frac{\partial^k \gamma(\cdot, 1)}{\partial y^k} \right\|_{L_\infty} \leqslant c \, \|\gamma\|_{L_{1,\text{unif}}}.$$

Proof. We make use of the following estimate for derivatives of the Poisson kernel $P(x, y) = y/(|x|^2 + y^2)^{(n+1)/2}$ (see Stein [1]):

$$\left| \frac{\partial^k P(x, y)}{\partial y^k} \right| \leqslant c(|x| + y)^{-n-k} \qquad (k = 1, 2, \ldots). \tag{1}$$

For $k > 0$ we obtain

$$\left| \frac{\partial^k \gamma(x, 1)}{\partial y^k} \right| = \left| \int \frac{\partial^k P(t, 1)}{\partial y^k} \gamma(x - t) \, dt \right|$$
$$\leqslant c \int \frac{|\gamma(x-t)| \, dt}{(|t| + 1)^{n+k}} \leqslant c\left(\int_{|t| \leqslant 1} |\gamma(x-t)| \, dt + \int_{|t| \geqslant 1} \frac{|\gamma(x-t)| \, dt}{|t|^{n+k}} \right)$$
$$\leqslant c\left(\|\gamma\|_{L_{1,\text{unif}}} + \int_1^\infty r^{-n-k-1} \, dr \int_{|t| \leqslant r} |\gamma(x-t)| \, dt \right).$$

We note that for $r \geqslant 1$ we have

$$\int_{|t| \leqslant r} |\gamma(x-t)| \, dt \leqslant c r^n \, \|\gamma\|_{L_{1,\text{unif}}}.$$

Therefore,

$$\left| \frac{\partial^k \gamma(x, 1)}{\partial y^k} \right| \leqslant c \, \|\gamma\|_{L_{1,\text{unif}}} \left(1 + \int_1^\infty r^{-k-1} \, dr \right).$$

Making use of the estimate $P(x, 1) \leqslant c(|x| + 1)^{-n-1}$, one can consider the case $k = 0$ in a similar way. □

Lemma 1. *Let $\gamma \in W_{1,\text{loc}}^{[l]}$, $\{l\} > 0$, $k = [l] + 1$. We have*

$$\left(\int_0^\infty \left| \frac{\partial^k \gamma(x, y)}{\partial y^k} \right|^2 y^{1-2\{l\}} \, dy \right)^{1/2} \leqslant c(S_l \gamma)(x).$$

Proof. First, consider the case $l \in (0, 1)$, $k = 1$. The identity

$$\int \frac{\partial P(t, y)}{\partial y} \, dt = 0$$

(see Stein [1]) implies

$$\frac{\partial \gamma(x, y)}{\partial y} = \int \frac{\partial P(t, y)}{\partial y} \gamma(x-t) \, dt = \int \frac{\partial P(t, y)}{\partial y} [\gamma(x-t) - \gamma(x)] \, dt.$$

Using (1) we get

$$\left| \frac{\partial \gamma(x, y)}{\partial y} \right| \leqslant c \int \frac{|\gamma(x-t) - \gamma(x)|}{(|t| + y)^{n+1}} \, dt.$$

Consequently,

$$\int_0^\infty \left| \frac{\partial \gamma(x, y)}{\partial y} \right|^2 y^{1-2l} \, dy$$

$$\leqslant c \int_0^\infty y^{1-2l} \, dy \left\{ \int \frac{|\gamma(x-t) - \gamma(x)|}{(|t| + y)^{n+1}} \, dt \right\}^2$$

$$\leqslant \int_0^\infty y^{-(1+2l+2n)} \, dy \left(\int_{|t| \leqslant y} |\gamma(x-t) - \gamma(x)| \, dt \right)^2$$

$$+ c \int_0^\infty y^{1-2l} \, dy \left(\int_{|t| \geqslant y} \frac{|\gamma(x-t) - \gamma(x)|}{|t|^{n+1}} \, dt \right)^2 = A_1 + A_2.$$

We have

$$A_1 = c \int_0^\infty y^{-(1+2l+2n)} \, dy \left(\int_{|\tau| \leqslant 1} |\gamma(x - \tau y) - \gamma(x)| \, y^n \, d\tau \right)^2$$

$$= c \int_0^\infty y^{-(1+2l)} \, dy \left(\int_{|\tau| \leqslant 1} |\gamma(x - \tau y) - \gamma(x)| \, d\tau \right)^2 \leqslant c[(S_l \gamma)(x)]^2.$$

To give a bound for A_2 we rewrite it as follows:

$$A_2 = c \int_0^\infty y^{1-2l} \, dy \left(\int_{r \geqslant y} r^{-n-2} \, dr \int_{y \leqslant |t| \leqslant r} |\gamma(x-t) - \gamma(x)| \, dt \right)^2.$$

Applying the Hardy inequality, we get

$$A_2 \leqslant c \int_0^\infty r^{-2n-2l-1} \, dr \left(\int_{|t| \leqslant r} |\gamma(x-t) - \gamma(x)| \, dt \right)^2$$

$$= c \int_0^\infty r^{-2l-1} \, dr \left(\int_{|\tau| \leqslant 1} |\gamma(x - \tau r) - \gamma(x)| \, d\tau \right)^2 \leqslant c[(S_l \gamma)(x)]^2.$$

Thus, for $l \in (0, 1)$,

$$\int_0^\infty \left|\frac{\partial \gamma(x, y)}{\partial y}\right|^2 y^{1-2l}\, dy \leq c[(S_l\gamma)(x)]^2.$$

Now let $k = 2m$ $(m = 1, 2, \ldots)$. Since $\gamma(x, y)$ is a harmonic function in R_+^{n+1},

$$\frac{\partial^2 \gamma(x, y)}{\partial y^2} = -\sum_{j=1}^n \frac{\partial^2 \gamma(x, y)}{\partial x_j^2}.$$

Therefore

$$\frac{\partial^k \gamma(x, y)}{\partial y^k} = (-1)^m \sum_{i_1,\ldots,i_m=1}^n \frac{\partial^{2m}\gamma(x, y)}{\partial x_{i_1}^2 \cdots \partial x_{i_m}^2}. \tag{2}$$

Using the identity $\int \partial P(t, y)/\partial t_i\, dt = 0$ $(i = 1, \ldots, n)$ and the estimate $|\partial P(x, y)/\partial x_i| \leq c(|x| + y)^{-n-1}$ (see Stein [1]), we get

$$\left|\frac{\partial^k \gamma(x, y)}{\partial y^k}\right| = \left|\sum_{i_1,\ldots,i_m=1}^n \int \frac{\partial P(t, y)}{\partial t_{i_1}}\right.$$
$$\left.\times \left[\frac{\partial^{[l]}\gamma(x-t)}{\partial x_{i_1}\,\partial x_{i_2}^2 \cdots \partial x_{i_m}^2} - \frac{\partial^{[l]}\gamma(x)}{\partial x_{i_1}\,\partial x_{i_2}^2 \cdots \partial x_{i_m}^2}\right] dt\right|$$
$$\leq c \int \frac{|\nabla_{[l]}\gamma(x-t) - \nabla_{[l]}\gamma(x)|}{(|t| + y)^{n+1}}\, dt.$$

We complete the proof in the same way as for $l \in (0, 1)$.

For $k = 2m + 1$ $(m = 1, 2, \ldots)$ we use the identity

$$\frac{\partial^k \gamma(x, y)}{\partial y^k} = (-1)^m \sum_{i_1,\ldots,i_m=1}^n \frac{\partial^{2m+1}\gamma(x, y)}{\partial y\, \partial x_{i_1}^2 \cdots \partial x_{i_m}^2}$$
$$= (-1)^m \sum_{i_1,\ldots,i_m=1}^n \int \frac{\partial P(t, y)}{\partial y}$$
$$\times \left[\frac{\partial^{[l]}\gamma(x-t)}{\partial x_{i_1}^2 \cdots \partial x_{i_m}^2} - \frac{\partial^{[l]}\gamma(x)}{\partial x_{i_1}^2 \cdots \partial x_{i_m}^2}\right] dt$$

which implies

$$\left|\frac{\partial^k \gamma(x, y)}{\partial y^k}\right| \leq c \int \frac{|\nabla_{[l]}\gamma(x-t) - \nabla_{[l]}\gamma(x)|}{(|t| + y)^{n+1}}\, dt.$$

Again we complete the proof in the same way as for $l \in (0, 1)$. □

We introduce the value

$$K = \sup_{x \in R^n, r \in (0,1]} r^{m-n/p} \|S_l\gamma; B_r(x)\|_{L_p}.$$

Lemma 2. *Let* $\gamma \in W^{[l]}_{1,\mathrm{loc}}$, $y \in (0, 1]$, $k = [l]+1$. *We have*

$$\left|\frac{\partial^k \gamma(x, y)}{\partial y^k}\right| \leqslant cKy^{\{l\}-m-1}.$$

Proof. Let $r/2 < y \leqslant r$, $r \in (0, 1]$. According to Lemma 1,

$$\int_{B_r(x)} \left(\int_0^\infty \left|\frac{\partial^k \gamma(t, y)}{\partial y^k}\right|^2 y^{1-2\{l\}}\, dy\right)^{p/2} dt \leqslant cK^p r^{n-mp}. \tag{3}$$

Applying the mean value theorem for harmonic functions and then the Cauchy inequality, we obtain

$$\begin{aligned}\left|\frac{\partial^k \gamma(x, y)}{\partial y^k}\right| &\leqslant cr^{-n-1} \int_{B_r(x)} \int_{r/2}^r \left|\frac{\partial^k \gamma(t, \eta)}{\partial \eta^k}\right| d\eta\, dt \\ &\leqslant cr^{-n-1/2} \int_{B_r(x)} \left(\int_{r/2}^r \left|\frac{\partial^k \gamma(t, \eta)}{\partial \eta^k}\right|^2 d\eta\right)^{1/2} dt \\ &\leqslant cr^{-n-1+\{l\}} \int_{B_r(x)} \left(\int_{r/2}^r \left|\frac{\partial^k \gamma(t, \eta)}{\partial \eta^k}\right|^2 \eta^{1-2\{l\}}\, d\eta\right)^{1/2} dt.\end{aligned}$$

Further, using (3) and the Hölder inequality we find

$$\left|\frac{\partial^k \gamma(x, y)}{\partial y^k}\right| \leqslant cr^{-n-1+\{l\}} r^{n(p-1)/p} K r^{(n-mp)/p} \leqslant cKy^{\{l\}-m-1}. \qquad \square$$

Lemma 3. *Let* $\gamma \in W^{[l]}_{1,\mathrm{loc}}$, $0 < l < m \leqslant n/p$. *Then, for all* $x \in R^n$,

$$|\gamma(x)| \leqslant c\{K^{l/m}[(S_l\gamma)(x)]^{(m-l)/m} + \|\gamma\|_{L_{1,\mathrm{unif}}}\}, \tag{4}$$

where K is the value introduced before Lemma 2.

Proof. Propositions 1 and 2 imply

$$|\gamma(x)| \leqslant c\left\{\|\gamma\|_{L_{1,\mathrm{unif}}} + \int_0^1 \left|\frac{\partial^{k+1}\gamma(x, y)}{\partial y^{k+1}}\right| y^k\, dy\right\}$$

with $k = [l]+1$. We put

$$\varphi(y) = \begin{cases} |\partial^{k+1}\gamma(x, y)/\partial y^{k+1}| & \text{for } 0 < y \leqslant 1, \\ 0 & \text{for } y > 1. \end{cases}$$

Then, for any $R > 0$,

$$\begin{aligned}\int_0^1 \left|\frac{\partial^{k+1}\gamma(x, y)}{\partial y^{k+1}}\right| y^k\, dy &= \int_0^\infty \varphi(y) y^k\, dy \\ &= \int_0^R \varphi(y) y^k\, dy + \int_R^\infty \varphi(y) y^k\, dy.\end{aligned}$$

Applying the Cauchy inequality to the first summand and using Lemma 2 to estimate the second one, we find

$$\int_0^\infty \varphi(y)y^k\,dy \leqslant c\Big\{\Big(\int_0^R \varphi(y)^2 y^{1-2\{l\}}\,dy\Big)^{1/2} \times\Big(\int_0^R y^{2k+2\{l\}-1}\,dy\Big)^{1/2} + K\int_R^\infty y^{k+\{l\}-1-m}\,dy\Big\}$$
$$\leqslant c\Big\{\Big(\int_0^\infty \varphi(y)^2 y^{1-2\{l\}}\,dy\Big)^{1/2} R^l + KR^{l-m}\Big\}.$$

Putting here

$$R = K^{1/m}\Big\{\int_0^\infty \varphi(y)^2 y^{1-2\{l\}}\,dy\Big\}^{-1/2m}$$

we arrive at

$$\int_0^\infty \varphi(y)y^k\,dy \leqslant cK^{l/m}\Big(\int_0^\infty \varphi(y)^2 y^{1-2\{l\}}\,dy\Big)^{(m-l)/2m}.$$

Consequently,

$$|\gamma(x)| \leqslant c\Big\{K^{l/m}\Big(\int_0^1 \Big|\frac{\partial^{k+1}\gamma(x,y)}{\partial y^{k+1}}\Big|^2 y^{1-2\{l\}}\,dy\Big)^{(m-l)/2m} + \|\gamma\|_{L_{1,\text{unif}}}\Big\}.$$

Applying Lemma 1, we complete the proof. □

Proof of Theorem 2.6 for $\{l\}>0$. By Theorem 2.2.8 it suffices to derive the estimate

$$\int_e |\gamma(x)|^{mp/(m-l)}\,dx \leqslant c(Q^{mp/(m-l)} + \|\gamma\|_{L_{1,\text{unif}}}^{mp/(m-l)})\,\text{cap}\,(e, H_p^m)$$

where

$$Q = \sup_e \frac{\|S_l\gamma; e\|_{L_p}}{[\text{cap}\,(e, H_p^m)]^{1/p}}.$$

By Lemma 3,

$$\int_e |\gamma(x)|^{mp/(m-l)}\,dx \leqslant c\Big\{K^{lp/(m-l)}\int_e |(S_l\gamma)(x)|^p\,dx + \|\gamma\|_{L_{1,\text{unif}}}^{mp/(m-l)}\,\text{mes}_n\,e\Big\}.$$

Using the obvious estimate $\text{mes}_n\,e \leqslant \text{cap}\,(e, H_p^m)$, hence we get

$$\left(\frac{\int_e |\gamma(x)|^{mp/(m-l)}\,dx}{\text{cap}\,(e, H_p^m)}\right)^{(m-l)/mp} \leqslant c\{K^{l/m}Q^{(m-l)/m} + \|\gamma\|_{L_{1,\text{unif}}}\}.$$

Taking into account that $K \leqslant Q$, we complete the proof. □

2.6.2 The Case of Integer l

The following assertion is due to Adams [2], p. 768.

Lemma. *Let $0<l<m\leq n/p$, $p>1$, and let I_l be the Riesz potential of order l with non-negative density $f\in L_p(R^n)$. Then*

$$(I_l f)(x)\leq c \sup_{r>0} (r^{m-n/p}\|f; B_r(x)\|_{L_p})^{l/m}[(Mf)(x)]^{(m-l)/m}$$

where M is the Hardy–Littlewood maximal operator.

Proof. Let $x=0$. We have

$$(I_l f)(0)=\int_{|z|\leq\delta}\frac{f(z)}{|z|^{n-l}}\,dz+\int_{|z|>\delta}\frac{f(z)}{|z|^{n-l}}\,dz.$$

The first integral can be estimated in the same way as in the proof of inequality (1.1.3/2):

$$\int_{|z|\leq\delta}\frac{f(z)}{|z|^{n-l}}\,dz\leq\sum_{k=0}^{\infty}\left(\frac{\delta}{2^k}\right)^{l-n}\int_{\delta/2^{k+1}<|z|\leq\delta/2^k} f(z)\,dz$$
$$\leq c\delta^l(Mf)(0)\sum_{k=0}^{\infty}2^{-kl}=c\delta^l[(Mf)(0)].$$

The second integral can be expressed in the form

$$\int_{|z|>\delta}\frac{f(z)}{|z|^{n-l}}\,dz=\sum_{j=0}^{\infty}\int_{B^{(j+1)}\setminus B^{(j)}}\frac{f(z)}{|z|^{n-l}}\,dz$$

where $B^{(j)}(z)=\{z:|z|\leq 2^j\delta\}$. For the integral over $B^{(j+1)}\setminus B^{(j)}$ we have

$$\int_{B^{(j+1)}\setminus B^{(j)}}\frac{f(z)}{|z|^{n-l}}\,dz\leq c(2^j\delta)^{l-n}\left(\int_{B^{(j)}}[f(z)]^p\,dz\right)^{1/p}(\text{mes}_n\, B^{(j)})^{(p-1)/p}$$
$$\leq ck(2^j\delta)^{l-m}$$

where $k=\sup_{r>0} r^{m-n/p}\|f; B_r\|_{L_p}$. Consequently,

$$\int_{|z|>\delta}\frac{f(z)}{|z|^{n-l}}\,dz\leq ck\delta^{l-m}\sum_{j=0}^{\infty}2^{j(l-m)}.$$

Finally,

$$\int\frac{f(z)}{|z|^{n-l}}\,dz\leq c[(Mf)(0)\delta^l+k\delta^{l-m}].$$

Putting here $\delta^{-m}=(Mf)(0)/k$, we get

$$\int\frac{f(z)}{|z|^{n-l}}\,dz\leq ck^{l/m}[(Mf)(0)]^{(m-l)/m}.\quad\square$$

Taking into account the inequality

$$|\gamma(x)| \leq c\left(\int_{|z|<2} \frac{|\nabla_l \gamma(x+z)|}{|z|^{n-l}}\, dz + \sup_x \|\gamma; B_1(x)\|_{L_1}\right)$$

(see (1.3.4/1)), from the lemma we obtain the following assertion:

Corollary. *Let $\gamma \in W^l_{1,\text{loc}}$, $0<l<m\leq n/p$. Then*

$$|\gamma(x)| \leq c\Big\{ \sup_{z\in R^n, r\in(0,1]} (r^{m-n/p} \|\nabla_l \gamma; B_r(z)\|_{L_p})^{l/m} \times [(M\, \nabla_l \gamma)(x)]^{(m-l)/m} + \|\gamma\|_{L_{1,\text{unif}}}\Big\}$$

for almost all $x \in R^n$.

With the notation

$$K = \sup_{x\in R^n, r\in(0,1]} r^{m-n/p} \|\nabla_l \gamma; B_r(x)\|_{L_p},$$

$$Q = \sup_e \frac{\|M\, \nabla_l \gamma; e\|_{L_p}}{[\text{cap}\,(e, H^m_p)]^{1/p}},$$

using the corollary and the theorem on the continuity of the maximal operator in $M(H^m_p \to L_p)$ presented in the next subsection, one can prove Theorem 2.6 in the same way as for fractional l.

2.6.3 Continuity of the Operator M in the Space $M(H^m_p \to L_p)$

Theorem (I. E. Verbitskii). *Let $1<p<\infty$, $0<m\leq n/p$, and let*

$$s(f) = \sup_e \|f; e\|_{L_p}/[\text{cap}\,(e, H^m_p)]^{1/p}. \tag{1}$$

We have

$$s(Mf) \leq c s(f) \tag{2}$$

where c is a constant depending only on n, m, p.

For the proof we need the following assertion which can be easily deduced from the results of Muckenhoupt [1], Theorem 9.

Lemma 1. *Let $p\in(1,\infty)$ and let φ be a measurable non-negative function with $M\varphi \leq c\varphi$ almost everywhere in R^n. Then, for any ball B and for any f vanishing outside B, the inequality*

$$\int_B |(Mf)(x)|^p\, \varphi(x)\, dx \leq A \int_B |f(x)|^p\, \varphi(x)\, dx, \tag{3}$$

with a constant A depending only on c, n, p, is valid.

Next we state two more lemmas on nonlinear Riesz potentials $U_{p,m}\mu = I_m(I_m\mu)^{p'-1}$ where μ is a measure, $mp<n$, $p>1$.

As in Remark 2.2.7/2, by h_p^m we denote a completion of the space D with respect to the norm $\|(-\Delta)^{m/2}u\|_{L_p}$ and by $\operatorname{cap}(e, h_p^m)$ we mean the capacity of a compactum e generated by the norm in the space h_p^m, i.e.

$$\operatorname{cap}(e, h_p^m) = \inf\{\|u\|_{h_p^m}^p : u \in D,\ u \geq 1 \text{ on } e\}.$$

It is well known (see Maz'ya, Havin [2]) that $u \in h_p^m$ if and only if $u = I_m f$ with $f \in L_p$.

All we said at the beginning of Subsection 2.1.2 up to Proposition 1 and in Subsection 2.1.3 on the capacity $\operatorname{cap}(\cdot, H_p^m)$ and on nonlinear Bessel potentials equally relates to the capacity $\operatorname{cap}(\cdot, h_p^m)$ and Riesz potentials.

Let $\mathscr{E}_{p,m}(\mu) = \|I_m\mu\|_{L_{p'}}^{p'}$ be the energy of the measure μ.

Lemma 2. *Let $mp<n$, $p>1$ and let μ be a measure in R^n with* $\operatorname{supp}\mu \subset B_1$.

1. *If $p \geq 2$, then*

$$\operatorname{cap}(\{x : (U_{p,m}\mu)(x) > t\}, h_p^m) \leq ct^{1-p}\mu(R^n) \tag{4}$$

for all $t>0$.

2. *If $p<2$, then*

$$\operatorname{cap}(\{x : (U_{p,m}\mu)(x) > t\}, h_p^m) \leq ct^{-1}[\mu(R^n)]^{p-1}[\mathscr{E}_{p,m}(\mu)]^{2-p} \tag{5}$$

for all $t>0$.

Proof. For (4), see Adams and Meyers [1]. This inequality is valid even for $p>2-m/n$ (Adams and Hedberg [1]). In the linear case $p=2$ this is a classical fact (see, for example, Carleson [1]).

Inequality (5) is due to Verbitskii. Here is the proof.

Let $E = \{x : (U_{p,m}\mu)(x) > t\}$ and let ν be the capacitary distribution of E. By Hölder's inequality with exponents $1/(p-1)$, $1/(2-p)$ we have

$$\begin{aligned}
t\operatorname{cap}(E, h_p^m) &\leq \int U_{p,m}\mu \, d\nu \\
&= \int\Big(\int \frac{d\mu(z)}{|z-y|^{n-m}}\Big)^{1/(p-1)} \int \frac{d\nu(x)}{|x-y|^{n-m}}\, dy \\
&\leq \Big[\int\Big(\int \frac{d\mu(z)}{|z-y|^{n-m}}\Big)^{p/(p-1)} dy\Big]^{2-p} \\
&\quad \times \Big[\int\int \frac{d\mu(z)}{|z-y|^{n-m}}\Big(\int \frac{d\nu(x)}{|x-y|^{n-m}}\Big)^{1/(p-1)} dy\Big]^{p-1} \\
&= (\mathscr{E}_{p,m}\mu)^{2-p}\Big(\int (U_{p,m}\nu)(z)\, d\mu(z)\Big)^{p-1} \\
&\leq c(\mathscr{E}_{p,m}\mu)^{2-p}[\mu(R^n)]^{p-1}. \quad \square
\end{aligned}$$

(The latter inequality holds by rough maximum principle for nonlinear potentials.)

Lemma 3 (Verbitskii). *If* $\alpha \in (1, n/(n-m))$ *for* $p \leqslant 2-m/n$ *and* $\alpha \in (1, (p-1)n/(n-mp))$ *for* $p > 2-m/n$, *then*

$$M(U_{p,m}\mu)^\alpha \leqslant c(U_{p,m}\mu)^\alpha. \tag{6}$$

Proof. In the case $1<p\leqslant 2-m/n$, the proof of this estimate is quite simple. Namely, let

$$g(y) = \left(\int \frac{d\mu(z)}{|y-z|^{n-m}}\right)^{1/(p-1)}.$$

By the Minkowski inequality,

$$\sup_{r>0} \frac{1}{\operatorname{mes}_n B_r} \int_{B_r(x)} \left(\int \frac{g(y)}{|z-y|^{n-m}}\, dy\right)^\alpha dz$$
$$\leqslant \left[\int g(y)\, dy \left(\sup_{r>0} \int_{B_r(x)} \frac{dz}{|y-z|^{(n-m)\alpha}}\right)^{1/\alpha}\right]^\alpha \leqslant c \int \frac{g(y)}{|y-x|^{n-m}}\, dy.$$

The result follows.

For $p>2-m/n$ the proof is less elementary. We consider here the case $p \geqslant 2$ which is only used in what follows (see Verbitskii [2], Lemma 3.5).

Since the function $\alpha \to (Mv^\alpha)^{1/\alpha}$ does not decrease, we may put $\alpha \geqslant p-1$. We have

$$M(U_{p,m}\mu)^\alpha = \sup_{r>0} \left\{\frac{1}{\operatorname{mes}_n B_r} \int_{B_r(x)} \left(\int \frac{g(y)\, dy}{|z-y|^{n-m}}\right)^\alpha dz\right\}.$$

First we note that

$$A_1(z) = \int_{|y-x|\geqslant 2|x-z|} \frac{g(y)\, dy}{|z-y|^{n-m}} \leqslant c \int \frac{g(y)\, dy}{|y-x|^{n-m}} = c(U_{p,m}\mu)(x).$$

Consequently,

$$\left\{\frac{1}{\operatorname{mes}_n B_r} \int_{B_r(x)} [A_1(z)]^\alpha\, dz\right\}^{1/\alpha} \leqslant c(U_{p,m}\mu)(x). \tag{7}$$

Next we derive an estimate for the integral

$$A_2(z) = \int_{|y-x|<2|x-z|} \frac{g(y)\, dy}{|z-y|^{n-m}}.$$

We put $\beta = n(p-2)/(p-1) - \varepsilon$ where ε is a small enough number which will be specified later. Applying Hölder's inequality with exponents $p-1$ and $(p-1)/(p-2)$ as well as the composition formula for the Riesz kernels, we obtain

$$A_2(z) \leqslant \left\{ \int_{|y-x|<2|x-z|} \frac{dy}{|z-y|^{\beta(p-1)/(p-2)}} \right\}^{(p-2)/(p-1)}$$

$$\times \left\{ \int \frac{[g(y)]^{p-1}\, dy}{|y-z|^{(n-m-\beta)(p-1)}} \right\}^{1/(p-1)}$$

$$\leqslant c\, |x-z|^{n(p-2)/(p-1)-\beta} \left\{ \int \frac{d\mu(u)}{|u-z|^{(n-m-\beta)(p-1)-m}} \right\}^{1/(p-1)}.$$

We note that the composition formula

$$\int \frac{du}{|x-u|^{n-a}\, |u-y|^{n-b}} = \frac{c}{|x-y|^{n-(a+b)}} \tag{8}$$

is valid for $a>0$, $b>0$, $a+b<n$. In our case, $a=n-(n-m-\beta)(p-1)$, $b=m$, so (8) is applicable provided that $\varepsilon<m$.

Since $r \geqslant |x-z|$, then

$$\left\{ \frac{1}{\operatorname{mes}_n B_r} \int_{B_r(x)} [A_2(z)]^\alpha\, dz \right\}^{(p-1)/p}$$

$$\leqslant c \left\{ \int \frac{dz}{|z-x|^{n-\alpha n(p-2)/(p-1)+\alpha\beta}} \left[\int \frac{d\mu(u)}{|u-z|^{(n-m-\beta)(p-1)-m}} \right]^{\alpha/(p-1)} \right\}^{(p-1)/\alpha}.$$

Applying the Minkowski inequality with exponent $\alpha/(p-1) \geqslant 1$, we find that the right-hand side does not exceed

$$c \int d\mu(u) \left\{ \int \frac{dz}{|x-z|^{\beta_1}\, |u-z|^{\beta_2}} \right\}^{(p-1)/\alpha}$$

where $\beta_1 = n - \alpha n(p-2)/(p-1) + \alpha\beta$, $\beta_2 = \alpha(n-m-\beta) - \alpha m/(p-1)$. If ε is subjected to the condition $\varepsilon < n/\alpha - (n-mp)/(p-1)$ then we may use formula (8) for the last integral. Then

$$\left\{ \frac{1}{\operatorname{mes}_n B_r} \int_{B_r(x)} [A_2(z)]^\alpha\, dz \right\}^{(p-1)/\alpha} \leqslant c \int \frac{d\mu(u)}{|u-x|^{n-mp}}.$$

Once again, applying formula (8) and the Minkowski inequality with exponent $1/(p-1) \leqslant 1$, we conclude that the last integral is dominated by

$$c \int d\mu(z) \left\{ \int \frac{dy}{|x-y|^{n-m}\, |z-y|^{(n-m)/(p-1)}} \right\}^{p-1} \leqslant c(U_{p,m}\mu)(x).$$

Hence the required inequality follows from (7). □

Proof of the theorem. It suffices to derive (2) for the 'truncated' maximal operator

$$(M_1 f)(x) = \sup_{r\in(0,1)} \frac{1}{\operatorname{mes}_n B_r(x)} \int_{B_r(x)} |f(y)|\,dy.$$

In fact, since $\operatorname{cap}(B_r, H_p^m) \leqslant c \operatorname{mes}_n B_r$ for $r \geqslant 1$, then, according to (1),

$$(M_2 f)(x) = \sup_{r\geqslant 1} \frac{1}{\operatorname{mes}_n B_r(x)} \int_{B_r(x)} |f(y)|\,dy$$
$$\leqslant \sup_{r\geqslant 1} \left(\frac{1}{\operatorname{mes}_n B_r(x)} \int_{B_r(x)} |f(y)|^p\,dy \right)^{1/p} \leqslant cs(f).$$

Therefore

$$\|M_2 f; e\|_{L_p} \leqslant cs(f)[\operatorname{mes}_n(e)]^{1/p} \leqslant cs(f)[\operatorname{cap}(e, H_p^m)]^{1/p}.$$

Further, we note that in (2) for the operator M_1 it suffices to limit consideration to compacta e lying inside B_1 (see Proposition 2.1.4/2).

Consider the case $mp < n$. The capacity $\operatorname{cap}(e, H_p^m)$ can be replaced by the equivalent $\operatorname{cap}(e, h_p^m)$. By μ we denote the capacitary measure of the compactum e. Then $\mu(R^n) = \mathscr{E}_{p,m}(\mu) = \operatorname{cap}(e, h_p^m)$ and $U_{p,m}\mu(x) \leqslant \mathfrak{M}$, where $\mathfrak{M}$ is the constant in the rough maximum principle for the potential $U_{p,m}$. Since $U_{p,m}\mu(x) \geqslant 1$ on e, then for $\alpha > 0$ we have

$$\int_e |M_1 f|^p\,dx \leqslant \int_{B_1} |(M_1 f)(x)|^p\,[(U_{p,m}\mu)(x)]^\alpha\,dx.$$

Let φ designate the restriction of f to the ball B_2 with radius 2 concentric with the ball B_1. It is clear that $(M_1 f)(x) \leqslant (M\varphi)(x)$ for all $x \in B_1$.

Let $\alpha \in (1, (p-1)n/(n-mp))$ for $p \geqslant 2$ and $\alpha \in (1, n/(n-m))$ for $p < 2$. By Lemma 3 we have

$$M[(U_{p,m}\mu)(x)]^\alpha \leqslant c[(U_{p,m}\mu)(x)]^\alpha$$

and by Lemma 1

$$\int_{B_1} |(M_1 f)(x)|^p\,[(U_{p,m}\mu)(x)]^\alpha\,dx \leqslant \int_{B_2} |M\varphi|^p\,[(U_{p,m}\mu)(x)]^\alpha\,dx$$
$$\leqslant c\int_{B_2} |\varphi(x)|^p\,[(U_{p,m}\mu)(x)]^\alpha\,dx = c\int_{B_2} |f(x)|^p\,[(U_{p,m}\mu)(x)]^\alpha\,dx.$$

Further,

$$\int_{B_2} |f(x)|^p\,[(U_{p,m}\mu)(x)]^\alpha\,dx = \int_0^{\mathfrak{M}} \left(\int_{E_t} |f(x)|^p\,dx \right) t^{\alpha-1}\,dt,$$

where $E_t = \{x \in B_2 : (U_{p,m}\mu)(x) > t\}$. Since the capacities $\operatorname{cap}(E_t, H_p^m)$ and cap (E_t, h_p^m) are equivalent for the sets contained in B_2, then

$$\int_{E_t} |f(x)|^p \, dx \leq c[s(f)]^p \operatorname{cap}(E_t, h_p^m).$$

Let $p \geq 2$. According to Lemma 2,

$$\operatorname{cap}(E_t, h_p^m) \leq ct^{1-p}\mu(R^n) = ct^{1-p} \operatorname{cap}(e, h_p^m).$$

Therefore, for $\alpha \in (p-1, n(p-1)/(n-mp))$,

$$\int_{B_2} |f(x)|^p \, [(U_{p,m}\mu)(x)]^\alpha \, dx \leq c \operatorname{cap}(e, h_p^m) \int_0^{\mathfrak{M}} t^{\alpha-p} \, dt$$
$$\leq c_1 \operatorname{cap}(e, h_p^m).$$

For $p<2$, by Lemma 2,

$$\operatorname{cap}(E_t, h_p^m) \leq ct^{-1}[\mu(R^n)]^{p-1}[\mathscr{E}_{p,m}(\mu)]^{2-p} = ct^{-1} \operatorname{cap}(e, h_p^m).$$

So, for $\alpha \in (1, n/(n-m))$,

$$\int_{B_2} |f(x)|^p \, [(U_{p,m}\mu)(x)]^\alpha \, dx \leq c \operatorname{cap}(e, h_p^m) \int_0^{\mathfrak{M}} t^{\alpha-2} \, dt$$
$$\leq c_1 \operatorname{cap}(e, h_p^m). \quad \square$$

Thus the theorem is proved for $mp<n$. For $mp=n$ the proof is analogous. However the necessity to use Bessel instead of Riesz potentials gives rise to certain technical difficulties which are not dealt with here.

3

The Space $M(W_p^m \to W_p^l)$ for Fractional m and l

The greater part of this chapter is devoted to the study of multipliers acting from W_p^m into W_p^l for fractional m and l. Section 3.1 contains auxiliary assertions. The description of the space $M(W_p^m \to W_p^l)$ for $p \in (1, \infty)$ is obtained in 3.2 and for $p = 1$ in 3.4. Sufficient conditions for a function to belong to the class $M(W_p^m \to W_p^l)$ in terms of Besov and Bessel potential spaces are given in 3.3. The theorem on a composite function $\varphi(\gamma)$ with $\gamma \in M(W_p^m \to W_p^l)$ is proved in 3.5.

The following three sections concern applications of multipliers. In the short Section 3.6 we apply the results on the space $M(W_2^m \to W_2^l)$ obtained earlier to translation invariant operators acting in pairs of weighted L_2-spaces. Section 3.7 contains upper and in some cases two-sided estimates for a norm of a differential operator acting from one Sobolev space into another. These estimates are formulated in terms of multiplier norms of coefficients. In 3.8 we show that coercive estimates in multiplier spaces are valid for solutions of elliptic equations and systems in R^n as well as for solutions of elliptic boundary value problems in R_+^n.

The last three sections contain the description of multipliers in the Besov space B_p^l, in the space BMO of functions with bounded mean oscillation and in certain spaces of analytic functions in a circle.

3.1 The Space B_p^l and its Properties

3.1.1 Definitions of Spaces and Auxiliary Assertions

We present an arbitrary positive number l in the form $l = k + \alpha$, where $\alpha \in (0, 1]$ and k is a non-negative integer. Let

$$(C_{p,l}u)(x) = \left(\int |\Delta_h^{(2)} \nabla_k u(x)|^p \, |h|^{-n-p\alpha} \, dh \right)^{1/p},$$

where $\Delta_h^{(2)} v(x) = v(x+2h) - 2v(x+h) + v(x)$.

We denote by B_p^l the completion of D with respect to the norm $\|C_{p,l}u\|_{L_p}+\|u\|_{L_p}$.

It is known that $\|u\|_{B_p^l}\sim\|C_{p,l}u\|_{L_p}+\|u\|_{W_p^{[l]}}$. We set $(D_{p,l}u)(x)=|\nabla_l u(x)|$ if $\{l\}=0$ and

$$(D_{p,l}u)(x)=\left(\int|\Delta_h\nabla_{[l]}u(x)|^p\,|h|^{-n-p\{l\}}\,dh\right)^{1/p}$$

if $\{l\}>0$. Here $\Delta_h v(x)=v(x+h)-v(x)$.

Let W_p^l be the space obtained by completion of D with respect to the norm $\|D_{p,l}u\|_{L_p}+\|u\|_{L_p}$.

The exposition of different aspects of the theory of the spaces B_p^l and W_p^l can be found in monographs by S. Nikol'skii [1], Stein [1], Triebel [4] and elsewhere.

For $\{l\}>0$ the spaces B_p^l and W_p^l have the same elements and their norms are equivalent. This follows from the estimates

$$(2-2^{\{l\}})D_{p,l}u\leqslant C_{p,l}u\leqslant(2+2^{\{l\}})D_{p,l}u \tag{1}$$

which in their turn can be deduced from the identity

$$\begin{aligned}2[\varphi(x+h)-\varphi(x)]=&-[\varphi(x+2h)-2\varphi(x+h)+\varphi(x)]\\&+[\varphi(x+2h)-\varphi(x)].\end{aligned}$$

We give some properties of the functions $C_{p,l}$ and $D_{p,l}$ which will be used henceforth.

Lemma 1. *If $m, l>0$, then $C_{p,l}u\leqslant J_mC_{p,l}\Lambda^m$ and $D_{p,l}u\leqslant J_mD_{p,l}\Lambda^m$, where $\Lambda^m=(-\Delta+1)^{m/2}$.*

Proof. Let $f=\Lambda^m u$ and let $l=k+\alpha$, $\alpha\in(0,1]$. It is clear that

$$\begin{aligned}(C_{p,l}u)(x)=(C_{p,l}J_mf)(x)&=\left(\int\left|\int[G_m(x-\xi+2h)-2G_m(x-\xi+h)\right.\right.\\&\left.\left.\quad+G_m(x-\xi)]\nabla_kf(\xi)\,d\xi\right|^p|h|^{-n-p\alpha}\,dh\right)^{1/p}\\&=\left(\int\left|\int G_m(x-\xi)(\nabla_kf(\xi+2h)-2\nabla_kf(\xi+h)\right.\right.\\&\left.\left.\quad+\nabla_kf(\xi))\,d\xi\right|^p|h|^{-n-p\alpha}\,dh\right)^{1/p}.\end{aligned}$$

Applying the Minkowski inequality, we obtain $C_{p,l}u\leqslant J_mC_{p,l}f$. An analogous inequality for $D_{p,l}u$ with $\{l\}>0$ needs no separate proof by virtue of (1). □

Lemma 2. *The following inequality holds:*

$$\|D_{p,\alpha}D_{p,\beta}u\|_{L_p}\leqslant c\,\|D_{p,\alpha+\beta}u\|_{L_p}, \tag{2}$$

where $\alpha,\beta>0$, $\alpha+\beta<1$, $1\leqslant p<\infty$.

Proof. Let $\chi \in R^n$ and $u_\chi(x) = u(x+\chi)$. It is clear that

$$|(D_{p,\beta}u)(x) - (D_{p,\beta}u_\chi)(x)| \leq \left(\int |\Delta_h[u(x) - u_\chi(x)]|^p \frac{dh}{|h|^{n+p\beta}}\right)^{1/p}.$$

Therefore

$$\|D_{p,\alpha}D_{p,\beta}u\|_{L_p}^p \leq \int\int\int |\Delta_h[u(x) - u_\chi(x)]|^p \frac{dh\, d\chi\, dx}{|h|^{n+p\beta}\, |\chi|^{n+\alpha p}}.$$

We represent the integral over R^{3n} in the right-hand side as the sum of two integrals, one of which is taken over $|h| \leq |\chi|$ and does not exceed

$$\int dx \int \frac{|\Delta_h u(x)|^p}{|h|^{n+\beta p}}\, dh \int_{|\chi| \geq |h|} \frac{d\chi}{|\chi|^{n+\alpha p}} + \int \frac{dh}{|h|^{n+\beta p}}$$
$$\times \int_{|\chi| \geq |h|} \frac{d\chi}{|\chi|^{n+\alpha p}} \int |\Delta_h u_\chi(x)|^p\, dx.$$

Obviously the second of these summands coincides with the first one which in its turn has the form

$$c \int dx \int |\Delta_h u(x)|^p\, |h|^{-n-(\alpha+\beta)p}\, dh = c\, \|D_{p,\alpha+\beta}u\|_{L_p}^p.$$

The integral over $|h| > |\chi|$ is estimated in the same way. □

Next we formulate some known assertions on the space B_p^l (Besov [1], Stein [1], Aronszajn, Mulla, Szeptycky [1], Szeptycky [1], Triebel [4], Uspenskii [1]).

Lemma 3. The following relation holds: $\|u\|_{B_p^l} \sim \|\Lambda^\mu u\|_{B_p^{l-\mu}}$, *where* $1 < p < \infty$ *and* $0 < \mu < l$.

Lemma 4. The following inequalities hold:

$$\|u\|_{H_p^l} \leq c\, \|u\|_{B_p^l} \quad \text{for} \quad 1 < p \leq 2,$$
$$\|u\|_{B_p^l} \leq c\, \|u\|_{H_p^l} \quad \text{for} \quad 2 \leq p < \infty.$$

In the next lemma we take $R^n = R^m \times R^k = \{x = (y, z) : y \in R^m,\ z \in R^k\}$.

Lemma 5. The following relation holds:

$$\|u\|_{B_p^l} \sim \|u\|_{L_p} + \left(\int_{R^k} \int_{R^m} \|\Delta_\eta^{(2)} u(\cdot, z); R^m\|_{L_p}^p \frac{d\eta\, dz}{|\eta|^{m+p\alpha}}\right)^{1/p}$$
$$+ \left(\int_{R^m} \int_{R^k} \|\Delta_\zeta^{(2)} u(y, \cdot); R^k\|_{L_p}^p \frac{d\zeta\, dy}{|\zeta|^{k+p\alpha}}\right)^{1/p},$$

where $l>0$, $p\in[1,\infty)$. Analogous norms appear under the decomposition of R^n into more than two factors.

Lemma 6. *Let $U\in D(R^{n+s})$ be any extension of $u\in D(R^n)$ onto the space $R^{n+s}=\{z=(x,y): x\in R^n,\ y\in R^s\}$.*

(i) *If $p\in(1,\infty)$, $l>0$, then*

$$\|u; R^n\|_{B_p^l}\sim\inf_{\{U\}}\|U; R^{n+s}\|_{H_p^{l+s/p}}. \tag{3}$$

(ii) *If $p\in[1,\infty)$, $l>0$, then*

$$\|u; R^n\|_{B_p^l}\sim\inf_{\{U\}}\|U; R^{n+s}\|_{B_p^{l+s/p}}. \tag{4}$$

(iii) *If $p\in[1,\infty)$, $l>0$, then*

$$\|u; R^n\|_{B_p^l}\sim\inf_{\{U\}}\left(\int_{R^{n+s}}|y|^{p(1-\{l\})-s}(|\nabla_{[l]+1}U|^p+|U|^p)\,dz\right)^{1/p}.$$

Lemma 7. *The following inequality holds: $\|u\|_{L_q}\leq c\,\|u\|_{B_p^l}$, where $p>1$ and*

$$q=pn/(n-pl)\quad\text{if}\quad n>pl,$$
$$q\in[p,\infty)\quad\text{if}\quad n=pl,$$
$$q\in[p,\infty]\quad\text{if}\quad n<pl.$$

The next assertion is a particular case of Lemma 2.3.1/1 by Polking [1]. To use his result it suffices to express $D_{p,l}u$ as

$$c\left(\int_0^\infty\int_{B_1}|\nabla_{[l]}u(x+\rho y)-\nabla_{[l]}u(x)|^p\,dy\frac{d\rho}{\rho^{1+p\{l\}}}\right)^{1/p}.$$

Lemma 8. *Let $\{l\}>0$, $p\in[2,\infty)$ and $r>np/(n+\{l\}p)$, $1<r<\infty$. Then $\|D_{p,l}u\|_{L_r}\leq c\,\|(-\Delta)^{l/2}u\|_{L_r}$.*

We now give a relation which follows from (3) and Theorem 2.1.1/3. However it can be proved directly.

Lemma 9. *Let $\{\eta_j\}_{j\geq0}$ be the same sequence as in Theorem 2.1.1/3. Then $\|u\|_{B_p^l}\sim(\sum_{j\geq0}\|u\eta_j\|^p_{B_p^l})^{1/p}$.*

The following lemma is the simple generalization of the Friedrichs inequality.

Lemma 10. *Let $u\in W_p^m$ and $\operatorname{supp}u\subset B_\delta$. Then, for $k\in(0,m]$, $\delta^k\,\|D_{p,k}u\|_{L_p}+\|u\|_{L_p}\leq c\,\delta^m\,\|D_{p,m}u\|_{L_p}$.*

From this inequality and Lemma 9 we have:

Corollary 1. *Let $\{B_\delta^{(j)}\}$ be a covering of R^n by open balls of radius $\delta \in (0,1)$ with finite multiplicity depending only on n. Further, let $u^{(j)} \in W_p^l$ and supp $u^{(j)} \subset B_\delta^{(j)}$. Then $\|\sum_j u^{(j)}\|_{W_p^l}^p \leqslant c \sum_j \|D_{p,l} u^{(j)}\|_{L_p}^p$.*

As usual we denote by $W_p^k(B_r)$ the space of functions with the finite norm

$$\|u; B_r\|_{W_p^k} = \begin{cases} \sum_{j=0}^{k} \|\nabla_j u; B_r\|_{L_p} \quad \text{for} \quad \{k\}=0, \\ \|u; B_r\|_{W_p^{[k]}} + \sum_{j=0}^{[k]} \left(\int_{B_r} \int_{B_r} |\nabla_j u(x) - \nabla_j u(y)|^p \frac{dx\,dy}{|x-y|^{n+p\{k\}}} \right)^{1/p} \\ \qquad \text{for} \quad \{k\}>0. \end{cases}$$

We introduce in $W_p^k(B_r)$ one more norm depending on r. Namely we set

$$|||u; B_r|||_{W_p^k} = \begin{cases} \sum_{j=0}^{k} r^{j-k} \|\nabla_j u; B_r\|_{L_p} \quad \text{for} \quad \{k\}=0, \\ r^{-\{k\}} |||u; B_r|||_{W_p^{[k]}} + \sum_{j=0}^{[k]} r^{j-[k]} \\ \times \left(\int_{B_r} \int_{B_r} |\nabla_j u(x) - \nabla_j u(y)|^p \frac{dx\,dy}{|x-y|^{n+p\{k\}}} \right)^{1/p} \\ \qquad \text{for} \quad \{k\}>0. \end{cases}$$

It is clear that the last norm is invariant under the similarity transformation.

We present certain properties of the norm $|||\cdot; B_r|||_{W_p^k}$ which will be used henceforth.

Lemma 11. *If k is positive fractional, then*

$$\|D_{p,k} u; B_r\|_{L_p} \leqslant c \sup_{x \in R^n} |||u; B_r(x)|||_{W_p^k}. \tag{5}$$

Proof. It suffices to estimate

$$\left(\int_{B_{r/2}(z)} dy \int_{R^n \setminus B_r(z)} |\nabla_{[k]} u(x) - \nabla_{[k]} u(y)|^p \frac{dx}{|x-y|^{n+p\{k\}}} \right)^{1/p}$$

where z is an arbitrary point of the ball B_r. This value does not exceed

$$\left(\int_{B_{r/2}(z)} |\nabla_{[k]}u(y)|^p \, dy \int_{R^n \setminus B_r(z)} \frac{dx}{|x-y|^{n+p\{k\}}}\right)^{1/p}$$

$$+\left(\int_{R^n \setminus B_r(z)} |\nabla_{[k]}u(x)|^p \, dx \int_{B_{r/2}(z)} \frac{dy}{|x-y|^{n+p\{k\}}}\right)^{1/p}$$

$$\leqslant cr^{-\{k\}} \|\nabla_{[k]}u; B_r(z)\|_{L_p} + c\left(r^n \int_{R^n \setminus B_r(z)} |\nabla_{[k]}u(x)|^p \, |x-z|^{-n-p\{k\}} \, dx\right)^{1/p}.$$

The second summand on the right is not more than

$$c\left(\int_{R^n \setminus B_r(z)} |\xi - x|^{-n-p\{k\}} \int_{B_r(\xi)} |\nabla_{[k]}u(x)|^p \, dx \, d\xi\right)^{1/p}$$

which does not exceed $cr^{-\{k\}} \sup_{x \in R^n} \|u; B_r(x)\|_{W_p^{[k]}}$. □

Next we formulate three well-known assertions on the norm $\|u; B_r\|_{W_p^m}$, leaving their proof to the reader as an exercise.

Lemma 12. *If $\varphi \in C_0^\infty(B_r)$ and $|\nabla_k \varphi| \leqslant cr^{-k}$, $k = 0, 1, \ldots, m$, then, for all $u \in W_p^k(B_{2r})$, $\|\varphi u; R^n\|_{W_p^m} \leqslant c\|u; B_{2r}\|_{W_p^m}$ for $r \leqslant 1$ and $\|\varphi u; R^n\|_{W_p^k} \leqslant c\|u; B_{2r}\|_{W_p^k}$ for $r > 1$.*

Lemma 13. *Let $u \in W_p^m(B_\delta)$. There exists a polynomial P of order $[m]$ having the form*

$$P(u; x) = \sum_\beta (x/\delta)^\beta \delta^{-n} \int_{B_\delta} \varphi_\beta(y/\delta) u(y) \, dy,$$

where $\varphi_\beta \in C_0^\infty(B_1)$, and such that

$$\|u - P(u; \cdot); B_\delta\|_{W_p^m}^p \leqslant c \int_{B_\delta} \int_{B_\delta} |\nabla_{[m]}u(x) - \nabla_{[m]}u(y)|^p \frac{dx \, dy}{|x-y|^{n+p\{m\}}}.$$

Lemma 14. *The inequality*

$$\|u; B_r\|_{W_p^s} \leqslant c \, \|u; B_r\|_{W_p^k}^{s/k} \|u; B_r\|_{L_p}^{(k-s)/k}, \qquad 0 < s < k, \tag{6}$$

is valid.

From (6) we immediately obtain

Corollary 2. *The following relations hold:*

$$\|u; B_r\|_{W_p^k} \sim r^{-k} \|u; B_r\|_{L_p} + \left(\int_{B_r} \int_{B_r} |\nabla_{[k]}u(x) - \nabla_{[k]}u(y)|^p \frac{dx \, dy}{|x-y|^{n+p\{k\}}}\right)^{1/p},$$

if k is fractional and

$$|||u; B_r|||_{W_p^k} \sim r^{-k}\, \|u; B_r\|_{L_p} + \|\nabla_k u; B_r\|_{L_p},$$

if k is integer.

From the known inequality, $\|u\,|x|^{-k}\|_{L_p} \leqslant c\,\|u\|_{W_p^k}$, $kp<n$, (see Lizorkin [2]) follows:

Lemma 15. *Inequality $|||u; B_r|||_{W_p^k} \leqslant c\,\|u; R^n\|_{W_p^k}$, where $kp<n$ and c is a constant independent on r, is valid.*

3.1.2. Estimates for Potentials

Let f be a non-negative function in $L_p(R^{n+s})$, $s \geqslant 0$, and let $R^{n+s} = \{z=(x, y): x \in R^n,\ y \in R^s\}$.

By $J_r^{(n+s)}f$ we denote the Bessel potential of order r with density f in R^{n+s}. In other words, $J_r^{(n+s)}f = G_r * f$, where $G_r(z) = cK_{(n+s-r)/2}(|z|)\,|z|^{(r-n-s)/2}$ (cf. 1.1.1). We recall asymptotic estimates for G_r.

For $|z| \to 0$,

$$G_r(z) \sim \begin{cases} |z|^{r-n-s}, & 0<r<n+s, \\ \log|z|^{-1}, & r=n+s, \\ 1, & r>n+s. \end{cases} \tag{1}$$

For $|z| \to \infty$,

$$G_r(z) \sim |z|^{(r-n-s-1)/2} e^{-|z|}. \tag{2}$$

Let $M^{(n)}$ be the Hardy–Littlewood maximal operator in R^n (cf. 1.1.1). Henceforth we need the following modification of Hedberg's inequality (1.1.3/1) obtained by authors [4].

Lemma 1. *For all $x \in R^n$,*

$$(J_{r\theta+s/p}^{(n+s)}f)(x, 0) \leqslant c[(J_{r+s/p}^{(n+s)}f)(x, 0)]^{\theta}[(M^{(n)}F)(x)]^{1-\theta}, \tag{3}$$

where $F(\xi) = \|f(\xi, \cdot); R^s\|_{L_p}$, $0<\theta<1$.

Proof. Let $\delta \in (0, 1]$ and let $E_\delta(x) = \{\zeta = (\xi, \eta): \xi \in R^n,\ \eta \in R^s,\ (x-\xi)^2 + \eta^2 > \delta^2\}$. We express the potential in the left-hand side of (3) as the sum of two integrals one of which is over $E_\delta(x)$. Let $r\theta < n + s/p'$;

then, by virtue of (1),

$$\int_{R^{n+s}\setminus E_\delta(x)} G_{r\theta+s/p}(x-\xi,\eta)f(\xi,\eta)\,d\xi\,d\eta$$
$$\leqslant c\int_{B_\delta(x)} F(\xi)\left(\int_{|\eta|<\delta}\frac{d\eta}{(|x-\xi|+|\eta|)^{p'q}}\right)^{1/p'} d\xi, \tag{4}$$

where $p'=p/(p-1)$ and $q=n-r\theta+s/p'$.

It is clear that

$$\left(\int_{|\eta|<\delta}\frac{d\eta}{(|x-\xi|+|\eta|)^{p'q}}\right)^{1/p'} \leqslant \begin{cases} c\,|x-\xi|, & r\theta<n,\\ c\log\dfrac{2\delta}{|x-\xi|}, & r\theta=n,\\ \delta^{r\theta-n}, & r\theta>n.\end{cases} \tag{5}$$

Consider, for example, the case $r\theta<n$. The right-hand side of (4) does not exceed

$$c\int_{B_\delta(x)}\frac{F(\xi)\,d\xi}{|x-\xi|^{n-r\theta}}=c\sum_{j=1}^{\infty}\int_{\delta 2^{-j}<|x-\xi|<\delta 2^{-j+1}}\frac{F(\xi)\,d\xi}{|x-\xi|^{n-r\theta}}$$
$$\leqslant c\sum_{j=1}^{\infty}2^{-j(n-r\theta)}\,\delta^{-n+r\theta}\int_{|x-\xi|<\delta 2^{-j+1}}F(\xi)\,d\xi$$
$$\leqslant c\,\delta^{r\theta}(M^{(n)}F)(x).$$

The same estimate follows from (4)–(5) for $r\theta\geqslant n$.

Now let $r\theta=n+s/p'$. By (1) the left-hand side of (4) is not more than

$$c\int_{B_\delta(x)}F(\xi)\left(\int_{|\eta|<\delta}|\log c(|x-\xi|+|\eta|)|^{p'}\,d\eta\right)^{1/p'}$$
$$\leqslant c(1+|\log\delta|)\,\delta^{s/p'}\int_{B_\delta(x)}F(\xi)\,d\xi\leqslant c(1+|\log\delta|)\,\delta^{r\theta}(M^{(n)}F)(x).$$

If $r\theta>n+s/p'$, then the left-hand side of (4) is bounded by

$$c\,\delta^{s/p'}\int_{B_\delta(x)}F(\xi)\,d\xi\leqslant c\,\delta^{n+s/p'}(M^{(n)}F)(x).$$

Thus, for $\delta\leqslant 1$,

$$\int_{R^{n+s}\setminus E_\delta(x)} G_{r\theta+s/p}(x-\xi,\eta)f(\xi,\eta)\,d\xi\,d\eta$$
$$\leqslant\begin{cases} c\delta^{r\theta}(M^{(n)}F)(x), & r\theta<n+s/p',\\ c(1+|\log\delta|)\,\delta^{r\theta}(M^{(n)}F)(x), & r\theta=n+s/p',\\ c\,\delta^{n+s/p'}(M^{(n)}F)(x), & r\theta>n+s/p'.\end{cases} \tag{6}$$

Next we estimate the integral over $E_\delta(x)$. In the case $r\theta < n+s/p'$ we have

$$\int_{E_\delta(x)} G_{r\theta+s/p}(x-\xi,\eta)f(\xi,\eta)\,d\xi\,d\eta \leqslant c\int_{E_\delta(x)\setminus E_1(x)} \frac{f(\xi,\eta)\,d\xi\,d\eta}{(|x-\xi|+|\eta|)^{n-r\theta+s/p'}}$$
$$+c\int_{E_1(x)} e^{-\sqrt{(x-\xi)^2+\eta^2}}\frac{f(\xi,\eta)\,d\xi\,d\eta}{(|x-\xi|+|\eta|)^{(n+1-r\theta+s/p')/2}}$$
$$\leqslant c\,\delta^{-r(1-\theta)}\int_{E_\delta(x)\setminus E_1(x)} \frac{f(\xi,\eta)\,d\xi\,d\eta}{(|x-\xi|+|\eta|)^{n-r+s/p'}}$$
$$+c\,\delta^{-r(1-\theta)/2}\int_{E_1(x)} e^{-\sqrt{(x-\xi)^2+\eta^2}}\frac{f(\xi,\eta)\,d\xi\,d\eta}{(|x-\xi|+|\eta|)^{(n+1-r+s/p')/2}}$$
$$\leqslant c\,\delta^{-r(1-\theta)}(J^{(n+s)}_{r+s/p}f)(x,0).$$

In the same way we obtain the estimate for $r\theta \geqslant n+s/p'$. Thus

$$\int_{E_\delta(x)} G_{r\theta+s/p}(x-\xi,\eta)f(\xi,\eta)\,d\xi\,d\eta \leqslant \begin{cases} c\,\delta^{-r(1-\theta)}(J^{(n+s)}_{r+s/p}f)(x,0) \\ \quad \text{for } r\theta<n+s/p', \\ c(1+|\log\delta|)(J^{(n+s)}_{r+s/p}f)(x,0) \\ \quad \text{for } r\theta=n+s/p', \\ c(J^{(n+s)}_{r+s/p}f)(x,0) \\ \quad \text{for } r\theta>n+s/p'. \end{cases} \tag{7}$$

Further, we notice that $G_r(z)=O(e^{-c|z|})$ implies

$$\int_{E_1(x)} G_{r\theta+s/p}(x-\xi,\eta)f(\xi,\eta)\,d\xi\,d\eta \leqslant c\int_{E_1(x)} F(\xi)e^{-c|x-\xi|}\,d\xi$$
$$= c\sum_{j=0}^{\infty}\int_{2^j<|x-\xi|<2^{j+1}} e^{-c|x-\xi|}F(\xi)\,d\xi$$
$$\leqslant c\sum_{j=0}^{\infty} e^{-c2^j}2^{nj}(M^{(n)}F)(x)$$
$$= c(M^{(n)}F)(x).$$

Combining this estimate with (6), where $\delta=1$, we arrive at

$$(J^{(n+s)}_{r\theta+s/p}f)(x,0)\leqslant c(M^{(n)}F)(x). \tag{8}$$

In the case $r\theta>n+s/p'$ we have

$$\int_{R^{n+s}\setminus E_1(x)} G_{r\theta+s/p}(x-\xi,\eta)f(\xi,\eta)\,d\xi\,d\eta \leqslant \int_{R^{n+s}\setminus E_1(x)} f(\xi,\eta)\,d\xi\,d\eta$$
$$\leqslant c(J^{(n+s)}_{r+s/p}f)(x,0)$$

which, together with (7), yields

$$(J_{r\theta+s/p}^{(n+s)}f)(x,0)\le c(J_{r+s/p}^{(n+s)}f)(x,0),\qquad r\theta>n+s/p'. \tag{9}$$

For $r\theta<n+s/p'$, estimates (6)–(8) imply

$$c(J_{r\theta+s/p}^{(n+s)}f)(x,0)\le\delta^{r\theta}(M^{(n)}F)(x)+\delta^{-r(1-\theta)}(J_{r+s/p}^{(n+s)}f)(x,0)$$

for all $\delta\in(0,\infty)$. By minimizing the right-hand side with respect to δ, we arrive at (3).

If $r\theta=n+s/p'$ then, by (6), (7),

$$(J_{r\theta+s/p}^{(n+s)}f)\le c(1+|\log\delta|)[\delta^{r\theta}(M^{(n)}F)(x)+(J_{r+s/p}^{(n+s)}f)(x,0)] \tag{10}$$

for all $\delta\in(0,1]$. If $(M^{(n)}F)(x)\le(J_{r+s/p}^{(n+s)}f)(x,0)$, then by (8)

$$(J_{r\theta+s/p}^{(n+s)}f)(x,0)\le c(J_{r+s/p}^{(n+s)}f)(x,0) \tag{11}$$

and (3) follows from (8) and (1.11). Let $(M^{(n)}F)(x)>(J_{r+s/p}^{(n+s)}f)(x,0)$. We put

$$\delta^{r\theta}=\frac{(J_{r+s/p}^{(n+s)}f)(x,0)}{(M^{(n)}F)(x)}$$

in (10). Then

$$\begin{aligned}(J_{r\theta+s/p}^{(n+s)}f)(x,0)&\le c(1+|\log\delta|)(J_{r+s/p}^{(n+s)}f)(x,0)\\&=c(1+|\log\delta|)\,\delta^{r\theta(1-\theta)}[(J_{r+s/p}^{(n+s)}f)(x,0)]^{\theta}[(M^{(n)}F)(x)]^{1-\theta}.\end{aligned} \tag{12}$$

Since $\delta\in(0,1]$, (3) is proved.

In the case $r\theta>n+s/p'$ the estimate (3) immediately follows from (8) and (9). □

Remark. Analogous, but simpler, arguments lead to the generalization of inequality (1.1.3/1):

$$(I_{r\theta+s/p}^{(n+s)}f)(x,0)\le c[(I_{r+s/p}^{(n+s)}f)(x,0)]^{\theta}[(M^{(n)}F)(x)]^{1-\theta}, \tag{13}$$

where $F(\xi)=\|f(\xi,\cdot);R^s\|_{L_p}$, $0<\theta<1$ and $r<n+s/p'$.

The following assertion is proved by Sjödin [1]. We use the same notation as in Lemma 1 and put $s=1$.

Lemma 2. *The inequality*

$$(J_{r+1/p}^{(n+1)}f)(x,0)\le c(J_r^{(n)}F)(x) \tag{14}$$

is valid.

Proof. From (1) follows, for $|x| \leq 1$,

$$\|G_{r+1/p}(x, \cdot); R^1\|_{L_{p'}} \leq \begin{cases} c\,|x|^{r-n} & \text{if} \quad r \leq n, \\ c & \text{if} \quad r > n. \end{cases} \tag{15}$$

Let $|x| > 1$. By virtue of (2) the norm on the left in (15) does not exceed

$$c\,|x|^{(r-n-1/p)/2}\left(\int_{R^1} (1+\tau^2)^{(r-n-1-1/p')p'/4} e^{-p'|x|\sqrt{(1+\tau^2)}}\,d\tau\right)^{1/p'}.$$

The main term of the asymptotics of the latter integral is equal to $c\,|x|^{-1/2}\exp(-p'\,|x|)$ (see, for example, Evgrafov [1]). So, for $|x| > 1$,

$$\|G_{r+1/p}(x, \cdot); R^1\|_{L_{p'}} \leq c\,|x|^{(r-n-1)/2}\exp(-|x|).$$

This and (2) imply $\|G_{r+1/p}(x, \cdot); R^1\|_{L_{p'}} \leq cG_r(x)$. It remains to apply Hölder's inequality to the potential $J^{(n+1)}_{r+1/p}f$. □

3.1.3. *The Capacity* cap $(\cdot, B_p^m)$

Let e be a compact set in R^n. We define the capacity by

$$\text{cap}\,(e, B_p^m) = \inf\{\|u\|^p_{B_p^m} : u \in D,\ u \geq 1 \quad \text{on} \quad e\},$$

where $m > 0$, $p \in (1, \infty)$. The capacity cap (e, W_p^m) is defined in the same way. It is clear that cap $(e, W_p^m) \sim$ cap (e, B_p^m) for $\{m\} > 0$.

The definitions of corresponding capacities together with part (i) of Lemma 3.1.1/6 imply:

Proposition 1. *The relation*

$$\text{cap}\,(e, B_p^m(R^n)) \sim \text{cap}\,(e, H_p^{m+1/p}(R^{n+1})) \tag{1}$$

is valid.

Proposition 2. (Sjödin [1]). *The estimate*

$$\text{cap}\,(e, H_p^m) \leq c\,\text{cap}\,(e, B_p^m) \tag{2}$$

holds.

Proof. By virtue of (2.1.2/3) there exists $f \in L_p(R^{n+1})$, $f \geq 0$, such that $J^{(n+1)}_{m+1/p}f \geq 1$ on e and

$$\|f; R^{n+1}\|^p_{L_p} \leq \text{cap}\,(e, H_p^{m+1/p}(R^{n+1})) + \varepsilon, \tag{3}$$

where $\varepsilon > 0$. This, together with Lemma 3.1.2/2, gives $J^{(n)}_m(cF) \geq 1$ on e, where $F(\xi) = \|f(\xi, \cdot); R^1\|_{L_p}$. Therefore

$$\text{cap}\,(e, H_p^m(R^n)) \leq c^p\,\|F; R^n\|^p_{L_p} = c^p\,\|f; R^{n+1}\|^p_{L_p}. \tag{4}$$

Combining (1), (3), (4), we complete the proof. □

This proposition and Lemma 3.1.1/4 give:

Proposition 3. *If* $p \geq 2$, *then* $\text{cap}(e, B_p^m) \sim \text{cap}(e, H_p^m)$.

From Proposition 2.1.2/1 and from (1) we directly obtain:

Proposition 4. *If* $mp > n$ *then for all non-empty compact sets* e *with* $d(e) \leq 1$ *the relation* $\text{cap}(e, B_p^m) \sim 1$ *is valid.*

The following assertion is a corollary of Propositions 2.1.2/2, 2.1.2/3 and estimate (2).

Proposition 5. (i) *If* $mp < n$, *then*

$$\text{cap}(e, B_p^m) \geq c(\text{mes}_n\, e)^{(n-mp)/n}. \tag{5}$$

(ii) *If* $mp = n$ *and* $d(e) \leq 1$, *then*

$$\text{cap}(e, B_p^m) \geq c\left(\log \frac{2^n}{\text{mes}_n\, e}\right)^{1-p}.$$

This together with (1) and Proposition 2.1.2/4 implies:

Proposition 6. (i) *If* $mp < n$ *and* $0 < r \leq 1$, *then* $\text{cap}(B_r, B_p^m) \sim r^{n-mp}$.
(ii) *If* $mp = n$, $0 < r \leq 1$, *then* $\text{cap}(B_r, B_p^m) \sim (\log 2/r)^{1-p}$.

An immediate corollary of Proposition 2.1.2/5 and formula (1) is:

Proposition 7. *For any compact set* e *the following relation holds*:

$$\text{cap}(e, B_p^m) \sim \sum_{j \geq 0} \text{cap}(e \cap B^{(j)}, B_p^m),$$

where $\{B^{(j)}\}_{j \geq 0}$ *is a covering of* R^n *by balls of unit diameter with a finite multiplicity depending only on* n.

3.2 Description of the Space $M(W_p^m \to W_p^l)$ for Fractional m and l and for $p > 1$

The exposition in this section follows the authors' paper [10].

3.2.1 Auxiliary Assertions

Since the spaces B_p^s are intermediate under real interpolation of spaces W_p^s (see, for example, Triebel [4]), then

$$\|\gamma\|_{M(B_p^{m-l+\mu} \to B_p^{\mu})} \leq c\, \|\gamma\|_{M(W_p^m \to W_p^l)}^{\mu/l}\, \|\gamma\|_{M(W_p^{m-l} \to L_p)}^{(l-\mu)/l}, \tag{1}$$

where $0 < \mu < l$.

Lemma 1. *Let γ_ρ be the mollification of $\gamma \in W_{p,\text{loc}}^l$ with kernel $K \geq 0$ and radius ρ. Then*

$$\|\gamma_\rho\|_{M(W_p^m \to W_p^l)} \leq \|\gamma\|_{M(W_p^m \to W_p^l)} \leq \varliminf_{\rho \to 0} \|\gamma_\rho\|_{M(W_p^m \to W_p^l)}. \tag{2}$$

The space $M(B_p^m \to B_p^l)$ has a similar property.

Proof. Let $u \in D$. It is clear that for $\{l\} > 0$

$$\|\gamma_\rho u\|_{W_p^l} = \left(\iint \left| \rho^{-n} \int K(\xi/\rho) \nabla_{[l],x}(Q(x+y, \xi) - Q(x, \xi))\, d\xi \right|^p \right.$$
$$\left. \times |y|^{-n-p\{l\}}\, dy\, dx \right)^{1/p} + \left(\int \left| \int \rho^{-n} K(\xi/\rho) Q(x, \xi)\, d\xi \right|^p dx \right)^{1/p},$$

where $Q(x, \xi) = \gamma(x - \xi)u(x)$. By the Minkowski inequality,

$$\|D_{p,l}(\gamma_\rho u)\|_{L_p} \leq \rho^{-n} \int K(\xi/\rho)\, \|D_{p,l} Q(\cdot, \xi)\|_{L_p}\, d\xi.$$

An analogous estimate holds for the norm $\|\gamma_\rho u\|_{L_p}$. Therefore

$$\|\gamma_\rho u\|_{W_p^l} \leq \rho^{-n} \int K(\xi/\rho)\, \|Q(\cdot, \xi)\|_{W_p^l}\, d\xi.$$

Since $\|Q(\cdot, \xi)\|_{W_p^l} \leq \|\gamma\|_{M(W_p^m \to W_p^l)} \|u\|_{W_p^m}$, the left inequality is proved for $\{l\} > 0$. The case $\{l\} = 0$ can be considered in the same way (see the proof of Lemma 1.3.1/1).

The right inequality follows from

$$\|\gamma u\|_{W_p^l} = \lim_{\rho \to 0} \|\gamma_\rho u\|_{W_p^l} \leq \varliminf_{\rho \to 0} \|\gamma_\rho\|_{M(W_p^m \to W_p^l)} \|u\|_{W_p^m}.$$

For the space $M(B_p^m \to B_p^l)$ the proof is similar. □

Lemma 2. *Let μ be a measure in R^n, $p \in (1, \infty)$, $k \in (0, \infty)$. The best constant C in*

$$\int |u|^p\, d\mu \leq C\, \|u\|_{W_p^k}^p, \qquad u \in D, \tag{3}$$

is equivalent to

$$\sup \frac{\mu(e)}{\text{cap}\,(e, W_p^k)}, \tag{4}$$

where e is an arbitrary compact set of positive capacity $\text{cap}\,(e, W_p^k)$. The same is valid when W_p^k is replaced by B_p^k.

Proof. For W_p^k with integer k the assertion was proved in Theorem 1.1.4.

Let $U \in D(R^{n+1})$ be an arbitrary extension of the function u to R^{n+1}. By Theorem 2.1.4/2,

$$\int_{R^n} |u|^p \, d\mu \leq c \sup_{e \subset R^n} \frac{\mu(e)}{\operatorname{cap}(e, H_p^{k+1/p}(R^{n+1}))} \|U; R^{n+1}\|_{H_p^{k+1/p}}.$$

Minimizing the right-hand side over all extensions of u and applying (3.1.3/1), we obtain the required upper bound for C. The lower bound is obvious. □

This implies the following:

Corollary. *Let $p \in (1, \infty)$, $k \in (0, \infty)$ and $\gamma \in L_{p,\mathrm{loc}}$. The relation*

$$\|\gamma\|_{M(W_p^k \to L_p)} \sim \sup_e \frac{\|\gamma; e\|_{L_p}}{[\operatorname{cap}(e, W_p^k)]^{1/p}} \tag{5}$$

is valid. The same relation for B_p^k holds.

Remark 1. From Proposition 3.1.3/7 and the corollary we get

$$\|\gamma\|_{M(W_p^k \to L_p)} \sim \sup_{\{e: d(e) \leq 1\}} \frac{\|\gamma; e\|_{L_p}}{[\operatorname{cap}(e, W_p^k)]^{1/p}}. \tag{6}$$

The similar relation for B_p^k is valid.

Remark 2. From the corollary and Propositions 3.1.3/2 and 3.1.3/3 the equality $M(B_p^k \to L_p) = M(H_p^k \to L_p)$ for $p \geq 2$ and the imbedding $M(H_p^k \to L_p) \subset M(B_p^k \to L_p)$ for $1 < p \leq 2$ follow.

Lemma 3. *Let $1 < p < \infty$, $0 < \lambda < \mu$ and let $\varphi \in L_{p\mu,\mathrm{loc}}$, $\varphi \geq 0$. Then*

$$\sup_e \left(\frac{\int_e \varphi^{\lambda p} \, dx}{\operatorname{cap}(e, W_p^\lambda)} \right)^{1/\lambda} \leq c \sup_e \left(\frac{\int_e \varphi^{\mu p} \, dx}{\operatorname{cap}(e, W_p^\mu)} \right)^{1/\mu}. \tag{7}$$

The same is true for the scale B_p^k.

Proof. Let $u \in D(R^n)$ and let $U \in D(R^{n+s})$ be an extension of u onto R^{n+s}. Here $s = 1, 2, \ldots$ for $\{\lambda\} > 0$ and $s = 0$ for integer λ. By f we denote the function $\Lambda^{\lambda + s/p} U$, where $\Lambda = (-\Delta + 1)^{1/2}$, Δ being the Laplace operator in R^{n+s}. Then

$$u(x) = (J_{\lambda + s/p}^{(n+s)} f)(x, 0).$$

Using Lemma 3.1.2/1 for $r = \mu$, $\theta = \lambda/\mu$ and the Hölder inequality, we obtain

$$\int \varphi^{\lambda p} |u|^p \, dx \leq c \int \varphi(x)^{\lambda p} [(J^{(n+s)}_{\mu+s/p} |f|)(x,0)]^{\lambda p/\mu} [(M^{(n)} F)(x)]^{(\mu-\lambda)p/\mu} \, dx$$

$$\leq c \left(\int \varphi(x)^{\mu p} [(J^{(n+s)}_{\mu+s/p} |f|)(x,0)]^p \, dx \right)^{\lambda/\mu}$$

$$\times \left(\int [(M^{(n)} F)(x)]^p \, dx \right)^{(\mu-\lambda)/\mu}.$$

By continuity of the operator $M^{(n)}$ in $L_p(R^n)$ and by Lemma 2 we have

$$\int \varphi^{\lambda p} |u|^p \, dx \leq c \, \|f; R^{n+s}\|_{L_p}^{(\mu-\lambda)p/\mu} \sup_e \left(\frac{\int_e \varphi^{\mu p} \, dx}{\text{cap}\,(e, W_p^{\mu})} \right)^{\lambda/\mu}$$

$$\times \|(J^{(n+s)}_{\mu+s/p} |f|)(\cdot, 0)\|_{W_p^{\mu}}^{\lambda p/\mu}.$$

(Here we have used the possibility of approximation of $(J^{(n+s)}_{\mu+s/p} |f|)(x,0)$ in W_p^{μ} by a sequence of functions in $D(R^n)$.) Since the last norm does not exceed

$$\|\Lambda^{\mu+s/p} J^{(n+s)}_{\mu+s/p} |f|; R^{n+s}\|_{L_p} = \|f; R^{n+s}\|_{L_p} = \|\Lambda^{\lambda+s/p} U; R^{n+s}\|_{L_p}$$

(see part (i) of Lemma 3.1.1/6), then

$$\int \varphi^{\lambda p} |u|^p \, dx \leq c \sup_e \left(\frac{\int_e \varphi^{\mu p} \, dx}{\text{cap}\,(e, W_p^{\mu})} \right)^{\lambda/\mu} \|\Lambda^{\lambda+s/p} U; R^{n+s}\|_{L_p}^p.$$

By minimizing the right-hand side over all extensions U of u, we obtain

$$\int \varphi^{\lambda p} |u|^p \, dx \leq c \sup_e \left(\frac{\int_e \varphi^{\mu p} \, dx}{\text{cap}\,(e, W_p^{\mu})} \right)^{\lambda/\mu} \|u\|_{W_p^{\lambda}}^p.$$

It remains to use Lemma 2.

The same arguments with $s \geq 1$ can be used for the scale B_p^k. □

3.2.2 *Imbedding* $M(W_p^m \to W_p^l) \subset M(W_p^{m-l} \to L_p)$ *for Fractional* m *and* l

Lemma. *Let* $p \in (1, \infty)$, $m \geq l > 0$ *and let* m *and* l *be fractional. Then*

$$\|\gamma\|_{M(W_p^{m-l} \to L_p)} \leq c \, \|\gamma\|_{M(W_p^m \to W_p^l)}. \tag{1}$$

Proof. Let γ_ρ be the mollification of $\gamma \in M(W_p^m \to W_p^l)$. Since $\gamma \in L_{p,\text{unif}}$, all the derivatives of γ_ρ are uniformly bounded and so γ_ρ is a multiplier in any space W_p^k. Next we consider three separate cases: $m \geq 2l$, $m = l$, $2l > m > l$.

1. Let $m \geq 2l$. We put $s = 1$ if $\{m - l\} > 0$ and $s = 0$ if $\{m - l\} = 0$. Let

$U \in D(R^{n+s})$ be an extension of $u \in D(R^n)$ to R^{n+s} for $s=1$ and $U=u$ for $s=0$. We express U as a potential $J^{(n+s)}_{m-l+s/p} f$ with density $f \in L_p(R^{n+s})$. According to Lemma 3.1.2/1,

$$|u(x)| \leq c[(J^{(n+s)}_{m+s/p}|f|)(x,0)]^{1-l/m}[(M^{(n)}F)(x)]^{l/m}.$$

Therefore

$$\|\gamma_\rho u\|_{L_p} \leq c\,\|f; R^{n+s}\|^{l/m}_{L_p}\,\|\gamma_\rho^{l/(m-l)}\gamma_\rho(J^{(n+s)}_{m+s/p}|f|)(\cdot,0)\|^{1-l/m}_{L_p}.$$

By virtue of Lemma 3.2.1/2 the right-hand side does not exceed

$$c\,\|f; R^{n+s}\|^{l/m}_{L_p}\,\|\gamma_\rho(J^{(n+s)}_{m+s/p}|f|)(\cdot,0)\|^{1-l/m}_{W_p^l}\sup_e\left[\frac{\int_e|\gamma_\rho|^{pl/(m-l)}\,dx}{\operatorname{cap}(e, W_p^l)}\right]^{(m-l)/pm}.$$

From Lemma 3.2.1/3 it follows that the last supremum is not more than

$$c\left(\sup_e\frac{\int_e|\gamma_\rho|^p\,dx}{\operatorname{cap}(e, W_p^{m-l})}\right)^{l/pm}$$

which can be changed to $c\,\|\gamma_\rho\|^{l/m}_{M(W_p^{m-l}\to L_p)}$ by Corollary 3.2.1.

Thus

$$\begin{aligned}\|\gamma_\rho u\|_{L_p} &\leq c\,\|f; R^{n+s}\|^{l/m}_{L_p}\,\|\gamma_\rho\|^{l/m}_{M(W_p^{m-l}\to L_p)}\,\|\gamma_\rho(J^{(n+s)}_{m+s/p}|f|)(\cdot,0)\|^{1-l/m}_{W_p^l}\\ &\leq c\,\|\gamma_\rho\|^{l/m}_{M(W_p^{m-l}\to L_p)}\,\|\gamma_\rho\|^{1-l/m}_{M(W_p^m\to W_p^l)}\,\|f; R^{n+s}\|^{l/m}_{L_p}\\ &\quad\times\|(J^{(n+s)}_{m+s/p}|f|)(\cdot,0)\|^{1-l/m}_{W_p^m}.\end{aligned}\tag{2}$$

If $s=0$ then (2) can be rewritten in the form

$$\|\gamma_\rho u\|_{L_p} \leq c\,\|\gamma_\rho\|^{l/m}_{M(W_p^{m-l}\to L_p)}\,\|\gamma_\rho\|^{1-l/m}_{M(W_p^m\to W_p^l)}\,\|u\|_{W_p^{m-l}}.\tag{3}$$

Let $s=1$. From Lemma 3.1.1/6 it follows that the last norm in (2) does not exceed

$$c\,\|f; R^{n+s}\|_{L_p} \sim \|U; R^{n+s}\|_{H_p^{m-l+s/p}}.$$

This and Lemma 3.1.1/6 give (3). The reference to Lemma 3.2.1/1 provides (1) in the case $m \geq 2l$.

2. Suppose $m=l$. For any integer $N>0$ we have

$$\|\gamma^N u\|^{1/N}_{L_p} \leq \|\gamma^N u\|^{1/N}_{W_p^l} \leq \|\gamma\|_{MW_p^l}\,\|u\|^{1/N}_{W_p^l}.$$

Passing to the limit as $N\to\infty$, we obtain $\|\gamma\|_{ML_p}=\|\gamma\|_{L_\infty}\leq\|\gamma\|_{MW_p^l}$.

3. Now let $m>l$ and let μ be a positive number such that $\mu<m-l$, $\mu<1$. Suppose that $m-l+\mu$ is fractional. By (3.2.1/1) the estimate

$$\|\gamma_\rho\|_{M(W_p^{m-l+\mu}\to W_p^\mu)} \leq c\,\|\gamma_\rho\|^{1-\mu/l}_{M(W_p^{m-l}\to L_p)}\,\|\gamma_\rho\|^{\mu/l}_{M(W_p^m\to W_p^l)}\tag{4}$$

holds. Since $m-l+\mu>2\mu$, then by the first part of the proof

$$\|\gamma_\rho\|_{M(W_p^{m-l}\to L_p)} \leq c\,\|\gamma_\rho\|_{M(W_p^{m-l+\mu}\to W_p^\mu)}$$

which, together with (4), gives (1) for γ_ρ. It remains to apply Lemma 3.2.1/1. □

3.2.3 Two-sided Estimate for the Norm in $M(W_p^m \to W_p^l)$, $0<l<1$

Lemma. *If $0<l\leq m$, $l<1$ and $\{m\}>0$, then*

$$\|\gamma\|_{M(W_p^m\to W_p^l)} \sim \|D_{p,l}\gamma\|_{M(W_p^m\to L_p)} + \|\gamma\|_{M(W_p^{m-l}\to L_p)}. \tag{1}$$

Proof. Let $t=0$ for $m=l$ and $t\in(0, m-l)$ for $m>l$. By virtue of Lemma 3.1.1/1,

$$(D_{p,l}u)(x) \leq c(J_t^{(n)} D_{p,l}\Lambda^t u)(x).$$

Therefore

$$\begin{aligned}\|\gamma D_{p,l}u\|_{L_p} &\leq c\,\|\gamma J_t^{(n)} D_{p,l}\Lambda^t u\|_{L_p}\\ &\leq c\,\|\gamma\|_{M(W_p^{m-l}\to L_p)}\,\|J_t^{(n)} D_{p,l}\Lambda^t u\|_{W_p^{m-l}}.\end{aligned}$$

Estimating the last norm by Lemma 3.1.1/3, we obtain

$$\begin{aligned}\|\gamma D_{p,l}u\|_{L_p} &\leq c\,\|\gamma\|_{M(W_p^{m-l}\to L_p)}\,\|D_{p,l}\Lambda^t u\|_{W_p^{m-l+t}}\\ &\leq c\,\|\gamma\|_{M(W_p^{m-l}\to L_p)}(\|D_{p,m-l-t}D_{p,l}\Lambda^t u\|_{L_p} + \|D_{p,l}\Lambda^t u\|_{L_p})\end{aligned}$$

which, together with Lemma 3.1.1/2, gives

$$\|\gamma D_{p,l}u\|_{L_p} \leq c\,\|\gamma\|_{M(W_p^{m-l}\to L_p)}(\|\Lambda^t u\|_{W_p^{m-t}} + \|\Lambda^t u\|_{W_p^t}).$$

This, Lemma 3.1.1/3 and the obvious inequality $\|D_{p,l}(\gamma u) - uD_{p,l}\gamma\|_{L_p} \leq \|\gamma D_{p,l}u\|_{L_p}$ imply

$$|\,\|D_{p,l}(\gamma u)\|_{L_p} - \|uD_{p,l}\gamma\|_{L_p}| \leq c\,\|\gamma\|_{M(W_p^{m-l}\to L_p)}\,\|u\|_{W_p^m}.$$

This leads to the inequality

$$\|D_{p,l}(\gamma u)\|_{L_p} \leq c(\|D_{p,l}\gamma\|_{M(W_p^m\to L_p)} + \|\gamma\|_{M(W_p^{m-l}\to L_p)})\,\|u\|_{W_p^m}$$

which provides the required upper estimate for the norm in $M(W_p^m \to W_p^l)$. On the other hand,

$$\|uD_{p,l}\gamma\|_{L_p} \leq (\|\gamma\|_{M(W_p^m\to W_p^l)} + c\,\|\gamma\|_{M(W_p^{m-l}\to L_p)})\,\|u\|_{W_p^m}$$

which, together with (3.2.2/1), gives the lower bound for the norm in $M(W_p^m \to W_p^l)$. □

3.2.4 Estimates for Derivatives of a Multiplier

Lemma 1. *If $\gamma\in M(W_p^m \to W_p^l)$, $\{m\}, \{l\}>0$, then $D^\alpha\gamma\in M(W_p^m \to W_p^{l-|\alpha|})$ for any multi-index α of order $|\alpha|\leq l$. The following inequality* *holds*:

$$\|D^\alpha \gamma\|_{M(W_p^m \to W_p^{l-|\alpha|})} \leqslant c\,\|\gamma\|_{M(W_p^m \to W_p^l)}. \tag{1}$$

This assertion can be proved in the same way as Lemma 2.2.3/1, but using interpolation inequality (3.2.1/1).

From (1) and (3.2.2/1) immediately follows:

Corollary. *The following estimate is valid*: $\|D^\alpha \gamma\|_{M(W_p^{m-l+|\alpha|} \to L_p)} \leqslant c\,\|\gamma\|_{M(W_p^m \to W_p^l)}$.

3.2.5 Lower Bound for the Norm of a Multiplier

Lemma. *Let m and l be fractional, $m \geqslant l > 0$. Then*

$$\|\gamma\|_{M(W_p^{m-l} \to L_p)} + \|D_{p,l}\gamma\|_{M(W_p^m \to L_p)} \leqslant c\,\|\gamma\|_{M(W_p^m \to W_p^l)}. \tag{1}$$

Proof. For $0 < l < 1$, inequality (1) is contained in Lemma 3.2.3.

For $l > 1$ it suffices to estimate only the second summand in the left-hand side of (1), since the first was estimated in Lemma 3.2.2. Let us assume that

$$\|D_{p,l}\gamma\|_{M(W_p^m \to L_p)} \leqslant c\,\|\gamma\|_{M(W_p^m \to W_p^l)}$$

for all fractional $l \in (0, N)$, where N is integer. Further, let $N < l < N+1$. We have

$$\|uD_{p,l}\gamma\|_{L_p} \leqslant c\Big(\|\gamma u\|_{W_p^l} + \sum_{j=0}^{N} \||\nabla_j \gamma|\, D_{p,l-j}u\|_{L_p} + \sum_{j=1}^{N} \||\nabla_j u|\, D_{p,l-j}\gamma\|_{L_p}\Big). \tag{2}$$

By t we denote any number in $(0, m-l+j)$ if $m > l$ or if $m = l$, $j > 0$. Further, let $t = 0$ if $m = l$, $j = 0$. According to Lemma 3.1.1/1,

$$(D_{p,l-j}u)(x) \leqslant c(J_t^{(n)} D_{p,l-j}\Lambda^t u)(x).$$

This and Corollary 3.2.4 imply

$$\begin{aligned}\||\nabla_j\gamma|\, D_{p,l-j}u\|_{L_p} &\leqslant \|\nabla_j \gamma\|_{M(W_p^{m-l+j} \to L_p)}\, \|J_t^{(n)} D_{p,l-j}\Lambda^t u\|_{W_p^{m-l+j}} \\ &\leqslant c\,\|\nabla_j\gamma\|_{M(W_p^{m-l+j} \to L_p)}\, \|D_{p,l-j}\Lambda^t u\|_{W_p^{m-l+j-t}}.\end{aligned}$$

In other words,

$$\begin{aligned}\||\nabla_j\gamma|\, D_{p,l-j}u\|_{L_p} \leqslant c\,\|\nabla_j\gamma\|_{M(W_p^{m-l+j} \to L_p)}(&\|D_{p,m-l+j-t}D_{p,\{l\}}\nabla_{[l-j]}\Lambda^t u\|_{L_p} \\ &+ \|D_{p,l-j}\Lambda^t u\|_{L_p}).\end{aligned}$$

If t is sufficiently close to $m - l + j$, then by Lemmas 3.1.1/2 and 3.1.1/3

$$\begin{aligned}\|D_{p,m-l+j-t}D_{p,\{l-j\}}\nabla_{[l-j]}\Lambda^t u\|_{L_p} &\leqslant c\,\|D_{p,m-t-[l-j]}\Lambda^t u\|_{L_p} \\ &\leqslant c_1\,\|\Lambda^t u\|_{W_p^{m-t}} \leqslant c_2\,\|u\|_{W_p^m}.\end{aligned}$$

Therefore

$$\||\nabla_j\gamma|\, D_{p,l-j}u\|_{L_p} \leqslant c\,\|\nabla_j\gamma\|_{M(W_p^{m-l+j}\to L_p)}\,\|u\|_{W_p^m} \tag{3}$$

and, by Corollary 3.2.4,

$$\||\nabla_j\gamma|\, D_{p,l-j}u\|_{L_p} \leqslant c\,\|\gamma\|_{M(W_p^m\to W_p^l)}\,\|u\|_{W_p^m}. \tag{4}$$

According to the induction hypothesis, for $j \geqslant 1$,

$$\begin{aligned}\||\nabla_j u|\, D_{p,l-j}\gamma\|_{L_p} &\leqslant \|D_{p,l-j}\gamma\|_{M(W_p^{m-j}\to L_p)}\,\|u\|_{W_p^m}\\ &\leqslant c_1\,\|\gamma\|_{M(W_p^{m-j}\to W_p^{l-j})}\,\|u\|_{W_p^m}.\end{aligned} \tag{5}$$

Applying interpolation inequality (3.2.1/1), we get

$$\||\nabla_j u|\, D_{p,l-j}\gamma\|_{L_p} \leqslant c\,\|\gamma\|^{j/l}_{M(W_p^{m-l}\to L_p)}\,\|\gamma\|^{1-j/l}_{M(W_p^m\to W_p^l)}\,\|u\|_{W_p^m}. \tag{6}$$

This, together with Lemma 3.2.2, implies

$$\||\nabla_j u|\, D_{p,l-j}\gamma\|_{L_p} \leqslant c_1\,\|\gamma\|_{M(W_p^m\to W_p^l)}\,\|u\|_{W_p^m}. \tag{7}$$

Estimating the right-hand side of (2) with the help of (4) and (7), we arrive at

$$\|uD_{p,l}\gamma\|_{L_p} \leqslant c\,\|\gamma\|_{M(W_p^m\to W_p^l)}\,\|u\|_{W_p^m}. \quad \square$$

3.2.6 Upper Bound for the Norm of a Multiplier

Lemma. *Let m and l be fractional, $m \geqslant l > 0$. The estimate*

$$\|\gamma\|_{M(W_p^m\to W_p^l)} \leqslant c(\|D_{p,l}\gamma\|_{M(W_p^m\to L_p)} + \|\gamma\|_{M(W_p^{m-l}\to L_p)}) \tag{1}$$

is valid.

Proof. Let the right-hand side of (1) be finite. As before, by γ_ρ we denote a mollification of γ with radius ρ. From the boundedness of the function γ_ρ with all its derivatives we obtain $\gamma_\rho \in M(W_p^m \to W_p^l)$. For any $u \in W_p^m$ we have

$$\|D_{p,l}(\gamma_\rho u)\|_{L_p} \leqslant c\left(\sum_{j=0}^{[l]} \||\nabla_j\gamma_\rho|\, D_{p,l-j}u\|_{L_p} + \sum_{j=0}^{[l]} \||\nabla_j u|\, D_{p,l-j}\gamma_\rho\|_{L_p}\right). \tag{2}$$

Let ε be any positive fractional number such that $\varepsilon < \{l\}$ and let $m-l+j+\varepsilon$ be fractional. By virtue of Corollary 3.2.4 and interpolation inequality (3.2.1/1),

$$\begin{aligned}\|\nabla_j\gamma_\rho\|_{M(W_p^{m-l+j}\to L_p)} &\leqslant c\,\|\gamma_\rho\|_{M(W_p^{m-l+j+\varepsilon}\to W_p^{j+\varepsilon})}\\ &\leqslant c_1\,\|\gamma_\rho\|^{(l-j-\varepsilon)/l}_{M(W_p^{m-l}\to L_p)}\,\|\gamma_\rho\|^{(j+\varepsilon)/l}_{M(W_p^m\to W_p^l)}.\end{aligned} \tag{3}$$

This and (3.2.5/3) show that the first sum in the right-hand side of (2) does not exceed

$$c\,\|u\|_{W_p^m} \sum_{j=0}^{[l]} \|\gamma_\rho\|_{M(W_p^{m-l}\to L_p)}^{(l-j-\varepsilon)/l} \|\gamma_\rho\|_{M(W_p^m\to W_p^l)}^{(j+\varepsilon)/l}.$$

By Lemma 3.2.5, for γ_ρ estimate (3.2.5/5) with $j \geq 1$ holds. Applying (3.2.1/1), we arive at (3.2.5/6) for γ_ρ. Thus, for any $\delta > 0$,

$$\|D_{p,l}(\gamma_\rho u)\|_{L_p} \leq (\delta\,\|\gamma_\rho\|_{M(W_p^m\to W_p^l)} + c(\delta)\,\|\gamma_\rho\|_{M(W_p^{m-l}\to L_p)})\,\|u\|_{W_p^m} + \|uD_{p,l}\gamma_\rho\|_{L_p}.$$

Consequently,

$$\|D_{p,l}(\gamma_\rho u)\|_{L_p} \leq (\delta\,\|\gamma_\rho\|_{M(W_p^m\to W_p^l)} + c_1(\delta)\,\|\gamma_\rho\|_{M(W_p^{m-l}\to L_p)} + c_2(\delta)\,\|D_{p,l}\gamma_\rho\|_{M(W_p^m\to L_p)})\,\|u\|_{W_p^m}$$

which gives (1) for γ_ρ. Referring to Lemma 3.2.1/1, we complete the proof. □

Remark. From (3) and Lemma 3.2.1/1 it follows that, for any $\varepsilon > 0$,

$$\|\nabla_j \gamma\|_{M(W_p^{m-l+j}\to L_p)} \leq \varepsilon\,\|\gamma\|_{M(W_p^m\to W_p^l)} + c(\varepsilon)\,\|\gamma\|_{M(W_p^{m-l}\to L_p)}.$$

3.2.7 *Description of the Space $M(W_p^m \to W_p^l)$ for Fractional m and l*

Combining Lemmas 3.2.5 and 3.2.6, we obtain the following result.

Theorem 1. *Let m and l be fractional, $m \geq l$, $p \in (1, \infty)$. The function γ belongs to $M(W_p^m \to W_p^l)$ if and only if $\gamma \in W_{p,\mathrm{loc}}^l$, $D_{p,l}\gamma \in M(W_p^m \to L_p)$ and $\gamma \in M(W_p^{m-l} \to L_p)$. The following relation holds:*

$$\|\gamma\|_{M(W_p^m\to W_p^l)} \sim \|D_{p,l}\gamma\|_{M(W_p^m\to L_p)} + \|\gamma\|_{M(W_p^{m-l}\to L_p)}.$$

This and Corollary 3.2.1 give:

Theorem 2. *The function γ belongs to the space $M(W_p^m \to W_p^l)$ with m and l fractional, $m \geq l$, $p \in (1, \infty)$ if and only if $\gamma \in W_{p,\mathrm{loc}}^l$ and, for any compact set $e \subset R^n$,*

$$\|D_{p,l}\gamma; e\|_{L_p}^p \leq \text{const cap}\,(e, W_p^m), \qquad \|\gamma; e\|_{L_p}^p \leq \text{const cap}\,(e, W_p^{m-l}).$$

The relation

$$\|\gamma\|_{M(W_p^m\to W_p^l)} \sim \sup_e \left[\frac{\|D_{p,l}\gamma; e\|_{L_p}}{[\text{cap}\,(e, W_p^m)]^{1/p}} + \frac{\|\gamma; e\|_{L_p}}{[\text{cap}\,(e, W_p^{m-l})]^{1/p}}\right] \tag{1}$$

is valid.

Remark 1. From Theorem 2 and Remark 3.2.1/1 it follows that

$$\|\gamma\|_{M(W_p^m\to W_p^l)} \sim \sup_{\{e:d(e)\leq 1\}} \left(\frac{\|D_{p,l}\gamma; e\|_{L_p}}{[\operatorname{cap}(e, W_p^m)]^{1/p}} + \frac{\|\gamma; e\|_{L_p}}{[\operatorname{cap}(e, W_p^{m-l})]^{1/p}}\right). \tag{2}$$

Remark 2. According to Corollary 3.2.4, the estimate

$$\sup_e \frac{\|\nabla_j\gamma; e\|_{L_p}}{[\operatorname{cap}(e, W_p^{m-l+j})]^{1/p}} \leq c\,\|\gamma\|_{M(W_p^m\to W_p^l)}, \qquad j=1,\ldots,[l]$$

is valid. From (3.2.1/1) and from the imbedding $M(W_p^m \to W_p^l) \subset M(W_p^{m-l} \to L_p)$ it follows that $M(W_p^m \to W_p^l)$ is imbedded continuously into $M(W_p^{m-j} \to W_p^{l-j})$, $j=0, 1, \ldots, [l]$. Consequently,

$$\sup_e \frac{\|D_{p,l-j}\gamma; e\|_{L_p}}{[\operatorname{cap}(e, W_p^{m-j})]^{1/p}} \leq c\,\|\gamma\|_{M(W_p^m\to W_p^l)}.$$

So relation (1) is equivalent to

$$\|\gamma\|_{M(W_p^m\to W_p^l)} \sim \sup_e \left(\sum_{j=0}^{[l]} \frac{\|D_{p,l-j}\gamma; e\|_{L_p}^p}{\operatorname{cap}(e, W_p^{m-j})} + \sum_{j=0}^{[l]} \frac{\|\nabla_j\gamma; e\|_{L_p}^p}{\operatorname{cap}(e, W_p^{m-l+j})}\right)^{1/p}. \tag{3}$$

It is clear that here one can restrict oneself to sets e satisfying the condition $d(e) \leq 1$.

3.2.8 *The Case* $mp > n$

Similarly to 1.3.3, where an analogous result was derived for integer m and l, we shall show that $M(W_p^m \to W_p^l) = W_{p,\mathrm{unif}}^l$, if m and l are fractional, $p > 1$ and $mp > n$.

First we note the following easily checked relation:

$$\|\gamma\|_{W_{p,\mathrm{unif}}^l} \sim \sup_{x\in R^n} \Bigg(\|\gamma; B_1(x)\|_{W_p^{[l]}} + \left(\int_{B_1(x)}\int_{B_1(x)} \frac{|\nabla_{[l]}\gamma(y) - \nabla_{[l]}\gamma(z)|^p}{|y-z|^{n+p\{l\}}}\,dy\,dz\right)^{1/p}\Bigg), \tag{1}$$

where $\{l\} > 0$.

Proposition. *If* $p > 1$, $mp > n$, *then*

$$\|\gamma\|_{M(W_p^m\to W_p^l)} \sim \|\gamma\|_{W_{p,\mathrm{unif}}^l}. \tag{2}$$

Proof. The lower bound for the norm in $M(W_p^m \to W_p^l)$ follows at once from (1) and (3.2.7/3).

We derive the upper bound. Since $\operatorname{cap}(e, W_p^m) \sim 1$ for $mp > n$, then

$$\sup_{\{e:d(e)\leq 1\}} \frac{\|D_{p,l}\gamma; e\|_{L_p}}{[\operatorname{cap}(e, W_p^l)]^{1/p}} \leq c \sup_{x\in R^n} \|D_{p,l}\gamma; B_{1/2}(x)\|_{L_p}.$$

According to Lemma 3.1.1/11, the right-hand side does not exceed $c\,\|\gamma\|_{W^l_{p,\mathrm{unif}}}$.

From Propositions 3.1.3/4–3.1.3/6 it follows that

$$\sup_{\{e:d(e)\leq 1\}} \frac{\|\gamma; e\|_{L_p}}{[\operatorname{cap}(e, W_p^{m-l})]^{1/p}} \leq c \sup_{x\in R^n} \|\gamma; B_1(x)\|_{L_q},$$

where $q=n/(m-l)$ if $p(m-l)<n$ and $q>p$ if $p(m-l)\geq n$. Since $W^l_{p,\mathrm{unif}}$ is imbedded into $L_{q,\mathrm{unif}}$ (cf. Lemma 3.1.1/7), then the latter norm does not exceed $c\,\|\gamma\|_{W^l_{p,\mathrm{unif}}}$. □

3.2.9 *One-sided Estimates for the Norm in* $M(W_p^m \to W_p^l)$

By analogy to 1.3.4 and 2.2.10 we state here lower and upper bounds for the norm in $M(W_p^m \to W_p^l)$, $mp\leq n$, $p>1$, which contain no capacity.

From Proposition 3.1.3/6 and Theorem 3.2.7/2 we obtain immediately:

Proposition 1. *The following estimates hold*:

$$\|\gamma\|_{M(W_p^m\to W_p^l)} \geq c \sup_{x\in R^n, r\in(0,1)} (r^{m-n/p}(\|D_{p,l}\gamma; B_r(x)\|_{L_p} + r^{-l}\|\gamma; B_r(x)\|_{L_p})),$$

where $mp<n$ *and*

$$\|\gamma\|_{M(W_p^m\to W_p^l)} \geq c \sup_{x\in R^n, r\in(0,1)} ((\log 2/r)^{1/p'}\|D_{p,l}\gamma; B_r(x)\|_{L_p} + r^{-l}\|\gamma; B_r(x)\|_{L_p}),$$

where $mp=n$.

Remark 3.2.7/1 and Proposition 3.1.3/5 give

Proposition 2. *The following estimates hold*:

$$\|\gamma\|_{M(W_p^m\to W_p^l)} \leq c \sup_{\{e:d(e)\leq 1\}} (\operatorname{mes}_n e)^{m/n-1/p}(\|D_{p,l}\gamma; e\|_{L_p} + (\operatorname{mes}_n e)^{-l/n}\|\gamma; e\|_{L_p}), \tag{1}$$

where $mp<n$;

$$\|\gamma\|_{M(W_p^m\to W_p^l)} \leq c \sup_{\{e:d(e)\leq 1\}} ((\log(2^n/\operatorname{mes}_n e))^{1/p'}\|D_{p,l}\gamma; e\|_{L_p} + (\operatorname{mes}_n e)^{-l/n}\|\gamma; e\|_{L_p}), \tag{2}$$

where $mp=n$.

3.3 Sufficient Conditions for Inclusion into $M(W_p^m \to W_p^l)$

It may be of use to compare the contents of this section with sufficient conditions for inclusion into the class $M(H_p^m \to H_p^l)$ obtained in 2.3. Here similar conditions are found for $M(W_p^m \to W_p^l)$, $\{m\}>0$, $\{l\}>0$. They are formulated in terms of spaces $B^{\mu}_{q,\infty}$ (cf. (3.3.1)), $B^l_{q,p}$ (cf. (3.3.2)) and $H^l_{n/m}$ (cf. (3.3.3)).

By $B^s_{q,\theta}$ we denote the space of functions in R^n having the finite norm

$$\|u\|_{B^s_{q,\theta}} = \left(\int \|\Delta_h \nabla_{[s]} u\|^{\theta}_{L_q} |h|^{-n-\theta\{s\}}\, dh\right)^{1/\theta} + \|u\|_{W_q^{[s]}},$$

where $\{s\}>0$, q, $\theta \geqslant 1$.

In this section we derive sufficient conditions for a function to belong to the space $M(W_p^m \to W_p^l)$, using Proposition 3.2.9/2. These conditions are formulated in terms of the space $B^s_{q,\theta,\mathrm{unif}}$.

It is clear that

$$\|u\|_{B^s_{q,\theta,\mathrm{unif}}} \sim \sup_{x\in R^n} \left[\left(\int_{B_1} \|\Delta_h \nabla_{[s]} u; B_1(x)\|^{\theta}_{L_q} |h|^{-n-\theta\{s\}}\, dh\right)^{1/\theta} + \|u; B_1(x)\|_{W_q^{[s]}}\right].$$

3.3.1 Conditions in Terms of the Space $B^{\mu}_{q,\infty}$

We prove an assertion, analogous to Theorem 2.3.3, borrowed from the authors' paper [13].

Theorem 1. *Let* $q \geqslant p > 1$, $\{m\}, \{l\} > 0$, $\mu = n/q - m + l$, $\mu > l$, $\{\mu\} > 0$.
(i) *If* $\gamma \in B^{\mu}_{q,\infty,\mathrm{unif}} \cap L_\infty$, *then* $\gamma \in MW_p^l$ *and*

$$\|\gamma\|_{MW_p^l} \leqslant c(\|\gamma\|_{B^{\mu}_{q,\infty,\mathrm{unif}}} + \|\gamma\|_{L_\infty}). \tag{1}$$

(ii) *If* $\gamma \in B^{\mu}_{q,\infty,\mathrm{unif}}$, then $\gamma \in M(W_p^m \to W_p^l)$ *and*

$$\|\gamma\|_{M(W_p^m \to W_p^l)} \leqslant c\, \|\gamma\|_{B^{\mu}_{q,\infty,\mathrm{unif}}}. \tag{2}$$

Proof. Let $m \geqslant l$. It suffices to assume that the difference $\varepsilon = \mu - l$ is small since the general case follows by interpolation between the pairs $\{W_p^{m-l}, L_p\}$ and $\{W_p^{n/q-\varepsilon}, W_p^{\mu-\varepsilon}\}$ (see (3.2.1/1)). So we assume that $1+[l] > \mu > l$.

Let e be a compact set in R^n, $d(e) \leqslant 1$, and let $|e| = \mathrm{mes}_n\, e$. We have

$$\|D_{p,l}\gamma; e\|^p_{L_p} = \int \frac{dh}{|h|^{n+p\{l\}}} \int_e |\nabla_{[l]}\gamma(x+h) - \nabla_{[l]}\gamma(x)|^p\, dx. \tag{3}$$

We express the integral over R^n as the sum of two integrals $i_1 + i_2$, the first being taken over the exterior of the ball $\{h : |h| < |e|^{1/n}\}$. Obviously,

$$i_1 \leq c \int_{|h|>|e|^{1/n}} \frac{dh}{|h|^{n+p\{l\}}} \left(\int_e |\nabla_{[l]}\gamma(x+h)|^p \, dx + \int_e |\nabla_{[l]}\gamma(x)|^p \, dx \right)$$

$$\leq c\,|e|^{1-p/q} \int_{|h|>|e|^{1/n}} \frac{dh}{|h|^{n+p\{l\}}} \left(\int_e |\nabla_{[l]}\gamma(x+h)|^q \, dx \right.$$

$$\left. + \int_e |\nabla_{[l]}\gamma(x)|^q \, dx \right)^{p/q}.$$

This and Corollary 2.3.2 imply

$$i_1 \leq c\,|e|^{1-p/q+\{\mu\}p/n} \int_{|h|>|e|^{1/n}} \frac{dh}{|h|^{n+p\{l\}}} N(\gamma)^p, \tag{4}$$

where $N(\gamma)$ is the right-hand side of either (1) or (2). Since $[\mu]=[l]$, $\{\mu\}-\{l\}=\mu-l$ and consequently

$$i_1 \leq c\,|e|^{1-mp/n} N(\gamma)^p.$$

By Hölder's inequality,

$$i_2 \leq |e|^{1-p/q} \int_{|h|<|e|^{1/n}} \frac{dh}{|h|^{n+p\{l\}}} \left(\int_e |\nabla_{[l]}\gamma(x+h) - \nabla_{[l]}\gamma(x)|^q \, dx \right)^{p/q}$$

$$\leq |e|^{1-p/q} \int_{|h|<|e|^{1/n}} \frac{dh}{|h|^{n+p(\{l\}-\{\mu\})}} N(\gamma)^p = c\,|e|^{1-pm/n} N(\gamma)^p.$$

This, together with (3) and (4), implies $\|D_{p,l}\gamma; e\|_{L_p} \leq c\,|e|^{1/p-m/n} N(\gamma)$. In addition, according to (2.3.3/5), for $m>l$

$$\|\gamma; e\|_{L_p} \leq c\,|e|^{1/p-(m-l)/n} N(\gamma). \tag{5}$$

It remains to use Proposition 3.2.9/2. □

Using the same arguments as in the proof of Corollary 2.3.4, we obtain:

Corollary 1. *Let $n=1$, $q \geq p$ and $lq<1$. If $\gamma \in L_\infty$ and $\mathrm{Var}_q(\gamma)<\infty$, then $\gamma \in MW_p^l$ and*

$$\|\gamma\|_{MW_p^l} \leq c(\|\gamma\|_{L_\infty} + \mathrm{Var}_q(\gamma)).$$

We give one more sufficient condition for a function to belong to the space $M(W_p^m \to W_p^l)$ for fractional m and l in the case $mp=n$. We introduce the semi-norm

$$\langle\gamma\rangle = \sup_{y \in R^n} \sup_{h \in B_{1/2}} |h|^{-\{l\}} \log(1/|h|) \, \|\Delta_h \nabla_{[l]}\gamma; B_1(y)\|_{L_p}.$$

Theorem 2. *Let* $\{m\}, \{l\} > 0$, $p > 1$.

(i) *If* $lp = n$, $\gamma \in L_\infty$ *and* $\langle\gamma\rangle < \infty$, *then* $\gamma \in MW_p^l$ *and*

$$\|\gamma\|_{MW_p^l} \leq c(\langle\gamma\rangle + \|\gamma\|_{L_\infty}). \tag{6}$$

(ii) *If* $mp = n$, $\gamma \in L_{p,\mathrm{unif}}$ *and* $\langle\gamma\rangle < \infty$, *then* $\gamma \in M(W_p^m \to W_p^l)$ *for* $l < m$ *and*

$$\|\gamma\|_{M(W_p^m \to W_p^l)} \leq c(\langle\gamma\rangle + \|\gamma\|_{L_{p,\mathrm{unif}}}). \tag{7}$$

Proof. By $Q(\gamma)$ we denote either of the right-hand sides of (6) and (7). We express the integral over R^n in (3) as the sum $i_1 + i_2$ of two integrals, the first being taken over the exterior of the ball

$$K = \{h : |h| < c_0 |e|^{1/n} (\log 2^n/|e|)^{(p-1)/p\{l\}}\},$$

where $|e| = \mathrm{mes}_n\, e$ and c_0 is a small positive constant depending on n and p. By virtue of Corollary 2.3.2 with $q = p$, $\mu = n/p = l$,

$$i_1 \leq c \int_{R^n \setminus K} \frac{dh}{|h|^{n+p\{l\}}} \left(\int_e |\nabla_{[l]}\gamma(x+h)|^p\, dx + \int_e |\nabla_{[l]}\gamma(x)|^p\, dx \right)$$

$$\leq c\, |e|^{\{l\}p/n} \int_{R^n \setminus K} \frac{dh}{|h|^{n+p\{l\}}} Q(\gamma)^p \leq c(\log 2^n/|e|)^{1-p} Q(\gamma)^p. \tag{8}$$

Moreover,

$$i_2 \leq \langle\gamma\rangle^p \int_K \frac{dh}{|h|^n (\log 1/|h|)^p}.$$

The integral in the right-hand side does not exceed $c(\log 2^n/|e|)^{1-p}$. This, together with (3) and (8), yields $(\log 2^n/|e|)^{(p-1)/p} \|D_{p,l}\gamma; e\|_{L_p} \leq cQ(\gamma)$. In addition, by (5), $|e|^{-l/n} \|\gamma; e\|_{L_p} \leq cQ(\gamma)$. Reference to Proposition 3.2.9/2 completes the proof. □

Verbitskii proved a certain analogue of Theorem 1(i) for multipliers in the space of number sequences with finite norm $(\sum_{k=-\infty}^{\infty} |a_k|^p (|k|+1)^\alpha)^{1/p}$ (oral communication). His result implies that a bounded function γ on ∂B_1 whose Fourier coefficients satisfy

$$a_n = \begin{cases} O(1/n) & \text{for } 2l < 1, \\ O(1/n \log n) & \text{for } 2l = 1, \end{cases}$$

belongs to $MW_2^l(\partial B_1)$. A general assertion of such a kind for integral operators in L_p in measure $(|x|+1)^\alpha dx$ with $-n < \alpha < n(p-1)$ can be found in Stein [1] § 6, Ch II. One can also derive similar results from Theorems 1 and 2. In fact, the next known lemma is valid.

Lemma 1 (see, for example, Triebel [4]). *The relation*

$$\|u\|_{B^s_{2,\infty}} \sim \sup_{R>1} R^s \|Fu; B_{2R}\backslash B_R\|_{L_2} + \|u\|_{L_2} \tag{9}$$

holds.

Proof. The lower bound for the norm in $B^s_{2,\infty}$ is obtained as follows:

$$\begin{aligned}\sup_h |h|^{-2\{s\}} \|\Delta_h \nabla_{[s]} u\|^2_{L_2} &= c \sup_{\rho>0} \rho^{-2\{s\}} \int_{\partial B_1} \|\Delta_{\rho\theta} \nabla_{[s]} u\|^2_{L_2} \, d\sigma_\theta \\ &= c \sup_{\rho>0} \rho^{-2\{s\}} \iint\limits_{\partial B_1} |\xi|^{2[s]} |Fu(\xi)|^2 \sin^2 \frac{\rho(\theta, \xi)}{2} \, d\sigma_\theta \, d\xi \\ &\geqslant c \sup_{\rho>0} \rho^{2(1-\{s\})} \int_{\rho^{-1}>|\xi|>(2\rho)^{-1}} |\xi|^{2[s]+1} |Fu(\xi)|^2 \, d\xi \\ &\geqslant c \sup_{R>0} R^{2s} \int_{B_{2R}\backslash B_R} |Fu(\xi)|^2 \, d\xi.\end{aligned}$$

On the other hand,

$$\begin{aligned}\sup_h |h|^{-2\{s\}} \|\Delta_h \nabla_{[s]} u\|^2_{L_2} \leqslant c \sup_h |h|^{-2\{s\}} &\int_{|\xi|\,|h|>1} |\xi|^{2[s]} |Fu(\xi)|^2 \, d\xi \\ + c \sup_h |h|^{2(1-\{l\})} &\int_{|\xi|\,|h|<1} |\xi|^{2[s]+2} |Fu(\xi)|^2 \, d\xi.\end{aligned} \tag{10}$$

The first summand in the right-hand side does not exceed

$$\begin{aligned}&c \sup_h |h|^{-2s} \sum_{j=1}^{\infty} 4^{j[s]} \int_{2^j>|\xi|\,|h|>2^{j-1}} |Fu(\xi)|^2 \, d\xi \\ &\leqslant c \sum_{j=1}^{\infty} 4^{-j\{s\}} \sup_R R^{2s} \int_{B_{2R}\backslash B_R} |Fu(\xi)|^2 \, d\xi.\end{aligned}$$

The second summand in the right-hand side of (10) is majorized by

$$\begin{aligned}&c \sup_h |h|^{-2s} \sum_{j=0}^{\infty} 4^{-j([s]+1)} \int_{2^{-j-1}<|\xi|\,|h|<2^{-j}} |Fu(\xi)|^2 \, d\xi \\ &\leqslant c \sum_{j=0}^{\infty} 4^{-j(1-\{s\})} \sup_R R^s \int_{B_{2R}\backslash B_R} |Fu(\xi)|^2 \, d\xi. \quad \square\end{aligned}$$

In the same way one may prove:

Lemma 2. *The norms*

$$\sup_{y \in R^n} \sup_{h \in B_{1/2}} |h|^{-\{l\}} \log(1/|h|) \, \|\Delta_h \nabla_{[l]} u; B_1(y)\|_{L_2} + \|u\|_{L_2},$$

$$\sup_{R>2} R^l \log R \, \|Fu; B_{2R} \backslash B_R\|_{L_2} + \|u\|_{L_2}$$

are equivalent.

Theorems 1, 2 and Lemmas 1, 2 immediately imply

Corollary 2. (i) *If* $1<p\leq 2$, $n/2>m$, m, l *are fractional*, $m>l$ *and* $(F\gamma)(\xi) = O((1+|\xi|)^{m-l-n})$, *then* $\gamma \in M(W_p^m \to W_p^l)$.

(ii) *If* $1<p\leq 2$, $n/2>l$, l *is fractional*, $\gamma \in L_\infty$ *and* $(F\gamma)(\xi) = O((1+|\xi|)^{-n})$, *then* $\gamma \in MW_p^l$.

(iii) *If* n *is odd*, $2l=n$, $\gamma \in L_\infty$ *and* $(F\gamma)(\xi) = O(|\xi|^{-n} (\log|\xi|)^{-1})$ *for* $|\xi| \geq 2$, *then* $\gamma \in MW_2^l$.

3.3.2 Conditions in Terms of the Space $B_{q,p}^l$

The condition $\gamma \in B^\mu_{q,\infty,\mathrm{unif}}$ in Theorem 3.3.1/1 requires the 'number of derivatives' μ to exceed l. In this subsection we obtain sufficient conditions for a function to belong to the class $M(W_p^m \to W_p^l)$ in terms of the space $B^l_{q,p,\mathrm{unif}}$. We recall that the diminishing of exponent θ leads to narrowing of $B^\mu_{q,\theta}$ and the diminishing of μ leads to expansion of this space. So new sufficient conditions are not comparable with the conditions of Theorem 3.3.1/1.

Theorem. *Let* $\{m\}, \{l\} > 0$, $p>1$.

(i) *Let* $q \in [n/l, \infty]$ *for* $pl<n$ *and* $q \in (p, \infty]$ *for* $lp=n$. *If* $\gamma \in B^l_{q,p,\mathrm{unif}} \cap L_\infty$, *then* $\gamma \in MW_p^l$ *and*

$$\|\gamma\|_{MW_p^l} \leq c\left(\sup_{x\in R^n}\left(\int_{B_1} \|\Delta_h \nabla_{[l]}\gamma; B_1(x)\|^p_{L_q} |h|^{-n-p\{l\}} \, dh\right)^{1/p} + \|\gamma\|_{L_\infty}\right). \qquad (1)$$

(ii) *Let* $m>l$, $q \in [n/m, \infty]$ *for* $mp<n$ *or* $q \in (p, \infty)$ *for* $mp=n$. *If* $\gamma \in B^l_{q,p,\mathrm{unif}}$, *then* $\gamma \in M(W_p^m \to W_p^l)$ *and*

$$\|\gamma\|_{M(W_p^m \to W_p^l)} \leq c \sup_{x \in R^n}\left(\left(\int_{B_1} \|\Delta_h \nabla_{[l]}\gamma; B_1(x)\|^p_{L_q} |h|^{-n-p\{l\}} \, dh\right)^{1/p} + \|\gamma; B_1(x)\|_{L_p}\right). \qquad (2)$$

Proof. Lemma 3.1.1/9 implies

$$\|\gamma\|_{M(W_p^m \to W_p^l)} \leq c \sup_{x \in R^n} \|\eta_x \gamma\|_{M(W_p^m \to W_p^l)},$$

where $\eta \in D(B_1)$, $\eta = 1$ on $B_{1/2}$ and $\eta_x(y) = \eta(x-y)$. Therefore it suffices to obtain (1), (2) under the assumption that the diameter of supp γ does not exceed 1.

Let $e \subset R^n$, $d(e) \leq 1$. We have

$$\|D_{p,l}\gamma; e\|_{L_p}^p \leq \int \|\Delta_h \nabla_{[l]}\gamma; e\|_{L_p}^p \frac{dh}{|h|^{n+p\{l\}}}$$

$$\leq (\text{mes}_n\, e)^{1-p/q} \int \|\Delta_h \nabla_{[l]}\gamma; e\|_{L_q}^p \frac{dh}{|h|^{n+p\{l\}}}$$

$$\leq (\text{mes}_n\, e)^{1-p/q} \sup_{x \in R^n} \int_{B_1} \|\Delta_h \nabla_{[l]}\gamma; B_1(x)\|_{L_q}^p \frac{dh}{|h|^{n+p\{l\}}}.$$

In the case $m > l$ we must estimate the norm of γ in $L_p(e)$. By Hölder's inequality,

$$\|\gamma; e\|_{L_p} \leq (\text{mes}_n\, e)^{1/p-(m-l)/n} \|\gamma; e\|_{L_{n/(m-l)}}.$$

Using the continuity of the imbedding operator: $B_{n/m,p}^l \to L_{n/(m-l)}$ (see Besov, Il'in, Nikol'skii [1], Theorem 18.10), we get

$$\|\gamma; e\|_{L_p} \leq c(\text{mes}_n\, e)^{1/p-(m-l)/n} \|\gamma\|_{B_{n/m,p,\text{unif}}^l}.$$

Reference to Proposition 3.2.9/2 completes the proof. □

Putting $q = \infty$ in (1), we obtain a simple condition for a function γ to belong to the class MW_p^l (and hence to $M(W_p^m \to W_p^l)$) in terms of the modulus of continuity ω of the vector-function $\nabla_{[l]}\gamma$:

$$\int_0 \left[\frac{\omega(t)}{t^{\{l\}+1/p}}\right]^p dt < \infty. \tag{3}$$

The last theorem contains the condition $lp \leq n$. Nevertheless, (3) ensures $\gamma \in MW_p^l$ for $lp > n$ since in that case $MW_p^l = W_{p,\text{unif}}^l \supset B_{\infty,p}^l$.

We show that even a rough condition (3) is the best possible in some sense.

Example. Let ω be a continuous increasing function on [0, 1] satisfying the inequalities

$$\delta \int_\delta^1 \frac{\omega(t)}{t^2} dt + \int_0^\delta \frac{\omega(t)}{t} dt \leq c\omega(\delta), \qquad 1 > \omega(\delta) \geq c\delta. \tag{4}$$

Further, let $\int_0^1 [\omega(t)/t^{\{l\}+1/p}]^p\, dt = \infty$. We construct a function γ on R^n such that:

1. the modulus of continuity of the vector-function $\nabla_{[l]}\gamma$ does not exceed $c\omega$, where $c = \text{const}$;

2. $\gamma \notin W^l_{p,\text{unif}}$ and hence $\gamma \notin M(W_p^m \to W_p^l)$.

We put

$$\gamma(x) = \prod_{i=1}^{n} \eta(x_i) \sum_{k=1}^{\infty} \mathrm{e}^{-[l]k}\omega(\mathrm{e}^{-k}) \sin(\mathrm{e}^k x_1), \tag{5}$$

where $\eta \in D(-2\pi, 2\pi)$, $\eta = 1$ on $(-\pi, \pi)$, $0 < \eta \leq 1$.

For small enough $|h|$ we have

$$|\nabla_{[l]}\gamma(x+h) - \nabla_{[l]}\gamma(x)| \leq c\Big(|h| \sum_{k=1}^{\infty} \omega(\mathrm{e}^{-k}) + |h| \sum_{k \leq \log|h|^{-1}} \omega(\mathrm{e}^{-k})\mathrm{e}^k + \sum_{k > \log|h|^{-1}} \omega(\mathrm{e}^{-k})\Big)$$

which, together with (4), gives

$$|\nabla_{[l]}\gamma(x+h) - \nabla_{[l]}\gamma(x)| \leq c\omega(|h|). \tag{6}$$

Further,

$$\|\gamma\|^p_{W_p^l} \geq c \int_{R^n} \int_{R^1} \left|\Delta_{te_1} \frac{\partial^{[l]}\gamma}{\partial x_1^{[l]}}\right|^p \frac{\mathrm{d}t}{t^{1+p\{l\}}}. \tag{7}$$

We set

$$f(x_1) = \sum_{k=1}^{\infty} \mathrm{e}^{-[l]k}\omega(\mathrm{e}^{-k})\mathrm{e}^{\mathrm{i}\mathrm{e}^k x_1}. \tag{8}$$

By virtue of (7) we have

$$\|f\|^p_{W_p^l} \geq c\, \|\mathrm{Im}\, f^{[l]}; (-\pi, \pi)\|^p_{W_p^{\{l\}}} \geq c\, \|f^{[l]}; (-\pi, \pi)\|^p_{W_p^{\{l\}}}.$$

It is clear that

$$\Delta_t f^{[l]}(x_1) = \sum_{k=1}^{\infty} \mathrm{e}^{-[l]k}\omega(\mathrm{e}^{-k})(\mathrm{e}^{\mathrm{i}\mathrm{e}^k t} - 1)\mathrm{e}^{\mathrm{i}\mathrm{e}^k x_1}.$$

According to a known property of lacunary trigonometric series (see Zygmund [1], v. 1, Theor. (8.20)),

$$\|\Delta_t f^{[l]}; (-\pi, \pi)\|^2_{L_p} \sim \sum_{k=1}^{\infty} \mathrm{e}^{-2[l]k}(\omega(\mathrm{e}^{-k}) \sin(\mathrm{e}^k t/2))^2.$$

Therefore

$$\|\gamma\|^p_{W_p^l} \geq c \int_0^{\pi} [\omega(\mathrm{e}^{-k(t)}) \sin(\mathrm{e}^{k(t)} t/2)]^p \frac{\mathrm{d}t}{t^{1+p\{l\}}},$$

where $k(t) = [\log 2t^{-1}]$. Finally,

$$\|\gamma\|_{W_p^l}^p \geq c \int_0^1 \left[\frac{\omega(t)}{t^{\{l\}+1/p}}\right]^p dt = \infty.$$

3.3.3 Conditions in Terms of the Space $H^l_{n/m}$

Here we obtain the upper bound for the norm in $M(W_p^m \to W_p^l)$, $p \geq 2$, $mp < n$, by the norm in the Bessel potential space $H^l_{n/m,\text{unif}}$.

Theorem. *The following estimates are valid:*

$$\|\gamma\|_{M(W_p^m \to W_p^l)} \leq c\, \|\gamma\|_{H^l_{n/m,\text{unif}}}, \tag{1}$$

where $m > l$, $\{m\}$, $\{l\} > 0$, $mp < n$ and $p \geq 2$;

$$\|\gamma\|_{MW_p^l} \leq c(\|\gamma\|_{H^l_{n/l,\text{unif}}} + \|\gamma\|_{L_\infty}), \tag{2}$$

where $\{l\} > 0$, $lp < n$ and $p \geq 2$.

Proof. It suffices to derive (1) and (2) under the assumption that the diameter of supp γ does not exceed 1 (cf. the beginning of the proof of Theorem 3.3.2).

By virtue of Lemma 3.1.1/8, $\|D_{p,l}\gamma\|_{L_{n/m}} \leq c\, \|\gamma\|_{H^l_{n/m}}$, where $p \geq 2$, $n > p(m - \{l\})$. Moreover, by Theorem 2.1.1/4, $\|\gamma\|_{L_{n/(m-l)}} \leq c\, \|\gamma\|_{H^l_{n/m}}$, where $m > l$. It remains to use the inequality $\|\gamma\|_{M(W_p^m \to W_p^l)} \leq c \sup_{x \in R^n} (\|D_{p,l}\gamma; B_1(x)\|_{L_{n/m}} + \|\gamma; B_1(x)\|_{L_{n/(m-l)}})$ which follows from (3.2.9/1). □

Possibly it is of interest to compare the last theorem with Proposition 2.3.1/1, according to which the right-hand sides of (1) and (2) majorize the norms of MH_p^l and $M(H_p^m \to H_p^l)$ for any $p \in (1, \infty)$.

We shall show that the condition $p \geq 2$ in the theorem of this subsection cannot be omitted.

Example. We consider the function γ defined by (3.3.2/5). Since, for $q \in (1, \infty)$,

$$\|\gamma\|_{H_q^l} \sim \sum_{j=1}^n \left\| \left(1 - \frac{\partial^2}{\partial x_j^2}\right)^{l/2} \gamma \right\|_{L_q}$$

(see the proof of Proposition 2.4/4), then

$$\|\gamma\|_{H_q^l} \sim \|\eta \operatorname{Im} f; R^1\|_{H_q^l},$$

where η and f are functions defined in Example 3.3.2.

It is known that $\eta \operatorname{Im} f \in H_q^l(R^1)$ if and only if the function $\operatorname{Im} f$ belongs to the space H_q^l on the unit circumference C and

$$\|\eta \operatorname{Im} f; R^1\|_{H_q^l} \sim \|\operatorname{Im} f; C\|_{H_q^l} = \left(\int_{-\pi}^{\pi} \left|\left(1-\frac{d^2}{d\theta^2}\right)^{l/2} \operatorname{Im} f(\theta)\right|^q d\theta\right)^{1/q}.$$

(We omit a standard but rather tedious proof of this fact.) Therefore

$$\|\gamma\|_{H_q^l} \sim \left(\sum_{u=1}^{\infty} e^{-2[l]k}(1+e^{2k})^{l/2}[\omega(e^{-k})]^2\right)^{1/2}$$

and consequently $\gamma \in H_q^l$ if and only if

$$\int_0^1 \left(\frac{\omega(t)}{t^{\{l\}}}\right)^2 \frac{dt}{t} < \infty. \tag{3}$$

From Proposition 2.3.1/1 and from the imbedding $M(H_p^m \to H_p^l) \subset H_{p,\text{unif}}^l$ we obtain that (3) is equivalent to $\gamma \in M(H_p^m \to H_p^l)$.

It was shown in Example 3.3.2 that $\gamma \in M(W_p^m \to W_p^l)$ if and only if

$$\int_0^1 \left(\frac{\omega(t)}{t^{\{l\}}}\right)^p \frac{dt}{t} < \infty. \tag{4}$$

If the theorem of this subsection were true for $p<2$, we would have implication (3) $\Rightarrow$ (4), which is obviously wrong.

What is more, we see that for $p<2$ one cannot give sufficient conditions for inclusion into $M(W_p^m \to W_p^l)$, $\{l\}>0$, in terms of H_q^l.

3.4 The Space $M(W_1^m \to W_1^l)$ for $\{m\}>0$ or $\{l\}>0$

3.4.1 An Imbedding Theorem

The basic result of this subsection is the following assertion, similar to Theorem 1.2.2/1.

Theorem 1. (Maz'ya [10]). *The best constant K in*

$$\int |u| \, d\mu \le K \, \|C_{1,l} u\|_{L_1} \tag{1}$$

is equivalent to

$$Q = \sup_{x \in R^n, \rho>0} \rho^{l-n} \mu(B_\rho(x)).$$

We shall use the following notation: $z=(x, y)$ and $\zeta=(\xi, \eta)$ are points of R^{n+s}, where $x, \xi \in R^n$, $y, \eta \in R^s$. Further, let $B_r^{(d)}(q)$ be a d-dimensional ball with centre $q \in R^d$.

To prove Theorem 1 we need some auxiliary assertions.

Lemma 1. *Let g be an open subset of R^{n+s} with compact closure and smooth boundary ∂g such that*

$$\int_{B_r^{(n+s)}(z)\cap g} |\eta|^\alpha \, d\zeta \Big/ \int_{B_r^{(n+s)}(z)} |\eta|^\alpha \, d\zeta = 1/2, \tag{2}$$

where $\alpha > -s$. Then

$$\int_{B_r^{(n+s)}(z)\cap \partial g} |\eta|^\alpha \, d\sigma(\zeta) \geqslant c r^{n+s-1}(r+|y|)^\alpha, \tag{3}$$

where σ is the $(n+s-1)$-dimensional area.

The proof is based on the following assertion.

Lemma 2. *Let $\alpha > -s$ for $s>1$ and $0 \geqslant \alpha > -1$ for $s=1$. Then for any $v \in C^\infty(\overline{B_r^{(n+s)}})$ there exists a constant V such that*

$$\int_{B_r^{(n+s)}} |v(\zeta)-V|\,|\eta|^\alpha \, d\zeta \leqslant cr \int_{B_r^{(n+s)}} |\eta|^\alpha |\nabla v(\zeta)| \, d\zeta. \tag{4}$$

Proof. It suffices to derive (4) for $r=1$. We put $B_1^{(n+s)} = B$ and $B_1^{(s)} \times B_1^{(n)} = Q$. By $R(\zeta)$ we denote the distance of the point $\zeta \in \partial Q$ from the origin, i.e. $R(\zeta) = (1+|\zeta|^2)^{1/2}$ for $|\eta|=1$, $|\xi|<1$, and $R(\zeta) = (1+|\eta|^2)^{1/2}$ for $|\xi|=1$, $|\eta|<1$. Taking into account that B is the bi-Lipschitz image of Q under the mapping $\zeta \to \zeta/R(\zeta)$, we can deduce (4) from the following inequality:

$$\int_Q |v(\zeta)-V|\,|\eta|^\alpha \, d\zeta \leqslant c \int_Q |\nabla v(\zeta)|\,|\eta|^\alpha \, d\zeta. \tag{5}$$

Let us show that this is really valid. Since $(s+\alpha)|\eta|^\alpha = \operatorname{div}(|\eta|^\alpha \eta)$, then integrating by parts in the left-hand side of (5) we find that it does not exceed

$$(s+\alpha)^{-1}\Big(\int_Q |\nabla v|\,|\eta|^{\alpha+1} \, d\zeta + \int_{B_1^{(n)}} d\xi \int_{\partial B_1^{(s)}} |v(\zeta)-V| \, ds(\eta)\Big). \tag{6}$$

We put $T = B_1^{(n)} \times (B_1^{(s)} \setminus B_{1/2}^{(s)})$. Let $s>1$. The second summand in (6) is not more than $c\int_T |\nabla v| \, d\zeta + c\int_T |v-V| \, d\zeta$. Hence, taking the mean value of v in T as V, we get (5) from (6). If $s=1$ then the set T has two components: $T_+ = B_1^{(n)} \times (1/2, 1)$ and $T_- = B_1^{(n)} \times (-1, -1/2)$. The same arguments as for the case $s>1$ lead to

$$\int_{B_1^{(n)}} |v(\xi, \pm 1) - V_\pm| \, d\xi \leqslant c \int_{T_\pm} |\nabla v(\zeta)| \, d\zeta \leqslant c \int_Q |\nabla v(\zeta)|\,|\eta|^\alpha \, d\zeta,$$

where $V_\pm$ is the mean value of v in $T_\pm$. It remains to be noted that

$$|V_+ - V_-| \leq c \int_{B_1^{(n)}} d\xi \int_{-1}^{1} \left|\frac{\partial v}{\partial \eta}\right| d\eta \leq c \int_Q |\nabla v(\zeta)|\, |\eta|^\alpha \, d\zeta$$

for $\alpha \leq 0$. Thus, for $s = 1$, inequality (5) follows with V_+ or V_- in place of V. □

Proof of Lemma 1. Let, for the sake of brevity, $B = B_r^{(n+s)}(z)$. We substitute for v in (4) the mollification χ_ρ of the characteristic function of g. Then the left-hand side is bounded from below by the sum $|1-V| \int_{e_1} |\eta|^\alpha \, d\zeta + |V| \int_{e_0} |\eta|^\alpha \, d\zeta$, where $e_i = \{z \in B : \chi_\rho(z) = i\}$, $i = 0, 1$.

Let ε be an arbitrarily small positive number. By (2), for sufficiently small ρ,

$$(\tfrac{1}{2} - \varepsilon)(|1-V| + |V|) \int_B |\eta|^\alpha \, d\zeta \leq cr \int_B |\eta|^\alpha \, |\nabla \chi_\rho(\zeta)| \, d\zeta.$$

Consequently,

$$\tfrac{1}{2} \int_B |y|^\alpha \, d\zeta \leq cr \varlimsup_{\rho \to +0} \int_B |\eta|^\alpha \, |\nabla_\zeta \chi_\rho(\zeta)| \, d\zeta = cr \int_{B \cap \partial g} |\eta|^\alpha \, d\sigma(\zeta).$$

(The latter equality can be derived from Corollary 1.2.1/1.) It remains to note that

$$\int_B |y|^\alpha \, d\zeta \geq c r^{n+s} (r + |y|)^\alpha. \quad \square$$

Lemma 3. Let ν be a measure in R^{n+s} and $\alpha > -s$. The best constant K_1 in

$$\int_{R^{n+s}} |U| \, d\nu \leq K_1 \int_{R^{n+s}} |y|^\alpha \, |\nabla_z U| \, dz, \qquad U \in D(R^{n+s}), \tag{7}$$

is equivalent to

$$Q_1 = \sup_{z;\rho > 0} (\rho + |y|)^{-\alpha} \rho^{1-n+s} \nu(B_\rho^{(n+s)}(z)). \tag{8}$$

Proof. 1. First let $m > 1$ or $0 \geq \alpha > -1$, $m = 1$. According to Proposition 1.2.1,

$$K_1 = \sup_g \frac{\nu(g)}{\int_{\partial g} |y|^\alpha \, d\sigma},$$

where g is an arbitrary open subset of R^{n+s} with compact closure and smooth boundary. We show that for any g there exists a covering by a sequence of balls $B_{\rho_i}^{(n+s)}(z_i)$, $i=1, 2, \ldots$, such that

$$\sum_i \rho_i^{n+s-1}(\rho_i+|y_i|)^\alpha \leq c\int_{\partial g} |y|^\alpha \, d\sigma. \tag{9}$$

Every point $z\in g$ is the centre of a ball $B_r^{(n+s)}(z)$ for which (2) holds. In fact, the ratio in the left-hand side of (2) is a continuous function in r, equal to unity for small values of r and tending to zero as $r\to\infty$. By Lemma 1.2.1/3 there exists a sequence of disjoint balls $B_{r_i}^{(n+s)}(z_i)$ such that $g\subset \bigcup_{i=1}^\infty B_{3r_i}^{(n+s)}(z_i)$. Lemma 1 implies

$$\int_{B_{r_i}^{(n+s)}(z_i)\cap \partial g} |y|^\alpha \, d\sigma \geq c r_i^{n+s-1}(r_i+|y_i|)^\alpha.$$

Consequently $\{B_{3r_i}^{(n+s)}(z_i)\}_{i\geq 1}$ is the required covering. Obviously,

$$\nu(g)\leq \sum_i \nu(B_{3r_i}^{(n+s)}(z_i)) \leq Q_1 \sum_i r_i^{n+s-1}(r_i+|y_i|^\alpha)$$
$$\leq cQ_1 \int_{\partial g} |y|^\alpha \, d\sigma.$$

Thus $K_1\leq cQ_1$.

2. Let $s=1$ and $\alpha>0$. We construct a covering of the set $\{\zeta: \eta\neq 0\}$ by balls $B^{(j)}$ such that the radius ρ_j of $B^{(j)}$ is equal to the distance from $B^{(j)}$ to the hyperplane $\{\zeta:\eta=0\}$. By $\{\varphi_j\}$ we denote a partition of unity subjected to the covering $\{B^{(j)}\}$ with $|\nabla\varphi_j|\leq c/\rho_j$ (see Stein [1], Ch. VI, §1). Using the present assertion for $\alpha=0$ we get

$$\int_{R^{n+1}} |\varphi_j u| \, d\nu \leq c \sup_{\rho;z} \rho^{-n}\nu_j(B_\rho^{(n+1)}(z)) \, \|\nabla(\varphi_j u); R^{n+1}\|_{L_1}$$

where ν_j is the restriction of the measure ν to $B^{(j)}$. It is clear that

$$\sup_{\rho;z} \rho^{-n}\nu_j(B_\rho^{(n+1)}(z)) \leq c \sup_{\rho\leq\rho_j; z\in B^{(j)}} \rho^{-n}\nu(B_\rho^{(n+1)}(z)).$$

Therefore

$$\int_{R^{n+1}} |\varphi_j u| \, d\nu$$
$$\leq c \sup_{\rho\leq\rho_j; z\in B^{(j)}} (\rho+\rho_j)^{-\alpha}\rho^{-n}\nu(B_\rho^{(n+1)}(z)) \int_{R^{n+1}} |\nabla(\varphi_j u)| \, |\eta|^\alpha \, d\zeta.$$

Summing over j, we find

$$\int_{R^{n+1}} |u| \, d\nu \leq cK_1\left(\int_{R^{n+1}} |\nabla u| \, |\eta|^\alpha \, d\zeta + \int_{R^{n+1}} |u| \, |\eta|^\alpha \, d\zeta\right).$$

Since

$$\int_{R^{n+1}} |u|\,|\eta|^{\alpha-1}\,d\zeta \leq \alpha^{-1}\int_{R^{n+1}} |\nabla u|\,|\eta|^{\alpha}\,d\zeta$$

for $\alpha > 0$, we also have $K_1 \leq cQ_1$ for $s=1$, $\alpha > 0$.

3. To obtain the converse estimate we put $U(\xi)=\varphi(\rho^{-1}(\zeta - z))$, where $\varphi \in D(B_2^{(n+s)}(0))$, $\varphi = 1$ on $B_1^{(n+s)}(0)$, into (7). We notice that

$$\int_{B_{2\rho}^{(n+s)}(z)} |\eta|^{\alpha}\,|\nabla_{\zeta} U|\,d\zeta \leq c\rho^{-1}\int_{B_{2\rho}^{(n+s)}(z)} |\eta|^{\alpha}\,d\zeta \leq c\rho^{n+s-1}(\rho+|y|)^{\alpha}. \quad \square$$

Corollary. *Let ν be a measure in R^n and let $\alpha > -s$. The best constant in (7) is equivalent to*

$$\sup_{x\in R^n, \rho>0} \rho^{1-n-s-\alpha}\nu(B_\rho^{(n+s)}(z)).$$

To prove this assertion it suffices to note that if supp $\nu \subset R^n$, then the value Q_1 defined in (8) is equivalent to the last supremum.

Proof of Theorem 1. The estimate $K \geq cQ$ can be obtained quite simply. It suffices to put $u(\xi)=\varphi(\rho^{-1}(x-\xi))$, where $\varphi \in D(B_2^{(n)}(0))$, $\varphi = 1$ on $B_1^{(n)}(0)$, into (1) and to note that

$$\int |u|\,d\mu \geq \mu(B_\rho^{(n)}(x)), \qquad \|C_{1,l}u\|_{L_1} = c\rho^{n-l}.$$

Now we obtain the estimate $K \leq cQ$. Let $l \in (0, 1)$. According to the corollary,

$$\int |u|\,d\mu \leq cQ\int_{R^{n+1}} |y|^{-l}\,|\nabla U|\,dz, \tag{10}$$

where $U \in D(R^{n+1})$ is an arbitrary extension of $u \in D(R^n)$ to R^{n+1}. If $l=1$, then by Theorem 1.2.2/1

$$\int |u|\,d\mu \leq cQ\int_{R^{n+1}} |\nabla_2 U|\,dz. \tag{11}$$

It is known (see Uspenskii [1]) that

$$\|C_{1,l}u\|_{L_1} \sim \inf_U \int_{R^{n+1}} |y|^{-l}\,|\nabla U|\,dz, \qquad l \in (0, 1),$$

$$\|C_{1,1}u\|_{L_1} \sim \inf_U \int_{R^{n+1}} |\nabla_2 U|\,dz.$$

So, by minimizing the right-hand sides of (10) and (11) over all extensions U, we arrive at

$$\int |u| \, d\mu \leq cQ \, \|C_{1,l}u\|_{L_1},$$

where $l \in (0, 1]$.

Suppose the estimate $K \leq cQ$ is proved under the condition $l \in (m-2, m-1]$, where m is integer, $m \geq 2$. Duplicating the argument used in part (ii) of Theorem 1.2.2/1, we obtain the required estimate for $l \in (m-1, m]$. □

Remark 1. From Theorem 1 it follows that for $l > n$ the inequality (1) is valid only in the trivial case $\mu = 0$, and for $l = n$ if and only if the measure μ is finite.

Theorem 2. *The best constant* K_0 *in* $\int |u| \, d\mu \leq K_0 \, \|u\|_{B_1^l}$ *is equivalent to*

$$Q_0 = \sup_{x \in R^n, \rho \in (0,1]} \rho^{l-n} \mu(B_\rho(x)).$$

Proof. The estimate $K_0 \geq cQ_0$ can be obtained in the same way as the estimate $K \geq cQ$ in Theorem 1. To prove the converse inequality we use the sequence $\{\eta_j\}_{j \geq 0}$ defined in Theorem 2.1.1/3. We apply Theorem 1 to the integral $\int |\eta_j u| \, d\mu_j$, where μ_j is the restriction of μ to the support of η_j. Then

$$\int |u| \, d\mu \leq c \sum_j \int |\eta_j u| \, d\mu_j \leq cQ_0 \sum_j \|C_{1,l}(\eta_j u)\|_{L_1}.$$

According to Lemma 3.1.1/9, the last sum does not exceed $c \, \|u\|_{B_1^l}$. □

Remark 2. It is clear that

$$Q_0 = \sup_{x \in R^n} \mu(B_1(x))$$

for $l \geq n$.

3.4.2 Description of the Space $M(W_1^m \to W_1^l)$ for $\{m\}>0$ or $\{l\}>0$

An obvious corollary of Theorem 3.4.1/2 and Remark 3.4.1/2 is

Lemma. *If* $m < n$, *then*

$$\|\gamma\|_{M(W_1^m \to L_1)} \sim \sup_{x \in R^n, r \in (0,1)} r^{m-n} \, \|\gamma; B_r(x)\|_{L_1}$$

and if $m \geqslant n$, *then*

$$\|\gamma\|_{M(W_1^m \to L_1)} \sim \sup_{x \in R^n} \|\gamma; B_1(x)\|_{L_1}.$$

We obtain necessary and sufficient conditions for a function to belong to the space $M(W_1^m \to W_1^l)$ for $\{l\} > 0$. The case of integer m and l was considered in 1.3.5.

In what follows we use the norm $|||\cdot; B_r|||_{W_1^l}$ introduced in 3.1.1: namely,

$$|||u; B_r|||_{W_1^l} = \sum_{j=0}^{l} r^{j-l} \|\nabla_j u; B_r\|_{L_1}$$

for integer l and

$$|||u; B_r|||_{W_1^l} = r^{-\{l\}} |||u; B_r|||_{W_1^{[l]}} + \sum_{j=0}^{[l]} r^{j-[l]} \int_{B_r} \int_{B_r} |\nabla_j u(x) - \nabla_j u(y)| \frac{dx\, dy}{|x-y|^{n+\{l\}}}$$

for fractional l.

According to Corollary 3.1.1/3, the following relations hold:

$$|||u; B_r|||_{W_1^l} \sim r^{-l} \|u; B_r\|_{L_1} + \|\nabla_l u; B_r\|_{L_1}$$

for integer l and

$$|||u; B_r|||_{W_1^l} \sim r^{-l} \|u; B_r\|_{L_1} + \int_{B_r} \int_{B_r} |\nabla_{[l]} u(x) - \nabla_{[l]} u(y)| \frac{dx\, dy}{|x-y|^{n+\{l\}}}$$

for fractional l.

The following assertion has been proved by the authors [11].

Theorem. *The function* γ *belongs to the space* $M(W_1^m \to W_1^l)$, $m \geqslant l \geqslant 0$, *if and only if* $\gamma \in W_{1,\mathrm{loc}}^l$ *and for any ball* $B_r(x)$, $0 < r < 1$,

$$|||\gamma; B_r(x)|||_{W_1^l} \leqslant \mathrm{const}\, r^{-m+n}.$$

Moreover,

$$\|\gamma\|_{M(W_1^m \to W_1^l)} \sim \sup_{x \in R^n, r \in (0,1)} r^{m-n} |||\gamma; B_r(x)|||_{W_1^l}. \tag{1}$$

For $m \geqslant n$ *the latter is equivalent to*

$$\|\gamma\|_{M(W_1^m \to W_1^l)} \sim \sup_{x \in R^n} |||\gamma; B_1(x)|||_{W_1^l} \sim \|\gamma\|_{W_{1,\mathrm{unif}}^l}. \tag{2}$$

Proof. By virtue of Lemma 3.1.1/15,

$$|||\varphi; B_r|||_{W_1^l} \leqslant c\, \|\varphi\|_{W_1^l}, \qquad l < n.$$

Let $u(y)=\eta((y-x)/r)$, where $r\in(0,1)$ for $m<n$ and $r=1$ for $m\geqslant n$, $\eta\in D(B_2(0))$, $\eta=1$ on $B_1(0)$. We put u into the inequality

$$\|\gamma u\|_{W_1^l}\leqslant\|\gamma\|_{M(W_1^m\to W_1^l)}\|u\|_{W_1^m}.$$

Then

$$\|\gamma;B_r(x)\|_{W_1^l}\leqslant c\,\|\gamma\|_{M(W_1^m\to W_1^l)}r^{n-m}$$

and the required lower bound for the norm $\|\gamma\|_{M(W_1^m\to W_1^l)}$ is obtained.

Now we turn to the proof of the upper bound. We have

$$\|D_{1,l}(\gamma u)\|_{L_1}\leqslant c_1\sum_{j=0}^{[l]}(\||\nabla_j u|\,D_{1,l-j}\gamma\|_{L_1}+\||\nabla_j\gamma|\,D_{1,l-j}u\|_{L_1}), \tag{3}$$

where $D_{1,k}\varphi=|\nabla_k\varphi|$ for integer k and

$$(D_{1,k}\varphi)(x)=\iint|\nabla_{[k]}\varphi(x)-\nabla_{[k]}\varphi(y)|\,\frac{dx\,dy}{|x-y|^{n+\{k\}}}$$

for fractional k. According to Theorems 1.2.2/2 and 3.4.1/2,

$$\||\nabla_j u|\,D_{1,l-j}\gamma\|_{L_1}\leqslant c\sup_{x\in R^n,r\in(0,1)}r^{m-j-n}\|D_{1,l-j}\gamma;B_r(x)\|_{L_1}\|\nabla_j u\|_{W_1^{m-j}}.$$

This, together with Lemma 3.1.1/11, implies

$$\||\nabla_j u|\,D_{1,l-j}\gamma\|_{L_1}\leqslant\sup_{x\in R^n,r\in(0,1)}r^{m-n}\|\gamma;B_r(x)\|_{W_1^l}\|u\|_{W_1^m} \tag{4}$$

for $m<n$ and

$$\||\nabla_j u|\,D_{1,l-j}\gamma\|_{L_1}\leqslant\sup_{x\in R^n}\|\gamma;B_1(x)\|_{W_1^l}\|u\|_{W_1^m} \tag{5}$$

for $m\geqslant n$.

If l is an integer, then

$$\||\nabla_j u|\,D_{1,l-j}u\|_{L_1}=\||\nabla_j\gamma|\,|\nabla_{l-j}u|\|_{L_1},$$

which, by virtue of Theorems 1.2.2/2 and 3.4.1/2, does not exceed

$$c\sup_{x\in R^n,r\in(0,1)}r^{m-l+j-n}\|\nabla_j\gamma;B_r(x)\|_{L_1}\|\nabla_{l-j}u\|_{W_1^{m-l+j}}.$$

Let l be a fractional number. We note that the proof of (3.2.5/3) is also valid for $p=1$. Consequently,

$$\||\nabla_j\gamma|\,D_{1,l-j}u\|_{L_1}\leqslant c\,\|\nabla_j\gamma\|_{M(W_1^{m-l+j}\to L_1)}\|u\|_{W_1^m}$$

which, together with the lemma, gives

$$\||\nabla_j\gamma|\,D_{1,l-j}u\|_{L_1}\leqslant c\sup_{x\in R^n,r\in(0,1)}r^{m-l+j-n}\|\nabla_j\gamma;B_r(x)\|_{L_1}\|u\|_{W_1^m}.$$

Therefore

$$\| |\nabla_j \gamma| \, D_{1,l} u \|_{L_1} \leq c \sup_{x \in R^n, r \in (0,1)} r^{m-n} \| \gamma; B_r(x) \|_{W_1^l} \|u\|_{W_1^m} \tag{6}$$

for $m < n$ and

$$\| |\nabla_j \gamma| \, D_{1,l} u \|_{L_1} \leq c \sup_{x \in R^n} \| \gamma; B_1(x) \|_{W_1^l} \|u\|_{W_1^m} \tag{7}$$

for $m \geq n$.

Substituting (4)–(7) into (3), we complete the proof. □

Corollary. *The inequality*

$$\|\gamma\|_{M(W_1^{m-j} \to W_1^{l-j})} \leq c \, \|\gamma\|_{M(W_1^m \to W_1^l)}^{1-j/l} \, \|\gamma\|_{M(W_1^{m-l} \to L_1)}^{j/l} \tag{8}$$

with $0 \leq j \leq l$ *holds.*

Proof. By Lemma 3.1.1/14,

$$\| \gamma; B_r(x) \|_{W_1^{l-j}} \leq c \, \| \gamma; B_r(x) \|_{W_1^l}^{1-j/l} \, \| \gamma; B_r(x) \|_{L_1}^{j/l}.$$

Thus

$$\sup_{x \in R^n, r \in (0,1)} r^{m-j-n} \| \gamma; B_r(x) \|_{W_1^{l-j}}$$

$$\leq c \Big(\sup_{x \in R^n, r \in (0,1)} r^{m-n} \| \gamma; B_r(x) \|_{W_1^l} \Big)^{1-j/l} \Big(\sup_{x \in R^n, r \in (0,1)} r^{m-l-n} \| \gamma; B_r(x) \|_{L_1} \Big)^{j/l}$$

and the result follows from the theorem. □

3.5 Functions of a Multiplier

According to a theorem by Hirschman [1], the composition $\varphi(\gamma)$ of $\varphi \in C^{0,\rho}$, $\rho \in (0, 1]$, and of a multiplier γ in the space W_2^l, $l \in (0, 1)$, represents a multiplier in W_2^r, where $r \in (0, l\rho)$ provided that $\rho < 1$ and $r = l$ provided that $\rho = 1$. The case $\rho = 1$ was considered by Beurling [1] earlier.

The purpose of this section is to present a generalization of Hirschman's result, obtained by the authors [10].

Theorem. *Let* $\gamma \in M(W_p^m \to W_p^l)$, $m \geq l$, $0 < l < 1$, $p > 1$. *Further, let* φ *be a function defined on* R^1 *if* $\operatorname{Im} \gamma = 0$ *or on* $\mathbb{C}^1$ *if* γ *is complex-valued. Suppose that* $\varphi(0) = 0$ *and for all* t *and* τ, $|\tau| < 1$, *the inequality* $|\varphi(t+\tau) - \varphi(t)| \leq A \, |\tau|^\rho$ *with* $\rho \in (0, 1]$ *is valid. Then* $\varphi(\gamma) \in M(W_p^{m-l+r} \to W_p^r)$, *where* $r \in (0, l\rho)$ *if* $\rho < 1$ *and* $r = l$ *if* $\rho = 1$. *The follow-*

ing estimate holds:

$$\|\varphi(\gamma)\|_{M(W_p^{m-l+r}\to W_p^r)}\leqslant cA(\|\gamma\|^{\rho}_{M(W_p^m\to W_p^l)}+\|\gamma\|_{M(W_p^m\to W_p^l)}).$$

Proof. First we note that for all t, τ

$$|\varphi(t+\tau)-\varphi(t)|\leqslant A(|\tau|^{\rho}+|\tau|). \tag{1}$$

Consider the case $\rho=1$. We have

$$\|\varphi(\gamma)u\|_{W_p^l}=\|D_{p,l}[\varphi(\gamma)u]\|_{L_p}+\|\varphi(\gamma)u\|_{L_p}$$
$$\leqslant\|uD_{p,l}\varphi(\gamma)\|_{L_p}+\|\varphi(\gamma)D_{p,l}u\|_{L_p}+\|\varphi(\gamma)u\|_{L_p}.$$

By (1) the latter sum does not exceed $2A(\|uD_{p,l}\gamma\|_{L_p}+\|\gamma D_{p,l}u\|_{L_p}+\|\gamma u\|_{L_p})$. It is clear that $\|\gamma D_{p,l}u\|_{L_p}\leqslant\|D_{p,l}(\gamma u)\|_{L_p}+\|uD_{p,l}\gamma\|_{L_p}$. So

$$\|\varphi(\gamma)u\|_{W_p^l}\leqslant 2A(2\,\|uD_{p,l}\gamma\|_{L_p}+\|\gamma u\|_{W_p^l}).$$

Applying Corollary 3.2.1, we get

$$\|\varphi(\gamma)u\|_{W_p^l}\leqslant 2A\left(c\sup_e\frac{\|D_{p,l}\gamma;e\|_{L_p}}{[\operatorname{cap}(e,W_p^m)]^{1/p}}+\|\gamma\|_{M(W_p^m\to W_p^l)}\right)\|u\|_{W_p^m}$$

which, together with Theorem 3.2.7/2, gives

$$\|\varphi(\gamma)\|_{M(W_p^m\to W_p^l)}\leqslant cA\,\|\gamma\|_{M(W_p^m\to W_p^l)}.$$

Now let $0<\rho<1$. Let us represent the integral

$$\int\int|\varphi(\gamma(x))u(x)-\varphi(\gamma(y))u(y)|^p\,|x-y|^{-n-pr}\,dy\,dx$$

as the sum of two integrals of which one is taken over the set $\mathfrak{M}=\{(x,y):|\gamma(y)|\leqslant|\gamma(x)|\}$. It is sufficient to estimate the integral over $\mathfrak{M}$ which obviously does not exceed

$$c\Big(\iint_{\mathfrak{M}}|u(x)|^p\,|\varphi(\gamma(x))-\varphi(\gamma(y))|^p\,|x-y|^{-n-pr}\,dy\,dx$$
$$+\iint_{\mathfrak{M}}|\varphi(\gamma(y))|^p\,|u(x)-u(y)|^p\,|x-y|^{-n-pr}\,dy\,dx. \tag{2}$$

We introduce two sets, $\mathfrak{M}_1(x)=\{y:|\gamma(y)|\leqslant|\gamma(x)|,\ |\gamma(x)-\gamma(y)|\leqslant 1\}$ and

$\mathfrak{M}_2(x)=\{y:|\gamma(y)|\leq|\gamma(x)|,\ |\gamma(x)-\gamma(y)|>1\}$. From (1) it follows that

$$\int_{\mathfrak{M}_1(x)} |\varphi(\gamma(x))-\varphi(\gamma(y))|^p\,|x-y|^{-n-pr}\,dy$$
$$\leq cA^p \int_{\mathfrak{M}_1(x)} |\gamma(x)-\gamma(y)|^{p\rho}\,|x-y|^{-n-pr}\,dy$$
$$\leq cA^p\Big(2^{p\rho}\,|\gamma(x)|^{p\rho}\int_{|x-y|\geq\delta_x} |x-y|^{-n-pr}\,dy$$
$$+\int_{|x-y|<\delta_x} |\gamma(x)-\gamma(y)|^{p\rho}\,|x-y|^{-n-pr}\,dy\Big), \tag{3}$$

where δ_x is a non-negative function which will be chosen henceforth. By estimating the last integral by means of the Hölder inequality, we obtain the majorant

$$\Big(\int_{|x-y|<\delta_x} |\gamma(x)-\gamma(y)|^p\,|x-y|^{-n-pl}\,dy\Big)^{\rho}$$
$$\times\Big(\int_{|x-y|<\delta_x} |x-y|^{-n-p(r-l\rho)/(1-\rho)}\,dy\Big)^{1-\rho}$$
$$\leq c(D_{p,l}\gamma(x))^{p\rho}\,\delta_x^{p(l\rho-r)}. \tag{4}$$

From (3) and (4) we have

$$\int_{\mathfrak{M}_1(x)} |\varphi(\gamma(x))-\varphi(\gamma(y))|^p\,|x-y|^{-n-pr}\,dy$$
$$\leq cA^p\Big(|\gamma(x)|^{p\rho}\,\delta_x^{-pr}+(D_{p,l}\gamma(x))^{p\rho}\,\delta_x^{p(\rho l-r)}\Big).$$

Minimizing the right-hand side over δ_x, we obtain

$$\int_{\mathfrak{M}_1(x)} |\varphi(\gamma(x))-\varphi(\gamma(y))|^p\,|x-y|^{-n-pr}\,dy$$
$$\leq cA^p\,|\gamma(x)|^{p(\rho-r/l)}(D_{p,l}\gamma(x))^{pr/l}.$$

Moreover, by (1),

$$\int_{\mathfrak{M}_2(x)} |\varphi(\gamma(x))-\varphi(\gamma(y))|^p\,|x-y|^{-n-pr}\,dy$$
$$\leq cA^p\int_{\mathfrak{M}_2(x)} |\gamma(x)-\gamma(y)|^p\,|x-y|^{-n-pr}\,dy\leq cA^p(D_{p,r}\gamma(x))^p.$$

Consequently, the first integral in (2) is not more than

$$\int |u(x)|^p\,|\gamma(x)|^{p(\rho-r/l)}(D_{p,l}\gamma(x))^{pr/l}\,dx+\int |u(x)|^p\,(D_{p,r}\gamma(x))^p\,dx. \tag{5}$$

Let $m \neq l$. We set $v = \Lambda^{-s} |\Lambda^s u|$, where $\Lambda = (-\Delta + 1)^{1/2}$, $s = m - l + r - \varepsilon$ and ε is a small positive number. The properties of the kernel G_r given in 1.1.1 and the Hölder inequality imply

$$v \leqslant c(\Lambda^r v)^{(l-r)/l} (\Lambda^{r-l} v)^{r/l}. \tag{6}$$

So the first summand in (5) does not exceed

$$c \int (\Lambda^{r-l} v D_{p,l}\gamma)^{pr/l} (\Lambda^r v)^{p(1-r/l)} |\gamma|^{p(\rho - r/l)}\, dx$$

$$\leqslant c\Big(\int (\Lambda^{r-l} v)^p (D_{p,l}\gamma(x))^p\, dx\Big)^{r/l} \Big(\int (\Lambda^r v)^p\, |\gamma|^{p(l\rho - r)/(l-r)}\, dx\Big)^{1-r/l}$$

$$\leqslant c_1 \Big(\sup_e \frac{\|D_{p,l}\gamma;\, e\|^p_{L_p}}{\operatorname{cap}(e, W^m_p)}\Big)^{r/l} \|\Lambda^{r-l} v\|^{pr/l}_{W^m_p}$$

$$\times \Big(\sup_e \frac{\||\gamma|^{(l\rho - r)/(l-r)};\, e\|^p_{L_p}}{\operatorname{cap}(e, W^{m-l}_p)}\Big)^{1-r/l} \|\Lambda^r v\|^{p(1-r/l)}_{W^{m-l}_p}.$$

Here we used the Hölder inequality and Corollary 3.2.1. Furthermore,

$$\|\Lambda^{r-l} v\|_{W^m_p} = \|\Lambda^{\varepsilon - m}\, |\Lambda^{m-l+r-\varepsilon} u|\, \|_{W^m_p} \leqslant c\, \||\Lambda^{m-l+r-\varepsilon} u|\|_{W^\varepsilon_p}$$
$$\leqslant c\, \|\Lambda^{m-l+r-\varepsilon} u\|_{W^\varepsilon_p} \leqslant c_1 \|u\|_{W^{m-l+r}_p}$$

(see Lemma 3.1.1/3) and similarly

$$\|\Lambda^r v\|_{W^{m-l}_p} = \|\Lambda^{\varepsilon - m + l}\, |\Lambda^{m-l+r-\varepsilon} u|\|_{W^{m-l}_p} \leqslant c\, \||\Lambda^{m-l+r-\varepsilon} u|\|_{W^\varepsilon_p}$$
$$\leqslant c_1 \|u\|_{W^{m-l+r}_p}.$$

Now the first term in (5) is estimated by

$$c\Big(\sup_e \frac{\|D_{p,l}\gamma;\, e\|^p_{L_p}}{\operatorname{cap}(e, W^m_p)}\Big)^{r/l} \Big(\sup_e \frac{\||\gamma|^{(l\rho - r)/(l-r)};\, e\|^p_{L_p}}{\operatorname{cap}(e, W^{m-l}_p)}\Big)^{1-r/l} \|u\|^p_{W^{m-l+r}_p}. \tag{7}$$

The same bound can be obtained for $m = l$, if we put $v = \Lambda^{-r} |\Lambda^r u|$ and apply $v \leqslant c(M\Lambda^r v)^{(l-r)/l} (\Lambda^{r-l} v)^{r/l}$ instead of (6) (see Lemma 1.1.3).

Taking into account that

$$\frac{\|\, |\gamma|^{(l\rho - r)/(l-r)};\, e\|^p_{L_p}}{\operatorname{cap}(e, W^{m-l}_p)} \leqslant \Big(\frac{\|\gamma;\, e\|^p_{L_p}}{\operatorname{cap}(e, W^{m-l}_p)}\Big)^{(l\rho - r)/(l-r)} \Big(\frac{\operatorname{mes}_n e}{\operatorname{cap}(e, W^{m-l}_p)}\Big)^{l(1-\rho)/(l-r)}$$
$$\leqslant \Big(\frac{\|\gamma;\, e\|^p_{L_p}}{\operatorname{cap}(e, W^{m-l}_p)}\Big)^{(l\rho - r)/(l-r)}$$

we obtain from Theorem 3.2.7/2 that (7), and consequently the first term in (5), does not exceed

$$c\, \|\gamma\|^{p\rho}_{M(W^m_p \to W^l_p)} \|u\|^p_{W^{m-l+r}_p}.$$

From Theorem 3.2.7/2, and by virtue of the imbedding of $M(W_p^m \to W_p^l)$ into $M(W_p^{m-l+r} \to W_p^r)$, the second summand in (5) is estimated in the following way:

$$\int |u(x)|^p \, (D_{p,r}\gamma(x))^p \, dx \leqslant c \sup_e \frac{\|D_{p,r}\gamma; e\|_{L_p}^p}{\text{cap}\,(e, W_p^{m-l+r})} \|u\|_{W_p^{m-l+r}}^p$$
$$\leqslant c_1 \|\gamma\|_{M(W_p^m \to W_p^l)}^p \|u\|_{W_p^{m-l+r}}^p. \quad \square$$

3.6 Estimates for Translation Invariant Operators in a Weighted L_2-space

3.6.1 *The Convolution Operator*

Let $K: u \to k * u$ be a convolution operator with the kernel k. The results of preceding sections for the case $p=2$ can be interpreted as the theorems on properties of K considered as a mapping from $L_2((1+|x|^2)^{m/2})$ into $L_2((1+|x|^2)^{l/2})$, $m \geqslant l \geqslant 0$. Here we have

$$\|u\|_{L_2((1+|x|^2)^{k/2})} = \left(\int |u|^2 \, (1+|x|^2)^k \, dx \right)^{1/2}.$$

For example, the operator K is continuous if and only if its symbol, i.e. the Fourier transform Fk, belongs to $M(W_2^m \to W_2^l)$. By virtue of Theorem 3.2.7/2 the latter is equivalent to the following properties: $Fk \in W_{2,\text{loc}}^l$ and, for every compactum $e \subset R^n$,

$$\int_e |Fk|^2 \, dx \leqslant \text{const cap}\,(e, W_2^{m-l}), \quad \int_e [D_{2,l}(Fk)]^2 \, dx \leqslant \text{const cap}\,(e, W_2^m),$$

where

$$(D_{2,l}u)(x) = \left(\int |\nabla_{[l],x}(u(x+h) - u(x))|^2 \, |h|^{-n-2\{l\}} \, dh \right)^{1/2}.$$

Moreover,

$$\|K\| \sim \sup_e \left(\frac{\|Fk; e\|_{L_2}}{[\text{cap}\,(e, W_2^{m-l})]^{1/2}} + \frac{\|D_{2,l}(Fk); e\|_{L_2}}{[\text{cap}\,(e, W_2^m)]^{1/2}} \right).$$

In the case $2m > n$,

$$\|K\| \sim \sup_{x \in R^n} (\|Fk; B_1(x)\|_{L_2} + \|D_{2,l}(Fk); B_1(x)\|_{L_2}) \sim \|Fk\|_{W_{2,\text{unif}}^l}.$$

Theorem 3.5 describes properties of a function of the operator $K: L_2((1+|x|^2)^{m/2}) \to L_2((1+|x|^2)^{l/2})$. Namely, let $0 < l < 1$ and φ be a complex-valued function of a complex argument, $\varphi(0) = 0$. By $\varphi(K)$ we denote the convolution operation with the symbol $\varphi(Fk)$. If φ satisfies the uniform Lipschitz condition, then $\varphi(K)$ is continuous in the same pair of spaces as K.

Replacing the Lipschitz condition by a weaker one, $|\varphi(t+\tau)-\varphi(t)| \leq A\,|\tau|^\rho$ where $|\tau|<1$, $\rho \in (0, 1)$, we derive the following assertion from Theorem 3.5. The operator

$$\varphi(K): L_2((1+|x|^2)^{(m-l+r)/2}) \to L_2((1+|x|^2)^{r/2})$$

with $r \in (0, l\rho)$ is continuous.

According to Corollary 2.5.2/2, a number λ belongs to the spectrum $\sigma(K)$ of an operator K continuous in $L_2((1+|x|^2)^{l/2})$ if and only if $(Fk-\lambda)^{-1} \notin L_\infty$.

Let $\lambda \in \sigma(K)$. By Theorem 2.5.2, λ is an eigenvalue of K if and only if, for all x in a set of positive measure,

$$\lim_{\rho\to 0} \rho^{-n} \operatorname{cap}(B_\rho(x)\backslash Z_\lambda, W_2^l) = 0$$

where $Z_\lambda = \{\xi \in R^n : (Fk)(\xi) = \lambda\}$. This condition is equivalent to $\operatorname{cap}(G\backslash Z_\lambda, W_2^l) < \operatorname{cap}(G, W_2^l)$ for some open set G. According to the same Theorem 2.5.2, λ belongs to residual spectrum $\sigma_r(K)$ if and only if any of the conditions stated above is not valid and $\operatorname{cap}(Z_\lambda, W_2^l) > 0$. Moreover, $\lambda \in \sigma_c(K) \Leftrightarrow \operatorname{cap}(Z_\lambda, W_2^l) = 0$.

If λ is a point of the spectrum of an operator K continuous in $L_2((1+|x|^2)^{-l/2})$, then Theorem 2.5.2 yields:

$$\lambda \in \sigma_p(K) \Leftrightarrow \operatorname{cap}(Z_\lambda, W_2^l) > 0,$$

$$\lambda \in \sigma_c(K) \Leftrightarrow \operatorname{cap}(Z_\lambda, W_2^l) = 0.$$

Consequently, $\sigma_r(K) = \varnothing$.

In Chapter 4 we obtain two-sided estimates for the essential norm of a multiplier. As a corollary, we derive necessary and sufficient conditions for the compactness of the operator of multiplication by $\gamma \in M(W_p^m \to W_p^l)$. It is clear that for $p=2$ these results yield two-sided estimates for the essential norm and conditions for the compactness of the operator $K: L_2((1+|x|^2)^{m/2}) \to L_2((1+|x|^2)^{l/2})$.

3.6.2 *Estimates for a Pair of Differential Operators in R^n*

The discussion of the preceding subsection is related to the problem of the dominance of differential operators with constant coefficients (see, for example, Hörmander [1], Treves [1], Gel'man and Maz'ya [1]). Here we consider one of the formulations of this problem for which the answer is due to Lemma 3.2.1/2.

Let $R(D)$ and $P(D)$ be differential operators in R^n with constant coefficients and let S be the space of infinitely differential functions defined on R^n and tending to zero at infinity with all derivatives faster than an arbitrary positive degree of $|x|^{-1}$ (see Schwartz [1], Gelfand and Shilov [1]). The following theorem holds:

Theorem 1. *The inequality*

$$\|R(D)u\|_{L_2((1+|x|^2)^{-l/2})} \leq C\,\|P(D)u\|_{L_2}, \qquad l>0, \tag{1}$$

is valid for every $u \in S$ if and only if $R/P \in M(W_2^l \to L_2)$ which is equivalent to

$$\sup_{\{e:d(e)\leq 1\}} \frac{\|R/P;\, e\|_{L_2}}{[\text{cap}\,(e,\, W_2^l)]^{1/2}} < \infty. \tag{2}$$

In particular, in the case $2l>n$ the condition (2) means that

$$\sup_{x\in R^n} \int_{B_1(x)} |R(\xi)/P(\xi)|^2\, d\xi < \infty. \tag{3}$$

Proof. Sufficiency: the left-hand side in (1) is equal to

$$\sup_{\varphi} \frac{|(R(D)u,\, \varphi)|}{\|\varphi\|_{L_2((1+|x|^2)^{l/2})}} = \sup_{\Phi} \frac{|(PFu,\, (\overline{R/P})\Phi)|}{\|\Phi\|_{W_2^l}},$$

where F is the Fourier transform in R^n. The right-hand side does not exceed

$$\|PFu\|_{L_2} \sup_{\Phi} \frac{\|(R/P)\Phi\|_{L_2}}{\|\Phi\|_{W_2^l}} = \|R/P\|_{M(W_2^l\to L_2)}\, \|P(D)u\|_{L_2}.$$

Necessity: let $Q_\varepsilon(\xi) = (|P(\xi)|^2+\varepsilon)^{1/2}$, where $\varepsilon = \text{const}>0$, and $Q_\varepsilon(D) = F^{-1}Q_\varepsilon F$. Since $FS=S$, the operator $Q_\varepsilon^{-1}(D) = F^{-1}Q_\varepsilon^{-1}F$ maps S into itself. We set $u = Q_\varepsilon^{-1}(D)f$, where f is an arbitrary function in S. By virtue of (1),

$$\|R(D)Q_\varepsilon^{-1}(D)f\|_{L_2((1+|x|^2)^{-l/2})} \leq C\,\|f\|_{L_2}$$

which is the same as

$$|(\Psi,\, (\bar{R}/Q_\varepsilon)\Phi)| \leq C\,\|\Psi\|_{L_2}\,\|\Phi\|_{W_2^l}.$$

Consequently, for all $\Phi \in W_2^l$, $\|(R/Q_\varepsilon)\Phi\|_{L_2} \leq C\,\|\Phi\|_{W_2^l}$. Passing to the limit as $\varepsilon \to +0$, we complete the proof. □

A rough corollary of Theorem 1 is the sufficiency of (3) for the validity of

$$\|R(D)u;\, K\|_{L_2} \leq C(K)\,\|P(D)u\|_{L_2}, \tag{4}$$

where $u \in S$, K is an arbitrary compact set in R^n and $C(K)$ is a constant independent of u. Namely, the following theorem holds:

Theorem 2 (Maz'ya [7]). *Inequality* (4) *is true if and only if the functions R and P satisfy* (3).

Proof. We need to prove only the necessity. Let x be a fixed point in R^n, let χ_x be the characteristic function of the ball $B_1(x)$ and let K be the cube $\{x \in R^n : |x_i| \leqslant 1, 1 \leqslant i \leqslant n\}$. We define a family of functions $\{u_{\varepsilon,h}\}$ by

$$Fu_{\varepsilon,h} = \left(\frac{\bar{R}\chi_x}{|P|^2+\varepsilon}\right)_h,$$

where $(\varphi)_h$ is the mollification of φ with radius h and ε is a positive number. Clearly, $u_{\varepsilon,h} \in S$ and it can be put into (4). We have

$$\lim_{h\to 0} \|P(D)u_{\varepsilon,h}\|_{L_2} = \left\|\frac{PR}{|P|^2+\varepsilon}; B_1(x)\right\|_{L_2} \leqslant \left\|\frac{R}{(|P|^2+\varepsilon)^{1/2}}; B_1(x)\right\|_{L_2}. \tag{5}$$

On the other hand,

$$\|R(D)u_{\varepsilon,h}; K\|_{L_2} = c\, \|\psi * RFu_{\varepsilon,h}\|_{L_2},$$

where $\psi(\xi) = \prod_{1\leqslant i\leqslant n} \xi_i^{-1} \sin \xi_i$. Therefore

$$\|R(D)u_{\varepsilon,h}; K\|_{L_2}^2 \geqslant c \int_{B_1(x)} \left| \int \psi(\xi-\eta) R(\eta)(Fu_{\varepsilon,h})(\eta)\, d\eta \right|^2 d\xi.$$

The right-hand side tends to

$$c \int_{B_1(x)} \left| \int_{B_1(x)} \psi(\xi-\eta) \frac{|R(\eta)|^2\, d\eta}{|P(\eta)|^2+\varepsilon} \right|^2 d\xi$$

as $h \to 0$. Here $|\xi-\eta|<2$ and consequently $\psi(\xi-\eta) \geqslant \text{const} > 0$. We arrive at the inequality

$$\varliminf_{h\to 0} \|R(D)u_{\varepsilon,h}; K\|_{L_2} \geqslant c \int_{B_1(x)} \frac{|R(\eta)|^2\, d\eta}{|P(\eta)|^2+\varepsilon}. \tag{6}$$

Now, from (4) for $u_{\varepsilon,h}$ and from (5), (6) we obtain

$$\|R(|P|^2+\varepsilon)^{-1/2}; B_1(x)\|_{L_2} \leqslant c.$$

By passing to the limit as $\varepsilon \to 0$, we get (3). □

3.7 The Norm of a Differential Operator

3.7.1 Operators Mapping W_p^h into W_p^{h-k}

In this subsection we discuss some simple applications of the space $M(W_p^m \to W_p^l)$ to the theory of differential operators, namely, to the question of the continuity of such operators in pairs of Sobolev spaces.

Lemma 1. (i) *The operator*

$$P(x, D_x)u = \sum_{|\alpha|\leq k} a_\alpha(x) D_x^\alpha u, \qquad x \in R^n, \tag{1}$$

represents a continuous mapping $W_p^h \to W_p^{h-k}$, $h \geq k$, *provided that* $a_\alpha \in M(W_p^{h-|\alpha|} \to W_p^{h-k})$ *for any multi-index* α. *The following estimate holds:*

$$\|P\|_{W_p^h \to W_p^{h-k}} \leq c \sum_{|\alpha|\leq k} \|a_\alpha\|_{M(W_p^{h-|\alpha|} \to W_p^{h-k})}. \tag{2}$$

(ii) *If* $p = 1$ *or if* $p(h-k) > n$ *and* $p > 1$, *then the relation*

$$\|P\|_{W_p^h \to W_p^{h-k}} \sim \sum_{|\alpha|\leq k} \|a_\alpha\|_{M(W_p^{h-|\alpha|} \to W_p^{h-k})} \tag{3}$$

is valid.

Proof. The estimate (2) is obvious. Let $x \in R^n$ and let $\eta \in D(B_2)$, $\eta = 1$ on B_1. Further, let $u(y) = \eta((x-y)/\delta)$, where $\delta \in (0, 1]$. By setting the function u into the inequality

$$\Big\| \sum_{|\alpha|\leq k} a_\alpha D^\alpha u \Big\|_{W_p^{h-k}} \leq c \, \|u\|_{W_p^h},$$

we obtain the estimate

$$\sup_{x \in R^n} \|a_o; B_\delta(x)\|_{W_p^{h-k}} \leq c\, \delta^{n/p-h}$$

which, together with Theorems 1.3.3, 1.3.5/1, 3.4.2 and with Proposition 3.2.8, shows that $a_o \in M(W_p^h \to W_p^{h-k})$.

Assume $a_\alpha \in M(W_p^{h-|\alpha|} \to W_p^{h-k})$ for $|\alpha| \leq \nu$, $\nu \leq k-1$ and

$$\sum_{|\alpha|\leq \nu} \|a_\alpha\|_{M(W_p^{h-|\alpha|} \to W_p^{h-k})} \leq c \, \|P\|_{W_p^h \to W_p^{h-k}}.$$

We show that the same holds when ν is replaced by $\nu + 1$. We have

$$\Big\| \sum_{|\alpha|\geq \nu+1} a_\alpha D^\alpha u \Big\|_{W_p^{h-k}} \leq \Big\| Pu - \sum_{|\alpha|\leq \nu} a_\alpha D^\alpha u \Big\|_{W_p^{h-k}}$$
$$\leq c\, \|P\|_{W_p^h \to W_p^{h-k}} \|u\|_{W_p^h}$$

for all $u \in W_p^h$. Putting here $u(y) = (x-y)^\alpha \eta((x-y)/\delta)$, $|\alpha| = \nu + 1$, we obtain

$$\sup_{x \in R^n} \|a_\alpha; B_\delta(x)\|_{W_p^{h-k}} \leq c\, \|P\|_{W_p^h \to W_p^{h-k}} \, \delta^{n/p-h+|\alpha|}.$$

This, together with Theorems 1.3.3, 1.3.5/1, 3.4.2 and Proposition 3.2.8, implies $a_\alpha \in M(W_p^{h-|\alpha|} \to W_p^{h-k})$. $\square$

Now we present an analogous result for matrix operators. Let $u(x) = \{u^1(x), u^2(x), \ldots, u^N(x)\}$ be an N-tuple vector-function. Consider the operator

$$Pu = \left\{ \sum_{k=1}^{N} P_{jk}(x, D_x)u^k \right\}_{j=1}^{M}, \qquad x \in R^n, \tag{4}$$

where $P_{jk}(x, D_x)u^k = \sum_{|\alpha| \leqslant s_j + t_k} a_{jk}^{(\alpha)}(x) D_x^\alpha u^k$ and s_j, t_k are integers.

Theorem 1. *Let $h \geqslant s = \max s_j$, $j = 1, \ldots, M$.*
(i) *The operator P is a continuous mapping*

$$P: \prod_{k=1}^{N} W_p^{t_k + h} \to \prod_{j=1}^{M} W_p^{h - s_j}, \tag{5}$$

if $a_{jk}^{(\alpha)} \in M(W_p^{t_k + h - |\alpha|} \to W_p^{h - s_j})$. The following estimate holds

$$\|P\| \leqslant c \sum_{k=1}^{N} \sum_{j=1}^{M} \sum_{|\alpha| \leqslant s_j + t_k} \|a_{jk}^{(\alpha)}\|_{M(W_p^{t_k + h - |\alpha|} \to W_p^{h - s_j})}. \tag{6}$$

(ii) *If $p(h - s) \geqslant n$, $p > 1$ or $p = 1$, then the relation*

$$\|P\| \sim \sum_{k=1}^{N} \sum_{j=1}^{M} \sum_{|\alpha| \leqslant s_j + t_k} \|a_{jk}^{(\alpha)}\|_{M(W_p^{t_k + h - |\alpha|} \to W_p^{h - s_j})}$$

holds.

Proof. Inequality (6) follows from Lemma 1. Let

$$\sum_{j=1}^{M} \left\| \sum_{k=1}^{N} P_{jk}(x, D_x)u^k \right\|_{W_p^{h - s_j}} \leqslant c\, \|P\| \sum_{k=1}^{N} \|u^k\|_{W_p^{h + t_k}}.$$

We set here $u^k = 0$ provided $k \neq i$. Then

$$\sum_{j=1}^{M} \|P_{ji}(x, D_x)u^i\|_{W_p^{h - s_j}} \leqslant c\, \|P\|\, \|u^i\|_{W_p^{h + t_i}}$$

and in the case $p(h - s) > n$, $p > 1$, or in the case $p = 1$, the lemma gives us the estimate

$$\sum_{j=1}^{M} \sum_{|\alpha| \leqslant s_j + t_i} \|a_{ji}^{(\alpha)}\|_{M(W_p^{h + t_i - |\alpha|} \to W_p^{h - s_j})} \leqslant c\, \|P\|,$$

where $i = 1, 2, \ldots, N$. □

Next we show that a relation similar to (3) with an arbitrary $p \in (1, \infty)$ can be obtained for operators of special form.

Theorem 2. *Let h and s be positive integers, $h \geqslant 2s$, $1 < p < \infty$ and*

$P(x, D_x)u = \sum_{j=0}^{s} b_j(x)\Delta^j u$, where Δ is the Laplace operator. Then P is a continuous mapping: $W_p^h \to W_p^{h-2s}$ if and only if $b_j \in M(W_p^{h-2j} \to W_p^{h-2s})$, $j = 0, \ldots, s$. Moreover, the relation

$$\|P\|_{W_p^h \to W_p^{h-2s}} \sim \sum_{j=0}^{s} \|b_j\|_{M(W_p^{h-2j} \to W_p^{h-2s})} \tag{7}$$

is valid.

Proof. The sufficiency as well as the upper bound for the norm of P follows from Lemma 1.

Suppose, for all $u \in W_p^h$, $\|Pu\|_{W_p^{h-2s}} \leqslant c\,\|u\|_{W_p^h}$. Let $u = 1$ in a neighbourhood of a compactum e with $d(e) \leqslant 1$. Then

$$\|\nabla_{h-2s} b_0; e\|_{L_p} + \|b_0; e\|_{L_p} \leqslant c\,\|u\|_{W_p^h}.$$

Consequently,

$$\|\nabla_{h-2s} b_0; e\|_{L_p} \leqslant c[\mathrm{cap}\,(e, W_p^h)]^{1/p}$$

and $\|b_0; B_1(x)\|_{L_p} \leqslant c$ for all $x \in R^n$. By Theorem 2.6 this means that $b_0 \in M(W_p^h \to W_p^{h-2s})$. Therefore, the operator $Q\Delta$ with $Qu = \sum_{j=0}^{s-1} b_{j+1}(x)\Delta^j u$ satisfies the inequality

$$\|Q\Delta u\|_{W_p^{h-2s}} \leqslant c\,\|u\|_{W_p^h}. \tag{8}$$

Let $\zeta \in W_p^{h-2}$, $\mathrm{supp}\,\zeta \subset \{x = (x_1, \ldots, x_n) : 0 < x_i < 1\}$. We put $w(x) = \zeta(x) - \zeta(-x)$ and $w_i(x) = w(x_1, \ldots, x_{i-1}, x_i/2, x_{i+1}, \ldots, x_n)$ for $i = 1, \ldots, n$. Further, let

$$v(x) = w(x) + \sum_{i=1}^{n} \alpha_i w_i(x - a_i) \tag{9}$$

where a_i are fixed points with $\mathrm{dist}\,(a_i, a_j) > (8n)^{1/2}$ and α_i are arbitrary constants. It is clear that all the functions on the right in (9) are orthogonal to unity and have disjoint supports. We show that coefficients α_i can be selected so that

$$\int x_j v(x)\,dx = 0, \qquad j = 1, \ldots, n.$$

The latter is equivalent to the following algebraic system with respect to $\alpha_1, \ldots, \alpha_n$:

$$2 \int x_j \zeta(x)\,dx \Big(1 + 2 \sum_{i=1}^{n} \alpha_i (1 + \delta_i^j)\Big) = 0, \qquad j = 1, \ldots, n,$$

where δ_i^j is the Kroneker delta. The system is solvable because $\det \|1 + \delta_i^j\|_{i,j=1}^n = n + 1$.

Let u be the harmonic (Newtonian for $n>2$ and logarithmic for $n=2$) potential with density v. Since v is orthogonal with $1, x_1, \ldots, x_n$ and the diameter of its support is bounded by a constant depending only on n, then

$$\|u\|_{W_p^h} \leq c\,\|v\|_{W_p^{h-2}} \leq c_1\,\|\zeta\|_{W_p^{h-2}}.$$

This and (8) imply

$$\|Q\zeta\|_{W_p^{h-2s}} \leq c\,\|\zeta\|_{W_p^{h-2}}.$$

By virtue of the arbitrariness of the origin, the latter inequality is valid for any function $\zeta \in W_p^{h-2}$ with support contained in any cube of the coordinate grid. Hence it is valid for any $\zeta \in W_p^{h-2}$, i.e. Q is a continuous operator $W_p^{h-2} \to W_p^{h-2s}$. Now the statement of the theorem follows by a successive reduction of order of the operator. □

Theorem 2 and item (ii) of Lemma 1 suggest the following hypothesis: relation (3) is valid for all p, h, k, i.e. the coefficients of any differential operator (1) mapping W_p^h into W_p^{h-k} are of necessity multipliers in corresponding Sobolev spaces. The next example disproves this suggestion.

Example. Let $x=(x', x_n)$, $x'=(x_1, \ldots, x_{n-1})$, $n \geq 3$. We show that the coefficient a of the bounded operator

$$a(x')\,\partial/\partial x_n : W_2^2 \to L_2 \tag{10}$$

need not be an element of the space $M(W_2^1 \to L_2)$.

Suppose

$$A = \sup_{y \in R^{n-1}, r \in (0,1)} \frac{\int_{B_r^{(n-1)}(y)} |a(x')|^2\,dx'}{r^{n-3}} < \infty$$

where $B_r^{(n-1)}(y)$ is an $(n-1)$-dimensional ball with centre y and radius r. By $\hat{u}(x', \lambda)$ we denote the Fourier transform of the function u with respect to x_n. According to Theorem 1.2.3/2,

$$\int_{R^{n-1}} |a(x')\hat{u}(x', \lambda)|^2\,dx' \leq A\,\|[\hat{u}(\cdot, \lambda)]^2; R^{n-1}\|_{W_1^2}$$

$$\leq cA \int_{R^{n-1}} (|\hat{u}(x', \lambda)|\,|\nabla_{2,x'}\hat{u}(x', \lambda)|$$

$$+|\nabla_{x'}\hat{u}(x', \lambda)|^2 + |\hat{u}(x', \lambda)|^2)\,dx'.$$

Consequently,

$$\int_{R^n} \left| a(x') \frac{\partial u}{\partial x_n} \right|^2 dx \leq cA \int_{R^1} d\lambda \int_{R^{n-1}} (|\lambda^2 \hat{u}(x', \lambda)|\, |\nabla_{2,x'} \hat{u}(x', \lambda)|$$

$$+ |\lambda \nabla_{x'} \hat{u}(x', \lambda)|^2 + |\lambda \hat{u}(x', \lambda)|^2)\, dx'$$

$$\leq c_1 A (\|\nabla_2 u\|_{L_2}^2 + \|\nabla u\|_{L_2}^2).$$

Thus the finiteness of the value A is sufficient for the continuity of operator (10). The necessity for the same condition results from the estimate

$$\|P\|_{W_p^h \to W_p^{h-k}} \geq c \sum_{|\alpha| \leq k} \sup_{x;\, r \in (0,1)} (r^{h-|\alpha|-n/p} \|\nabla_{h-k} a_\alpha;\, B_r(x)\|_{L_p}$$

$$+ r^{k-|\alpha|-n/p} \|a_\alpha;\, B_r(x)\|_{L_p}) \tag{11}$$

derived in the proof of the second part of Lemma 1 for all $p \in [1, \infty)$.

If (3) is valid for all p, h, k then the continuity of operator (10) would imply $a \in M(W_2^1 \to L_2)$.

We choose the coefficient a as follows:

$$a(x') = \rho^{-1} |\log \rho|^{\varepsilon - 1} \eta(x_1, x_2) \zeta(x_3, \ldots, x_{n-1})$$

where $\rho^2 = x_1^2 + x_2^2$, $0 < \varepsilon < 1/2$, $\eta \in C_0^\infty(B_1^{(2)})$, $\zeta \in C_0^\infty(B_1^{(n-3)})$. It is clear that for any $y \in R^{n-1}$ and for all $r \in (0, 1/2)$ we have

$$\int_{B_r^{(n-1)}(y)} |a(x')|^2\, dx' \leq c r^{n-3} \int_{B_r^{(2)}} \rho^{-2} |\log \rho|^{2(\varepsilon - 1)}\, dx_1\, dx_2$$

$$= c r^{n-3} |\log r|^{2\varepsilon - 1}.$$

Therefore $A < \infty$. Suppose $\|au;\, R^n\|_{L_2} \leq c\, \|u;\, R^n\|_{W_2^1}$ for any $u \in C_0^\infty(R^n)$. Then, for all $v \in C_0^\infty(R^2)$,

$$\|\eta \rho^{-1} |\log \rho|^{\varepsilon - 1} v;\, R^2\|_{L_2} \leq c\, \|v;\, R^2\|_{W_2^1}.$$

The latter implies

$$\int_{\rho < r} \rho^{-2} |\log \rho|^{2(\varepsilon - 1)}\, dx_1\, dx_2 \leq c \operatorname{cap}(B_r^{(2)};\, W_2^1(R^2))$$

for any $r \in (0, 1/2)$, which contradicts the relation $\operatorname{cap}(B_r^{(2)};\, W_2^1(R^2)) \sim |\log r|^{-1}$. Thus, operator (10) with function a under consideration is continuous although $a \notin M(W_2^1(R^n) \to L_2(R^n))$.

A necessary and sufficient condition for continuity of operator (10) derived in this example can be generalized to a certain class of operators with coefficients depending on a part of variables.

Theorem 3. *Let $y \in R^s$, $z \in R^{n-s}$, $s \leq n$; let h and k be integers, $h \geq k$,* *and*

$$P(y, D_y, D_z)u = \sum_{0 \leq |\beta|+|\gamma| \leq k} a_{\beta\gamma}(y) D_y^\beta D_z^\gamma u.$$

The operator $P(y, D_y, D_z)$ *is a continuous mapping:* $W_2^h(R^n) \to W_2^{h-k}(R^n)$ *if and only if the operator* $P(y, D_y, 0)$: $W_2^h(R^s) \to W_2^{h-k}(R^s)$ *is continuous and, for any multi-indices* β, γ,

$$\sup_{y \in R^s, r \in (0,1)} r^{h-|\beta|-|\gamma|-s/2}(\|\nabla_{h-k} a_{\beta\gamma}; B_r^{(s)}(y)\|_{L_2} + r^{k-h} \|a_{\beta\gamma}; B_r^{(s)}(y)\|_{L_2}) < \infty \tag{12}$$

where $B_r^{(s)}(y)$ *is an s-dimensional ball with centre y and radius r.*

The proof of this theorem is based on the following assertion (see Maz'ya [14], Ch. I).

Lemma 2. *Let* μ *be a measure in* R^s *such that*

$$K = \sup_{y \in R^s, r \in (0,1)} r^{-\sigma} \mu(B_r^{(s)}(y)) < \infty \tag{13}$$

for a certain $\sigma \in [0, s]$. *Further, let l and m be integers,* $0 \leq l < m$, $\sigma > s - 2(m-l)$. *Then, for all* $u \in D$,

$$\left(\int_{R^s} |\nabla_l u|^2 \, d\mu \right)^{1/2} \leq cK \|u; R^s\|_{W_2^m}^{\tau} \|u; R^s\|_{L_2}^{1-\tau} \tag{14}$$

where $\tau = (2l + s - \sigma)/2m$ *and c is a constant independent of u and* μ. *Moreover, condition* (13) *is necessary for the validity of* (14).

Proof of Theorem 3. Sufficiency: first we prove the continuity of the operator

$$a_{\beta\gamma}(y) D_y^\beta D_z^\gamma : W_2^h(R^n) \to W_2^{h-k}(R^n) \quad \text{for} \quad |\gamma| > 0.$$

With this aim in view we show the continuity of the operator

$$(D_y^\mu a_{\beta\gamma}(y)) D_y^{\rho+\beta} D_z^{\gamma+\theta} : W_2^h(R^n) \to L_2(R^n)$$

where $0 \leq |\mu| + |\rho| + |\theta| \leq h - k$. By $\hat{u}(y, \lambda)$ we denote the Fourier transform of the function u with respect to variable z. Putting $l = |\rho| + |\beta|$, $\sigma = 2|\beta| + 2|\gamma| + s - 2k - 2|\mu|$, $m = k + |\mu| + |\rho| + |\theta|$ in Lemma 2, we conclude that

$$\int_{R^s} |(D_y^\mu a_{\beta\gamma}(y)) D_y^{\rho+\beta} \hat{u}(y, \lambda)|^2 \, dy$$
$$\leq c A_{\beta,\gamma,\mu}^2 \|\hat{u}(\cdot, \lambda); R^s\|_{W_2^m}^{2\tau} \|\hat{u}(\cdot, \lambda); R^s\|_{L_2}^{2(1-\tau)}$$

where $\tau = (|\mu|+|\rho|+k-|\gamma|)/(|\mu|+|\rho|+|\theta|+k)$ and

$$A_{\beta,\gamma,\mu} = \sup_{y\in R^s, r\in(0,1)} r^{-\sigma/2} \|D_y^\mu a_{\beta\gamma}; B_r^{(s)}(y)\|_{L_2}.$$

(We note that the condition $\sigma > s-2(m-l)$ in Lemma 2 is equivalent to $|\gamma|>0$.) Multiplying the latter inequality by $|\lambda|^{2(|\gamma|+|\theta|)}$ and integrating over λ, we get

$$\|(D_y^\mu a_{\beta\gamma})D_y^{\rho+\beta}D_z^{\gamma+\theta}u; R^n\|_{L_2}^2$$

$$\leq cA_{\beta,\gamma,\mu}^2 \int_{R^{n-s}} \|\hat{u}(\cdot,\lambda); R^s\|_{W_2^m}^{2\tau} \||\lambda|^m \hat{u}(\cdot,\lambda); R^s\|_{L_2}^{2(1-\tau)}\, d\lambda$$

$$\leq cA_{\beta,\gamma,\mu}^2 \|u; R^n\|_{W_2^m}^2.$$

Since $m \leq h$ and $A_{\beta,\gamma,\mu}$ does not exceed the value (12), the operator $P(y, D_y, D_z) - P(y, D_y, 0): W_2^h(R^n) \to W_2^{h-k}(R^n)$ is continuous.

In order to derive the continuity of the operator $P(y, D_y, 0)$ in the same pair of spaces we prove the inequality

$$\|P(y, D_y, 0)v; R^s\|_{L_2} \leq c\,\|v; R^s\|_{W_2^h}^{k/h} \|v; R^s\|_{L_2}^{1-k/h}. \tag{15}$$

Applying Lemma 2 with $l=|\beta|$, $m=h$ for any multi-index β with $|\beta|\leq k$, we get

$$\|a_{\beta 0}D_y^\beta v; R^s\|_{L_2}$$

$$\leq c \sup_{y\in R^s, r\in(0,1)} r^{k-|\beta|-s/2} \|a_{\beta 0}; B_r^{(s)}(y)\|_{L_2} \|v; R^s\|_{W_2^h}^{k/h} \|v; R^s\|_{L_2}^{1-k/h}$$

which entails (15). Using (15) together with (12), we find

$$\|P(\cdot, D_y, 0)u; R^n\|_{W_2^{h-k}}^2 \leq c \int_{R^{n-s}} (\|P(\cdot, D_y, 0)\hat{u}(\cdot,\lambda); R^s\|_{W_2^{h-k}}^2$$

$$+ |\lambda|^{2(h-k)} \|P(\cdot, D_y, 0)\hat{u}(\cdot,\lambda); R^s\|_{L_2}^2)\, d\lambda$$

$$\leq c \int_{R^{n-s}} (\|\hat{u}(\cdot,\lambda); R^s\|_{W_2^h}^2$$

$$+ \|\hat{u}(\cdot,\lambda); R^s\|_{W_2^h}^{2k/h} \||\lambda|^h \hat{u}(\cdot,\lambda); R^s\|_{L_2}^{2(1-k/h)})\, d\lambda$$

$$\leq c\,\|u; R^n\|_{W_2^h}^2$$

for all $u \in W_2^h(R^n)$. The sufficiency follows.

Necessity: we note that by replacing s by n in (12) we obtain an equivalent condition. Then the finiteness of (12) for all multi-indices β,γ follows from (11). Thus the first part of this theorem implies the continuity of the operator $P(y, D_y, D_z) - P(y, D_y, 0): W_2^h(R^n) \to W_2^{h-k}(R^n)$. Since the operator $P(y, D_y, D_z)$: $W_2^h(R^n) \to W_2^{h-k}(R^n)$ is continuous, then for all $v \in W_2^h(R^n)$

$$\|P(y, D_y, 0)v; R^n\|_{W_2^{h-k}} \leq c\,\|v; R^n\|_{W_2^h}.$$

Substituting here $v(x) = \eta(z)u(y)$, where η is a fixed function in $D(R^{n-s})$ and u is an arbitrary element of $W_2^h(R^s)$, we complete the proof of Theorem 3. □

3.7.2 Operators in Pairs of Multiplier Spaces

We now turn to theorems on the boundedness of differential operators mapping a space of multipliers into another one.

Lemma 1. *Let $m \leq n$ and let $\gamma \in W_1^l \cap M(W_1^m \to W_1^l)$. Then*

$$\|\gamma_\delta\|_{M(W_1^m \to W_1^l)} \leq c\,\delta^{m-l}(\|\gamma\|_{M(W_1^m \to W_1^l)} + \|\gamma\|_{W_1^l}),$$

where $\gamma_\delta(y) = \gamma(y/\delta)$, $0 < \delta < 1$.

Proof. By Theorems 1.3.5/1 and 3.4.2,

$$\|\gamma_\delta\|_{M(W_1^m \to W_1^l)} \leq c \sup_{x \in R^n, r \in (0,1)} r^{m-n} \||\gamma_\delta; B_r(x)\||_{W_1^l}.$$

Since

$$\||\gamma_\delta; B_r(x)\||_{W_1^l} \leq \delta^{n-1}\, \||\gamma; B_{r/\delta}\||_{W_1^l},$$

then

$$\|\gamma_\delta\|_{M(W_1^m \to W_1^l)} \leq c\,\delta^{m-l}\Big(\sup_{x \in R^n, r \in (0,\delta]} (r/\delta)^{m-n}\, \||\gamma; B_{r/\delta}(x)\||_{W_1^l}$$

$$+ \sup_{x \in R^n, r > \delta} (r/\delta)^{m-n}\, \||\gamma; B_{r/\delta}(x)\||_{W_1^l}\Big).$$

By using Theorems 1.3.5/1 and 3.4.2 once more, we complete the proof. □

Lemma 2. *Let P be the operator defined by* (3.7.1/1) *and let $r \geq h \geq k \geq 0$, where r and h are simultaneously either integers or fractional numbers.*

(i) *If $a_\alpha \in M(W_p^{h-|\alpha|} \to W_p^{h-k})$ for any multi-index α, then the mapping $P: M(W_p^r \to W_p^h) \to M(W_p^r \to W_p^{h-k})$ is continuous and the following estimate holds:*

$$\langle P \rangle \leq c \sum_{|\alpha| \leq k} \|a_\alpha\|_{M(W_p^{h-|\alpha|} \to W_p^{h-k})}. \tag{1}$$

Here by $\langle \cdot \rangle$ we denote the norm of an operator mapping $M(W_p^r \to W_p^h)$ into $M(W_p^r \to W_p^{h-k})$.

(ii) *If* $p(h-k)>n$, $p>1$ *or* $p=1$, *then the relation* $\langle P\rangle \sim \sum_{|\alpha|\le k} \|a_\alpha\|_{M(W_p^{h-|\alpha|}\to W_p^{h-k})}$ *is valid.*

Proof. (i) As a consequence of Lemmas 1.3.2/1 and 3.2.4/1, we have the inclusion $D^\alpha u \in M(W_p^r \to W_p^{h-|\alpha|})$. Since $a_\alpha \in M(W_p^{h-|\alpha|} \to W_p^{h-k})$, then $a_\alpha D^\alpha u \in M(W_p^r \to W_p^{h-k})$ and

$$\langle a_\alpha D^\alpha\rangle \le c\, \|a_\alpha\|_{M(W_p^{h-|\alpha|}\to W_p^{h-k})}.$$

The first assertion is proved.

(ii) By setting $u=1$ into the estimate

$$\|P\|_{M(W_p^r\to W_p^{h-k})} \le c\langle P\rangle\, \|u\|_{M(W_p^r\to W_p^h)}$$

we obtain $\|a_0\|_{M(W_p^r\to W_p^{h-k})} \le c\langle P\rangle$. Suppose $a_\alpha \in M(W_p^{h-|\alpha|}\to W_p^{h-k})$ for $|\alpha|\le\nu$, $\nu\le k-1$, and let

$$\sum_{|\alpha|\le\nu} \|a_\alpha\|_{M(W_p^{h-|\alpha|}\to W_p^{h-k})} \le c\langle P\rangle.$$

We show that the same is true after the change from ν to $\nu+1$. We have

$$\Big\|\sum_{|\alpha|\ge\nu+1} a_\alpha D^\alpha u\Big\|_{M(W_p^r\to W_p^{h-k})} = \Big\|Pu - \sum_{|\alpha|\le\nu} a_\alpha D_u^\alpha\Big\|_{M(W_p^r\to W_p^{h-k})}$$
$$\le c\langle P\rangle\, \|u\|_{M(W_p^r\to W_p^h)}. \tag{2}$$

Let $p=1$, $r\le n$. We put in (2) $u(y)=(x-y)^\beta\eta((x-y)/\delta)$, where $|\beta|=\nu+1$ and η is the function defined in the proof of Lemma 1. By using Theorems 1.3.5/1 and 3.4.2 we obtain

$$\delta^{r-n}\, \|a_\beta D^\beta u; B_\delta(x)\|_{W_1^{h-k}} = \delta^{r-n} \Big\|\sum_{|\alpha|\ge\nu+1} a_\alpha D^\alpha u; B_\delta(x)\Big\|_{W_1^{h-k}}$$
$$\le c\langle P\rangle\, \|u\|_{M(W_1^r\to W_1^h)}.$$

According to Lemma 1, $\|u\|_{M(W_1^r\to W_1^h)} \le c\,\delta^{|\beta|+r-h}$. Therefore

$$\|a_\beta\|_{M(W_1^{h-|\beta|}\to W_1^{h-k})} \le c \sup_{x\in R^n,\delta\in(0,1)} \delta^{h-|\beta|-n}\, \|a_\beta; B_\delta(x)\|_{W_1^{h-k}} \le c\langle P\rangle.$$

In the case $p(h-k)>n$, $p>1$ we set $u(y)=(x-y)^\beta\eta(x-y)$ in (2). Then, as a consequence of Proposition 3.2.8 and Theorem 1.3.3, we have

$$\|a_\beta\|_{M(W_p^{h-|\beta|}\to W_p^{h-k})} \le c \sup_{x\in R^n} \|a_\beta; B_1(x)\|_{W_p^{h-k}}$$
$$\le c_1 \sup_{x\in R^n} \Big\|\sum_{|\alpha|\ge\nu+1} a_\alpha D^\alpha u; B_1(x)\Big\|_{W_p^{h-k}}$$
$$\le c_2\langle P\rangle\, \|u\|_{M(W_p^r\to W_p^h)} \le c_3\langle P\rangle.$$

The following theorem is derived from Lemma 2 in the same way as

Theorem 3.7.1/1 is obtained from Lemma 3.7.1/1. We shall use the same notation as in Theorem 3.7.1/1.

Theorem. *Let $r \geq h \geq s$, where r and h are simultaneously integers or fractional numbers.* (i) *The operator P represents a continuous mapping*

$$P: \prod_{k=1}^{N} M(W_p^r \to W_p^{h+t_k}) \to \prod_{j=1}^{M} M(W_p^r \to W_p^{h-s_j}) \tag{3}$$

provided that $a_{jk}^{(\alpha)} \in M(W_p^{h+t_k-|\alpha|} \to W_p^{h-s_j})$. The following estimate holds:

$$\langle P \rangle \leq c \sum_{k=1}^{N} \sum_{j=1}^{M} \sum_{|\alpha| \leq s_j + t_k} \| a_{jk}^{(\alpha)} \|_{M(W_p^{h+t_k-|\alpha|} \to W_p^{h-s_j})},$$

where $\langle \cdot \rangle$ is the norm of an operator acting in the pair of spaces (3).

(ii) *If $p(h-s) \geq n$, $p > 1$ or $p = 1$, then the relation*

$$\langle P \rangle \sim \sum_{k=1}^{N} \sum_{j=1}^{M} \sum_{|\alpha| \leq s_j + t_k} \| \alpha_{jk}^{(\alpha)} \|_{M(W_p^{h+t_k-|\alpha|} \to W_p^{h-s_j})}$$

is valid.

3.8 Coercive Estimates for Solutions of Elliptic Equations in Spaces of Multipliers

It is well known that the solutions of elliptic boundary value problems satisfy coercive estimates in Sobolev spaces. The purpose of this section is to show that similar estimates are valid for norms in classes of multipliers acting in a Sobolev space or in a pair of Sobolev spaces.

3.8.1 The Case of Operators in R^n

Theorem. *Let P be an elliptic (in the sense of Douglis and Nirenberg) operator (3.7.1/4), where $M = N$. Let the coefficients of P be constant. Further, let $\gamma = \{\gamma^1, \ldots, \gamma^N\}$ be a vector-function in the space $\prod_k W_{p,\text{loc}}^{h+t_k} \cap \prod_k M(W_p^{r-h-t_k} \to L_p)$ and let $P\gamma \in \prod_j M(W_p^r \to W_p^{h-s_j})$, where $r \geq h + t_k \geq 0$, $r \geq h - s_j \geq 0$, $1 \leq j, k \leq N$. Then $\gamma \in \prod_k M(W_p^r \to W_p^{h+t_k})$ and the estimate*

$$\|\gamma\|_{\prod_k M(W_p^r \to W_p^{h+t_k})} \leq C(\|P\gamma\|_{\prod_j M(W_p^r \to W_p^{h-s_j})} + \|\gamma\|_{\prod_k M(W^{r-h+t_k} \to L_p)}) \tag{1}$$

holds.

Proof. It is known that, for all $u \in \prod_k W_p^{t_k+h}$,

$$\|u\|_{\prod_k W_p^{t_k+h}} \leq C_1(\|Pu\|_{\prod_j W_p^{h-s_j}} + \|u\|_{L_p}).$$

Consequently, for all $\varphi \in C_0^\infty$,

$$\|\gamma_\rho \varphi\|_{\prod_k W_p^{t_k+h}} \leqslant C_1(\|\varphi P \gamma_\rho\|_{\prod_j W_p^{h-s_j}} + \|\gamma_\rho \varphi\|_{L_p} + \|[\varphi, P]\gamma_\rho\|_{\prod_j W_p^{h-s_j}}), \tag{2}$$

where $[\varphi, P]$ is a commutator of P and the operator of multiplication by φ. As usual, by γ_ρ we denote a mollification of γ.

It is clear that

$$\|\varphi P \gamma_\rho\|_{\prod_j W_p^{h-s_j}} \leqslant \|P\gamma_\rho\|_{\prod_j M(W_p^r \to W_p^{h-s_j})} \|\varphi\|_{W_p^r}, \tag{3}$$

$$\|\gamma_\rho \varphi\|_{L_p} \leqslant \|\gamma_\rho\|_{\prod_k M(W_p^{r-h-t_k} \to L_p)} \|\varphi\|_{W_p^{r-h-\min_k t_k}}. \tag{4}$$

It remains to estimate the third summand in (2). For any multi-index α, $|\alpha| \leqslant s_j + t_k$, we have

$$\begin{aligned}\|[\varphi, D^\alpha]\gamma_\rho\|_{\prod_j W_p^{h-s_j}} &\leqslant c \sum_{0<\beta\leqslant\alpha} \|D^\beta \varphi D^{\alpha-\beta}\gamma_\rho\|_{\prod_k W_p^{h-|\alpha|+t_k}} \\ &\leqslant c_1 \sum_{0<\beta\leqslant\alpha} \|D^{\alpha-\beta}\gamma_\rho\|_{\prod_k M(W_p^{r-|\beta|} \to W_p^{h-|\alpha|+t_k})} \|D^\beta \varphi\|_{W_p^{r-|\beta|}}\end{aligned}$$

which, together with Lemma 1.3.2/1, yields

$$\begin{aligned}\|[\varphi, D^\alpha]\gamma_\rho\|_{\prod_j W_p^{h-s_j}} &\leqslant c \sum_{1\leqslant\nu\leqslant|\alpha|} \|\gamma_\rho\|_{\prod_k M(W_p^{r-\nu} \to W_p^{h-\nu+t_k})} \|\varphi\|_{W_p^r} \\ &\leqslant (\varepsilon \|\gamma_\rho\|_{\prod_k M(W_p^r \to W_p^{h+t_k})} \\ &\quad + c(\varepsilon) \|\gamma_\rho\|_{\prod_k M(W_p^{r-h-t_k} \to L_p)} \|\varphi\|_{W_p^r}.\end{aligned}$$

This and inequalities (2)–(4) imply

$$\begin{aligned}\|\gamma_\rho \varphi\|_{\prod_k W_p^{h+t_k}} &\leqslant C_2(\|P\gamma_\rho\|_{\prod_j M(W_p^r \to W_p^{h-s_j})} \\ &\quad + \varepsilon \|\gamma_\rho\|_{\prod_k M(W_p^r \to W_p^{h+t_k})} + c(\varepsilon) \|\gamma_\rho\|_{\prod_k M(W_p^{r-h-t_k} \to L_p)}) \|\varphi\|_{W_p^r}.\end{aligned}$$

Consequently, function γ_ρ satisfies (1). Since $P\gamma_\rho = (P\gamma)_\rho$, then, applying Lemma 1.3.1/1, we complete the proof. □

Remark 1. If $r - h = t_k$ for all $k = 1, \ldots, N$, then the additional assumption $\gamma \in \prod_k M(W_p^{r-h-t_k} \to L_p)$ is equivalent to $\gamma \in L_\infty$ and estimate (1) takes the form

$$\|\gamma\|_{MW_p^r} \leqslant C(\|P\gamma\|_{\prod_j M(W_p^r \to W_p^{h-s_j})} + \|\gamma\|_{L_\infty}).$$

In particular, the inequality

$$\|\gamma\|_{MW_p^r} \leqslant C(\|P\gamma\|_{M(W_p^r \to W_p^{r-2\sigma})} + \|\gamma\|_{L_\infty}) \tag{5}$$

is valid for a scalar elliptic operator P of order 2σ with constant coefficients. Here we *a priori* assume that $\gamma \in W_{p,\mathrm{loc}}^r \cap L_\infty$ and $P\gamma \in M(W_p^r \to W_p^{r-2\sigma})$, $r \geqslant 2\sigma$.

Remark 2. We shall show that the norm $\|\gamma\|_{L_\infty}$ in the right-hand side of (5) cannot be omitted. Let $P=-\Delta+1$, $r=2$, $2p<n$, and let $\gamma_0(x)=\eta(x)\log|x|$, where $\eta\in C_0^\infty(B_1)$, $\eta(0)=1$. Then $P\gamma_0=O(|x|^{-2})$ and $\operatorname{supp} P\gamma_0\subset B_1$. Therefore $P\gamma_0\in M(W_p^2\to L_p)$. Suppose $\|\gamma\|_{MW_p^2}\leq c\,\|P\gamma\|_{M(W_p^2\to L_p)}$ for all $\gamma\in W_{p,\mathrm{loc}}^2\cap L_\infty$. Let us substitute a mollification $(\gamma_0)_\rho$ of the function γ_0 into the latter estimate. Then, according to Proposition 1.3.8/4 and Lemma 1.3.1/4,

$$\|(\gamma_0)_\rho\|_{L_\infty}\leq C\,\|(P\gamma_0)_\rho\|_{M(W_p^2\to L_p)}\leq C\,\|P\gamma_0\|_{M(W_p^2\to L_p)}.$$

Passing to the limit as $\rho\to 0$, we obtain a contradiction.

3.8.2. *Auxiliary Assertions on the Space $M(W_p^m(R_+^{n+1})\to W_p^l(R_+^{n+1}))$ for Integer m and l*

In this subsection we assemble some facts on the space of multipliers $M(W_p^m(R_+^{n+1})\to W_p^l(R_+^{n+1}))$ which will be used in 3.8.3 for the proof of coercive estimates for multiplier norms of solutions of elliptic boundary value problems in the half-space $R_+^{n+1}=\{z=(x,x_{n+1}): x\in R^n,\ x_{n+1}>0\}$.

Let π be the Hestenes extension operator, defined by

$$\pi(v)(z)=\begin{cases} v(z) & \text{for}\quad x_{n+1}>0,\\ \sum_{j=1}^{l}\alpha_j v(x,-jx_{n+1}) & \text{for}\quad x_{n+1}<0.\end{cases}$$

where α_j satisfy the conditions

$$\sum_{j=1}^{l}(-1)^k j^k\alpha_j=1,\qquad 0\leq k\leq l-1.$$

Lemma 1. If $\gamma\in M(W_p^m(R_+^{n+1})\to W_p^l(R_+^{n+1}))$, then

$$\pi(\gamma)\in M(W_p^m(R^{n+1})\to W_p^l(R^{n+1}))$$

and the estimate

$$\|\pi(\gamma);R^{n+1}\|_{M(W_p^m\to W_p^l)}\leq c\,\|\gamma;R_+^{n+1}\|_{M(W_p^m\to W_p^l)}\tag{1}$$

is valid.

Proof. Since $\gamma\in W_{p,\mathrm{loc}}^l(\overline{R_+^{n+1}})$, then, using the known fact that $\pi(\gamma)\in W_{p,\mathrm{loc}}^l(R^{n+1})$, we obtain $\pi(\gamma)u\in W_p^l(R^{n+1})$ for any $u\in C_0^\infty(R^{n+1})$. We have

$$\|\pi(\gamma)u;R^{n+1}\|_{W_p^l}^p=\|\gamma u;R_+^{n+1}\|_{W_p^l}+\|\pi(\gamma)u;R_-^{n+1}\|_{W_p^l}^p$$

$$\leq\|\gamma u;R_+^{n+1}\|_{W_p^l}^p+c\sum_{j=1}^{l}\|\gamma u_j;R_+^{n+1}\|_{W_p^l}^p,$$

where $u_j(x, x_{n+1}) = u(x, -x_{n+1}/j)$ and $R_-^{n+1} = \{z = (x, x_{n+1}) : x \in R^n, x_{n+1} < 0\}$. Therefore

$$\|\pi(\gamma)u; R^{n+1}\|_{W_p^l} \leq c\,\|\gamma; R_+^{n+1}\|_{M(W_p^m \to W_p^l)} \times \left(\|u; R_+^{n+1}\|_{W_p^m} + \sum_{j=1}^{l} \|u_j; R_+^{n+1}\|_{W_p^m}\right)$$

and since $\|u_j; R_+^{n+1}\|_{W_p^m} \leq c\,\|u; R_-^{n+1}\|_{W_p^m}$, then

$$\|\pi(\gamma)u; R^{n+1}\|_{W_p^l} \leq c\,\|\gamma; R_+^{n+1}\|_{M(W_p^m \to W_p^l)}\,\|u; R^{n+1}\|_{W_p^m}. \quad \square$$

Theorem. *The function γ belongs to $M(W_p^m(R_+^{n+1}) \to W_p^l(R_+^{n+1}))$ if and only if $\gamma \in W_{p,\mathrm{loc}}^l(R_+^{n+1})$ and for any compactum $e \subset \overline{R_+^{n+1}}$ the following inequalities hold:*

$$\|\gamma; e\|_{L_p}^p \leq \text{const cap}\,(e, W_p^{m-l}(R^{n+1})),$$
$$\|\nabla_l \gamma; e\|_{L_p}^p \leq \text{const cap}\,(e, W_p^m(R^{n+1})).$$

The estimates

$$\|\gamma; R^{n+1}\|_{M(W_p^m \to W_p^l)}^p \geq c_1 \sup_e \sum_{k=0}^{l} \frac{\|\nabla_k \gamma; e\|_{L_p}^p}{\text{cap}\,(e, W_p^{m-l+k}(R^{n+1}))}, \tag{2}$$

$$\|\gamma; R^{n+1}\|_{M(W_p^m \to W_p^l)}^p \leq c_2 \sup_e \left(\frac{\|\gamma; e\|_{L_p}^p}{\text{cap}\,(e, W_p^{m-l}(R^{n+1}))} + \frac{\|\nabla_l \gamma; e\|_{L_p}^p}{\text{cap}\,(e, W_p^m(R^{n+1}))}\right) \tag{3}$$

are valid.

Proof. As a consequence of (1.3.2/11) and (1.3.2/12), we find

$$\sup_{e \subset \overline{R_+^{n+1}}} \frac{\|\nabla_k \gamma; e\|_{L_p}^p}{\text{cap}\,(e, W_p^{m-l+k}(R^{n+1}))} \leq \sup_{e \subset R^{n+1}} \frac{\|\nabla_k \pi(\gamma); e\|_{L_p}^p}{\text{cap}\,(e, W_p^{m-l+k}(R^{n+1}))} \leq c\,\|\pi(\gamma); R^{n+1}\|_{M(W_p^m \to W_p^l)}^p.$$

Now to obtain the lower bound for the norm in $M(W_p^m(R_+^{n+1}) \to W_p^l(R_+^{n+1}))$ it suffices to use Lemma 1.

We turn to the proof of the upper bound. Let $u \in C_0^\infty(\overline{R_+^{n+1}})$ and $\gamma \in M(W_p^m(R_+^{n+1}) \to W_p^l(R_+^{n+1}))$. We have

$$\|\gamma u; R_+^{n+1}\|_{W_p^l}^p \leq \|\pi(\gamma)\pi(u); R^{n+1}\|_{W_p^l}^p.$$

By Theorem 1.3.2/2 the right-hand side does not exceed

$$c \sup_{e \subset R^{n+1}} \sum_{k=0}^{l} \frac{\|\nabla_k \pi(\gamma); e\|_{L_p}^p}{\text{cap}\,(e, W_p^{m-l+k}(R^{n+1}))} \|\pi(u); R^{n+1}\|_{W_p^m}^p.$$

Since the latter norm is not more than $c\,\|u; R^{n+1}_+\|_{W^m_p}$, it remains to prove the inequality

$$\sup_{e\subset R^{n+1}} \frac{\|\nabla_k \pi(\gamma); e\|^p_{L_p}}{\operatorname{cap}(e, W^{m-l+k}_p(R^{n+1}))} \leqslant c \sup_{e\subset \overline{R^{n+1}_+}} \frac{\|\nabla_k \gamma; e\|^p_{L_p}}{\operatorname{cap}(e, W^{m-l+k}_p(R^{n+1}))}. \quad (4)$$

Let $e_\pm = e \cap \overline{R^{n+1}_\pm}$. The left-hand side of (4) does not exceed the sum

$$\frac{\|\nabla_k \gamma; e_+\|^p_{L_p}}{\operatorname{cap}(e_+, W^{m-l+k}_p(R^{n+1}))} + \frac{\|\nabla_k \pi(\gamma); e_-\|^p_{L_p}}{\operatorname{cap}(e_-, W^{m-l+k}_p(R^{n+1}))}. \quad (5)$$

Put $e_j = \{z : (x, -x_{n+1}/j) \in e_-\}$, $j = 1, \ldots, l$. Since $\|\nabla_k \pi(\gamma); e_-\|_{L_p} \leqslant c\,\|\nabla_k \gamma; e_j\|_{L_p}$ and

$$\operatorname{cap}(e_-, W^{m-l+k}_p(R^{n+1})) \sim \operatorname{cap}(e_j, W^{m-l+k}_p(R^{n+1})),$$

the second term in (5) is majorized up to a constant factor by the right-hand side of (3). □

An immediate corollary of the theorem is

Lemma 2. *If $\gamma \in M(W^m_p(R^{n+1}_+) \to W^l_p(R^{n+1}_+))$, then*

$$D^\alpha \gamma \in M(W^m_p(R^{n+1}_+) \to W^{l-|\alpha|}_p(R^{n+1}_+))$$

for any multi-index α of order $|\alpha| \leqslant l$. The estimate

$$\|D^\alpha \gamma; R^{n+1}_+\|_{M(W^m_p \to W^{l-|\alpha|}_p)} \leqslant c\,\|\gamma; R^{n+1}_+\|_{M(W^m_p \to W^l_p)}$$

holds.

Lemma 3. *Let $[\gamma]_\rho$ be a mollification of γ in R^{n+1}_+ with respect to variables x with non-negative kernel and radius ρ. The following inequalities are valid:*

$$\|[\gamma]_\rho; R^{n+1}_+\|_{M(W^m_p \to W^l_p)} \leqslant \|\gamma; R^{n+1}_+\|_{M(W^m_p \to W^l_p)} \leqslant \varliminf_{\rho\to 0} \|[\gamma]_\rho; R^{n+1}_+\|_{M(W^m_p \to W^l_p)}, \quad (6)$$

$$\|[\gamma]_\rho; R^n\|_{M(W^{m-1/p}_p \to W^{l-1/p}_p)} \leqslant \|\gamma; R^n\|_{M(W^{m-1/p}_p \to W^{l-1/p}_p)} \leqslant \varliminf_{\rho\to 0} \|[\gamma]_\rho; R^n\|_{M(W^{m-1/p}_p \to W^{l-1/p}_p)}. \quad (7)$$

The proof of (6) is analogous to that of Lemma 1.3.1/1 and estimates (7) were proven in Lemma 3.2.1/1.

Lemma 4. *If $\gamma \in W^l_{p,\mathrm{unif}}(R^{n+1}_+)$ and $[\gamma]_\rho$ is a mollification of γ in R^{n+1} with respect to variables x, then $[\gamma]_\rho \in MW^l_p(R^{n+1}_+)$.*

Proof. Obviously all the derivatives with respect to x of the function $\psi = [\gamma]_\rho$ belong to $W^l_{p,\mathrm{unif}}(R^{n+1}_+)$. Therefore all the derivatives of ψ up to the order l, but $\partial^l\psi/\partial y^l$, are bounded. What is more, for all $y \in R^1_+$,

$$\sup_x \int_y^{y+1} |\partial^l \psi(x,t)/\partial t^l|^p \, dt < \infty.$$

Using these properties and the estimate

$$|u(x,y)|^p \leqslant c \sum_{j=0}^{l} \int_y^{y+1} |\partial^j u(x,t)/\partial t^j|^p \, dt,$$

we easily arrive at $\|\psi u\|_{W^l_p} \leqslant C \|u\|_{W^l_p}$. □

3.8.3 A Boundary Value Problem in a Half-Space

Let the operator $\{P, P_1, \ldots, P_\sigma\}$ of the boundary value problem be considered in R^{n+1}_+, where P is a differential operator of order 2σ and P_j are operators of boundary conditions, generated by differential operators of orders σ_j. Suppose coefficients of operators P and P_j are constant and operators $\{P, P_j\}$ form the elliptic boundary value problem (see Agmon, Douglis, Nirenberg [1]).

Theorem. Let $\gamma \in W^h_{p,\mathrm{loc}}(R^{n+1}_+) \cap M(W^{r-h}_p(R^{n+1}_+) \to L_p(R^{n+1}_+))$, *where* r *and* h *are integers,* $r \geqslant h - \sigma_j > 0$, $r \geqslant h - 2\sigma$. *Further, let* $P\gamma \in M(W^r_p(R^{n+1}_+) \to W^{h-2\sigma}_p(R^{n+1}_+))$, $P_j\gamma \in M(W^{r-1/p}_p(R^n) \to W^{h-\sigma_j-1/p}_p(R^n))$. *Then* $\gamma \in M(W^r_p(R^{n+1}_+) \to W^h_p(R^{n+1}_+))$ *and*

$$\|\gamma; R^{n+1}_+\|_{M(W^r_p \to W^h_p)} \leqslant C\Big(\|P\gamma; R^{n+1}_+\|_{M(W^r_p \to W^{h-2\sigma}_p)} + \sum_{j=1}^{\sigma} \|P_j\gamma; R^n\|_{M(W^{r-1/p}_p \to W^{h-\sigma_j-1/p}_p)} + \|\gamma; R^{n+1}_+\|_{M(W^{r-h}_p \to L_p)}\Big). \tag{1}$$

Proof. It is known (see Agmon, Douglis, Nirenberg [1]) that, for all $u \in W^h_p(R^{n+1}_+)$,

$$\|u; R^{n+1}_+\|_{W^h_p} \leqslant C_1\Big(\|Pu; R^{n+1}_1\|_{W^{h-2\sigma}_p} + \sum_{j=1}^{\sigma} \|P_j u; R^n\|_{W^{h-\sigma_j-1/p}_p} + \|u; R^{n+1}_+\|_{L_p}\Big). \tag{2}$$

First we assume that γ belongs to $M(W^r_p(R^{n+1}_+) \to W^h_p(R^{n+1}_+))$. By virtue of (2), for all $\varphi \in C_0^\infty(\overline{R_+^{n+1}})$,

$$\|\gamma\varphi; R_+^{n+1}\|_{W_p^h} \leq C_1\Big(\|\varphi P\gamma; R_+^{n+1}\|_{W_p^{h-2\sigma}} + \sum_{j=1}^{\sigma} \|\varphi P_j\gamma; R^n\|_{W_p^{h-\sigma_j-1/p}}$$
$$+\|\varphi\gamma; R_+^{n+1}\|_{L_p} + \|[\varphi, P]\gamma; R_+^{n+1}\|_{W_p^{h-2\sigma}}$$
$$+\sum_{j=1}^{\sigma} \|[\varphi, P_j]\gamma; R^n\|_{W_p^{h-\sigma_j-1/p}}\Big), \tag{3}$$

where $[\varphi, P]$, $[\varphi, P_j]$ are commutators of P, P_j and the operator of multiplication by φ. It is clear that

$$\|\varphi P\gamma; R_+^{n+1}\|_{W_p^{h-2\sigma}} \leq \|P\gamma; R_+^{n+1}\|_{M(W_p^r \to W_p^{h-2\sigma})} \|\varphi; R_+^{n+1}\|_{W_p^r},$$
$$\|\varphi\gamma; R_+^{n+1}\|_{L_p} \leq \|\gamma; R_+^{n+1}\|_{M(W_p^{r-h} \to L_p)} \|\varphi; R_+^{n+1}\|_{W_p^{r-h}}, \tag{4}$$
$$\|\varphi P_j\gamma; R^n\|_{W_p^{h-\sigma_j-1/p}} \leq \|P_j\gamma; R^n\|_{M(W_p^{r-1/p} \to W_p^{h-\sigma_j-1/p})} \|\varphi; R^n\|_{W_p^{r-1/p}}.$$

Let us estimate the norm $\|[\varphi, P]\gamma; R_+^{n+1}\|_{W_p^{h-2\sigma}}$. For any multi-index α, $|\alpha| \leq 2\sigma$, we have

$$\|[\varphi, D^\alpha]\gamma; R_+^{n+1}\|_{W_p^{h-2\sigma}} \leq c \sum_{0<\beta\leq\alpha} \|D^\beta\varphi D^{\alpha-\beta}\gamma; R_+^{n+1}\|_{W_p^{h-|\alpha|}}$$
$$\leq c_1 \sum_{0<\beta\leq\alpha} \|D^{\alpha-\beta}\gamma; R_+^{n+1}\|_{M(W_p^{r-|\beta|} \to W_p^{h-|\alpha|})}$$
$$\times \|D^\beta\varphi; R_+^{n+1}\|_{W_p^{r-|\beta|}}.$$

This and Lemma 3.8.2/2 yield

$$\|[\varphi, D^\alpha]\gamma; R_+^{n+1}\|_{W_p^{h-2\sigma}} \leq c \sum_{0<\beta\leq\alpha} \|\gamma; R_+^{n+1}\|_{M(W_p^{r-|\beta|} \to W_p^{h-|\beta|})} \|\varphi; R_+^{n+1}\|_{W_p^r}.$$

By (1.3.1/1) the expression on the right does not exceed

$$c \sum_{0<\beta\leq\alpha} \|\gamma; R_+^{n+1}\|^{1-|\beta|/h}_{M(W_p^r \to W_p^h)} \|\gamma; R_+^{n+1}\|^{|\beta|/h}_{M(W_p^{r-h} \to L_p)} \|\varphi; R_+^{n+1}\|_{W_p^r}.$$

So, for any $\varepsilon > 0$,

$$\|[\varphi, P]\gamma; R_+^{n+1}\|_{W_p^{h-2\sigma}} \leq (\varepsilon \|\gamma; R_+^{n+1}\|_{M(W_p^r \to W_p^h)}$$
$$+ c(\varepsilon) \|\gamma; R_+^{n+1}\|_{M(W_p^{r-h} \to L_p)}) \|\varphi; R_+^{n+1}\|_{W_p^r}. \tag{5}$$

We turn to the estimate for the norm $\|[\varphi, P_j]\gamma; R^n\|_{W_p^{h-\sigma_j-1/p}}$. For any multi-index α, $|\alpha| \leq \sigma_j$, we have

$$\|[\varphi, D^\alpha]\gamma; R^n\|_{W_p^{h-\sigma_j-1/p}} \leq c \sum_{0<\beta\leq\alpha} \|D^\beta\varphi D^{\alpha-\beta}\gamma; R^n\|_{W_p^{h-|\alpha|-1/p}}$$
$$\leq c_1 \sum_{0<\beta\leq\alpha} \|D^{\alpha-\beta}\gamma; R^n\|_{M(W_p^{r-|\beta|-1/p} \to W_p^{h-|\alpha|-1/p})}$$
$$\times \|D^\beta\varphi; R^n\|_{W_p^{r-|\beta|-1/p}}.$$

It is clear that

$$\|D^{\alpha-\beta}\gamma; R^n\|_{M(W_p^{r-|\beta|-1/p} \to W_p^{h-|\alpha|-1/p})} \leq c\,\|D^{\alpha-\beta}\gamma; R_+^{n+1}\|_{M(W_p^{r-|\beta|} \to W_p^{h-|\alpha|})}.$$

This and Lemma 3.8.2/2 imply

$$\|[\varphi, D^\alpha]\gamma; R^n\|_{W_p^{h-\sigma_j-1/p}} \leq c \sum_{0<\beta\leq\alpha} \|\gamma; R_+^{n+1}\|_{M(W_p^{r-|\beta|} \to W_p^{h-|\beta|})} \|\varphi\|_{W_p^r}.$$

The right-hand side of this inequality was estimated earlier. Therefore for any $\varepsilon > 0$ the norm $\|[\varphi, P_j]\gamma; R^n\|_{W_p^{h-\sigma_j-1/p}}$ is majorized by the right-hand side in (5). This, together with (3)–(5), implies

$$\|\gamma\varphi; R_+^{n+1}\|_{W_p^h} \leq C_1\Big(\|P\gamma; R_+^{n+1}\|_{M(W_p^r \to W_p^{h-2\sigma})} + \sum_{j=1}^{\sigma} \|P_j\gamma; R^n\|_{M(W_p^{r-1/p} \to W_p^{h-\sigma_j-1/p})} + \varepsilon\,\|\gamma; R_+^{n+1}\|_{M(W_p^r \to W_p^h)} + c(\varepsilon)\,\|\gamma; R_+^{n+1}\|_{M(W_p^{r-h} \to L_p)}\Big)\,\|\varphi\|_{W_p^r}.$$

Consequently estimate (1) is valid.

Let us get rid of the assumption $\gamma \in M(W_p^r(R_+^{n+1}) \to W_p^h(R_+^{n+1}))$.

Let η_y be a function in $C_0^\infty(R^{n+1})$ defined by $\eta_y(z) = \eta(z-y)$, $y \in R^{n+1}$. Since

$$\|\Gamma\eta_y; R_+^{n+1}\|_{W_p^l} \leq \|\Gamma; R_+^{n+1}\|_{M(W_p^m \to W_p^l)} \|\eta_y; R_+^{n+1}\|_{W_p^m} \leq \text{const}\,\|\Gamma; R_+^{n+1}\|_{M(W_p^m \to W_p^l)}$$

for all $y \in \overline{R_+^{n+1}}$, then

$$M(W_p^m(R_+^{n+1}) \to W_p^l(R_+^{n+1})) \subset W_{p,\text{unif}}^l(R_+^{n+1}).$$

The inclusion

$$M(W_p^{m-1/p}(R^n) \to W_p^{l-1/p}(R^n)) \subset W_{p,\text{unif}}^{l-1/p}(R^n)$$

can be derived in a similar way. This and the conditions of the theorem imply $P\gamma \in W_{p,\text{unif}}^{h-2\sigma}(R_+^{n+1})$ and $P_j\gamma \in W_{p,\text{unif}}^{h-\sigma_j-1/p}(R^n)$. The last, together with a local coercive estimate (see Agmon, Douglis, Nirenberg [1]), leads to $\gamma \in W_{p,\text{unif}}^h(R_+^{n+1})$. It remains to substitute the mollification of γ with respect to variables x into (1) and to use Lemmas 3.8.2/3 and 3.8.2/4. □

Remark 1. In the same way one can prove the generalization of the theorem to the elliptic boundary value problem for a system, elliptic in the sense of Douglis–Nirenberg (cf. Theorem 3.8.1).

Remark 2. If $r = h$, then (1) takes the form

$$\|\gamma; R_+^{n+1}\|_{MW_p^r} \leq C\Big(\|P\gamma; R_+^{n+1}\|_{M(W_p^r \to W_p^{r-2\sigma})} + \sum_{j=1}^{\sigma} \|P_j\gamma; R^n\|_{M(W_p^{r-1/p} \to W_p^{r-\sigma_j-1/p})} + \|\gamma; R_+^{n+1}\|_{L_\infty}\Big). \quad (6)$$

Here we *a priori* assume that $\gamma \in W^r_{p,\mathrm{loc}}(R^{n+1}_+) \cap L_\infty(R^{n+1}_+)$ and $P\gamma \in M(W^r_p(R^{n+1}_+) \to W^{r-2\sigma}_p(R^{n+1}_+))$, $r \geq 2\sigma$.

Remark 3. We shall show that the norm $\|\gamma; R^{n+1}_+\|_{L_\infty}$ in the right-hand side of (6) cannot be omitted even if the operator $\{P, P_1, \ldots, P_\sigma\}$ satisfies (2) without the norm $\|u; R^{n+1}_+\|_{L_p}$ (cf. Remark 3.8.1/2).

Let $P = -\Delta + 1$, $P_1 = \partial/\partial x_{n+1}|_{R^n}$, $r = 2$, $2p < n$. By η we denote a function in $C_0^\infty(R^{n+1})$ with support in the unit ball centred at the origin. Let $\eta(0) = 1$ and let $\partial\eta/\partial x_{n+1} = 0$ for $x_{n+1} = 0$. We put $\Gamma(z) = \eta(z)\log|z|$. It is clear that $\Gamma \in W^2_p(R^{n+1}_+)$, $P\Gamma = O(|z|^{-2})$, $P\Gamma = 0$ for $|z| > 1$. Therefore $P\Gamma \in M(W^2_p(R^{n+1}_+) \to L_p(R^{n+1}_+))$. Further, we notice that $P_1\Gamma = 0$. Suppose

$$\|\gamma; R^{n+1}_+\|_{MW^2_p} \leq C(\|P\gamma; R^{n+1}_+\|_{M(W^2_p \to L_p)} + \|P_1\gamma; R^n\|_{M(W^{2-1/p}_p \to W^{1-1/p}_p)})$$

for all $\gamma \in W^2_p(R^{n+1}_+) \cap L_\infty(R^{n+1}_+)$ and substitute a mollification $[\Gamma]_\rho$ of Γ with respect to variables x into the latter inequality. Since $MW^r_p(R^{n+1}_+) \subset L_\infty(R^{n+1}_+)$, then

$$\|[\Gamma]_\rho; R^{n+1}_+\|_{L_\infty} \leq C\,\|P[\Gamma]_\rho; R^{n+1}_+\|_{M(W^2_p \to L_p)}.$$

The right-hand side of this inequality is uniformly bounded with respect to ρ by virtue of Lemma 3, although the left-hand side tends to infinity as $\rho \to 0$.

3.9 Multipliers in the Space B^l_p

In this section we study multipliers on the space B^l_p. We limit consideration to the case $p \in (1, \infty)$ and to multipliers which preserve B^l_p.

The basic result contains necessary and sufficient conditions for a function to belong to the space MB^l_p. For fractional l such conditions were obtained in Theorem 3.2.7/1, according to which

$$\|\gamma\|_{MB^l_p} \sim \sup_e \frac{\|C_{p,l}\gamma; e\|_{L_p}}{[\mathrm{cap}\,(e, B^l_p)]^{1/p}} + \|\gamma\|_{L_\infty}.$$

(We recall that functions $C_{p,l}\gamma$ and $D_{p,l}\gamma$ are equivalent for $\{l\} > 0$.)

We show below that the same result holds for an arbitrary positive l. The exposition follows the authors' paper [4].

3.9.1 An Auxiliary Estimate

Lemma. Let $u \in D$ and let v be a measurable function on R^n. Then

$$\iint |\Delta_h v(x)\,\Delta_h u(x)|^p\,|x|^{-n-p}\,dh\,dx \leq c \sup_e \frac{\|D_{p,\delta}v; e\|^p_{L_p}}{\mathrm{cap}\,(e, W^{k-1+\delta}_p)} \|u\|^p_{B^k_p}, \qquad (1)$$

where $k = 1, 2, \ldots$ and $\delta \in (0, 1)$.

Proof. Let $U \in D(R^{n+s})$ be an extension of the function u to R^{n+s}, where s is such that $k < n + s/p'$. By f we denote the function $\Lambda^{k+s/p}U$, where $\Lambda = (-\Delta+1)^{1/2}$, Δ is the Laplace operator in R^{n+s}. Then

$$u(x) = \int_{R^{n+s}} G_{k+s/p}(x-\xi, \eta) f(\xi, \eta)\, d\xi\, d\eta.$$

The function G_r is introduced at the beginning of Subsection 3.1.2, where its properties are also listed. Let

$$\begin{aligned}
\mathfrak{M}_1 &= \{(\xi, \eta) : 4|h| < |x-\xi| + |\eta| < 1\},\\
\mathfrak{M}_2 &= \{(\xi, \eta) : |x-\xi| + |\eta| < \min(1, 4|h|)\},\\
\mathfrak{M}_3 &= \{(\xi, \eta) : |\xi - x| + |\eta| > \max(1, 4|h|)\},\\
\mathfrak{M}_4 &= \{(\xi, \eta) : 4|h| > |\xi - x| + |\eta| > 1\}.
\end{aligned}$$

It is clear that

$$|\Delta_h u(x)| \leqslant \int_{R^{n+s}} |G_{k+s/p}(x-\xi+h, \eta) - G_{k+s/p}(x-\xi, \eta)|\, |f(\xi, \eta)|\, d\xi\, d\eta.$$

We represent the latter integral as the sum of four integrals over $\mathfrak{M}_1, \ldots, \mathfrak{M}_4$. We have

$$\int_{\mathfrak{M}_1} \leqslant c\,|h| \int_{\mathfrak{M}_1} t_\theta^{-(n-k+1+s/p')} |f(\xi, \eta)|\, d\xi\, d\eta,$$

where $t_\theta^2 = (x+\theta h - \xi)^2 + \eta^2$, $\theta \in (0, 1)$. Obviously $t_\theta^2 \geqslant c\ ((x-\xi)^2 + \eta^2)$ on $\mathfrak{M}_1$. Hence

$$\begin{aligned}
\int_{\mathfrak{M}_1} &\leqslant c\,|h|^{1-\delta} \int_{\mathfrak{M}_1} t_0^{-(n-k+1+s/p'-\delta)} |f(\xi, \eta)|\, d\xi\, d\eta\\
&\leqslant c\,|h|^{1-\delta} (\Lambda^{-(k-1+s/p+\delta)} |f|)(x, 0). \qquad (2)
\end{aligned}$$

For the integral over $\mathfrak{M}_2$ we obtain

$$\begin{aligned}
\int_{\mathfrak{M}_2} &\leqslant \int_{\mathfrak{M}_2} (t_1^{-(n-k+s/p')} + t_0^{-(n-k+s/p')})\, |f(\xi, \eta)|\, d\xi\, d\eta\\
&\leqslant c\,|h|^{1-\delta} \int_{\mathfrak{M}_2} (t_1^{-(n-k+s/p'+1-\delta)} + t_0^{-(n-k+s/p'+1-\delta)})\, |f(\xi, \eta)|\, d\xi\, d\eta.
\end{aligned}$$

Consequently,

$$\int_{\mathfrak{M}_2} \leqslant c\,|h|^{1-\delta} [(\Lambda^{-(k-1+s/p+\delta)} |f|)(x+h, 0) + (\Lambda^{-(k-1+s/p+\delta)} |f|)(x, 0)].$$

By using the asymptotics (3.1.2/2), we derive

$$\int_{\mathfrak{M}_3} \leqslant c\,|h| \int_{\mathfrak{M}_3} \mathrm{e}^{-t_0/2}\,|f(\xi,\eta)|\,\mathrm{d}\xi\,\mathrm{d}\eta.$$

Therefore

$$\int_{\mathfrak{M}_3} \leqslant c\,|h|^{1-\delta} \int_{\mathfrak{M}_3} t_0^{(k-n-s/p'-1+\delta)/2}\,\mathrm{e}^{-t_0/4}\,|f(\xi,\eta)|\,\mathrm{d}\xi\,\mathrm{d}\eta$$

$$\leqslant c\,|h|^{1-\delta}\,(\Lambda^{-(k-1+s/p+\delta)}\,|F|)\Big(\frac{x}{4},0\Big), \tag{3}$$

where $F(\xi,\eta)=f(4\xi,4\eta)$. In the same way,

$$\int_{\mathfrak{M}_4} \leqslant c \int_{\mathfrak{M}_4} (\mathrm{e}^{-t_1/2}+\mathrm{e}^{-t_0/2})\,|f(\xi,\eta)|\,\mathrm{d}\xi\,\mathrm{d}\eta$$

$$\leqslant c\,|h|^{1-\delta} \int_{\mathfrak{M}_4} [t_1^{(k-n-s/p'-1+\delta)/2}\mathrm{e}^{-t_1/4} + t_0^{(k-n-s/p'-1+\delta)/2}\mathrm{e}^{-t_0/4}]\,|f(\xi,\eta)|\,\mathrm{d}\xi\,\mathrm{d}\eta$$

and consequently,

$$\int_{\mathfrak{M}_4} \leqslant c\,|h|^{1-\delta}\Big[(\Lambda^{-(k-1+s/p+\delta)}\,|F|)\Big(\frac{x+h}{4},0\Big) + (\Lambda^{-(k-1+s/p+\delta)}\,|F|)\Big(\frac{x}{4},0\Big)\Big]. \tag{4}$$

Summing (2)–(4) and noting that $G_r(az)\geqslant G_r(z)$ for $a<1$, we obtain

$$|\Delta_h u(x)| \leqslant c\,|h|^{1-\delta}\Big[(\Lambda^{-(k-1+s/p+\delta)}\,|F|)\Big(\frac{x+h}{4},0\Big) + (\Lambda^{-(k-1+s/p+\delta)}\,|F|)\Big(\frac{x}{4},0\Big)\Big].$$

Hence

$$\iint |\Delta_h v(x)\,\Delta_h u(x)|^p\,|h|^{-n-p}\,\mathrm{d}h\,\mathrm{d}x$$

$$\leqslant c \int \Big[(\Lambda^{-(k-1+s/p+\delta)}\,|F|)\Big(\frac{x}{4},0\Big)\Big]^p \int |\Delta_h v(x)|^p\,|h|^{-n-p\delta}\,\mathrm{d}h\,\mathrm{d}x$$

$$= c_1 \int [(\Lambda^{-(k-1+s/p+\delta)}\,|F|)(x,0)]^p \int |\Delta_h v(4x)|^p\,|h|^{-n-p\delta}\,\mathrm{d}h\,\mathrm{d}x.$$

Lemma 3.2.1/2 gives

$$\int\int |\Delta_h v(x)\,\Delta_h u(x)|^p\,|h|^{-n-p}\,dh\,dx$$

$$\leqslant c\sup_e \frac{\int_e \int |\Delta_h v(4x)|^p\,|h|^{-n-p\delta}\,dh\,dx}{\operatorname{cap}(e, W_p^{k-1+\delta})}\,\|(\Lambda^{-(k-1+s/p+\delta)}\,|F|)(\cdot,0)\|^p_{W_p^{k-1+\delta}}. \tag{5}$$

It is clear that

$$\int_e \int |\Delta_h v(4x)|^p\,|h|^{-n-p\delta}\,dh\,dx = c\int_{4e}\int |\Delta_h v(x)|^p\,|h|^{-n-p\delta}\,dh\,dx, \tag{6}$$

$$\operatorname{cap}(e, W_p^{k-1+\delta}) \geqslant c\operatorname{cap}(4e, W_p^{k-1+\delta}), \tag{7}$$

where $4e = \{x : x/4 \in e\}$. We note also that

$$\|(\Lambda^{-(k-1+s/p+\delta)}\,|F|)(\cdot,0)\|_{W_p^{k-1+\delta}} \leqslant \|F; R^{n+s}\|_{L_p} = 4^{-(n+s)}\,\|f; R^{n+s}\|_{L_p}$$
$$= 4^{-(n+s)}\,\|\Lambda^{k+s/p}U; R^{n+s}\|_{L_p}.$$

This, together with (5)–(7), yields

$$\int\int |\Delta_h v(x)\,\Delta_h u(x)|^p\,|h|^{-n-p}\,dh\,dx$$

$$\leqslant c\sup_e \frac{\|D_{p,\delta}v; e\|^p_{L_p}}{\operatorname{cap}(e, W_p^{k-1+\delta})}\,\|\Lambda^{k+s/p}U; R^{n+s}\|^p_{L_p}.$$

By minimizing the latter norm over all extensions U, we complete the proof. □

3.9.2 Interpolation Inequalities

We shall use the following known assertion on the interpolation property of the scale B_p^l (see, for example, Triebel [4]).

Proposition. *Let* $p \in (1, \infty)$, $\theta \in (0, 1)$, $l_0 > l_1 > 0$, $l = \theta l_0 + (1-\theta)l_1$. *Further, let L be a linear continuous operator in* $B_p^{l_0}$, *admitting an extension to a continuous operator in* $B_p^{l_1}$. *Then L can be extended to a continuous operator in* B_p^l *and*

$$\|L\|_{B_p^l \to B_p^l} \leqslant c\,\|L\|^{\theta}_{B_p^{l_0} \to B_p^{l_0}}\,\|L\|^{1-\theta}_{B_p^{l_1} \to B_p^{l_1}}.$$

This, in particular, implies

$$\|\gamma\|_{MB_p^l} \leqslant c\,\|\gamma\|^{\theta}_{MB_p^{l_0}}\,\|\gamma\|^{1-\theta}_{MB_p^{l_1}}, \tag{1}$$

where $p \in (1, \infty)$, $l_0, l_1 > 0$ *and* $l = \theta l_0 + (1-\theta)l_1$.

We prove two more interpolation inequalities for multipliers in B_p^l.

Lemma. *Let $\gamma \in MB_p^l$, $l>0$, $p\in(1,\infty)$. Then $\gamma\in L_\infty \cap B_{p,\mathrm{unif}}^l$ and the following inequalities hold:*

$$\|\gamma\|_{MB_p^\sigma} \leqslant c\,\|\gamma\|_{L_\infty}^{1-\sigma/l}\,\|\gamma\|_{MB_p^l}^{\sigma/l}, \qquad 0<\sigma\leqslant l, \tag{2}$$

$$\sup_e \frac{\|C_{p,\sigma}\gamma;\, e\|_{L_p}}{[\mathrm{cap}\,(e,\, B_p^\sigma)]^{1/p}} \leqslant c\,\|\gamma\|_{L_\infty}^{1-\sigma/l}\,\|\gamma\|_{MB_p^l}^{\sigma/l}. \tag{3}$$

Proof. Obviously, $\gamma\in B_{p,\mathrm{unif}}^l$. Let $u\in D$ and let N be an arbitrary positive number. It is clear that

$$\|\gamma^N u\|_{L_p}^{1/N} \leqslant \|\gamma^N u\|_{B_p^l}^{1/N} \leqslant \|\gamma\|_{MB_p^l}\,\|u\|_{B_p^l}^{1/N}.$$

Consequently,

$$\|\gamma\|_{L_\infty} \leqslant \|\gamma\|_{MB_p^l}. \tag{4}$$

We perform the proof in three steps.

1. Let $l=1$. One can easily check the identity

$$\Delta_h^{(2)}(\gamma_\rho u) = \gamma_\rho\,\Delta_h^{(2)} u + u\,\Delta_h^{(2)}\gamma_\rho + \Delta_{2h}\gamma_\rho\,\Delta_{2h} u - 2\,\Delta_h\gamma_\rho\,\Delta_h u, \tag{5}$$

where γ_ρ is a mollification of γ with radius ρ. Therefore

$$\begin{aligned}
\|uC_{p,1}\gamma_\rho\|_{L_p} &\leqslant \|\gamma_\rho u\|_{B_p^1} + \|\gamma_\rho C_{p,1} u\|_{L_p}\\
&\quad + \left(\int\!\!\int |\Delta_{2h}\gamma_\rho(x)\,\Delta_{2h}u(x)|^p\,|h|^{-n-p}\,dh\,dx\right)^{1/p}\\
&\quad + 2\left(\int\!\!\int |\Delta_h\gamma_\rho(x)\,\Delta_h u(x)|^p\,|h|^{-n-p}\,dh\,dx\right)^{1/p}\\
&\leqslant \|\gamma_\rho u\|_{B_p^1} + \|\gamma_\rho C_{p,1} u\|_{L_p}\\
&\quad + 4\left(\int\!\!\int |\Delta_h\gamma_\rho(x)\,\Delta_h u(x)|^p\,|h|^{-n-p}\,dh\,dx\right)^{1/p}.
\end{aligned}$$

Applying Lemma 3.9.1 and the estimate $\|\gamma_\rho\|_{L_\infty} \leqslant \|\gamma_\rho\|_{MB_p^1}$, we obtain

$$\|uC_{p,1}\gamma_\rho\|_{L_p} \leqslant c\left(\|\gamma_\rho\|_{MB_p^1} + \sup_e \frac{\|D_{p,\sigma}\gamma_\rho;\, e\|_{L_p}}{[\mathrm{cap}\,(e,\, W_p^\sigma)]^{1/p}}\right)\|u\|_{B_p^1},$$

where $\sigma\in(0,1)$. This immediately implies

$$\sup_e \frac{\|C_{p,1}\gamma_\rho;\, e\|_{L_p}}{[\mathrm{cap}\,(e,\, B_p^1)]^{1/p}} \leqslant c(\|\gamma_\rho\|_{MB_p^1} + \|\gamma_\rho\|_{MW_p^\sigma}). \tag{6}$$

Since $W_p^t = B_p^t$ for fractional t, then from (1) we get

$$\|\gamma_\rho\|_{MW_p^\sigma} \leqslant c\,\|\gamma_\rho\|_{MB_p^1}^{(\sigma-\varepsilon)/(1-\varepsilon)}\,\|\gamma_\rho\|_{MW_p^\varepsilon}^{(1-\sigma)/(1-\varepsilon)}, \qquad 0<\varepsilon<\delta.$$

By (3.2.1/1),

$$\|\gamma_\rho\|_{MB_p^\varepsilon} \leqslant c\, \|\gamma_\rho\|_{MW_p^\sigma}^{\varepsilon/\sigma}\, \|\gamma_\rho\|_{L_\infty}^{(\sigma-\varepsilon)/\sigma}.$$

So the last two estimates imply

$$\|\gamma_\rho\|_{MW_p^\sigma} \leqslant c\, \|\gamma_\rho\|_{MB_p^1}^{\sigma}\, \|\gamma_\rho\|_{L_\infty}^{1-\sigma}. \tag{7}$$

The latter, together with (6), yields

$$\sup_e \frac{\|C_{p,1}\gamma_\rho;\, e\|_{L_p}}{[\mathrm{cap}\,(e, B_p^1)]^{1/p}} \leqslant c\, \|\gamma_\rho\|_{MB_p^1}. \tag{8}$$

Applying Lemma 3.2.1/1, from (7) we obtain (2) with $l=1$. Moreover we get

$$\overline{\lim_{\rho\to 0}}\, \sup_e \frac{\|C_{p,1}\gamma_\rho;\, e\|_{L_p}}{[\mathrm{cap}\,(e, B_p^1)]^{1/p}} \leqslant c\, \|\gamma\|_{MB_p^1}$$

from (8). Noting that $\|C_{p,1}(\gamma_\rho-\gamma)\|_{L_p} \to 0$ as $\rho\to+0$, we arrive at (3) for $\sigma=l=1$. Inequality (3) for fractional σ and $l=1$ follows from (7) and

$$\|\gamma\|_{MW_p^\sigma} \sim \sup_e \frac{\|C_{p,\sigma}\gamma;\, e\|_{L_p}}{[\mathrm{cap}\,(e, B_p^\sigma)]^{1/p}} + \|\gamma\|_{L_\infty} \tag{9}$$

which is contained in Theorem 3.2.7/1.

2. Let the lemma be proved for all positive integers up to $l-1$. By (5) we have

$$\|uC_{p,l}\gamma_\rho\|_{L_p} \leqslant \|\gamma_\rho u\|_{B_p^l} + c \sum_{|\alpha|+|\beta|=l-1} \|C_{p,1}(D^\alpha\gamma_\rho D^\beta u)\|_{L_p} + \||\nabla_{l-1}\gamma_\rho|\, C_{p,1}u\|_{L_p}$$

$$+4\Big(\int\int |\Delta_h \nabla_{l-1}\gamma_\rho(x)|^p\, |\Delta_h u(x)|^p\, |h|^{-n-p}\, dh\, dx\Big)^{1/p}.$$

Applying (5) once more, we obtain

$$\|uC_{p,l}\gamma_\rho\|_{L_p} \leqslant \|\gamma_\rho u\|_{B_p^l} + c\sum_{j=0}^{l-1} \||\nabla_j\gamma_\rho|\, C_{p,l-j}u\|_{L_p} + c\sum_{j=1}^{l-1} \||\nabla_j u C_{p,l-j}\gamma_\rho\|_{L_p}$$
$$+c\sum_{j=0}^{l-1}\Big(\int\int |\Delta_h\nabla_j\gamma_\rho(x)|^p\, |\Delta_h\nabla_{l-1-j}u|^p\, |h|^{-n-p}\, dh\, dx\Big)^{1/p}. \tag{10}$$

According to Lemma 3.1.1/1, $(C_{p,l-j}u)(x) \leqslant (J_j^{(n)} C_{p,l-j}\Lambda^j u)(x)$. This and Lemma 3.2.1/2 imply

$$\||\nabla_j\gamma_\rho|\, C_{p,l-j}u\|_{L_p} \leqslant c \sup_e \frac{\|\nabla_j\gamma_\rho;\, e\|_{L_p}}{[\mathrm{cap}\,(e, W_p^j)]^{1/p}}\, \|C_{p,l-j}\Lambda^j u\|_{L_p}. \tag{11}$$

Since $j<l$, then, by virtue of Lemma 3.1.1/3,

$$\|C_{p,l-j}\Lambda^j u\|_{L_p} \leq \|\Lambda^j u\|_{B_p^{l-j}} \leq c\,\|u\|_{B_p^l}. \tag{12}$$

Taking into account that the family of spaces B_p^σ forms an interpolation scale and $W_p^{j+\varepsilon}=B_p^{j+\varepsilon}$ for $\varepsilon\in(0,1)$, we obtain

$$\|\gamma_\rho\|_{MW_p^{j+\varepsilon}} \leq c\,\|\gamma_\rho\|_{MB_p^l}^{(j+\varepsilon-1)/(l-1)}\,\|\gamma_\rho\|_{MB_p^1}^{(l-j-\varepsilon)/(l-1)}, \tag{13}$$

$$\|\gamma_\rho\|_{MB_p^1} \leq c\,\|\gamma_\rho\|_{MB_p^l}^{(1-\sigma)/(l-\sigma)}\,\|\gamma_\rho\|_{MB_p^\sigma}^{(l-1)/(l-\sigma)}. \tag{14}$$

Unifying (7) with (14), we arive at

$$\|\gamma_\rho\|_{MB_p^1} \leq c\,\|\gamma_\rho\|_{MB_p^l}^{1/l}\,\|\gamma_\rho\|_{L_\infty}^{(l-1)/l}.$$

Substituting the last inequality into (13), we find

$$\|\gamma_\rho\|_{MW_p^{j+\varepsilon}} \leq c\,\|\gamma_\rho\|_{MB_p^l}^{(j+\varepsilon)/l}\,\|\gamma_\rho\|_{L_\infty}^{(l-j-\varepsilon)/l}. \tag{15}$$

Applying (3.2.7/3), we conclude

$$\sup_e \frac{\|\nabla_j\gamma_\rho;e\|_{L_p}}{[\mathrm{cap}\,(e,W_p^j)]^{1/p}} \leq c\,\|\gamma_\rho\|_{L_\infty}^{\varepsilon/(j+\varepsilon)}\,\|\gamma_\rho\|_{MB_p^{j+\varepsilon}}^{j/(j+\varepsilon)}$$

which, together with (15), gives

$$\sup_e \frac{\|\nabla_j\gamma_\rho;e\|_{L_p}}{[\mathrm{cap}\,(e,W_p^j)]^{1/p}} \leq c\,\|\gamma_\rho\|_{L_\infty}^{(l-j)/l}\,\|\gamma_\rho\|_{MB_p^l}^{j/l}.$$

This and (11), (12) imply

$$\||\nabla_j\gamma_\rho|\,C_{p,l-j}u\|_{L_p} \leq c\,\|\gamma_\rho\|_{L_\infty}^{(l-j)/l}\,\|\gamma_\rho\|_{MB_p^l}^{j/l}\,\|u\|_{B_p^l}. \tag{16}$$

Using Lemma 3.2.1/2, we get

$$\||\nabla_j u|\,C_{p,l-j}\gamma_\rho\|_{L_p} \leq c\,\sup_e \frac{\|C_{p,l-j}\gamma_\rho;e\|_{L_p}}{[\mathrm{cap}\,(e,B_p^{l-j})]^{1/p}}\,\|u\|_{B_p^l}, \qquad 1\leq j\leq l-1.$$

This and the induction hypothesis lead to

$$\||\nabla_j u|\,C_{p,l-j}\gamma_\rho\|_{L_p} \leq c\,\|\gamma_\rho\|_{L_\infty}^{j/l}\,\|\gamma_\rho\|_{MB_p^l}^{(l-j)/l}\,\|u\|_{B_p^l}. \tag{17}$$

To estimate the third sum in the right-hand side of (10) we make use of Lemma 3.9.1 for $k=1+j$ and $\delta=\varepsilon$. Then

$$\left(\int\int|\Delta_h\nabla_j\gamma_\rho(x)|^p\,|\Delta_h\nabla_{l-1-j}u|^p\,|h|^{-n-p}\,dh\,dx\right)^{1/p}$$

$$\leq c\,\sup_e \frac{\|D_{p,j+\varepsilon}\gamma_\rho;e\|_{L_p}}{[\mathrm{cap}\,(e,W_p^{j+\varepsilon})]^{1/p}}\,\|u\|_{B_p^l}.$$

Further, (15) and Theorem 3.2.7/1 imply that the right-hand side of the latter inequality does not exceed $c\,\|\gamma_\rho\|_{MB_p^l}^{(j+\varepsilon)/l}\,\|\gamma_\rho\|_{L_\infty}^{(l-j-\varepsilon)/l}\,\|u\|_{B_p^l}$. Unifying this with (16) and (17), we obtain from (10)

$$\|uC_{p,l}\gamma_\rho\|_{L_p} \leq c\Big(\|\gamma_\rho\|_{MB_p^l} + \sum_{j=1}^{l-1} \|\gamma_\rho\|_{L_\infty}^{j/l} \|\gamma_\rho\|_{MB_p^l}^{(l-j)/l}$$
$$+ \sum_{j=0}^{l-1} \|\gamma_\rho\|_{L_\infty}^{(l-j-\varepsilon)/l} \|\gamma_\rho\|_{MB_p^l}^{(j+\varepsilon)/l}\Big) \|u\|_{B_p^l}.$$

Consequently,

$$\sup_e \frac{\|C_{p,l}\gamma_\rho; e\|_{L_p}}{[\text{cap}(e, B_p^l)]^{1/p}} \leq c\, \|\gamma_\rho\|_{MB_p^l}. \tag{18}$$

We put $j+\varepsilon=\sigma$ into (15) and use the last inequality for $l=\sigma$. Then we obtain estimates (2), (3) for the function γ_ρ with $\sigma \in (0, l]$. Applying Lemma 3.2.1/1 and the convergence of $C_{p,l}\gamma_\rho$ to $C_{p,l}\gamma$ in L_p-norm, we complete the proof of (2), (3) for integer l.

3. Let l be a positive fractional number. If σ is also fractional, $\sigma \leq l$, then (2) and (3) follow from the coincidence of spaces MW_p^l and MB_p^l combined with (9). If σ is an integer, then for any $\varepsilon \in (0, 1)$, $\varepsilon < l-\sigma$, we have $MB_p^{\sigma+\varepsilon} = MW_p^{\sigma+\varepsilon}$. Hence

$$\|\gamma\|_{MB_p^\sigma} \leq c\, \|\gamma\|_{L_\infty}^{\varepsilon/(\sigma+\varepsilon)} \|\gamma\|_{MB_p^{\sigma+\varepsilon}}^{\sigma/(\sigma+\varepsilon)} \leq c\, \|\gamma\|_{L_\infty}^{(l-\sigma)/l} \|\gamma\|_{MB_p^l}^{\sigma/l}.$$

So inequality (2) follows for fractional l.

Applying Lemma 3.2.1 and the convergence of $C_{p,l}\gamma_\rho$ to $C_{p,l}\gamma$ in L_p-norm, we obtain from (18), with $l=\sigma$,

$$\sup_e \frac{\|C_{p,\sigma}\gamma; e\|_{L_p}}{[\text{cap}(e, B_p^\sigma)]^{1/p}} \leq c\, \|\gamma\|_{MB_p^\sigma}.$$

The latter, together with (2), yields (3) for integer σ and fractional l. □

3.9.3 Description of the Space MB_p^l

Lemma. *Let l be a positive integer, $1<p<\infty$, $\gamma \in L_\infty \cap B_{p,\text{loc}}^l$ and let*

$$\sup_e \frac{\|C_{p,l}\gamma; e\|_{L_p}}{[\text{cap}(e, B_p^l)]^{1/p}} < \infty.$$

Then $\gamma \in MB_p^l$ and

$$\|\gamma\|_{MB_p^l} \leq c\Big(\|\gamma\|_{L_\infty} + \sup_e \frac{\|C_{p,l}\gamma; e\|_{L_p}}{[\text{cap}(e, B_p^l)]^{1/p}}\Big). \tag{1}$$

Proof. Using identity (3.9.2/5), we obtain

$$\|\gamma_\rho u\|_{B_p^l} \leq c\Big(\sum_{j=0}^{l-1} \||\nabla_j\gamma_\rho|\, C_{p,l-j}u\|_{L_p} + \sum_{j=0}^{l-1} \||\nabla_j u|\, C_{p,l-j}\gamma_\rho\|_{L_p}$$
$$+ \sum_{j=0}^{l-1}\Big(\int\int |\Delta_h \nabla_j \gamma_\rho(x)|^p\, |\Delta_h \nabla_{l-1-j} u|^p\, |h|^{-n-p} dh\, dx\Big)^{1/p}\Big)$$

(cf. (3.9.2/10)). Each of the summands on the right was estimated in the proof of Lemma 3.9.2. Therefore

$$\|\gamma_\rho u\|_{B^l_p} \leqslant c \sum_{j=0}^{l-1} (\|\gamma_\rho\|^{j/l}_{L_\infty} \|\gamma_\rho\|^{(l-j)/l}_{MB^l_p} + \|\gamma_\rho\|^{(l-j-\varepsilon)/l}_{L_\infty} \|\gamma_\rho\|^{(j+\varepsilon)/l}_{MB^l_p}) \|u\|_{B^l_p},$$

where $\varepsilon \in (0, 1)$. Using (3.9.2/4) and Lemma 3.2.1/1, we complete the proof. □

Unifying the above Lemma with Lemma 3.9.2, we arrive at the basic result of the present section.

Theorem. *A function γ belongs to the space MB^l_p if and only if $\gamma \in L_\infty \cap B^l_{p,\mathrm{loc}}$ and, for any compact set $e \subset R^n$, $\|C_{p,l}\gamma; e\|^p_{L_p} \leqslant \mathrm{const}\ \mathrm{cap}\,(e, B^l_p)$. The relation*

$$\|\gamma\|_{MB^l_p} \sim \sup_e \frac{\|C_{p,l}\gamma; e\|_{L_p}}{[\mathrm{cap}\,(e, B^l_p)]^{1/p}} + \|\gamma\|_{L_\infty}$$

holds.

3.10 Multipliers in BMO

The space BMO of functions with bounded mean oscillation (see John and Nirenberg [1], Campanato [1], Fefferman [1], Stegenga [1], Janson [1], [2] and elsewhere) plays an important role in modern analysis. This space is defined as follows. Let $Q(x, r)$ be the cube $\{y \in R^n : |y_i - x_i| < r/2, i = 1, \ldots, n\}$. By $f(Q)$ we denote the mean value of f on the cube Q, that is

$$f(Q) = \frac{1}{\mathrm{mes}_n\, Q} \int_Q f(x)\, dx.$$

Further, we introduce the mean oscillation of f on Q by

$$O(f, Q) = \frac{1}{\mathrm{mes}_n\, Q} \int_Q |f(x) - f(Q)|\, dx.$$

By BMO we denote the space of functions, summable in R^n and such that

$$\sup_{x \in R^n, 0<r<1} O(f, Q(x, r)) < \infty.$$

Provided with the norm

$$\|f\|_{\mathrm{BMO}} = \|f; R^n\|_{L_1} + \sup_{x \in R^n, 0<r<1/2} O(f, Q(x, r)),$$

BMO becomes a Banach space.

We can include BMO in the family of spaces BMO_φ which contain functions with the finite norm

$$\|f; R^n\|_{L_1} + \sup_{x\in R^n, 0<r<1/2} \frac{O(f, Q(x,r))}{\varphi(r)},$$

where φ is a positive non-decreasing function on $(0, 1/2)$.

Stegenga [1] and Janson [1] obtained the description of the space $M(\mathrm{BMO})$ of multipliers in BMO. We present their theorem and its proof.

Theorem 1. *The space $M(\mathrm{BMO})$ coincides with $\mathrm{BMO}_{|\log r|^{-1}} \cap L_\infty$.*

Proof. We begin with the derivation of the following useful estimate for functions in BMO:

$$|f(Q_r) - f(Q_{1/2})| \leq c\, \|f\|_{\mathrm{BMO}}\, |\log r|, \tag{1}$$

where $Q_\rho = Q(x, \rho)$. It is clear that the left-hand side of this inequality does not exceed

$$\int_r^{1/2} \left|\frac{d}{d\rho} f(Q_\rho)\right| d\rho \leq c \int_r^{1/2} \frac{d\rho}{\rho^n} \int_{\partial Q_\rho} |f(y) - f(Q_\rho)|\, ds_y$$
$$\leq c_1 \int_r^{1/2} \frac{\partial \rho}{\rho^{n+1}} \int_{\rho/2}^{\rho} dr \int_{\partial Q_r} |f(y) - f(Q_\rho)|\, ds_y.$$

Therefore

$$|f(Q_r) - f(Q_{1/2})| \leq c \int_r^{1/2} \frac{d\rho}{\rho^{n+1}} \int_{Q_\rho} |f(y) - f(Q_\rho)|\, dy$$

which immediately yields (1). By virtue of (1),

$$|f(Q_r)| \leq 2^n \|f\|_{L_1} + c\, \|f\|_{\mathrm{BMO}}\, |\log r|. \tag{2}$$

Let $\gamma \in \mathrm{BMO}_{|\log r|^{-1}} \cap L_\infty$. Then

$$r^{-n} \int_{Q_r} |(\gamma f)(y) - \gamma(Q_r) f(Q_r)|\, dy$$
$$\leq r^{-n} \int_{Q_r} |\gamma(x)|\, |f(x) - f(Q_r)|\, dy + r^{-n} \int_{Q_r} |f(Q_r)|\, |\gamma(y) - \gamma(Q_r)|\, dy$$
$$\leq \|\gamma\|_{L_\infty} O(f, Q_r) + |f(Q_r)|\, O(\gamma, Q_r).$$

From (2) it follows that the last sum is not more than

$$c\, |\log r|\, (\|\gamma\|_{L_\infty} + \|\gamma\|_{\mathrm{BMO}_{|\log r|^{-1}}})\, \|f\|_{\mathrm{BMO}}.$$

It remains to note that, for any number a,

$$O(f, Q_r) \leqslant \frac{2}{\text{mes}_n\, Q_r} \int_{Q_r} |f(y) - a|\, dy.$$

Hence $\gamma \in M(\text{BMO})$.

Now let us show that $M(\text{BMO}) \subset L_\infty \cap \text{BMO}_{|\log r|^{-1}}$. Obviously,

$$\|\gamma\|_{L_\infty} = \lim_{N\to\infty} \|\gamma^N f\|_{L_1}^{1/N} \leqslant \varlimsup_{N\to\infty} \|\gamma^N f\|_{\text{BMO}}^{1/N} \leqslant \|\gamma\|_{M(\text{BMO})}. \tag{3}$$

We see that

$$\frac{1}{\text{mes}_n\, Q_r} \int_{Q_r} |\gamma(y) - \gamma(Q_r)|\, |f(y)|\, dy \leqslant O(\gamma f, Q_r) + 2\, \|\gamma\|_{L_\infty}\, O(f, Q_r).$$

This, together with (3) and the inequality $O(\gamma f, Q_r) \leqslant \|\gamma\|_{M(\text{BMO})}\, \|f\|_{\text{BMO}}$, implies

$$\frac{1}{\text{mes}_n\, Q_r} \int_{Q_r} |\gamma(y) - \gamma(Q_r)|\, |f(y)|\, dy \leqslant 3\, \|\gamma\|_{M(\text{BMO})}\, \|f\|_{\text{BMO}}.$$

By substituting here $f(y) = \eta(y) \log|x - y|$, where $\eta \in C_0^\infty(Q_1)$, $\eta = 1$ on $Q_{1/2}$, we obtain

$$O(\gamma, Q_r) \leqslant c\, |\log r|^{-1}. \quad \square$$

A little more complicated is the proof of the following general result.

Theorem 2. (Janson [1]). *Let the function $\varphi(r)r^{-1}$ be 'almost decreasing' in the sense that $\varphi(\rho)\rho^{-1} \leqslant \text{const}\, \varphi(r)r^{-1}$ provided that $\rho \geqslant r$. Then $M(\text{BMO}_\varphi) = \text{BMO}_\psi \cap L_\infty$, where $\psi(r) = \varphi(r)/(\int_r^1 \varphi(t)t^{-1}\,dt)$.*

By using the duality of the Hardy space H^1 and BMO (see Stein [1]), Janson [1] obtained the coincidence of spaces of multipliers in H^1 and BMO. In other words, $MH^1 = \text{BMO}_{|\log r|^{-1}} \cap L_\infty$.

3.11 Multipliers in Certain Spaces of Analytic Functions

The content of the present chapter as well as that of Chapter 5 is connected with some new results on multipliers in spaces of analytic functions on the disc B_1 with boundary values in B_p^l. Here we restrict ourselves to a brief outline.

Let φ be a positive continuous and summable function given on B_1. By $A_p(\varphi)$ we denote the space of functions which are analytic in B_1 and have the finite norm

$$|f(O)|+\left(\int_{B_1} |f'(z)|^p \varphi(z)\,dx\,dy\right)^{1/p}, \qquad z=x+iy,\ 1\leq p<\infty.$$

One can easily show (see Proposition 1.6/4) that the space $MA_p(\varphi)$ of multipliers in $A_p(\varphi)$ is imbedded into the space H^∞ of analytic and bounded functions on B_1. This immediately implies that $\gamma\in MA_p(\varphi)$ if and only if $\gamma\in H^\infty$ and, for all $g\in A_p(\varphi)$,

$$\int_{B_1} |g(z)|^p\,|\gamma'(z)|^p\,\varphi(z)\,dx\,dy<\infty.$$

This last means that $A_p(\varphi)$ is imbedded into the space $L_p(B_1,\mu)$ of functions defined on B_1 and summable with order p with respect to the measure μ with density $|\gamma'|^p\,\varphi$.

Thus, to describe the space $MA_p(\varphi)$ it suffices to characterize all Borel measures in B_1 for which the imbedding $A_p(\varphi)\subset L_p(B_1,\mu)$ holds.

The results in question concern the case $\varphi(z)=(1-|z|)^{p(1-l)-1}$, $p\in[1,2]$, $l\in[0,1/p]$. We put $A_p(\varphi)=A_p^l$.

The space A_p^l can be naturally identified with the space $\{f\in B_p^l(\partial B_1):\hat f(n)=0,\ n=-1,-2,\dots\}$, where

$$\hat f(n)=(2\pi)^{-1}\int_{-\pi}^{\pi} f(e^{it})e^{-int}\,dt.$$

According to the well-known theorem of Carleson [1] (see also Stein [1] § 4 ch. 7, N. K. Nikol'skii [1] Lecture 7), the Hardy space $H^p(B_1)$ is imbedded into $L_p(B_1,\mu)$ if and only if

$$\sup_{\zeta\in\partial B_1, r>0} r^{-1}\mu(B_1\cap B_r(\zeta))<\infty. \tag{1}$$

This condition can be rewritten as

$$\sup_{\lambda\in B_1}\int_{B_1}\frac{1-|\lambda|^2}{|1-\bar\lambda z|^2}\,d\mu(z)<\infty \tag{2}$$

(see N. K. Nikol'skii [1], p. 198–200).

Vinogradov attracted our attention to the fact that each of conditions (1), (2) is necessary and sufficient for the imbedding $A_p^0\subset L_p(B_1,\mu)$, $1\leq p\leq 2$. The proof of sufficiency follows from the mentioned Carleson theorem and from the imbedding $H^p\subset A_p^0$ for $p\in[1,2]$ (for $p=2$ the spaces H^p and A_p^0 coincide). The necessity follows from the boundedness of the function $\|\psi_\lambda\|_{A_p^0}$, $\lambda\in B_1$, where

$$\psi_\lambda(z)=\frac{(1-|\lambda|^2)^{1/p}}{(1-\bar\lambda z)^{2/p}}.$$

Taking into account the properties of the space $MA_p(\varphi)$, discussed above, we obtain the following result.

Function γ belongs to space MA_p^0 if and only if $\gamma \in H^\infty$ and the measure μ with density $|\gamma'(z)|^p (1-|z|)^{p-1}$ satisfies either (1) or (2).

Vinogradov [1] noted that this criterion admits one more formulation,

$$\gamma \in MA_p^0 \Leftrightarrow \sup_{\lambda \in B_1} \|\gamma \circ \omega_\lambda\|_{A_p^0} < \infty,$$

where $\omega_\lambda(z) = (\lambda - z)/(1 - \bar{\lambda} z)$.

The space $M(A_2^m \to A_2^l)$, even for negative values of m and l, was studied by Stegenga [2]. His principal result is presented in the following theorem, the formulation of which contains the capacity cap $(E, W_2^l(\partial B_1))$ of a subset E of the circumference ∂B_1, being defined in the same manner as the capacity of a subset of R^1 (see 3.1.3).

Theorem 1 (Stegenga [2]). *Let $0 < l \leq 1/2$. The space A_2^l is imbedded into $L_2(B_1, \mu)$ if and only if*

$$\mu(U_j S(I_j)) \leq c \operatorname{cap}(U_j I_j, W_2^l(\partial B_1)). \tag{3}$$

Here $\{I_j\}$ is any finite collection of disjoint arcs on ∂B_1, $S(I) = \{z : z/|z| \in I, 1 - |z| < \operatorname{mes}_1(I)/2\pi\}$.

This implies that an analytic function γ defined on B_1 is a multiplier in A_2^l, $0 < l \leq 1/2$, if and only if $\gamma \in L_\infty(B_1)$ and

$$\int_{U_j S(I_j)} |\gamma'(z)|^2 (1-|z|)^{1-2l} \, dx\, dy \leq c \operatorname{cap}(U_j I_j, W_2^l(\partial B_1)). \tag{4}$$

Stegenga [2] constructed the example which shows that the validity of (3) for only one arbitrary arc is insufficient for the inclusion of a bounded analytic function into $MA_2^{1/2}$. The same paper contains references to earlier literature on spaces $M(A_2^m \to A_2^l)$.

Theorem 3.4.1/2 implies the following result relating to the case $p = 1$.

Theorem 2. *Let $0 < l < 1$. The space A_1^l is imbedded into $L(B_1, \mu)$ if and only if*

$$\sup_{\zeta \in \partial B_1, r > 0} r^{l-1} \mu(B_1 \cap B_r(\zeta)) < \infty. \tag{5}$$

Proof. The necessity follows by substitution of the test function into inequalities for corresponding norms in the same way as for $l = 0$.

Now we prove sufficiency. Let u be an arbitrary function in $W_1^l(\partial B_1)$

and let Pu be its Poisson integral. We have

$$\int_{B_1} |(Pu)(z)|\, \mathrm{d}\mu(z) \leq \int_{\partial B_1} |u(\tau)|\, \mathrm{d}\nu(\tau), \tag{6}$$

where ν is the absolute continuous measure on ∂B_1 with derivative

$$(2\pi)^{-1} \int_{B_1} \frac{1-|z|^2}{|\tau - z|^2}\, \mathrm{d}\mu(z).$$

According to Theorem 3.4.1/2, the right-hand side of (6) does not exceed

$$c\, \|u; \partial B_1\|_{W_1^l} \sup_{\zeta \in \partial B_1, r>0} r^{l-1} \nu(\partial B_1 \cap B_r(\zeta)).$$

We put $C_0(\zeta) = B_1 \cap B_{2r}(\zeta)$ and

$$C_j(\zeta) = \{z \in B_1 : 2^j r \leq |z - \zeta| \leq 2^{j+1} r\}, \qquad j = 1, 2, \ldots.$$

Obviously,

$$\nu(\partial B_1 \cap B_r(\zeta)) = \sum_{j=0}^{\infty} (2\pi)^{-1} \int_{\partial B_1 \cap B_r(\zeta)} |\mathrm{d}\tau| \int_{C_j(\zeta)} \frac{1-|z|^2}{|\tau - z|^2}\, \mathrm{d}\mu(z).$$

The summand corresponding to $j = 0$ is equal to

$$\int_{B_1 \cap B_{2r}(\zeta)} (P\chi_{\partial B_1 \cap B_r(\zeta)})(z)\, \mathrm{d}\mu(z),$$

where χ_e is the characteristic function of the set e, and so it does not exceed $\mu(B_1 \cap B_{2r}(\zeta))$. If $z \in C_j(\zeta)$, $j \geq 1$, then $(1-|z|^2)\,|\tau - z|^{-2} = O(2^{-j} r^{-1})$. Consequently,

$$\int_{\partial B_1 \cap B_r(\zeta)} |\mathrm{d}\tau| \int_{C_j(\zeta)} \frac{1-|z|^2}{|\tau - z|^2}\, \mathrm{d}\mu(z) \leq c 2^{-j} \mu(C_j(\zeta)).$$

Thus

$$\begin{aligned} r^{l-1} \nu(\partial B_1 \cap B_r(\zeta)) &\leq r^{l-1} \mu(B_1 \cap B_{2r}(\zeta)) + c \sum_{j=1}^{\infty} (2^j r)^{l-1} \mu(C_j(\zeta)) 2^{-jl} \\ &\leq c \sup_{r>0} r^{l-1} \mu(B_1 \cap B_{2r}(\zeta)). \quad \square \end{aligned}$$

Now we conclude that $\gamma \in MA_1^l$ if and only if $\gamma \in H^\infty$ and

$$\sup_{\zeta \in \partial B_1, r>0} r^{l-1} \int_{B_1 \cap B_r(\zeta)} |\gamma'(z)| \frac{\mathrm{d}x\, \mathrm{d}y}{(1-|z|)^l} < \infty.$$

Verbitskii noticed that a certain space wider than A_1^l is imbedded into $L(B_1, \mu)$ provided that measure μ satisfies (5). By AH_p^l $(1 \leq p < \infty,\ 0 < l < 1)$ we denote the space of analytic functions which can be naturally identified with $\{f \in H_p^l(\partial B_1) : \hat{f}(n) = 0,\ n = -1, -2, \ldots\}$.

Theorem 3 (Verbitskii). *Let $0<l<1$. Then $AH_1^l \subset L(B_1, \mu)$ if and only if (5) is valid.*

Proof. The necessity follows from Theorem 2, since $AH_p^l \supset H_p^l$ for $p \leq 2$ (see Peetre [1]). The sufficiency results from the duality of the Hardy space $H^1(B_1)$ and BMO (see Fefferman, Stein [1]). □

The following assertion, more general than Theorem 2, is due to S. A. Vinogradov. With any function φ, analytic in B_1, we associate the convolution operator $\varphi *$:

$$\varphi * f(z) = \sum_{k \geq 0} \hat{\varphi}(k) \hat{f}(k) z^k$$

where $\hat{f}(k)$ $(k=0, 1, \ldots)$ are Taylor coefficients of an analytic function f.

Theorem 4 (Vinogradov [1]). *Let $0<l<1$. The operator $\varphi *$ maps A_1^l continuously into $L(B_1, \mu)$ if and only if*

$$\sup_{\lambda \in B_1} \int_{B_1} (1-|\lambda|)^{n+l} \, |\varphi^{(n)}(z\lambda)| \, d\mu(z) < \infty$$

for some integer n with $n+l>0$.

This condition, with $\varphi(z) = (1-z)^{-1}$, is equivalent to (5).

Next we state a result similar to Theorem 3 for a wider class of convolution operators in the Hardy space $H^1(B_1)$.

Theorem 5 (Vinogradov [1]). *Let ν be a finite Borel measure on ∂B_1. The convolution operator $f * \nu$ maps $H^1(B_1)$ continuously into $L(B_1, \mu)$ provided that*

$$\sup_{\lambda \in B_1} \int_{|z| \geq |\lambda|} \int_{\partial B_1} \frac{1-|\lambda|^2 \, |z|^2}{|1-\bar{\xi}\lambda z|^2} \, d\nu(\xi) \, d\mu(z) < \infty.$$

In the case $d\nu(\xi) = |1-\xi|^{l-1} \, |d\xi|$, the condition

$$\sup_{\lambda \in B_1} \int_{|z| \geq |\lambda|} \frac{d\mu(z)}{|1-\lambda z|^{1-l}} < \infty$$

is sufficient for the imbedding $AH_1^l \subset L(B_1, \mu)$.

It can easily be seen that the latter condition is not equivalent to (5).

By using his approach to the Carleson theorem, Vinogradov (see N. K. Nikol'skii [1] pp. 195–209) obtained the following imbedding theorem for the space A_p^l, $1 \leq p \leq 2$, $0<l<1/p$.

Theorem 6 (Vinogradov [1]). *If Borel measure μ defined on B_1 satisfies*

$$\sup_{\lambda \in B_1} \int_{|\lambda|<|z|<1} \frac{d\mu(z)}{|1-\bar{\lambda} z|^{1-pl}} < \infty, \tag{7}$$

then $A_p^l \subset L_p(B_1, \mu)$.

This immediately implies that γ is a multiplier in A_p^l provided that $\gamma \in H^\infty$ and

$$\sup_{\lambda \in B_1} \int_{|\lambda|<|z|<1} \frac{|\gamma'(z)|^p (1-|z|)^{p(1-l)-1}}{|1-\bar{\lambda} z|^{1-pl}} \, dx \, dy < \infty. \tag{8}$$

Verbitskii [1, 2] found a necessary and sufficient condition for so-called inner functions in B_1 (see N. K. Nikol'skii [1]) to belong to MA_p^l.

By definition, a function a in H^∞ is inner if $\lim_{|z|\to 1} |a(z)| = 1$ almost everywhere on ∂B_1. By $\{a_k\}_{k \geq 0}$ we denote a sequence of zeroes of a in B_1.

In the case $lp \geq 1$ the class MA_p^l contains only the Blaschke products with a finite number of factors:

$$z^n \prod_{k \geq 0} \frac{|a_k|}{a_k} \frac{a_k - z}{1 - \bar{a}_k z}.$$

If $0 < lp < 1$ then we connect with $\{a_k\}_{k \geq 0}$ a measure defined for $e \subset B_1$ by the formula

$$\mu_{lp}(e) = \prod_{a_k \in e} (1-|a_k|)^{1-lp}.$$

Theorem 7 (Verbitskii [1]). *Let $1 < p < \infty$ and $0 < lp < 1$. An inner funtion belongs to MA_p^l if and only if it is a Blaschke product satisfying*

$$\mu_{lp}(U_j S(I_j)) \leq c \operatorname{cap}(U_j I_j, W_p^l(\partial B_1)) \tag{9}$$

for any finite collection $\{I_j\}$ of disjoint arcs on ∂B_1.

Inner functions are convenient for the construction of various examples in the theory of multipliers. In particular, using Theorem 7, one can easily verify that the validity of inequalities similar to (4) for any arc is not sufficient for a bounded analytic function to belong to MA_p^l, $0 < lp < 1$, $p > 1$.

In Verbitskii [2] condition (9) is checked for the Blaschke product γ with zeroes $a_k = (1 - k^{-2-\beta}) \exp(2\pi i/k)$ with $\beta > 0$. It is shown that $\gamma \in MA_p^l$ if and only if $0 < lp \leq \beta/(\beta+1)$. The author notes that known sufficient conditions for $\gamma \in MA_p^l$ whose formulations do not contain a capacity fail to derive this result in the whole generality. For example, (8) is applicable only for $lp \in (0, \beta/(\beta+1))$ and, consequently, it is unnecessary for $\gamma \in MA_p^l$.

Verbitskii [2] developed one more approach to imbedding theorems for spaces of analytic and harmonic functions which permits the obtaining of necessary and sufficient conditions similar to D. R. Adams' criterion (1.1.4/6). We give this result for the class AH_p^l of analytic functions in H^p with boundary values in the Bessel potential space.

Theorem 8 (Verbitskii [2]). *Let $1<p<\infty$, $0<lp<1$ and let μ be a positive Borel measure in B_1. The space AH_p^l is imbedded into $L_p(B_1, \mu)$ if and only if*

$$\int_0^{2\pi} d\theta \left(\int_e |1 - z \exp(-i\theta)|^{\alpha-1} \, d\mu(z) \right)^{p/(p-1)} \leq c\mu(e) \tag{10}$$

for any compactum $e \subset B_1$.

Although the notion of capacity is not used in the formulation of Theorem 8, its proof is based on some variant of nonlinear potential theory with kernels of potentials being convolutions of Riesz and Poisson kernels.

Next we state three corollaries of Theorem 8. The first immediately follows from the known relations between the W_p^l and H_p^l capacities (see Sjödin [1], D. R. Adams [4]).

Corollary 1. *Let $1<p<\infty$, $0<lp<1$. Condition (10) is sufficient and for $p \geq 2$ is necessary for the imbedding $A_p^l \subset L_p(B_1, \mu)$.*

Corollary 2. *Let $1<p<\infty$ $0<lp<1$, $l>2-p$. The spaces AH_p^l and A_p^l are imbedded into $L_p(B_1, \mu)$ if*

$$\sup_{\lambda \in B_1} \int_0^{2\pi} |1 - \lambda \exp(it)|^{\alpha-1} \, dt \left\{ \int_{|z| \geq |\lambda|} |1 - \bar{\lambda} z|^{\alpha-1} \, d\mu(z) \right\}^{p'} < \infty.$$

Corollary 3. *Let $1<p\leq 2$, $0<lp<1$. The space AH_p^l is imbedded into $L_p(B_1, \mu)$ if (7) is valid.*

Since $A_p^l \subset AH_p^l$ for $p \leq 2$, Corollary 3 implies Theorem 6.

4

The Essential Norm in $M(W_p^m \to W_p^l)$

Here we continue the study of elements of the space $M(W_p^m \to W_p^l)$, where $p \geq 1$ and m, l are simultaneously either integers or fractional numbers.

By $\operatorname{ess} \|\gamma\|_{M(W_p^m \to W_p^l)}$ we denote the essential norm of the operator of multiplication by $\gamma \in M(W_p^m \to W_p^l)$, i.e. the value

$$\inf_{\{T\}} \|\gamma - T\|_{W_p^m \to W_p^l},$$

where $\{T\}$ is the totality of compact operators: $W_p^m \to W_p^l$.

In 4.2 for $m > l$ and in 4.3 for $m = l$ we give two-sided estimates for the essential norm in $M(W_p^m \to W_p^l)$. The main results can be combined in the following assertion.

Theorem. *Let $\gamma \in M(W_p^m \to W_p^l)$, $m \geq l \geq 0$ and let m and l be simultaneously either integers or non-integers.*

(i) *If $p > 1$ and $mp \leq n$, then*

$$\operatorname{ess} \|\gamma\|_{M(W_p^m \to W_p^l)}$$

$$\sim \lim_{\delta \to 0} \sup_{\{e : d(e) \leq \delta\}} \left(\frac{\|\gamma; e\|_{L_p}}{[\operatorname{cap}(e, W_p^{m-l})]^{1/p}} + \frac{\|D_{p,l}\gamma; e\|_{L_p}}{[\operatorname{cap}(e, W_p^m)]^{1/p}} \right)$$

$$+ \lim_{r \to \infty} \sup_{\{e \subset R^n \backslash B_r : d(e) \leq 1\}} \left(\frac{\|\gamma; e\|_{L_p}}{[\operatorname{cap}(e, W_p^{m-l})]^{1/p}} + \frac{\|D_{p,l}\gamma; e\|_{L_p}}{[\operatorname{cap}(e, W_p^m)]^{1/p}} \right).$$

In particular,

$$\operatorname{ess} \|\gamma\|_{MW_p^l} \sim \|\gamma\|_{L_\infty} + \lim_{\delta \to 0} \sup_{\{e : d(e) \leq \delta\}} \frac{\|D_{p,l}\gamma; e\|_{L_p}}{[\operatorname{cap}(e, W_p^l)]^{1/p}}$$

$$+ \lim_{r \to \infty} \sup_{\{e \subset R^n \backslash B_r : d(e) \leq 1\}} \frac{\|D_{p,l}\gamma; e\|_{L_p}}{[\operatorname{cap}(e, W_p^l)]^{1/p}}.$$

(ii) *If $m<n$, then*

$$\mathrm{ess}\,\|\gamma\|_{M(W_1^m\to W_1^l)}\sim\overline{\lim_{\delta\to 0}}\,\delta^{m-n}\sup_{x\in R^n}(\delta^{-l}\|\gamma;B_\delta(x)\|_{L_1}+\|D_{1,l}\gamma;B_\delta(x)\|_{L_1})$$

$$+\overline{\lim_{|x|\to\infty}}\sup_{r\in(0,1)}r^{m-n}(r^{-l}\|\gamma;B_r(x)\|_{L_1}+\|D_{1,l}\gamma;B_r(x)\|_{L_1}).$$

In particular,

$$\mathrm{ess}\,\|\gamma\|_{MW_1^l}\sim\|\gamma\|_{L_\infty}+\overline{\lim_{\delta\to 0}}\sup_{x\in R^n}\delta^{l-n}\|D_{1,l}\gamma;B_\delta(x)\|_{L_1}$$

$$+\overline{\lim_{|x|\to\infty}}\sup_{r\in(0,1)}r^{l-n}\|D_{1,l}\gamma;B_r(x)\|_{L_1}.$$

(iii) *If either $mp>n$, $p>1$ or $m\geqslant n$, $p=1$, then*

$$\mathrm{ess}\,\|\gamma\|_{M(W_p^m\to W_p^l)}\sim\overline{\lim_{|x|\to\infty}}\|\gamma;B_1(x)\|_{W_p^l}\quad\text{for } m>l,$$

$$\mathrm{ess}\,\|\gamma\|_{MW_p^l}\sim\|\gamma\|_{L_\infty}+\overline{\lim_{|x|\to\infty}}\|\gamma;B_1(x)\|_{W_p^l}.$$

From this theorem we immediately obtain the description of the space $\mathring{M}(W_p^m\to W_p^l)$, $m>l$, of compact multipliers (see 4.2.6). In 4.3.1 we show that the maximum modulus of a multiplier in W_p^l does not exceed its essential norm. The estimates for the essential norm of a differential operator formulated in terms of essential norms of its coefficients are given in 4.4. The concluding section 4.5 concerns some applications of multipliers to the theory of singular integral operators in Sobolev spaces.

4.1 Auxiliary Assertions

First we give the definitions of three 'truncating' functions $\eta_{\delta,x}$, $\mu_{\delta,x}$ and ζ_r which are often used in this chapter.

Definition 1. Let $x\in R^n$, $0<\delta<1$ and let η be a function in $C_0^\infty[0,2)$ equal to 1 on $[0,1]$. Moreover, let $0\leqslant\eta\leqslant 1$.
We introduce the function $R^n\ni y\to\eta_{\delta,x}(y)=\eta(|y-x|/\delta)$.

Definition 2. Let η be the same function as in Definition 1. We put

$$\mu_{\delta,x}(y)=\eta\left(\frac{2\log\delta}{\log|y-x|}\right),\qquad \delta\in(0,1/2).$$

We shall also use the notation $\eta_\delta=\eta_{\delta,0}$ and $\mu_\delta=\mu_{\delta,0}$.

Definition 3. Let $\zeta_r(y) = \zeta(y/r)$, where $r > 1$, $\zeta \in C^\infty(R^n)$, $\zeta(y) = 0$ for $y \in B_1$ and $\zeta(y) = 1$ for $y \in R^n \setminus B_2$. Moreover, let $0 \leq \zeta \leq 1$.

We prove some technical lemmas.

Lemma 1. *Let* $\varphi \in D(B_\delta)$, $|\nabla_k \varphi| \leq c\delta^{-k}$, $k = 0, 1, \ldots, [l]+1$. *Then the following estimate holds:*

$$\|\varphi\|_{MW_p^l} \leq c, \tag{1}$$

where $lp < n$, $p \geq 1$ *or* $l = n$, $p = 1$.

Proof. Let $u \in D$. According to Lemma 3.1.1/12 and Corollary 3.1.1/3,

$$\begin{aligned}\|\varphi u\|_{W_p^l} &\leq c\, \|u; B_{2\delta}\|_{W_p^l} \\ &\leq c_1\left(\left(\int_{B_{2\delta}} \int_{B_{2\delta}} |\nabla_{[l]} u(x) - \nabla_{[l]} u(y)|^p \frac{dx\, dy}{|x-y|^{n+p\{l\}}}\right)^{1/p} \right.\\ &\quad \left. + \|u; B_{2\delta}\|_{L_{pn/(n-p)}}\right).\end{aligned} \tag{2}$$

Now the result follows from Lemma 3.1.1/7. □

Lemma 2. *If* $u \in D$ *and* $lp < n$, *then*

$$\sup_{x \in R^n} \|\eta_{\delta,x} u\|_{W_p^l} \xrightarrow[\delta \to 0]{} 0, \tag{3}$$

where $\eta_{\delta,x}$ *is the function from Definition* 1.

This assertion follows from estimate (2).

Lemma 3. *Let* $\{l\} > 0$. *Then*

$$|\nabla_j \mu_\delta(z)| \leq c\, |\log|z||^{-1}\, |z|^{-j}, \tag{4}$$

$$\int_{B_\delta} \frac{|\nabla_j \mu_\delta(z) - \nabla_j \mu_\delta(y)|^p}{|z-y|^{n+p\{l\}}}\, dy \leq c_j\, |\log|z||^{-p}\, |z|^{-p(\{l\}+j)}, \tag{5}$$

where $z \in B_\delta$, $j = 0, 1, \ldots,$

Proof. Estimate (4) is obvious. Let us prove (5). Since

$$|D^\alpha \mu_\delta(z)| \leq |z|^{-|\alpha|} \sum_{k=1}^{|\alpha|} \sigma_k\left(\frac{2\log\delta}{\log|z|}\right)(2\log\delta)^{-k},$$

where $\sigma_k \in D(-1, 1)$, then we have

$$|\nabla_j\mu_\delta(z)-\nabla_j\mu_\delta(y)|\leqslant\begin{cases} c_j\,|\log\delta|^{-1}\,|z-y|\,|z|^{-j-1} \\ \quad \text{if } |z|/2\leqslant|y|\leqslant|z|; \\ c_j\,|\log\delta|^{-1}(\max\{|z|,|y|\})^{-1} \\ \quad \text{if } j>0 \text{ and either } |y|<|z|/2 \text{ or } |z|<|y|/2; \\ c_j\,|\log\delta|^{-1}\left|\log\dfrac{|z|}{|y|}\right| \\ \quad \text{if } j=0 \text{ and either } |y|<|z|/2 \text{ or } |z|<|y|/2. \end{cases}$$

From these estimates we obtain

$$\int_{B_\delta}\frac{|\nabla_j\mu_\delta(z)-\nabla_j\mu_\delta(y)|^p}{|z-y|^{n+p\{l\}}}\,dy\leqslant c_j\,|\log\delta|^{-p}\,|z|^{-p(\{l\}+j)}$$

which is equivalent to (5) for $z\in B_\delta\backslash B_{\delta^3}$. Let $z\in B_{\delta^3}$. Then

$$c_j^{-1}\int_{B_\delta\backslash B_{\delta^3}}\frac{|\delta_{0j}-\nabla_j\mu_\delta(y)|^p}{|y|^{n+p\{l\}}}\,dy\leqslant\int_{B_\delta}\frac{|\nabla_j\mu_\delta(z)-\nabla_j\mu_\delta(y)|^p}{|z-y|^{n+p\{l\}}}\,dy$$
$$\leqslant c_j\int_{B_\delta\backslash B_{\delta^2}}\frac{|\delta_{0j}-\nabla_j\mu_\delta(y)|^p}{|y|^{n+p\{l\}}}\,dy,$$

where $c_j>1$ and δ_{0j} is the Kroneker delta. Consequently,

$$\int_{B_\delta\backslash B_{\delta^2}}\frac{|\delta_{0j}-\nabla_j\mu_\delta(y)|^p}{|y|^{n+p\{l\}}}\,dy\leqslant c\,|\log\delta|^{-p}\,|z|^{-p(\{l\}+j)}.$$

By putting here $|z|=\delta^3$ and taking into account that $t^{3(\{l\}+j)}\,|\log t|$ increases near $t=0$, we arrive at (5). □

Lemma 4. *Let $\nabla\psi\in D(B_1)$, $0\leqslant\psi\leqslant1$ and $\psi_r(y)=\psi(y/r)$, $r\geqslant1$. Then $\|\psi_r\|_{MW_p^l}\leqslant1+cr^{-\sigma}$, where $\sigma=l$ provided that $0<l<1$ and $\sigma=1$ provided that $l\geqslant1$.*

Proof. The assertion is obvious for integer l. Let $\{l\}>0$. Then

$$\|\psi_r u\|_{W_p^l}\leqslant\|\psi_r D_{p,l}u\|_{L_p}+\|\psi_r u\|_{L_p}$$
$$+\sum_{\substack{|\alpha|+|\beta|=[l],\\ |\alpha|>0}}\|D^\alpha\psi_r D^\beta u\|_{W_p^{\{l\}}}+\|\nabla_{[l]}u D_{p,\{l\}}\psi_r\|_{L_p}.$$

We notice that $D^\alpha\psi_r=r^{-|\alpha|}(D^\alpha\psi)_r$ and $D_{p,\{l\}}\psi_r=r^{-\{l\}}(D_{p,\{l\}}\psi)_r$. Since the function $(D^\alpha\psi)_r$ is bounded together with all its derivatives and since the function $(D_{p,\{l\}}\psi)_r$ is uniformly bounded with respect to r, then

$$\|D^\alpha \psi_r D^\beta u\|_{W_p^{\{l\}}} \leq c r^{-|\alpha|} \|u\|_{W_p^{|\beta|+\{l\}}},$$

$$\|\nabla_{[l]} u D_{p,\{l\}} \psi_r\|_{L_p} \leq c r^{-\{l\}} \|u\|_{W_p^l}. \quad \square$$

Lemma 5. Let ψ and ψ_r be the functions defined in Lemma 4. Let, besides, $\psi = 0$ in $B_{1/2}$. Then, for any $u \in W_p^l$, $\lim_{r\to\infty} \|\psi_r u\|_{W_p^l} = 0$.

The proof follows from the inequality

$$\|\psi_r u\|_{W_p^l} \leq \|\psi_r D_{p,l} u\|_{L_p} + \|\psi_r u\|_{L_p} + c r^{-\sigma} \|u\|_{W_p^l}$$

established in the proof of Lemma 4.

We shall use the following simple property of capacity.

Lemma 6. Let $mp < n$, $k \in [0, m]$. Further, let e be a compact subset of the ball B_r, $r \in (0, 1)$. Then

$$\text{cap}\,(e, W_p^k) \leq c r^{(m-k)p}\, \text{cap}\,(e, W_p^m). \tag{6}$$

Proof. Let $u \in D$, $u \geq 1$ on e. We have

$$[\text{cap}\,(e, W_p^k)]^{1/p} \leq \|\eta_r u\|_{W_p^k} = \|D_{p,k}(\eta_r u)\|_{L_p} + \|\eta_r u\|_{L_p}.$$

By Lemma 3.1.1/10,

$$\|\eta_r u\|_{L_p} \leq c r^m \|D_{p,m}(\eta_r u)\|_{L_p}$$

and therefore

$$[\text{cap}\,(e, W_p^k)]^{1/p} \leq c r^{m-k} \|\eta_r u\|_{W_p^m}.$$

From this and from Lemmas 1 and 4 we obtain

$$[\text{cap}\,(e, W_p^k)]^{1/p} \leq c_1 r^{m-k} \|u\|_{W_p^m}.$$

Minimizing the right-hand side, we arrive at (6). $\square$

4.2 Two-sided Estimates for the Essential Norm in the Case $m > l$

4.2.1 Estimates in Terms of Truncating Functions

Lemma. The following estimate is valid:

$$\overline{\lim_{r\to\infty}} \|\zeta_r \gamma\|_{M(W_p^m \to W_p^l)} \leq c \text{ ess } \|\gamma\|_{M(W_p^m \to W_p^l)},$$

where $m \geq l$, $p \geq 1$.

Proof. Let $\varepsilon>0$ and let $T=T(\gamma, \varepsilon)$ be a compact operator such that

$$\|\gamma-T\| \leq \text{ess}\, \|\gamma\|_{M(W_p^m \to W_p^l)}+\varepsilon.$$

Then, for all $u \in W_p^m$,

$$\|\gamma u-Tu\|_{W_p^l} \leq (\text{ess}\, \|\gamma\|_{M(W_p^m \to W_p^l)}+\varepsilon)\, \|u\|_{W_p^m}. \tag{1}$$

Let S be the unit ball in W_p^m centred at the origin and let $\{v_k\}$ be a finite ε-net in TS. Without loss of generality we can assume $v_k \in D$.

It is clear that $\zeta_r v_k=0$ for sufficiently large r. Therefore

$$\|\zeta_r \gamma u\|_{W_p^l}=\|\zeta_r(\gamma u-v_k)\|_{W_p^l} \leq \|\zeta_r(\gamma u-Tu)\|_{W_p^l}+\|\zeta_r(Tu-v_k)\|_{W_p^l}.$$

From this and from Lemma 4.1/4 we obtain

$$\|\zeta_r \gamma u\|_{W_p^l} \leq c(\text{ess}\, \|\gamma\|_{M(W_p^m \to W_p^l)}+\varepsilon)\, \|u\|_{W_p^m}.$$

The result follows. □

Theorem. *Let $m>1$ and let either $lp<n$, $p>1$ or $l \leq n$, $p=1$. Then the relation*

$$\text{ess}\, \|\gamma\|_{M(W_p^m \to W_p^l)} \sim \overline{\lim_{\delta \to 0}} \sup_{x \in R^n} \|\eta_{\delta,x}\gamma\|_{M(W_p^m \to W_p^l)}+\overline{\lim_{r \to \infty}} \|\zeta_r \gamma\|_{M(W_p^m \to W_p^l)} \tag{2}$$

is valid.

Proof. (i) The lower bound for the essential norm: we use the notation introduced in the proof of the lemma. For any $u \in S$,

$$\begin{aligned}\|\eta_{\delta,x}\gamma u\|_{W_p^l} &\leq \|\eta_{\delta,x}(\gamma u-v_k)\|_{W_p^l}+\|\eta_{\delta,x}v_k\|_{W_p^l}\\ &\leq \|\eta_{\delta,x}(\gamma u-Tu)\|_{W_p^l}+\|\eta_{\delta,x}(Tu-v_k)\|_{W_p^l}+\|\eta_{\delta,x}v_k\|_{W_p^l}.\end{aligned}$$

From this and from Lemma 4.1/1 we get

$$\|\eta_{\delta,x}\gamma u\|_{W_p^l} \leq c(\text{ess}\, \|\gamma\|_{M(W_p^m \to W_p^l)}+2\varepsilon)+\varepsilon$$

which, together with reference to the lemma, completes the proof of the lower bound for $\text{ess}\, \|\gamma\|_{M(W_p^m \to W_p^l)}$.

(ii) The upper bound for the essential norm: we choose δ and r so that the following estimates hold:

$$\begin{aligned}&\sup_{x \in R^n} \|\eta_{2\delta,x}\gamma\|_{M(W_p^m \to W_p^l)} \leq \overline{\lim_{\delta \to 0}} \sup_{x \in R^n} \|\eta_{\delta,x}\gamma\|_{M(W_p^m \to W_p^l)}+\varepsilon,\\ &\|\zeta_r\gamma\|_{M(W_p^m \to W_p^l)} \leq \overline{\lim_{r \to \infty}} \|\zeta_r\gamma\|_{M(W_p^m \to W_p^l)}+\varepsilon.\end{aligned} \tag{3}$$

By $\{K_\delta^{(j)}\}$ we denote a finite covering of the ball B_{2r} by open balls with radius δ and centres x_j. We can choose the balls $K_\delta^{(j)}$ so that the multiplicity of the covering of B_{2r} by balls $K_{2\delta}^{(j)}$ depends only on n. Let

$\{\varphi^{(j)}\}$ be a partition of unity subjected to $\{K_\delta^{(j)}\}$ and such that $|\nabla_k \varphi^{(j)}| \leq c\delta^{-k}$, $k=0,1,\dots$. We introduce the polynomials $P^{(j)} = P^{(j)}(u;\cdot)$ specified for the balls $K_{2\delta}^{(j)}$ by Lemma 3.1.1/13.

Further, let $\Gamma = (1-\zeta_r)\gamma$ and let T_* be the finite-dimensional operator defined by

$$(T_* u)(x) = \Gamma(x) \sum_j \varphi^{(j)}(x) P^{(j)}(u;x). \tag{4}$$

We have

$$\|(\gamma - T_*)u\|_{W_p^l} \leq \|(\Gamma - T_*)u\|_{W_p^l} + \|\zeta_r \gamma u\|_{W_p^l}. \tag{5}$$

Since $(\Gamma - T_*)u = \sum_j \Gamma \eta_{2\delta,x_j} \phi^{(j)}(u - P^{(j)})$, then, by Corollary 3.1.1/1,

$$\|(\Gamma - T_*)u\|_{W_p^l}^p \leq c \sum_j \|\Gamma \eta_{2\delta,x_j} \varphi^{(j)}(u-P^{(j)})\|_{W_p^l}^p$$

$$\leq c \sup_j \|\Gamma \eta_{2\delta,x_j}\|_{M(W_p^m \to W_p^l)}^p \sum_j \|\varphi^{(j)}(u-P^{(j)})\|_{W_p^m}^p. \tag{6}$$

Using Lemmas 3.1.1/12, 3.1.1/13, we obtain that the last sum does not exceed

$$c \sum_j \|u - P^{(j)}; K_{2\delta}^{(j)}\|_{W_p^m}^p \leq c_1 \|D_{p,m} u\|_{L_p}^p. \tag{7}$$

Further we note that, by Lemma 4.1/4,

$$\|\Gamma \eta_{2\delta,x_j}\|_{M(W_p^m \to W_p^l)} \leq c\, \|\gamma \eta_{2\delta,x_j}\|_{M(W_p^m \to W_p^l)}.$$

This and inequalities (3), (5)–(7) imply

$$\|\gamma - T_*\|_{W_p^m \to W_p^l}$$

$$\leq c\Big(\overline{\lim_{\delta\to 0}} \sup_{x\in R^n} \|\eta_{\delta,x}\gamma\|_{M(W_p^m \to W_p^l)} + \overline{\lim_{r\to\infty}} \|\zeta_r \gamma\|_{M(W_p^m \to W_p^l)} + \varepsilon\Big). \quad \square \tag{8}$$

Remark. Relation (2) fails for $lp > n$. In fact, let $\gamma = 1$; it is clear that $1 \in M(W_p^m \to W_p^l)$. On the other hand, Proposition 3.2.8 implies $\lim_{\delta\to 0} \|\eta_{\delta,x}\|_{M(W_p^m \to W_p^l)} = \infty$.

4.2.2. *Estimates in Terms of Capacity (the Case $mp < n$, $p > 1$)*

The following theorem presents one more relation for the essential norm.

Theorem. *If $mp<n$ and $p\in(1,\infty)$, then*

$$\operatorname{ess}\|\gamma\|_{M(W_p^m\to W_p^l)}\sim\lim_{\delta\to 0}\sup_{\{e:d(e)\leqslant\delta\}}\left(\frac{\|\gamma;e\|_{L_p}}{[\operatorname{cap}(e,W_p^{m-l})]^{1/p}}+\frac{\|D_{p,l}\gamma;e\|_{L_p}}{[\operatorname{cap}(e,W_p^m)]^{1/p}}\right)$$
$$+\lim_{r\to\infty}\sup_{\{e\subset R^n\backslash B_r:d(e)\leqslant 1\}}\left(\frac{\|\gamma;e\|_{L_p}}{[\operatorname{cap}(e,W_p^{m-l})]^{1/p}}+\frac{\|D_{p,l}\gamma;e\|_{L_p}}{[\operatorname{cap}(e,W_p^m)]^{1/p}}\right).\tag{1}$$

Proof. We limit consideration to the case of fractional m and l, since for integer m and l the proof is analogous and slightly simpler.

(i) The lower bound for the essential norm: we introduce the notation

$$f_k(\gamma;e)=\frac{\|D_{p,k}\gamma;e\|_{L_p}}{[\operatorname{cap}(e,W_p^{m-l+k})]^{1/p}},\qquad 0\leqslant k\leqslant l.$$

Clearly, for any compact set e with $d(e)\leqslant\delta$,

$$f_0(\gamma;e)\leqslant\sup_{x\in R^n}\sup_{e\subset B_\delta(x)}f_0(\gamma;e)\leqslant\sup_{x\in R^n}\sup_{e\subset B_\delta(x)}f_0(\eta_{\delta,x}\gamma;e).$$

This, together with Theorem 3.2.7/2, implies

$$f_0(\gamma;e)\leqslant c\sup_{x\in R^n}\|\eta_{\delta,x}\gamma\|_{M(W_p^m\to W_p^l)}.$$

Applying Theorem 4.2.1, we obtain

$$f_0(\gamma;e)\leqslant c(\operatorname{ess}\|\gamma\|_{M(W_p^m\to W_p^l)}+\varepsilon)\tag{2}$$

for δ small enough.

Now we turn to the bound for $f_l(\gamma;e)$. For $e\subset B_\delta(x)$ we have

$$\|D_{p,l}\gamma;e\|_{L_p}^p=\int_e dy\int_{B_{2\delta}(x)}\frac{|\nabla_{[l]}(\gamma(y)\eta_{2\delta,x}(y))-\nabla_{[l]}(\gamma(z)\eta_{2\delta,x}(z))|^p}{|y-z|^{n+p\{l\}}}dz$$
$$+\int_e dy\int_{R^n\backslash B_{2\delta}(x)}|\nabla_{[l]}\gamma(y)-\nabla_{[l]}\gamma(z)|^p\frac{dz}{|y-z|^{n+p\{l\}}}.$$

Therefore

$$\|D_{p,l}\gamma;e\|_{L_p}\leqslant c\Bigg[\|D_{p,l}(\eta_{2\delta,x}\gamma);e\|_{L_p}$$
$$+\left(\int_e|\nabla_{[l]}\gamma(y)|^p\,dy\int_{R^n\backslash B_{2\delta}(x)}\frac{dz}{|y-z|^{n+p\{l\}}}\right)^{1/p}$$
$$+\left(\int_{R^n\backslash B_{2\delta}(x)}|\nabla_{[l]}\gamma(y)|^p\,dy\int_e\frac{dz}{|y-z|^{n+p\{l\}}}\right)^{1/p}\Bigg].\tag{3}$$

The second summand on the right does not exceed

$$c_1 \delta^{-\{l\}} \|\nabla_{[l]}\gamma; e\|_{L_p} \le c_2 \delta^{-\{l\}} \|\nabla_{[l]}(\gamma\eta_{2\delta,x}); e\|_{L_p}. \tag{4}$$

Since $e \subset B_\delta(x)$, the third summand is not more than

$$c(\text{mes}_n\, e)^{1/p}\left(\int_{R^n\setminus B_{2\delta}(x)} \frac{|\nabla_{[l]}\gamma(z)|^p\, dz}{|z-x|^{n+p\{l\}}}\right)^{1/p}$$

$$\le c_1(\text{mes}_n\, e)^{1/p}\left(\int_{R^n\setminus B_{2\delta}(x)} |z-x|^{-n-p\{l\}}\delta^{-n}\int_{B_\delta(z)} |\nabla_{[l]}\gamma(\xi)|^p\, d\xi\, dz\right)^{1/p}$$

$$\le c_2(\text{mes}_n\, e)^{1/p} \sup_{z\in R^n} \|\nabla_{[l]}(\eta_{2\delta,z}\gamma); B_\delta(z)\|_{L_p}\delta^{-\{l\}-n/p}. \tag{5}$$

Therefore

$$f_l(\gamma; e) \le c\left[f_l(\eta_{2\delta,x}\gamma; e) + \delta^{-\{l\}}\left(\frac{\text{cap}\,(e, W_p^{m-\{l\}})}{\text{cap}\,(e, W_p^m)}\right)^{1/p} f_{[l]}(\eta_{2\delta,x}\gamma; e)\right.$$

$$\left. + \left(\frac{\delta^{-mp}\, \text{mes}_n\, e}{\text{cap}\,(e, W_p^m)}\right)^{1/p} \sup_{x\in R^n}\, \sup_{e\subset B_\delta(x)} f_{[l]}(\eta_{2\delta,x}\gamma; e)\right]. \tag{6}$$

Now from (6) and from Lemma 4.1/6 we obtain

$$f_l(\gamma; e) \le c \sup_{x\in R^n}\, \sup_{e\subset B_\delta(x)} [f_l(\eta_{2\delta,x}\gamma; e) + f_{[l]}(\eta_{2\delta,x}\gamma; e)]$$

which, together with (3.2.7/3) and Theorem 4.2.1, implies

$$f_l(\gamma; e) \le \sup_{x\in R^n} \|\eta_{2\delta,x}\gamma\|_{M(W_p^m\to W_p^l)} \le c_1(\text{ess}\, \|\gamma\|_{M(W_p^m\to W_p^l)} + \varepsilon). \tag{7}$$

Combining (2) and (7), we arrive at the inequality

$$\lim_{\delta\to 0}\, \sup_{\{e: d(e)\le\delta\}} \left(\frac{\|\gamma; e\|_{L_p}}{[\text{cap}\,(e, W_p^{m-l})]^{1/p}} + \frac{\|D_{p,l}\gamma; e\|_{L_p}}{[\text{cap}\,(e, W_p^m)]^{1/p}}\right)$$

$$\le c\, \text{ess}\, \|\gamma\|_{M(W_p^m\to W_p^l)}. \tag{8}$$

Now let $e \subset R^n\setminus B_{3r}$. It is clear that $f_0(\gamma; e) \le f_0(\zeta_r\gamma; e)$ and, by virtue of Theorems 3.2.7/2 and 4.2.1, estimate (2) holds provided that r is sufficiently large.

Let us estimate $f_l(\gamma; e)$. We have

$$\|D_{p,l}\gamma; e\|_{L_p}^p = \int_e dy \int_{R^n\setminus B_{2r}} \frac{|\nabla_{[l]}(\gamma(y)\zeta_r(y)) - \nabla_{[l]}(\gamma(z)\zeta_r(z))|^p}{|y-z|^{n+p\{l\}}}\, dz$$

$$+ \int_e dy \int_{B_{2r}} \frac{|\nabla_{[l]}\gamma(y) - \nabla_{[l]}\gamma(z)|^p}{|y-z|^{n+p\{l\}}}\, dz.$$

Consequently,

$$\|D_{p,l}\gamma; e\|_{L_p} \leqslant c\left[\|D_{p,l}(\zeta_r\gamma); e\|_{L_p} + \left(\int_e |\nabla_{[l]}\gamma(y)|^p \, dy \int_{B_{2r}} \frac{dz}{|y-z|^{n+p\{l\}}}\right)^{1/p}\right.$$
$$\left. + \left(\int_{B_{2r}} |\nabla_{[l]}\gamma(z)|^p \, dz \int_e \frac{dy}{|y-z|^{n+p\{l\}}}\right)^{1/p}\right]. \tag{9}$$

The second summand on the right in (9) does not exceed

$$cr^{-\{l\}}\left(\int_e |\nabla_{[l]}\gamma(y)|^p \, dy\right)^{1/p} \leqslant c_1 r^{-\{l\}} f_{[l]}(\gamma; e)[\text{cap}\,(e, W_p^{m-\{l\}})]^{1/p}$$

which, by Lemma 4.1/6 and Theorem 3.2.7/2, is not more than $c_2 r^{-\{l\}} \|\gamma\|_{M(W_p^m \to W_p^l)}[\text{cap}\,(e, W_p^m)]^{1/p}$.

Let us get a similar estimate for the third summand in the right-hand side of (9). We have

$$\left(\int_{B_{2r}} |\nabla_{[l]}\gamma(z)|^p \, dz \int_e \frac{dy}{|y-z|^{n+p\{l\}}}\right)^{1/p} \leqslant cf_{[l]}(\gamma; B_{2r})(\text{mes}_n\, e)^{1/p} r^{-m}.$$

Therefore the third summand is majorized by

$$cr^{-m} \|\gamma\|_{M(W_p^m \to W_p^l)} \,\text{cap}\,[(e, W_p^m)]^{1/p}.$$

Finally,

$$f_l(\gamma; e) \leqslant c[f_l(\zeta_r\gamma; e) + (r^{-\{l\}} + r^{-m}) \|\gamma\|_{M(W_p^m \to W_p^l)}]. \tag{10}$$

Turning to the supremum with respect to e in both parts of this inequality and making $r \to \infty$, we arrive at

$$\lim_{r\to\infty} \sup_{e \subset R^n \setminus B_r} f_l(\gamma; e) \leqslant c \lim_{r\to\infty} \sup_{e} f_l(\zeta_r\gamma; e).$$

Combining this estimate with (2) and Theorem 4.2.1, we conclude

$$\lim_{r\to\infty} \sup_{e \subset R^n \setminus B_r} \left(\frac{\|\gamma; e\|_{L_p}}{[\text{cap}\,(e, W_p^{m-l})]^{1/p}} + \frac{\|D_{p,l}\gamma; e\|_{L_p}}{[\text{cap}\,(e, W_p^m)]^{1/p}}\right) \leqslant c \,\text{ess}\, \|\gamma\|_{M(W_p^m \to W_p^l)}. \tag{11}$$

Summing (8) and (11), we obtain the required estimate for the essential norm.

(ii) The upper bound for the essential norm: let e be an arbitrary compactum in R^n. We have

$$\|D_{p,l}(\eta_{\delta,x}\gamma); e\|_{L_p} \leq c\Big[\sum_{j=0}^{[l]} \| \,|\nabla_j(\eta_{2\delta,x}\gamma)|\, D_{p,l-j}\eta_{\delta,x}; e\|_{L_p}$$

$$+\sum_{j=0}^{[l]} \Big(\int_{R^n} |\nabla_j \eta_{\delta,x}(y)|^p \, dy$$

$$\times \int_e \frac{|\nabla_{[l]-j}(\eta_{2\delta,x}(y)\gamma(y)) - \nabla_{[l]-j}(\eta_{2\delta,x}(z)\gamma(z))|^p}{|y-z|^{n+p\{l\}}} \, dz\Big)^{1/p}$$

$$+\|D_{p,l}(\eta_{\delta,x}\eta_{2\delta,x}\gamma); e\setminus B_{4\delta}(x)\|_{L_p}\Big]. \tag{12}$$

The obvious estimate $D_{p,l-j}\eta_{\delta,x} \leq c\delta^{j-l}$ and Lemma 4.1/6 imply

$$\| \,|\nabla_j(\eta_{2\delta,x}\gamma)|\, D_{p,l-j}\eta_{\delta,x}; e\|_{L_p} \leq c\delta^{j-l} f_j(\eta_{2\delta,x}\gamma; e)[\text{cap}\,(e, W_p^{m-l+j})]^{1/p}$$
$$\leq c f_j(\eta_{2\delta,x}\gamma; e)[\text{cap}\,(e, W_p^m)]^{1/p}.$$

By Theorem 3.2.7/2,

$$f_j(\eta_{2\delta,x}\gamma; e) \leq c\, \|\eta_{2\delta,x}\gamma\|_{M(W_p^{m-l+j} \to W_p^j)}. \tag{13}$$

Hence, from Lemma 4.1/1 and Remark 3.2.6, we get

$$f_j(\eta_{2\delta,x}\gamma; e) \leq c \sup_{\xi\in R^n} \|\eta_{\delta,\xi}\eta_{2\delta,x}\gamma\|_{M(W_p^{m-l+j} \to W_p^j)}$$

$$\leq c \sup_{\xi\in R^n} \|\eta_{\delta,\xi}\gamma\|_{M(W_p^{m-l+j} \to W_p^j)}$$

$$\leq \varepsilon \sup_{\xi\in R^n} \|\eta_{\delta,\xi}\gamma\|_{M(W_p^m \to W_p^l)} + c(\varepsilon) \sup_{\xi\in R^n} \|\eta_{\delta,\xi}\gamma\|_{M(W_p^{m-l} \to L_p)}.$$

So the first sum in (12) does not exceed

$$(\varepsilon\, \|\eta_{\delta,x}\gamma\|_{M(W_p^m \to W_p^l)} + c(\varepsilon)\, \|\eta_{\delta,x}\gamma\|_{M(W_p^{m-l} \to L_p)})[\text{cap}\,(e, W_p^m)]^{1/p}. \tag{14}$$

The j-th summand in the second sum on the right in (12) is majorized by

$$c\delta^{-j}\, \|D_{p,l-j}(\eta_{2\delta,x}\gamma); e\|_{L_p} \leq c\delta^{-j} f_{l-j}(\eta_{2\delta,x}\gamma; e)[\text{cap}\,(e, W_p^{m-j})]^{1/p}$$
$$\leq c f_{l-j}(\eta_{2\delta,x}\gamma; e)[\text{cap}\,(e, W_p^m)]^{1/p}.$$

Using the same arguments as when estimating $f_j(\gamma; e)$, we get

$$f_{l-j}(\eta_{2\delta,x}\gamma; e) \leq \varepsilon \sup_{\xi\in R^n} \|\eta_{\delta,\xi}\gamma\|_{M(W_p^m \to W_p^l)} + c(\varepsilon) \sup_{\xi\in R^n} \|\eta_{\delta,\xi}\gamma\|_{M(W_p^{m-l} \to L_p)}$$

for $j = 1, \ldots, [l]$. Therefore the second sum on the right in (12) does not exceed

$$\left[\varepsilon \sup_{x\in R^n} \|\eta_{\delta,x}\gamma\|_{M(W_p^m\to W_p^l)} + c(\varepsilon) \sup_{\xi\in R^n} \|\eta_{\delta,\xi}\gamma\|_{M(W_p^{m-l}\to L_p)} + cf_l(\gamma; e)\right] \times[\operatorname{cap}(e, W_p^m)]^{1/p}.$$

Now we give a bound for $\|D_{p,l}(\eta_{\delta,x}\eta_{2\delta,x}\gamma); e\backslash B_{4\delta}(x)\|_{L_p}$. By Hölder's inequality and the known estimate for the capacity, we find for $z\in B_{2\delta}(x)$

$$\int_{e\backslash B_{4\delta}(x)} |y-z|^{-n-p\{l\}}\, dy \leqslant (\operatorname{mes}_n e)^{(n-mp)/n} \times\left(\int_{e\backslash B_{4\delta}(x)} |y-z|^{-(n+p\{l\})n/mp}\, dy\right)^{mp/n} \leqslant c \operatorname{cap}(e, W_p^m)\delta^{p(m-\{l\})-n}.$$

Consequently,

$$\begin{aligned}&\|D_{p,l}(\eta_{\delta,x}\eta_{2\delta,x}\gamma); e\backslash B_{4\delta}(x)\|_{L_p}^p\\ &\leqslant c\sum_{j=0}^{[l]} \int_{B_\delta(x)} |\nabla_j(\eta_{2\delta,x}(z)\gamma(z))|^p\, dz\, \delta^{(j-[l])p} \int_{e\backslash B_{4\delta}(x)} |y-z|^{-p\{l\}-n}\, dy\\ &\leqslant c \operatorname{cap}(e, W_p^m) \sum_{j=0}^{[l]} \delta^{(m-j+l)p-n} \int_{B_{2\delta}(x)} |\nabla_j(\eta_{2\delta,x}(z)\gamma(z))|^p dz\\ &\leqslant c \operatorname{cap}(e, W_p^m) \sum_{j=0}^{[l]} f_j(\eta_{2\delta,x}\gamma; B_{2\delta}(x)).\end{aligned}$$

Following the same lines as when estimating $f_j(\eta_{2\delta,x}\gamma; e)$, we conclude that the third summand on the right in (12) does not exceed (14). Substituting the derived estimates into (12), we get

$$f_l(\eta_{\delta,x}\gamma; e) \leqslant c\left(\varepsilon \sup_{\xi\in R^n} \|\eta_{\delta,\xi}\gamma\|_{M(W_p^m\to W_p^l)} + c(\varepsilon) \sup_{\xi\in R^n} \|\eta_{\delta,\xi}\gamma\|_{M(W_p^{m-l}\to L_p)} + f_l(\gamma; e)\right)$$

which implies

$$\begin{aligned}&\|\eta_{\delta,x}\gamma\|_{M(W_p^m\to W_p^l)}\\ &\leqslant c\left(\varepsilon \sup_{\xi\in R^n} \|\eta_{\delta,\xi}\gamma\|_{M(W_p^m\to W_p^l)} + c(\varepsilon) \sup_{\xi\in R^n} \|\eta_{\delta,\xi}\gamma\|_{M(W_p^{m-l}\to L_p)}\right.\\ &\left.\quad + \sup_{e\subset B_{2\delta}(x)} f_0(\gamma; e) + \sup_{e\subset B_{4\delta}(x)} f_l(\gamma; e)\right).\end{aligned}$$

Taking the supremum over x in both sides, we find

$$\|\eta_{\delta,x}\gamma\|_{M(W_p^m\to W_p^l)} \leqslant c\Big(\sup_{\{e:d(e)\leqslant 4\delta\}} f_0(\gamma;e) + \sup_{\{e:d(e)\leqslant 8\delta\}} f_l(\gamma;e)\Big). \tag{15}$$

Now let $e \subset R^n$, $d(e) \leqslant 1$ and let r be a large enough positive number. We have

$$\begin{aligned}\|D_{p,l}(\zeta_{3r}\gamma); e\|_{L_p} &\leqslant c\Bigg[\sum_{j=0}^{[l]} \| |\nabla_j\gamma|\, D_{p,l-j}\zeta_{3r}; e\backslash B_r\|_{L_p} \\ &+ \sum_{j=0}^{[l]}\Big(\int_{R^n} |\nabla_j\zeta_{3r}(y)|^p\, dy \int_{e\backslash B_r} \frac{|\nabla_{[l]-j}\gamma(y)-\nabla_{[l]-j}\gamma(z)|^p}{|y-z|^{n+p\{l\}}}\, dy\, dz\Big)^{1/p} \\ &+ \|D_{p,l}(\zeta_{3r}\gamma); e\cap B_r\|_{L_p}\Bigg].\end{aligned} \tag{16}$$

The first sum on the right does not exceed

$$\begin{aligned}&c\sum_{j=0}^{[l]} r^{j-l} f_j(\gamma; e\backslash B_r)[\operatorname{cap}(e\backslash B_r, W_p^{m-l+j})]^{1/p} \\ &\leqslant cr^{-\{l\}}\|\gamma\|_{M(W_p^m\to W_p^l)}[\operatorname{cap}(e, W_p^m)]^{1/p}.\end{aligned}$$

The j-th summand on the right in (16) is majorized by

$$cr^{-j}\|D_{p,l-j}\gamma; e\backslash B_r\|_{L_p} \leqslant cr^{-j} f_{l-j}(\gamma; e\backslash B_r)[\operatorname{cap}(e\backslash B_r, W_p^{m-j})]^{1/p}$$

and so the sum in question is majorized by

$$c(f_l(\gamma; e\backslash B_r) + r^{-1}\|\gamma\|_{M(W_p^m\to W_p^l)})[\operatorname{cap}(e, W_p^m)]^{1/p}.$$

Further, we estimate the last summand on the right in (16). We have

$$\begin{aligned}\|D_{p,l}(\zeta_{3r}\gamma); e\cap B_r\|_{L_p}^p &\leqslant \sum_{j=0}^{[l]} \int_{R^n\backslash B_{3r}} |\nabla_j\gamma(z)|^p\, dz\, r^{(j-[l])p} \\ &\quad\times \int_{e\cap B_r} |y-z|^{-n-p\{l\}}\, dy \\ &\leqslant c\, \operatorname{mes}_n e \sum_{j=0}^{[l]} \int_{R^n\backslash B_{3r}} \frac{|\nabla_j\gamma(z)|^p}{|z|^{n+p\{l\}}}\, dz\, r^{(j-[l])p}.\end{aligned}$$

We note that

$$\begin{aligned}\int_{R^n\backslash B_{3r}} |\nabla_j\gamma(z)|^p \frac{dz}{|z|^{n+p\{l\}}} &\leqslant r^{-p\{l\}} \sup_{\xi\in R^n\backslash B_{2r}} \|\nabla_j\gamma; B_1(\xi)\|_{L_p}^p \\ &\leqslant cr^{-p\{l\}}\|\gamma\|_{M(W_p^m\to W_p^l)}^p.\end{aligned}$$

Therefore

$$\|D_{p,l}(\zeta_{3r}\gamma); e\cap B_r\|_{L_p} \leqslant c(\operatorname{mes}_n e)^{1/p} r^{-\{l\}}\|\gamma\|_{M(W_p^m\to W_p^l)}.$$

Consequently,

$$\|D_{p,l}(\zeta_{3r}\gamma);\, e\|_{L_p} \leqslant c(f_l(\gamma;\, e\backslash B_r)+r^{-\{l\}}\|\gamma\|_{M(W_p^m\to W_p^l)})[\text{cap}\,(e,\, W_p^m)]^{1/p}.$$

By using Theorem 3.2.7/2 we obtain the estimate

$$\|\zeta_{3r}\gamma\|_{M(W_p^m\to W_p^l)} \leqslant c_1 \sup_{e\subset R^n\backslash B_r, d(e)\leqslant 1} (f_l(\gamma;\, e)+f_0(\gamma;\, e))$$
$$+c_2 r^{-\{l\}}\|\gamma\|_{M(W_p^m\to W_p^l)}. \tag{17}$$

From this, together with (15) and Theorem 4.2.1, we derive the required upper bound for the essential norm. □

4.2.3 Estimates in Terms of Capacity (the Case $mp=n,\ p>1$)

Theorem. *The relation* (4.2.2/1) *is also valid for* $mp=n,\ p>1$.

Proof. As in the proof of Theorem 4.2.2, we need to consider only the more difficult case of fractional m and l.

(i) *The lower bound for the essential norm.* Let e be a compactum in R^n, $d(e)\leqslant\delta;\ \delta\in(0,1/2)$. The argument leading to (4.2.2/2) applies equally well to the obtaining of the required estimate for $f_0(\gamma;\, e)$.

For $f_l(\gamma;\, e)$ we have estimates (4.2.2/3)–(4.2.2/5). The right-hand side of (4.2.2/4) does not exceed

$$c\delta^{-\{l\}}(\text{mes}_n\, e)^{\{l\}/n}\|\nabla_{[l]}(\gamma\eta_{2\delta,x});\, e\|_{L_{pn/(n-p\{l\})}}$$
$$\leqslant c\frac{|\log\delta|^{(p-1)/p}}{|\log \text{mes}_n\, e|^{(p-1)/p}}\|\nabla_{[l]}(\gamma\eta_{2\delta,x});\, e\|_{L_{pn/(n-p\{l\})}}.$$

The expression on the right is not more than

$$c[\text{cap}\,(e,\, W_p^m)]^{1/p}\sup_{x\in R^n}\frac{\|D_{p,l}(\gamma\eta_{2\delta,x});\, B_{4\delta}(x)\|_{L_p}}{[\text{cap}\,(B_{4\delta}(x),\, W_p^m)]^{1/p}}. \tag{1}$$

Similarly, the right-hand side of (4.2.2/5) does not exceed

$$c(\text{mes}_n\, e)^{1/p}\,|\log \text{mes}_n e|^{(p-1)/p}[\text{cap}\,(e,\, W_p^m)]^{1/p}\delta^{-n/p}$$
$$\times\sup_{z\in R^n}\|D_{p,l}(\gamma\eta_{2\delta,z});\, B_{4\delta}(z)\|_{L_p}$$

which, in turn, is majorized by (1). Thus

$$f_l(\gamma;\, e)\leqslant c\sup_{x\in R^n} f_l(\gamma\eta_{2\delta,x};\, B_{4\delta}(x)).$$

Combining this with (3.2.7/3) and Theorem 4.2.1, we get (4.2.2/7). Consequently we arrive at (4.2.2/8).

The proof of inequality (4.2.2/10) is also valid for $mp=n$. Thus the required lower bound for the essential norm is obtained.

(ii) *The upper bound for the essential norm.* We take the relation (4.2.1/2) as the basis of the proof. Since $\eta_{\delta^2,x}\mu_{\delta,x}=\eta_{\delta^2,x}$, then, by Lemma 4.1/1,

$$\|\eta_{\delta^2,x}\gamma\|_{M(W_p^m\to W_p^l)}\leqslant c\,\|\mu_{\delta,x}\gamma\|_{M(W_p^m\to W_p^l)}.$$

Our aim is to prove the estimate

$$\|\mu_{\delta,x}\gamma\|_{M(W_p^m\to W_p^l)}\leqslant c \sup_{e\subset B_{6\delta}(x)} (f_0(\gamma;e)+f_l(\gamma;e)). \tag{2}$$

Let e be a compact set in R^n with $d(e)<1/2$. We have

$$\begin{aligned}&\|D_{p,l}(\mu_{\delta,x}\gamma);e\|_{L_p}\\ &\leqslant c\Bigg[\sum_{j=0}^{[l]}\|\,|\nabla_j\gamma|\,D_{p,l-j}\mu_{\delta,x};e\cap B_{2\delta}(x)\|_{L_p}\\ &+\sum_{j=1}^{[l]}\left(\int|\nabla_j\mu_{\delta,x}(y)|^p\,dy\int_{e\cap B_{2\delta}(x)}\frac{|\nabla_{[l]-j}\gamma(y)-\nabla_{[l]-j}\gamma(z)|^p}{|y-z|^{n+p\{l\}}}\,dz\right)^{1/p}\\ &+\|D_{p,l}(\mu_{\delta,x}\gamma);e\setminus B_{2\delta}(x)\|_{L_p}+\|\mu_{\delta,x}D_{p,l}\gamma;e\|_{L_p}\Bigg].\end{aligned} \tag{3}$$

By applying Lemma 4.1/3, we find

$$\|\,|\nabla_j\gamma|\,D_{p,l-j}\mu_{\delta,x};e\cap B_{2\delta}(x)\|_{L_p}\leqslant c\,\|\,|\nabla_j\gamma|\,|\log r|^{-1}\,r^{j-l};e\cap B_{2\delta}(x)\|_{L_p}, \tag{4}$$

where $r(z)=|z-x|$. By Hölder's inequality the right-hand side does not exceed

$$c\,\|\nabla_j\gamma;B_{2\delta}(x)\|_{L_{pn/(n-p(l-j))}}\|\,|\log r|^{-1}\,r^{j-l};e\cap B_{2\delta}(x)\|_{L_{n/(l-j)}}.$$

Since the function $|\log t|^{-1}\,t^{j-l}$ decreases near $t=0$, the maximum value of the integral

$$\int_E\frac{dz}{|z|^n\,|\log|z|\,|^{n/(l-j)}}$$

over all sets E with prescribed small $\text{mes}_n\,E$ is attained at the ball centred at $z=0$. Consequently,

$$\|\,|\log r|^{-1}\,r^{j-l};e\cap B_{2\delta}(x)\|_{L_{n/(l-j)}}\leqslant c\,|\log \text{mes}_n(e\cap B_{2\delta}(x))|^{(l-j-n)/n}. \tag{5}$$

Further, we note that

$$\|\nabla_j\gamma;B_{2\delta}(x)\|_{L_{pn/(n-p(l-j))}}\leqslant c(\|D_{p,l}\gamma;B_{2\delta}(x)\|_{L_p}+\delta^{-l}\,\|\gamma;B_{2\delta}(x)\|_{L_p}).$$

From (4) and the two last inequalities we obtain

$$\| |\nabla_j \gamma| \, D_{p,l-j}\mu_{\delta,x}; e\cap B_{2\delta}(x)\|_{L_p} \leq c\, |\log \delta|^{(p(l-j)-n)/np}(|\log \delta|^{(1-p)/p} f_l(\gamma; B_{2\delta}(x)) + f_0(\gamma; B_{2\delta}(x)) \times |\log \mathrm{mes}_n\,(e\cap B_{2\delta}(x))|^{(1-p)/p}.$$

By applying Proposition 3.1.3/5, we arrive at

$$\| |\nabla_j \gamma| \, D_{p,l-j}\mu_{\delta,x}; e\cap B_{2\delta}(x)\|_{L_p} \leq c(f_l(\gamma; B_{2\delta}(x)) + f_0(\gamma; B_{2\delta}(x))[\mathrm{cap}\,(e, W_p^m)]^{1/p}. \tag{6}$$

The general term of the second sum on the right in (3) is equal to

$$\left(\int \frac{dh}{|h|^{n+p\{l\}}} \int_{e\cap B_{2\delta}(x)} |\Delta_h \nabla_{[l]-j}\gamma(z)|^p \, |\nabla_j \mu_{\delta,x}(z+h)|^p \, dz\right)^{1/p}.$$

Since supp $\mu_{\delta,x}\subset B_\delta(x)$, the latter expression does not exceed

$$\| |\nabla_j \mu_{\delta,x}; e\cap B_{2\delta}(x)\|_{L_{n/j}} \left(\int_{B_{3\delta}(x)} \|\Delta_h \nabla_{[l]-j}\gamma; B_{2\delta}(x)\|^p_{L_{np/(n-jp)}} \frac{dh}{|h|^{n+p\{l\}}}\right)^{1/p}.$$

By using Lemma 4.1/3 and applying the same argument as in the proof of (5), we conclude

$$\|\nabla_j \mu_{\delta,x}; e\cap B_{2\delta}(x)\|_{L_{n/j}} \leq c\, |\log \mathrm{mes}_n\,(e\cap B_{2\delta}(x))|^{(j-n)/n}. \tag{7}$$

It is known that the space $B^{s_1}_{q_1,p}$ is imbedded continuously into $B^{s_2}_{q_2,p}$ with $s_1 - n/q_1 = s_2 - n/q_2$, $1<q_1<q_2<\infty$ (see Besov [1]) This, in particular, implies

$$\int_{B_{3\delta}(x)} \|\Delta_h \nabla_{[l]-j}\gamma; B_{2\delta}(x)\|^p_{L_{np/(n-jp)}} \frac{dh}{|h|^{n+p\{l\}}} \leq c\, \||\gamma; B_{6\delta}(x)\||^p_{W_p^l} \sim c(\|D_{p,l}\gamma; B_{6\delta}(x)\|_{L_p} + \delta^{-l}\, \|\gamma; B_{6\delta}(x)\|_{L_p})^p. \tag{8}$$

From (7) and (8) we obtain that the general term of the second sum on the right in (3) is not more than

$$c\, |\log \delta|^{(pj-n)/np}(|\log \delta|^{(1-p)/p} f_l(\gamma; B_{6\delta}(x)) + f_0(\gamma; B_{6\delta}(x)))\, |\log \mathrm{mes}_n (e\cap B_{2\delta}(x))|^{(1-p)/p}.$$

This, together with Proposition 3.1.3/5, yields

$$\left(\int |\nabla_j \mu_{\delta,x}(y)|^p \, dy \int_{e\cap B_{2\delta}(x)} \frac{|\nabla_{[l]-j}\gamma(y) - \nabla_{[l]-j}\gamma(z)|^p}{|y-z|^{n+p\{l\}}} dz\right)^{1/p} \leq c(f_l(\gamma; B_{6\delta}(x)) + f_0(\gamma; B_{6\delta}(x)))[\mathrm{cap}\,(e, W_p^m)]^{1/p}. \tag{9}$$

Let us estimate the norm $\|D_{p,l}(\mu_{\delta,x}\gamma); e\backslash B_{2\delta}(x)\|_{L_p}$, which is obviously equal to

$$\left(\int_{e\setminus B_{2\delta}(x)} dy \int_{B_\delta(x)} \frac{|\nabla_{[l]}(\mu_{\delta,x}\gamma)(z)|^p}{|y-z|^{n+p\{l\}}}\,dz\right)^{1/p}$$

$$=\left(\int_{B_\delta(x)} |\nabla_{[l]}(\mu_{\delta,x}\gamma)(z)|^p\,dz \int_{e\setminus B_{2\delta}(x)} \frac{dy}{|y-z|^{n+p\{l\}}}\right)^{1/p}.$$

It is clear that

$$\int_{e\setminus B_{2\delta}(x)} \frac{dy}{|y-z|^{n+p\{l\}}} \leqslant \min\{\delta^{-n-p\{l\}}\,\mathrm{mes}_n\, e, \delta^{-p\{l\}}\}$$

$$\leqslant \frac{|\log \delta|^{p-1}}{|\log \mathrm{mes}_n\, e|^{p-1}\,\delta^{p\{l\}}}.$$

Moreover, by virtue of Lemma 3.1.1/10,

$$\int_{B_\delta(x)} |\nabla_{[l]}(\mu_{\delta,x}\gamma)(z)|^p\,dz$$

$$\leqslant \delta^{p\{l\}} \int_{B_\delta(x)} \int_{B_\delta(x)} \frac{|\nabla_{[l]}(\mu_{\delta,x}\gamma)(y) - \nabla_{[l]}(\mu_{\delta,x}\gamma)(z)|^p}{|y-z|^{n+p\{l\}}}\,dz\,dy.$$

Consequently,

$$\|D_{p,l}(\mu_{\delta,x}\gamma);\, e\setminus B_{2\delta}(x)\|^p_{L_p}$$

$$\leqslant c\frac{|\log \delta|^{p-1}}{|\log \mathrm{mes}_n\, e|^{p-1}}\Bigg(\|D_{p,l}\gamma;\, B_\delta(x)\|^p_{L_p} + \sum_{j=0}^{[l]} \|\, |\nabla_j \gamma|\, D_{p,l-j}\mu_{\delta,x};\, B_\delta(x)\|^p_{L_p}$$

$$+\sum_{j=1}^{[l]} \int_{B_\delta(x)} |\nabla_j \mu_{\delta,x}(y)|^p\,dy \int_{B_\delta(x)} \frac{|\nabla_{[l]-j}\gamma(y) - \nabla_{[l]-j}\gamma(z)|^p}{|y-z|^{n+p\{l\}}}\,dz\Bigg).$$

Putting $e = B_\delta(x)$ into (6) and (9), we see that either of the two last sums may be estimated from above by

$$c\,|\log \delta|^{1-p}\,(f_l(\gamma;\, B_{6\delta}(x)) + f_0(\gamma;\, B_{6\delta}(x)))^p.$$

Therefore

$$\|D_{p,l}(\mu_{\delta,x}\gamma);\, e\setminus B_{2\delta}(x)\|^p_{L_p} \leqslant c(f_l(\gamma;\, B_{6\delta}(x)) + f_0(\gamma;\, B_{6\delta}(x)))^p\,\mathrm{cap}\,(e, W^m_p).$$

By setting this, together with (6) and (9), into (3), we arrive at (2).

The estimate

$$\overline{\lim_{r\to\infty}}\, \|\zeta_r \gamma\|_{M(W^m_p \to W^l_p)} \leqslant c \lim_{r\to\infty} \sup_{e\subset R^n\setminus B_r, d(e)\leqslant 1} (f_l(\gamma;\, e) + f_0(\gamma;\, e))$$

is obtained at the end of the proof of Theorem 4.2.2. □

4.2.4 Sharpening of the Lower Bound for the Essential Norm in the Case $m>l$, $mp\leq n$, $p>1$

In addition to Theorems 4.2.2, 4.2.3 we prove the following.

Theorem. *If $m>l$, $mp\leq n$, $p>1$, then the following estimate holds:*

$$\begin{aligned}
&\operatorname{ess}\|\gamma\|_{M(W_p^m\to W_p^l)}\\
&\geq c\lim_{\delta\to 0}\sup_{\{e:d(e)\leq\delta\}}\left(\sum_{j=0}^{[l]}\frac{\|D_{p,l-j}\gamma;e\|_{L_p}^p}{\operatorname{cap}(e,W_p^{m-j})}+\sum_{j=0}^{[l]}\frac{\|\nabla_j\gamma;e\|_{L_p}^p}{\operatorname{cap}(e,W_p^{m-l+j})}\right)^{1/p}\\
&\quad+c\lim_{r\to\infty}\sup_{\{e\subset R^n\setminus B_r:d(e)\leq 1\}}\left(\sum_{j=0}^{[l]}\frac{\|D_{p,l-j}\gamma;e\|_{L_p}^p}{\operatorname{cap}(e,W_p^{m-j})}+\sum_{j=0}^{[l]}\frac{\|\nabla_j\gamma;e\|_{L_p}^p}{\operatorname{cap}(e,W_p^{m-l+j})}\right)^{1/p}.
\end{aligned}$$

To prove this theorem we need the following auxiliary assertion:

Lemma. *For any multi-index α, $|\alpha|\leq l<m$,*

$$\operatorname{ess}\|D^\alpha\gamma\|_{M(W_p^m\to W_p^{l-|\alpha|})}\leq c\sum_{j=0}^{|\alpha|}\operatorname{ess}\|\gamma\|_{M(W_p^{m-j}\to W_p^{l-j})}.$$

Proof. It suffices to limit consideration to $|\alpha|=1$. By T_j, $j=0,1$, we denote compact operators from W_p^{m-j} into W_p^{l-j} such that

$$\|\gamma u-T_ju\|_{W_p^{l-j}}\leq(\operatorname{ess}\|\gamma\|_{M(W_p^{m-j}\to W_p^{l-j})}+\varepsilon)\|u\|_{W_p^{m-j}}.$$

For any function $u\in W_p^m$ we have

$$\|\gamma\nabla u+u\nabla\gamma-\nabla T_0u\|_{W_p^{l-1}}\leq(\operatorname{ess}\|\gamma\|_{M(W_p^m\to W_p^l)}+\varepsilon)\|u\|_{W_p^m}$$

and therefore

$$\begin{aligned}
&\|u\nabla\gamma-\nabla T_0u+T_1\nabla u\|_{W_p^{l-1}}\\
&\quad\leq(\operatorname{ess}\|\gamma\|_{M(W_p^m\to W_p^l)}+\varepsilon)\|u\|_{W_p^m}+\|\gamma\nabla u-T_1\nabla u\|_{W_p^{l-1}}\\
&\quad\leq(\operatorname{ess}\|\gamma\|_{M(W_p^m\to W_p^l)}+\operatorname{ess}\|\gamma\|_{M(W_p^{m-1}\to W_p^{l-1})}+2\varepsilon)\|u\|_{W_p^m}.
\end{aligned}$$

Since $T=\nabla T_0-T_1\nabla$ is a compact operator from W_p^m into W_p^{l-1}, then

$$\operatorname{ess}\|\nabla\gamma\|_{M(W_p^m\to W_p^{l-1})}\leq\operatorname{ess}\|\gamma\|_{M(W_p^m\to W_p^l)}+\operatorname{ess}\|\gamma\|_{M(W_p^{m-1}\to W_p^{l-1})}.\quad\square$$

Proof of the Theorem. From interpolation inequality (3.2.1/1) and from Theorem 4.2.1, we find

$$\operatorname{ess}\|\gamma\|_{M(W_p^{m-j}\to W_p^{l-j})}\leq c\operatorname{ess}\|\gamma\|_{M(W_p^m\to W_p^l)}^{(l-j)/l}\operatorname{ess}\|\gamma\|_{M(W_p^{m-l}\to L_p)}^{j/l}$$

which, together with Theorems 1.3.2/1 and 3.2.7/1, gives

$$\operatorname{ess}\|\gamma\|_{M(W_p^{m-j}\to W_p^{l-j})}\leq c\operatorname{ess}\|\gamma\|_{M(W_p^m\to W_p^l)}.$$

From this and the lemma we conclude that

$$\text{ess}\,\|\nabla_j \gamma\|_{M(W_p^m \to W_p^{l-j})} \leq c\ \text{ess}\,\|\gamma\|_{M(W_p^m \to W_p^l)}.$$

It remains to make use of Theorems 4.2.2 and 4.2.3. □

4.2.5 Estimates for the Essential Norm for $mp > n$, $p > 1$ and for $p = 1$

In the two cases mentioned we can neglect the capacity. The simplest formulation is for $mp > n$, $p \geq 1$ and for $m = n$, $p = 1$.

Theorem 1. *If $mp > n$, $p \geq 1$ or $m = n$, $p = 1$ and $\gamma \in M(W_p^m \to W_p^l)$, $m > l$, then*

$$\text{ess}\,\|\gamma\|_{M(W_p^m \to W_p^l)} \sim \overline{\lim_{|x| \to \infty}}\, \|\gamma; B_1(x)\|_{W_p^l}. \tag{1}$$

Proof. Applying Proposition 3.2.8 and Theorem 3.4.2 to the multiplier $\zeta_r \gamma$, from Lemma 4.2.1, we obtain

$$\overline{\lim_{r \to \infty}} \sup_{x \in R^n} \|\zeta_r \gamma; B_1(x)\|_{W_p^l} \leq c\ \text{ess}\,\|\gamma\|_{M(W_p^m \to W_p^l)}$$

which is equivalent to

$$\overline{\lim_{|x| \to \infty}}\, \|\gamma; B_1(x)\|_{W_p^l} \leq c\ \text{ess}\,\|\gamma\|_{M(W_p^m \to W_p^l)}.$$

Let us prove the opposite estimate. Proposition 3.2.8 and Theorem 3.4.2 imply $\|\gamma\|_{MW_p^l} \sim \|\gamma\|_{M(W_p^m \to W_p^l)} < \infty$. This and Lemma 3.1.1/12 yield

$$\|(1-\zeta_r)\gamma u\|_{W_p^l} \leq \|\gamma\|_{MW_p^l} \|(1-\zeta_r)u\|_{W_p^l} \leq \text{const}\,\|u; B_{4r}\|_{W_p^l}.$$

Since any bounded subset of W_p^m is compact in $W_p^l(B_{4r})$, the operator $(1-\zeta_r)\gamma : W_p^m \to W_p^l$ is compact. Consequently, for any $r > 0$,

$$\text{ess}\,\|\gamma\|_{M(W_p^m \to W_p^l)} = \text{ess}\,\|\zeta_r \gamma\|_{M(W_p^m \to W_p^l)} \leq \|\zeta_r \gamma\|_{M(W_p^m \to W_p^l)}.$$

By estimating the last norm with the help of Proposition 3.2.8 and passing to the limit as $r \to \infty$, we complete the proof. □

Theorem 2. *If $l < m < n$, then*

$$\text{ess}\,\|\gamma\|_{M(W_1^m \to W_1^l)} \sim \overline{\lim_{\delta \to 0}}\, \delta^{m-n} \sup_{x \in R^n} \|\|\gamma; B_\delta(x)\|\|_{W_1^l}$$

$$+ \overline{\lim_{|x| \to \infty}} \sup_{r \in (0,1)} r^{m-n}\, \|\|\gamma; B_r(x)\|\|_{W_1^l}. \tag{2}$$

Proof. According to Theorems 3.4.2 and 4.2.1,

$$\text{ess}\,\|\gamma\|_{M(W_1^m\to W_1^l)}\sim\overline{\lim_{\delta\to 0}}\sup_{x,y\in R^n}\sup_{r\in(0,1)} r^{m-n}\,|||\eta_{\delta,x}\gamma;B_r(y)|||_{W_1^l}$$

$$+\overline{\lim_{\rho\to\infty}}\sup_{y\in R^n}\sup_{r\in(0,1)} r^{m-n}\,|||\zeta_\rho\gamma;B_r(y)|||_{W_1^l}.$$

Let the first summand on the right be denoted by A_1 and the second by A_2. We have

$$A_2\geqslant\overline{\lim_{\rho\to\infty}}\sup_{y\in B_{2\rho}}\sup_{r\in(0,1)} r^{m-n}\,|||\gamma;B_r(y)|||_{W_1^l}$$

$$=\overline{\lim_{|y|\to\infty}}\sup_{r\in(0,1)} r^{m-n}\,|||\gamma;B_r(y)|||_{W_1^l}.$$

An analogous upper bound for A_2 can be obtained as follows:

$$A_2\leqslant\overline{\lim_{\rho\to\infty}}\sup_{y\in B_{\rho/2}}\sup_{r\in(0,1)} r^{m-n}\,|||\zeta_\rho\gamma;B_r(y)|||_{W_1^l}$$

$$\leqslant c\,\overline{\lim_{\rho\to\infty}}\sup_{y\in B_{\rho/2}}\sup_{r\in(0,1)} r^{m-n}\,|||\gamma;B_{2r}|||_{W_1^l}$$

$$\leqslant c_1\,\overline{\lim_{|y|\to\infty}}\sup_{r\in(0,1)} r^{m-n}\,|||\gamma;B_r(y)|||_{W_1^l}.$$

Now we turn to estimates for A_1. We have

$$A_1\geqslant\overline{\lim_{\delta\to 0}}\sup_{x\in R^n}\delta^{m-n}\,|||\eta_{\delta,x}\gamma;B_\delta(x)|||_{W_1^l}$$

$$\geqslant c\,\overline{\lim_{\delta\to 0}}\sup_{x\in R^n}\delta^{m-n}\,|||\gamma;B_{\delta/2}(x)|||_{W_1^l}.$$

On the other hand,

$$\sup_{r\in(0,1)} r^{m-n}\,|||\eta_{\delta,x}\gamma;B_r(y)|||_{W_1^l}$$

$$\leqslant\sup_{r\in(0,\delta/2)} r^{m-n}\,|||\eta_{\delta,x}\gamma;B_r(y)|||_{W_1^l}+(2\delta)^{m-n}\sup_{r\in(\delta/2,1)}|||_{\eta_{\delta,x}}\gamma;B_r(y)|||_{W_1^l}.$$

The first summand on the right does not exceed

$$c\sup_{r\in(0,\delta/2)} r^{m-n}\,|||\gamma;B_{2r}(y)|||_{W_1^l}$$

and the second one is not more than

$$c\delta^{m-n}\,|||\gamma;B_{2\delta}(x)|||_{W_1^l}.$$

Consequently,

$$A_1\leqslant c\,\overline{\lim_{\delta\to 0}}\,\delta^{m-n}\sup_{x\in R^n}|||\gamma;B_\delta(x)|||_{W_1^l}.\quad\square$$

Remark 1. From Lemma 3.1.1/11 it follows that (2) can be rewritten as

$$\text{ess}\,\|\gamma\|_{M(W_1^m\to W_1^l)} \sim \overline{\lim_{\delta\to 0}}\, \delta^{m-n} \sup_{x\in R^n} (\delta^{-l}\,\|\gamma; B_\delta(x)\|_{L_1} + \|D_{1,l}\gamma; B_\delta(x)\|_{L_1})$$

$$+ \overline{\lim_{|x|\to\infty}} \sup_{r\in(0,1)} r^{m-n}(r^{-l}\,\|\gamma; B_r(x)\|_{L_1} + \|D_{1,l}\gamma; B_r(x)\|_{L_1}).$$

Remark 2. Let T_* be the operator defined by (4.2.1/4) for $pm \leq n$, $p>1$ and $m<n$, $p=1$. For other values of p, m we put $T_* = (1-\zeta_r)\gamma$. In the proofs of Theorems 4.2.1, 4.2.2, 4.2.3, 4.2.5/1 and 4.2.5/2 we proved in passing the following estimates for the norm of $\gamma - T_*$ for fixed δ and r.

(i) If $mp<n$, $p>1$, $m>l$, then

$$\|\gamma - T_*\|_{W_p^m\to W_p^l}$$
$$\leq c \sup_{\{e:d(e)\leq 8\delta\}} (f_l(\gamma; e) + f_0(\gamma; e))$$
$$+ c\left(\sup_{\{e\subset R^n\setminus B_{r/2}:d(e)\leq 1\}} (f_l(\gamma; e) + f_0(\gamma; e)) + r^{-\{l\}}\,\|\gamma\|_{M(W_p^m\to W_p^l)}\right).$$

(ii) If $mp=n$, $p>1$, $m>l$, then the latter estimate remains valid with $\{e: \mathrm{d}(e) \leq 8\delta\}$ being changed to $\{e: \mathrm{d}(e) \leq \delta^{1/2}\}$.

(iii) If $l<m<n$, then

$$\|\gamma - T_*\|_{W_1^m\to W_1^l} \leq c\delta^{m-n} \sup_{x\in R^n} (\delta^{-l}\,\|\gamma; B_\delta(x)\|_{L_1} + \|D_{1,l}\gamma; B_\delta(x)\|_{L_1})$$

$$+ c \sup_{x\in R^n\setminus B_{r/2}} \sup_{\rho\in(0,1)} (\rho^{-l}\,\|\gamma; B_\rho(x)\|_{L_1} + \|D_{1,l}\gamma; B_\rho(x)\|_{L_1}).$$

(iv) If $mp>n$, $p>1$ or $m\geq n$, $p=1$, $m>l$, then

$$\|\gamma - T_*\|_{W_p^m\to W_p^l} \leq c \sup_{x\in R^n\setminus B_{r/2}} \|\gamma; B_1(x)\|_{W_p^l}.$$

From (i)–(iv), together with lower estimates for the essential norm proved in the theorems mentioned, it follows that, given any ε, one can find so large r and so small δ that

$$\|\gamma - T_*\|_{W_p^m\to W_p^l} \leq c(\text{ess}\,\|\gamma\|_{M(W_p^m\to W_p^l)} + \varepsilon). \tag{3}$$

4.2.6 The Space of Compact Multipliers

By $\mathring{M}(W_p^m \to W_p^l)$, $m>l$, we denote the set of functions γ such that the operator of multiplication by γ is a compact operator acting from W_p^m into W_p^l.

Obviously, $\gamma \in \mathring{M}(W_p^m \to W_p^l)$ if and only if $\text{ess}\,\|\gamma\|_{M(W_p^m\to W_p^l)} = 0$. So Theorems 4.2.2, 4.2.3, 4.2.5/1, 4.2.5/2 imply the following necessary and sufficient conditions for a function $\gamma \in M(W_p^m \to W_p^l)$ to belong to the class $\mathring{M}(W_p^m \to W_p^l)$.

Theorem 1. (i) *If $mp \leq n$, $p>1$, then $\gamma \in \mathring{M}(W_p^m \to W_p^l)$ if and only if*

$$\lim_{\delta\to 0} \sup_{\{e: d(e)\leq\delta\}} \left(\frac{\|\gamma; e\|_{L_p}}{[\text{cap}\,(e, W_p^{m-l})]^{1/p}} + \frac{\|D_{p,l}\gamma; e\|_{L_p}}{[\text{cap}\,(e, W_p^m)]^{1/p}} \right) = 0, \tag{1}$$

$$\lim_{r\to\infty} \sup_{\{e\subset R^n\setminus B_r: d(e)\leq 1\}} \left(\frac{\|\gamma; e\|_{L_p}}{[\text{cap}\,(e, W_p^{m-l})]^{1/p}} + \frac{\|D_{p,l}\gamma; e\|_{L_p}}{[\text{cap}\,(e, W_p^m)]^{1/p}} \right) = 0. \tag{2}$$

(ii) *Let either $mp>n$, $p\geq 1$ or $m=n$, $p=1$. Then $\gamma \in \mathring{M}(W_p^m \to W_p^l)$ if and only if $\gamma \in W^l_{p,\text{unif}}$ and*

$$\lim_{|x|\to\infty} \|\gamma; B_1(x)\|_{W_p^l} = 0. \tag{3}$$

(iii) *In the case $m<n$ for $\gamma \in \mathring{M}(W_1^m \to W_1^l)$ it is necessary and sufficient that, together with (3), the following equality holds:*

$$\lim_{\delta\to 0} \delta^{m-n} \sup_{x\in R^n} \||\gamma; B_\delta(x)\||_{W_1^l} = 0. \tag{4}$$

From Theorem 4.2.1 we immediately obtain:

Theorem 2. *Let $lp<n$, $m>l$ and $p\geq 1$. Then $\gamma \in \mathring{M}(W_p^m \to W_p^l)$ if and only if*

$$\lim_{\delta\to 0} \sup_{x\in R^n} \|\eta_{\delta,x}\gamma\|_{M(W_p^m\to W_p^l)} = 0,$$

$$\lim_{r\to\infty} \|\zeta_r\gamma\|_{M(W_p^m\to W_p^l)} = 0.$$

From this, together with 3.2.9 and 3.3, we can get various necessary or sufficient conditions for $\gamma \in \mathring{M}(W_p^m \to W_p^l)$ which do not contain capacity.

The following theorem gives one more description of $\mathring{M}(W_p^m \to W_p^l)$.

Theorem 3. *The space $\mathring{M}(W_p^m \to W_p^l)$ is the completion of D with respect to the norm of the space $M(W_p^m \to W_p^l)$.*

Proof. By Theorem 1, $D \subset \mathring{M}(W_p^m \to W_p^l)$. Therefore any function in $M(W_p^m \to W_p^l)$, approximated by a sequence in D in the norm of $M(W_p^m \to W_p^l)$, generates a compact operator of multiplication: $W_p^m \to W_p^l$.

Further, we prove the converse assertion. Let $\gamma \in \mathring{M}(W_p^m \to W_p^l)$. According to items (ii), (iii) of Theorem 1 it suffices to consider the case

$mp \geq n$, $p > 1$. By Theorem 4.2.1 we have

$$\lim_{r\to\infty} \|\gamma - (1-\zeta_r)\gamma\|_{M(W_p^m \to W_p^l)} = 0. \tag{5}$$

Let $\Gamma = (1-\zeta_r)\gamma$ and let Γ_ρ be the mollification of Γ with radius ρ. By T_* and $T_*^{(\rho)}$ we denote the operators specified by (4.2.1/4) for functions Γ and Γ_ρ respectively. From (4.2.1/8) follows

$$\lim_{\delta\to 0} \|\Gamma - T_*\|_{W_p^m \to W_p^l} = 0. \tag{6}$$

Further, by (4.2.1/8) and Theorem 4.2.1,

$$\|\Gamma_\rho - T_*^{(\rho)}\|_{W_p^m \to W_p^l} \leq c \varlimsup_{\delta\to 0} \sup_{x\in R^n} \|\eta_{\delta,x}\Gamma_\rho\|_{M(W_p^m \to W_p^l)} + \varepsilon$$

$$\leq c \operatorname{ess} \|\Gamma_\rho\|_{M(W_p^m \to W_p^l)} + \varepsilon = \varepsilon. \tag{7}$$

The last equality is valid since $\Gamma_\rho \in D$.

From the definition of operators T_* and $T_*^{(\rho)}$ we get

$$\|(T_* - T_*^{(\rho)})u\|_{W_p^l} \leq c(\delta, r)\|u\|_{L_p} \sum_j \|(\Gamma - \Gamma_\rho)\varphi^{(j)}\|_{W_p^l}$$

and hence

$$\|T_* - T_*^{(\rho)}\|_{W_p^m \to W_p^l} \leq c(\delta, r)\, \|\Gamma - \Gamma_\rho\|_{W_p^l}. \tag{8}$$

The right-hand side of this inequality tends to zero as $\rho \to 0$. Since

$$\|\Gamma - \Gamma_\rho\|_{M(W_p^m \to W_p^l)} \leq \|\Gamma - T_*\|_{W_p^m \to W_p^l} + \|\Gamma_\rho - T_*^{(\rho)}\|_{W_p^m \to W_p^l}$$
$$+ \|T_* - T_*^{(\rho)}\|_{W_p^m \to W_p^l},$$

then by (6)–(8) $\lim_{r\to\infty, \rho\to 0} \|\gamma - \Gamma_\rho\|_{M(W_p^m \to W_p^l)} = 0$. Since $\Gamma_\rho \in D$, the result follows. □

4.3 Two-sided Estimates for the Essential Norm in the Case $m = l$

4.3.1 *Estimate for the Maximum Modulus of a Multiplier in W_p^l by its Essential Norm*

Theorem. *If $l > 0$ and $1 \leq p < \infty$, then* $\|\gamma\|_{L_\infty} \leq \operatorname{ess} \|\gamma\|_{MW_p^l}$.

Proof. Let T be a compact operator in W_p^l such that, for all $u \in W_p^l$,

$$\|(\gamma - T)u\|_{W_p^l} \leq (\operatorname{ess} \|\gamma\|_{MW_p^l} + \varepsilon)\, \|u\|_{W_p^l}. \tag{1}$$

Let η be an arbitrary function in $D(Q_k)$, where Q_k is the cube $\{y : |y_j| < \pi k\}$ and k is integer. We consider the sequence

$$u_N(y) = N^{-l} \exp(iNy_1)\eta(y), \quad N = 1, 2, \ldots.$$

Obviously, for integer l we have $\|u_N\|_{W_p^l}=\|\eta\|_{L_p}+O(N^{-1})$. Let l be fractional, $0<l<1$. Then

$$\|u_N\|_{W_p^l}=N^{-l}\,\|e^{iNy_1}\eta\|_{W_p^l}=N^{-l}\,\|D_{p,l}e^{iNy_1}\eta\|_{L_p}+O(N^{-1}).$$

Clearly, $|D_{p,l}(e^{iNy_1}\eta)-|\eta|\,D_{p,l}e^{iNy_1}|\leq D_{p,l}\eta$. Since $D_{p,l}e^{iNy_1}=a_lN^l$, where $a_l=\text{const}>0$, then

$$\|D_{p,l}(e^{iNy_1}\eta)-a_lN^l\,|\eta|\,\|_{L_p}=O(1). \tag{2}$$

Let $l>1$. Then $D_{p,l}(e^{iNy_1}\eta)=D_{p,\{l\}}(\nabla_{[l]}e^{iNy_1})$. We have

$$|D_{p,\{l\}}(\nabla_{[l]}(e^{iNy_1}\eta))-N^{[l]}D_{p,\{l\}}(e^{iNy_1}\eta)|\leq c\sum_{j=0}^{[l]}N^jD_{p,\{l\}}\left(e^{iNy_1}\frac{\partial^{[l]-j}\eta}{\partial y_1^{[l]-j}}\right).$$

This and (2) yield $\|D_{p,l}(e^{iNy_1}\eta)-N^la_{\{l\}}\|_{L_p}\leq cN^{[l]}$. So in the case $\{l\}>0$ we obtain

$$\|u_N\|_{W_p^l}=a_{\{l\}}\,\|\eta\|_{L_p}+O(N^{-\{l\}}). \tag{3}$$

We shall show that $\{u_N\}$ weakly converges to zero in $\mathring{W}_p^l(Q_k)$. Let f be any linear functional on $\mathring{W}_p^l(Q_k)$. If $p\leq 2$, then the restriction of f to $\mathring{W}_2^l(Q_k)$ is a linear functional on $\mathring{W}_2^l(Q_k)$. Consequently, $f(u_N)=\int \Lambda^lu_N\Lambda^l\psi\,dx$, where $\psi\in\mathring{W}_2^l(Q_k)$. Since $\|\Lambda^lu_N-e^{iNy_1}\eta\|_{L_2}=O(N^{-1})$ and the sequence $\int_{Q_k}e^{iNy_1}\eta(y)\Lambda^l\psi\,dy$ tends to zero, being a sequence of Fourier coefficients of a function in $L_2(Q_k)$, then $f(u_N)\xrightarrow[N\to\infty]{}0$.

Let $p>2$. Taking into account the imbedding of H_p^l into W_p^l, we get $|f(u_N)|\leq c\,\|u_N\|_{H_p^l}$. Therefore $f(u_N)=\int g\Lambda^lu_N\,dy$, where $g\in L_{p'}$. Since

$$\|\Lambda^lu_N-e^{iNy_1}\eta\|_{L_p}=O(N^{-1}),$$

then

$$f(u_N)=\int_{Q_k}e^{iNy_1}\eta(y)g\,dy+O(N^{-1})\,\|g\|_{L_{p'}}.$$

Applying the Hausdorff-Young theorem (see Zygmund [1]) to the function $\eta g\in L_{p'}(Q_k)$, $p'<2$, we conclude that $f(u_N)\xrightarrow[N\to\infty]{}0$.

By φ we denote a function in $D(Q_1)$ which is equal to unity on the cube $Q_{1-\delta}$, $\delta>0$, and we then set $\varphi_k(y)=\varphi(y/k)$. The compactness of the operator φ_kT in $\mathring{W}_p^l(Q_k)$ implies $\varphi_kTu_N\xrightarrow[N\to\infty]{}0$ in $\mathring{W}_p^l(Q_k)$. Now from Lemma 4.1/4 and (1) follows

$$\begin{aligned}\varlimsup_{N\to\infty}\|\varphi_k\gamma u_N\|_{W_p^l}&=\varlimsup_{N\to\infty}\|\varphi_k(\gamma-T)u_N\|_{W_p^l}\\&\leq(1+O(k^{-\delta}))\varlimsup_{N\to\infty}\|(\gamma-T)u_N\|_{W_p^l}\\&\leq(1+O(k^{-\delta}))\varlimsup_{N\to\infty}\|u_N\|_{W_p^l}(\operatorname{ess}\|\gamma\|_{MW_p^l}+\varepsilon)\end{aligned}$$

which, together with (3), yields

$$\overline{\lim_{N\to\infty}} \|\varphi_k \gamma u_N\|_{W_p^l} \leqslant (1+O(k^{-\delta}))a_{\{l\}} \|\eta\|_{L_p} (\text{ess} \|\gamma\|_{MW_p^l} + \varepsilon).$$

With the same arguments as in the proof of (3), we obtain

$$\lim_{N\to\infty} \|\varphi_k \gamma u_N\|_{W_p^l} = a_{\{l\}} \|\varphi_k \gamma\eta\|_{L_p}.$$

Thus

$$\overline{\lim_{k\to\infty}} \|\varphi_k \gamma\eta\|_{L_p} \leqslant \|\eta\|_{L_p} \text{ ess} \|\gamma\|_{MW_p^l}.$$

Since $\varphi_k\eta = \eta$ for large values of k and η is an arbitrary function in D, the result follows. □

4.3.2 Estimates for the Essential Norm in Terms of Truncating Functions (the Case $lp \leqslant n$, $p > 1$)

Theorem 1. *For $lp < n$, $p \geqslant 1$ the following relations hold:*

$$\text{ess} \|\gamma\|_{MW_p^l} \sim \overline{\lim_{\delta\to 0}} \sup_{x\in R^n} \|\eta_{\delta,x}\gamma\|_{MW_p^l} + \overline{\lim_{r\to\infty}} \|\zeta_r\gamma\|_{MW_p^l}. \tag{1}$$

The proof of this relation can be obtained by duplicating the proof of Theorem 4.2.1, where $m = l$.

Theorem 2. *If $0 < l \leqslant 1$, $lp = n$, $p > 1$, then*

$$\text{ess} \|\gamma\|_{MW_p^l} \sim \|\gamma\|_{L_\infty} + \overline{\lim_{\delta\to 0}} \sup_{x\in R^n} \|\eta_{\delta,x} D_{p,l}\gamma\|_{M(W_p^l\to L_p)} + \overline{\lim_{r\to\infty}} \|\zeta_r\gamma\|_{MW_p^l}. \tag{2}$$

Proof. (i) The upper bound for the essential norm: we choose δ and r so that

$$\sup_{x\in R^n} \|\eta_{\delta,x} D_{p,l}\gamma\|_{M(W_p^l\to L_p)} \leqslant \overline{\lim_{\delta\to 0}} \sup_{x\in R^n} \|\eta_{\delta,x} D_{p,l}\gamma\|_{M(W_p^l\to L_p)} + \varepsilon,$$

$$\|\zeta_r\gamma\|_{MW_p^l} \leqslant \overline{\lim_{r\to\infty}} \|\zeta_r\gamma\|_{MW_p^l} + \varepsilon.$$

Let Γ and T_* be the function and the operator introduced in the second part of Theorem 4.2.1. By (4.2.1/5) it suffices to get the estimate for $\|(\Gamma - T_*)u\|_{W_p^l}$. We obtain

$$\|(\Gamma - T_*)u\|_{W_p^l}^p \leqslant A + B + C, \tag{3}$$

where

$$A = \Big\|\Gamma \sum_j \varphi^{(j)}(u - P^{(j)})\Big\|_{L_p}^p, \qquad B = \Big\|\sum_j (D_{p,l}\Gamma)\eta_{2\delta,x_j}\varphi^{(j)}(u - P^{(j)})\Big\|_{L_p}^p,$$

$$C = \Big\|\Gamma D_{p,l} \sum_j \varphi^{(j)}(u - P^{(j)})\Big\|_{L_p}^p.$$

By Lemma 3.1.1/13,

$$A \leqslant \|\gamma\|_{L_\infty}^p \sum_j \|u-P^{(j)}; K_\delta^{(j)}\|_{L_p}^p \leqslant c\, \|\gamma\|_{L_\infty}^p \|D_{p,l}u\|_{L_p}^p. \tag{4}$$

From Lemmas 3.1.1/12 and 3.1.1/13 it follows that

$$\begin{aligned} B &\leqslant c \sum_j \|(D_{p,l}\Gamma)\eta_{2\delta,x_j}\varphi^{(j)}(u-P^{(j)})\|_{L_p}^p \\ &\leqslant c \sup_j \|\eta_{2\delta,x_j}D_{p,l}\Gamma\|_{M(W_p^l\to L_p)}^p \sum_j \|\varphi^{(j)}(u-P^{(j)})\|_{L_p}^p \\ &\leqslant c_1 \sup_j \|\eta_{2\delta,x_j}D_{p,l}\Gamma\|_{M(W_p^l\to L_p)}^p \|D_{p,l}u\|_{L_p}^p. \end{aligned} \tag{5}$$

By using Lemmas 3.1.1/11–3.1.1/13, we deduce

$$\begin{aligned} C &\leqslant c\, \|\gamma\|_{L_\infty}^p \sum_j \|\varphi^{(j)}(u-P^{(j)})\|_{W_p^l}^p \\ &\leqslant c_1 \|\gamma\|_{L_\infty}^p \sum_j \|u-P^{(j)}; K_\delta^{(j)}\|_{W_p^l}^p \leqslant c_2 \|\gamma\|_{L_\infty}^p \|u\|_{W_p^l}^p \end{aligned}$$

which, together with (3)–(5), implies

$$\|(\Gamma - T_*)u\|_{W_p^l} \leqslant c\left(\|\gamma\|_{L_\infty} + \sup_j \|\eta_{2\delta,x_j}D_{p,l}\Gamma\|_{M(W_p^l\to L_p)}\right)\|u\|_{W_p^l}.$$

Lemma 4.1/4 enables one to replace Γ by γ in the right-hand side. The required estimate for the essential norm is obtained.

(ii) The lower bound for the essential norm: let T be a compact operator in W_p^l such that

$$\|D_{p,l}(\gamma u) - D_{p,l}(Tu)\|_{L_p} \leqslant (\text{ess}\,\|\gamma\|_{MW_p^l} + \varepsilon)\,\|u\|_{W_p^l}.$$

Hence

$$\|uD_{p,l}\gamma - D_{p,l}Tu\|_{L_p} \leqslant (\text{ess}\,\|\gamma\|_{MW_p^l} + \varepsilon)\,\|u\|_{W_p^l} + \|\gamma D_{p,l}u\|_{L_p}.$$

From the inequality $\|D_{p,l}v_1 - D_{p,l}v_2\|_{L_p} \leqslant \|D_{p,l}(v_1-v_2)\|_{L_p}$ it follows that the compactness of the set $\{Tu : u\in S\}$ in the space W_p^l, where S is the unit ball in W_p^l, implies the compactness of the set $\{D_{p,l}Tu : u\in S\}$ in L_p. Let functions $\{w_k\}$ form a finite ε-net in the latter set. Then, for $u\in S$,

$$\begin{aligned} \|u\eta_{\delta,x}D_{p,l}\gamma\|_{L_p} &\leqslant \|\eta_{\delta,x}(D_{p,l}Tu - w_k)\|_{L_p} + \|\eta_{\delta,x}w_k\|_{L_p} \\ &\quad + \text{ess}\,\|\gamma\|_{MW_p^l} + \varepsilon + \|\gamma\|_{L_\infty}. \end{aligned}$$

Consequently, for any $x\in R^n$ and for small enough $\delta>0$,

$$\|u\eta_{\delta,x}D_{p,l}\gamma\|_{L_p} \leqslant c(\text{ess}\,\|\gamma\|_{MW_p^l} + \varepsilon).$$

(Here we have used Theorem 4.3.1.) This, together with Lemma 4.2.1 and Theorem 4.3.1, implies the required lower bound for the norm ess $\|\gamma\|_{MW_p^l}$. □

Theorem 3. *If* $l \geq 1$, $lp = n$, $p > 1$, *then*

$$\text{ess}\,\|\gamma\|_{MW_p^l} \sim \|\gamma\|_{L_\infty} + \overline{\lim_{\delta\to 0}} \sup_{x\in R^n} \|\eta_{\delta,x}\nabla_k\gamma\|_{M(W_p^l \to W_p^{l-k})} + \overline{\lim_{r\to\infty}} \|\zeta_r\gamma\|_{MW_p^l}$$

where $k = 1, \ldots, [l]$.

Proof. (i) Upper bound for the essential norm. We have

$$\|(\Gamma - T_*)u\|_{W_p^l}^p \leq c\left(\left\|\nabla_k\left(\Gamma \sum_j \varphi^{(j)}(u - P^{(j)})\right)\right\|_{W_p^{l-k}}^p + \|(\Gamma - T_*)u\|_{L_p}^p\right). \tag{6}$$

The second summand on the right does not exceed $c\|\gamma\|_{L_\infty}^p \|u\|_{W_p^l}^p$ (see estimate (4)). The first summand is not more than

$$c \sum_{|\alpha|+|\beta|=k} \left\|D^\alpha\Gamma \sum_j D^\beta[\varphi^{(j)}(u - P^{(j)})]\right\|_{W_p^{l-k}}^p$$

$$\leq c \sum_{|\alpha|=0}^{k} \sup_j \|\eta_{2\delta,x_j} D^\alpha\Gamma\|_{M(W_p^{l-k+|\alpha|} \to W_p^{l-k})}^p \sum_j \|\varphi^{(j)}(u - P^{(j)})\|_{W_p^l}^p. \tag{7}$$

Since $p(l-k) < n$, then, by making use of Theorems 4.2.1 and 1, we conclude that the expression on the right in the latter inequality is majorized by

$$c\left(\sup_j \|\eta_{2\delta,x_j}\nabla_k\Gamma\|_{M(W_p^l \to W_p^{l-k})} + \sum_{j=0}^{k-1} \text{ess}\,\|\nabla_j\Gamma\|_{M(W_p^{l-k+j} \to W_p^{l-k})}\right)^p \|u\|_{W_p^l}^p$$

which, by virtue of Lemma 4.2.4, does not exceed

$$c\left(\sup_j \|\eta_{2\delta,x_j}\nabla_k\Gamma\|_{M(W_p^l \to W_p^{l-k})} + \sum_{i=1}^{k} \text{ess}\,\|\Gamma\|_{MW_p^{l-i}}\right)^p \|u\|_{W_p^l}^p.$$

From the interpolation theorem we get

$$\begin{aligned} \text{ess}\,\|\Gamma\|_{MW_p^{l-i}} &\leq \|\Gamma - T_*\|_{W_p^{l-i} \to W_p^{l-i}} \leq c\,\|\Gamma - T_*\|_{W_p^l \to W_p^l}^{(l-i)/l} \|\Gamma - T_*\|_{L_p \to L_p}^{i/l} \\ &\leq \varepsilon\,\|\Gamma - T_*\|_{W_p^l \to W_p^l} + c(\varepsilon)\,\|\gamma\|_{L_\infty}, \end{aligned} \tag{8}$$

where ε is an arbitrarily small positive number. Therefore the right-hand side of (6) does not exceed

$$c\left(\sup_j \|\eta_{2\delta,x_j}\nabla_k\Gamma\|_{M(W_p^l \to W_p^{l-k})} + \varepsilon\,\|\Gamma - T_*\|_{W_p^l \to W_p^l} + c(\varepsilon)\,\|\gamma\|_{L_\infty}\right)^p \|u\|_{W_p^l}^p.$$

Choosing ε small enough and applying Lemma 4.1/4, we obtain from the last inequality and from (6) that

$$\|(\Gamma - T_*)\|_{W_p^l \to W_p^l} \leq c\left(\sup_j \|\eta_{2\delta,x_j} \nabla_k \Gamma\|_{M(W_p^l \to W_p^{l-k})} + \|\gamma\|_{L_\infty}\right) \tag{9}$$

which, together with (4.2.1/5) and Lemma 4.1/4, gives the required upper bound for the essential norm.

(ii) The lower bound for the essential norm. By Lemma 4.2.1 and Theorem 4.3.1 it suffices to show that

$$\|\eta_{\delta,x} \nabla_k \gamma\|_{M(W_p^l \to W_p^{l-k})} \leq c(\text{ess}\, \|\gamma\|_{MW_p^l} + \varepsilon) \tag{10}$$

for all $x \in R^n$ and small enough $\delta > 0$.

Let T be a compact operator for which (4.3.1/1) holds. Then, for all $u \in W_p^l$,

$$\|\nabla_k[(\gamma - T)u]\|_{W_p^{l-k}} \leq (\text{ess}\, \|\gamma\|_{MW_p^l} + \varepsilon)\, \|u\|_{W_p^l}.$$

Since $p(l-k) < n$,

$$\|\eta_{\delta,x} \nabla_k[(\gamma - T)u]\|_{W_p^{l-k}} \leq c(\text{ess}\, \|\gamma\|_{MW_p^l} + \varepsilon)\, \|u\|_{W_p^l}.$$

Let S be the unit ball in W_p^l. The set $\{v = D^\alpha Tu, |\alpha| = k : u \in S\}$ is compact in W_p^{l-k}. Let $\{v_\nu\}$ be an ε-net in W_p^{l-k} for the latter set. Without loss of generality we may assume that $v_\nu \in D$.

Since $p(l-k) < n$, then by Lemma 4.1/2, for small δ,

$$\sup_{x \in R^n} \|\eta_{\delta,x} v_\nu\|_{W_p^{l-k}} < \varepsilon$$

and so

$$\sup_{x \in R^n} \|\eta_{\delta,x} \nabla_k (Tu)\|_{W_p^{l-k}} < c\varepsilon.$$

Thus, for all $u \in S$,

$$\begin{aligned}
&\|u\eta_{\delta,x} \nabla_k \gamma\|_{W_p^{l-k}} \\
&\quad \leq c\left(\text{ess}\, \|\gamma\|_{MW_p^l} + \sum_{\substack{|\alpha|+|\beta|=k,\\ |\alpha|>0}} \|\eta_{\delta,x} D^\alpha u D^\beta \gamma\|_{W_p^{l-k}} + \varepsilon\right) \\
&\quad \leq c\left(\text{ess}\, \|\gamma\|_{MW_p^l} + \sum_{\substack{|\alpha|+|\beta|=k,\\ |\alpha|>0}} \|\eta_{\delta,x} D^\beta \gamma\|_{M(W_p^{l-|\alpha|} \to W_p^{l-k})} \|D^\alpha u\|_{W_p^{l-|\alpha|}} + \varepsilon\right).
\end{aligned} \tag{11}$$

From Theorems 4.2.1 and 1, it follows that for small δ

$$\|\eta_{\delta,x} D^\beta \gamma\|_{M(W_p^{l-|\alpha|} \to W_p^{l-k})} \leq c \text{ ess}\, \|D^\beta \gamma\|_{M(W_p^{l-|\alpha|} \to W_p^{l-k})}.$$

Making use of Theorem 1 and the interpolation theorem (see (8)), we obtain

$$\operatorname{ess} \|D^\beta \gamma\|_{M(W_p^{l-|\alpha|} \to W_p^{l-k})} \leq c \sum_{j=0}^{|\beta|} \operatorname{ess} \|\gamma\|_{MW_p^{l-|\alpha|-j}}$$
$$\leq \varepsilon \|\gamma - T_*\|_{W_p^l \to W_p^l} + c(\varepsilon) \|\gamma\|_{L_\infty}.$$

From this, together with (11) and the following estimate

$$\|\gamma - T_*\|_{W_p^l \to W_p^l} \leq c\left(\sup_j \|\eta_{2\delta,x_j} \nabla_k \gamma\|_{M(W_p^l \to W_p^{l-k})} + \|\gamma\|_{L_\infty} + \|\zeta_r \gamma\|_{MW_p^l}\right)$$

established in the first part of the proof (see (9)), we get for $u \in S$

$$\|u\eta_{\delta,x} \nabla_k \gamma\|_{W_p^{l-k}} \leq c(\varepsilon) \operatorname{ess} \|\gamma\|_{MW_p^l} + \varepsilon \sup_j \|\eta_{2\delta,x_j} \nabla_k \gamma\|_{M(W_p^l \to W_p^{l-k})} + \varepsilon.$$

Inequality (10) is obtained, which completes the proof of the theorem. □

4.3.3 *Estimates for the Essential Norm in Terms of Capacities (the Case $lp \leq n$, $p > 1$)*

For the theorem of this subsection we need the following assertion.

Lemma. *If $lp \leq n$, $p > 1$, then*

$$\operatorname{ess} \|\gamma\|_{MW_p^\sigma} \leq c \operatorname{ess} \|\gamma\|_{MW_p^l}^{\sigma/l} \|\gamma\|_{L_\infty}^{1-\sigma/l}, \qquad 0 < \sigma < l.$$

Proof. When proving any of Theorems 4.3.2/1–4.3.2/3 it was shown in passing that for some r and δ from the definition of T_* we have

$$\|\gamma - T_*\|_{W_p^l \to W_p^l} \leq c \operatorname{ess} \|\gamma\|_{MW_p^l} + \varepsilon \tag{1}$$

(cf. Remark 4.2.5/2). Moreover,

$$\|\gamma - T_*\|_{L_p \to L_p} \leq c \|\gamma\|_{L_\infty}.$$

So, interpolating between W_p^l and L_p, we get

$$\operatorname{ess} \|\gamma\|_{MW_p^\sigma} \leq \|\gamma - T_*\|_{W_p^\sigma \to W_p^\sigma} \leq c(\operatorname{ess} \|\gamma\|_{MW_p^l} + \varepsilon)^{\sigma/l} \|\gamma\|_{L_\infty}^{1-\sigma/l}. \quad \square$$

Theorem. *Let $lp \leq n$, $p > 1$. We have*

$$\operatorname{ess} \|\gamma\|_{MW_p^l} \sim \|\gamma\|_{L_\infty} + \lim_{\delta \to 0} \sup_{\{e : d(e) \leq \delta\}} \frac{\|D_{p,l}\gamma; e\|_{L_p}}{[\operatorname{cap}(e, W_p^l)]^{1/p}}$$
$$+ \lim_{r \to \infty} \sup_{\{e \subset R^n \setminus B_r : d(e) \leq 1\}} \frac{\|D_{p,l}\gamma; e\|_{L_p}}{[\operatorname{cap}(e, W_p^l)]^{1/p}}. \tag{2}$$

Proof. For $lp<n$ it suffices to duplicate the proof of Theorem 4.2.2, putting $m=l$.

Inequalities (4.2.2/10) and (4.2.2/17), which are also valid for $m=l\leqslant n/p$, imply

$$\overline{\lim_{r\to\infty}}\, \|\zeta_r\gamma\|_{MW_p^l} \sim \lim_{r\to\infty}\left(\|\gamma; R^n\setminus B_r\|_{L_\infty} + \sup_{\{e\subset R^n\setminus B_r : d(e)\leqslant 1\}} \frac{\|D_{p,l}\gamma; e\|_{L_p}}{[\operatorname{cap}(e, W_p^l)]^{1/p}}\right). \tag{3}$$

Hence, from Theorem 4.3.2/2 and Remark 3.2.1/1, we have (2) for $0<l\leqslant 1$, $lp=n$.

Let $lp=n$, $l>1$. In the proof of Theorem 4.2.3 it is shown that

$$\|\eta_{\delta^2,x}\nabla\gamma\|_{M(W_p^l\to W_p^{l-1})}\leqslant c\left(\sup_{\{e : d(e)\leqslant 2\delta\}} \frac{\|D_{p,l-1}(\nabla\gamma); e\|_{L_p}}{[\operatorname{cap}(e, W_p^l)]^{1/p}} + \frac{\|\nabla\gamma; e\|_{L_p}}{[\operatorname{cap}(e, W_p^1)]^{1/p}}\right),$$

$$\|\zeta_r\gamma\|_{MW_p^l}\leqslant c\left(\sup_{\{e\subset R^n\setminus B_{r/2} : d(e)\leqslant 1\}}\left(\frac{\|D_{p,l-1}(\nabla\gamma); e\|_{L_p}}{[\operatorname{cap}(e, W_p^l)]^{1/p}} + \frac{\|\nabla\gamma; e\|_{L_p}}{[\operatorname{cap}(e, W_p^1)]^{1/p}}\right)+\|\gamma\|_{L_\infty}\right).$$

Using the estimate for $\operatorname{ess}\|\gamma\|_{MW_p^1}$, from the latter and Theorem 4.3.2/3, we get

$$\operatorname{ess}\|\gamma\|_{MW_p^l}\leqslant c\left(\|\gamma\|_{L_\infty} + \lim_{\delta\to 0}\sup_{\{e : d(e)\leqslant\delta\}} \frac{\|D_{p,l}\gamma; e\|_{L_p}}{[\operatorname{cap}(e, W_p^l)]^{1/p}} + \lim_{r\to\infty}\sup_{\{e\subset R^n\setminus B_r : d(e)\leqslant 1\}} \frac{\|D_{p,l}\gamma; e\|_{L_p}}{[\operatorname{cap}(e, W_p^l)]^{1/p}} + \operatorname{ess}\|\gamma\|_{MW_p^1}\right).$$

It remains to note that, by the lemma,

$$\operatorname{ess}\|\gamma\|_{MW_p^1}\leqslant c\operatorname{ess}\|\gamma\|_{MW_p^l}^{1/l}\,\|\gamma\|_{L_\infty}^{(l-1)/l}.$$

Let us verify the lower bound for the essential norm. By the lemma and Theorem 4.3.1,

$$\operatorname{ess}\|\gamma\|_{MW_p^l}\geqslant c\operatorname{ess}\|\gamma\|_{MW_p^{l-1}}$$

which, together with Lemma 4.2.4, gives the estimate

$$\operatorname{ess}\|\gamma\|_{MW_p^l}\geqslant c\operatorname{ess}\|\nabla\gamma\|_{M(W_p^l\to W_p^{l-1})}.$$

Taking into account Theorem 4.2.3, we obtain that the right-hand side of this inequality is majorized by

$$c\lim_{\delta\to 0}\sup_{\{e : d(e)\leqslant\delta\}} \frac{\|D_{p,l}\gamma; e\|_{L_p}}{[\operatorname{cap}(e, W_p^l)]^{1/p}}.$$

It remains to use (3) and Theorem 4.3.2/2. The result follows. □

4.3.4 Two-sided Estimates for the Essential Norm in the Cases $lp > n$, $p > 1$ and $p = 1$

Theorem 1. *If $lp > n$, $p > 1$, then*

$$\operatorname{ess} \|\gamma\|_{MW_p^l} \sim \|\gamma\|_{L_\infty} + \overline{\lim_{|x|\to\infty}} \|\gamma; B_1(x)\|_{W_p^l}. \tag{1}$$

Proof. From Proposition 3.2.8 and Lemma 4.1/4 we obtain

$$\overline{\lim_{|x|\to\infty}} \|\gamma; B_1(x)\|_{W_p^l} \sim \overline{\lim_{r\to\infty}} \|\zeta_r \gamma\|_{MW_p^l}. \tag{2}$$

So the required lower bound for the essential norm follows from Lemma 4.2.1 and Theorem 4.3.1.

Now we establish the upper bound. Let Γ and T_* be the function and the operator specified in the second part of Theorem 4.2.1. With the function Γ_ρ, obtained as the mollification of Γ with radius ρ, we associate the operator $T_*^{(\rho)}$ by the same rule. Using Proposition 3.2.8 for sufficiently small ρ, we find

$$\|\Gamma - \Gamma_\rho\|_{MW_p^l} \leqslant c\, \|\Gamma - \Gamma_\rho\|_{W_p^l} < \varepsilon. \tag{3}$$

Next we note that the proof of inequality (4.3.2/9) equally suits the case $lp > n$. Replacing the numbers l and k by an integer s, $s > np$ in (4.3.2/9) and using Proposition 3.2.8, we arrive at

$$\|\Gamma_\rho - T_*^{(\rho)}\|_{W_p^s \to W_p^s} \leqslant c\left(\sup_j \|\eta_{2\delta, x_j} \nabla_s \Gamma_\rho\|_{L_p} + \|\gamma\|_{L_\infty}\right).$$

So for small δ,

$$\|\Gamma_\rho - T_*^{(\rho)}\|_{W_p^s \to W_p^s} \leqslant c\, \|\gamma\|_{L_\infty}. \tag{4}$$

The same inequality obviously holds for $s = 0$. By interpolating between L_p and W_p^s, we obtain (4) for $s = l$ which, together with (3), yields

$$\|\Gamma - T_*^{(\rho)}\|_{W_p^l \to W_p^l} \leqslant c\, \|\gamma\|_{L_\infty} + \varepsilon.$$

The result follows from the latter inequality and (4.2.1/5).

Further, we note that one may replace the operator $T_*^{(\rho)}$ by T_* in the last inequality. In fact, from Lemmas 3.1.1/12, 3.1.1/13 and Corollary 3.1.1/1 it follows that

$$\left\|u - \sum_j \varphi^{(j)} P^{(j)}\right\|_{W_p^l}^p \leqslant c \sum_j \|\varphi^{(j)}(u - P^{(j)})\|_{W_p^l}^p \leqslant c_1\, \|u\|_{W_p^l}^p.$$

Therefore

$$\|(T_* - T_*^{(\rho)})u\|_{W_p^l} \leq \|\Gamma - \Gamma_\rho\|_{MW_p^l} \left\|\sum_j \varphi^{(j)} P^{(j)}\right\|_{W_p^l} \leq c\, \|\Gamma - \Gamma_\rho\|_{MW_p^l}\, \|u\|_{W_p^l}$$

and it remains to use inequality (3). □

Theorem 2. *If $l<n$, then*

$$\text{ess}\, \|\gamma\|_{MW_1^l} \sim \overline{\lim_{\delta\to 0}}\, \delta^{l-n} \sup_{x\in R^n} |||\gamma; B_\delta(x)|||_{W_1^l}$$

$$+ \overline{\lim_{|x|\to\infty}} \sup_{r\in(0,1)} r^{l-n}\, |||\gamma; B_r(x)|||_{W_1^l}. \tag{5}$$

The proof runs in the same way as that of Theorem 4.2.5/2, in which one should put $m=l$ and use Theorem 4.3.2/1 instead of Theorem 4.2.1.

Remark 1. By virtue of Lemma 3.1.1/11, relation (5) can be rewritten as

$$\text{ess}\, \|\gamma\|_{MW_1^l} \sim \|\gamma\|_{L_\infty} + \overline{\lim_{\delta\to 0}}\, \delta^{l-n} \sup_{x\in R^n} \|D_{1,l}\gamma; B_\delta(x)\|_{L_1}$$

$$+ \overline{\lim_{|x|\to\infty}} \sup_{r\in(0,1)} r^{l-n}\, \|D_{1,l}\gamma; B_r(x)\|_{L_1}.$$

Theorem 3. *If $l\geq n$, then*

$$\text{ess}\, \|\gamma\|_{MW_1^l} \sim \|\gamma\|_{L_\infty} + \overline{\lim_{|x|\to\infty}}\, \|\gamma; B_1(x)\|_{W_1^l}. \tag{6}$$

Proof. The lower bound for the essential norm directly follows from Lemma 4.2.1 and Theorems 3.4.2, 4.3.1.

Next we obtain the upper bound. Let $k=[l]+1-n$. We have

$$\|(\Gamma - T_*)u\|_{W_1^l} \leq c\left(\left\|\nabla_k\left(\Gamma \sum_j \varphi^{(j)}(u-P^{(j)})\right)\right\|_{W_1^{l-k}} + \|(\Gamma - T_*)u\|_{L_1}\right).$$

The second summand on the right does not exceed $c\,\|\gamma\|_{L_\infty}\|u\|_{W_1^l}$ (see estimate (4.3.2/4)). The first one is not more than

$$c \sum_{|\alpha|+|\beta|=k} \left\|D^\alpha \Gamma \sum_j D^\beta[\varphi^{(j)}(u-P^{(j)})]\right\|_{W_1^{l-k}}$$

$$\leq c \sum_{|\alpha|=0}^{k} \sup_j \|\eta_{2\delta,x_j} D^\alpha \Gamma\|_{M(W_1^{l-k+|\alpha|}\to W_1^{l-k})} \sum_j \|\varphi^{(j)}(u-P^{(j)})\|_{W_1^l}.$$

With the help of Lemmas 3.1.1/11–3.1.1/13 we obtain that the last norm is majorized by $c\,\|u\|_{W_1^l}$.

Now we show that

$$\overline{\lim_{\delta\to 0}} \sup_{x\in R^n} \|\eta_{\delta,x}\Gamma\|_{MW_1^{l-k}} \leqslant c\,\|\Gamma\|_{L_\infty}. \tag{7}$$

Since $l-k<n$, then by Theorems 4.3.2/1 and 2 the left-hand side of (7) does not exceed

$$c \operatorname{ess} \|\Gamma\|_{MW_1^{l-k}} \sim \overline{\lim_{\delta\to 0}}\, \delta^{l-k-n} \sup_{x\in R^n} \|\Gamma; B_\delta(x)\|_{W_1^{l-k}}$$

$$\leqslant \overline{\lim_{\delta\to 0}}\, \delta^{l-n} \sup_{x\in R^n} \|\Gamma; B_\delta(x)\|_{W_1^l} = \|\Gamma\|_{L_\infty}.$$

Thus, (7) follows.

To complete the proof it suffices to establish the equality

$$\lim_{\delta\to 0} \sup_{x\in R^n} \|\eta_{\delta,x}\nabla_j\Gamma\|_{M(W_1^{l-k+j}\to W_1^{l-k})} = 0, \tag{8}$$

where $j=1,\dots,k$. By Theorem 4.2.1 the left-hand side is equivalent to the essential norm of $\nabla_j\Gamma$ in $M(W_1^{l-k+j}\to W_1^{l-k})$. Since $\nabla_j\Gamma\in W_1^{l-j}$, $\operatorname{supp}\nabla_j\Gamma\subset B_{2r}$, and $l-k+j>n$, then by Theorem 4.2.6/1, part (ii), $\nabla_j\Gamma\in \mathring{M}(W_1^{l-k+j}\to W_1^{l-k})$. The equality (8) is proved, and so is the theorem. □

Remark 2. In addition to Lemma 4.3.3 we note that, by Theorems 2, 3 and estimate (3.1.1/6), the following interpolation inequality is valid:

$$\operatorname{ess}\|\gamma\|_{MW_1^\sigma} \leqslant c(\operatorname{ess}\|\gamma\|_{MW_1^l})^{\sigma/l}\,\|\gamma\|_{L_\infty}^{1-\sigma/l}, \quad 0<\sigma<l.$$

4.3.5 The Essential Norm of an Element in the Space $\mathring{M}W_p^l$

According to Theorem 4.2.6/3, the space of compact multipliers $\mathring{M}(W_p^m\to W_p^l)$, $m>l$, coincides with the completion of D with respect to the norm of the space $M(W_p^m\to W_p^l)$. Similarly, $\mathring{M}W_p^l$ denotes the completion of D with respect to the norm of the space MW_p^l. The following theorem shows that the essential norm in $\mathring{M}W_p^l$ is equivalent to the norm in L_∞.

Theorem 1. If $\gamma\in\mathring{M}W_p^l$, $l\geqslant 0$, $p\geqslant 1$, then

$$\|\gamma\|_{L_\infty} \leqslant \operatorname{ess}\|\gamma\|_{MW_p^l} \leqslant c\,\|\gamma\|_{L_\infty}. \tag{1}$$

Proof. The left-hand estimate was obtained in Theorem 4.3.1. Let us establish the upper bound for the essential norm. Without loss of generality one may assume that $\gamma\in D$.

Let $p>1$, $lp\geqslant n$. Since $\operatorname{cap}(e, W_p^l)\geqslant c(\operatorname{mes}_n e)^\nu$, where $\nu\in(0,1)$ and

$d(e)\leq\delta$ (see Proposition 3.1.3/5), then

$$\frac{\|D_{p,l}\gamma;e\|_{L_p}}{[\text{cap}(e,W_p^l)]^{1/p}}\leq\text{const}(\text{mes}_n\,e)^{(1-\nu)/p}$$

and

$$\lim_{\delta\to 0}\sup_{\{e:d(e)\leq\delta\}}\frac{\|D_{p,l}\gamma;e\|_{L_p}}{[\text{cap}(e,W_p^l)]^{1/p}}=0.$$

For any compact set $e\subset R^n\backslash B_r$ with $d(e)\leq 1$,

$$\frac{\|D_{p,l}\gamma;e\|_{L_p}}{[\text{cap}(e,W_p^l)]^{1/p}}\leq\text{const}\,r^{-\{l\}-n/p}(\text{mes}_n\,e)^{(1-\nu)/p}.$$

Therefore

$$\lim_{r\to\infty}\sup_{\{e\subset R^n\backslash B_r:d(e)\leq 1\}}\frac{\|D_{p,l}\gamma;e\|_{L_p}}{[\text{cap}(e,W_p^l)]^{1/p}}=0.$$

Now the right-hand inequality in (1) follows from Theorem 4.3.3.

For $lp>n$, $p>1$ or $l\geq n$, $p=1$, the result is the corollary of the equality $\lim_{|x|\to\infty}\|\gamma;B_1(x)\|_{W_p^l}=0$ and Theorems 4.3.4/1, 4.3.4/3.

Finally, for $l<n$, $p=1$, the desired estimate for the essential norm immediately follows from Remark 4.3.4/1, since for $\gamma\in D$

$$\lim_{\delta\to 0}\delta^{l-n}\sup_{x\in R^n}\|D_{1,l}\gamma;B_\delta(x)\|_{L_1}=0,$$

$$\lim_{|x|\to\infty}\sup_{r\in(0,1)}r^{l-n}\|D_{1,l}\gamma;B_r(x)\|_{L_1}=0.\quad\square$$

We can describe the space $\mathring{M}W_p^l$ making no use of approximation by functions in D. The following assertion, supplementing Theorem 4.2.6/1, is valid.

Theorem 2. *Function γ belongs to $\mathring{M}W_p^l$ if and only if γ is a continuous function converging to zero at infinity and satisfying one of the conditions:*

(i) *If $lp\leq n$, $p>1$, then*

$$\sup_{\{e:d(e)\leq\delta\}}\frac{\|D_{p,l}\gamma;e\|_{L_p}}{[\text{cap}(e,W_p^l)]^{1/p}}=o(1)\quad as\quad \delta\to 0,\tag{2}$$

$$\sup_{\{e\subset R^n\backslash B_r:d(e)\leq 1\}}\frac{\|D_{p,l}\gamma;e\|_{L_p}}{[\text{cap}(e,W_p^l)]^{1/p}}=o(1)\quad as\quad r\to\infty.\tag{3}$$

(ii) If $l\leq n$, then

$$\sup_{x\in R^n}\delta^{l-n}\|D_{1,l}\gamma;B_\delta(x)\|_{L_1}=o(1)\quad as\quad \delta\to 0,\tag{4}$$

$$\sup_{r\in(0,1)}r^{l-n}\|D_{1,l}\gamma;B_r(x)\|_{L_1}=o(1)\quad as\quad |x|\to\infty.$$

(iii) *If* $lp>n$, $p>1$ *or* $l \geq n$, $p=1$, *then*

$$\|D_{p,l}\gamma; B_1(x)\|_{L_p} = o(1) \quad as \quad |x| \to \infty. \tag{6}$$

Proof. Necessity: let $\gamma \in \mathring{M}W_p^l$ and let $\{\gamma_j\}$ be a sequence of functions in D approximating γ in MW_p^l. From the expressions for equivalent norms in MW_p^l derived in Chapter 3 it follows that the left-hand sides of (2)–(6), with γ being replaced by $\gamma - \gamma_j$, are arbitrarily small for sufficiently large j. On the other hand, in the proof of Theorem 1 it has been shown that relations (2)–(6) are valid for $\gamma_j \in D$. Consequently, they hold for γ as well.

Sufficiency: let a function $\gamma \in C \cap MW_p^l$ satisfy one of conditions (2)–(6) and let $\gamma(x) \to 0$ as $|x| \to \infty$. For $lp>n$, $p>1$ and for $p=1$ the possibility of approximation of γ by mollifications of functions $\zeta_r\gamma$, $r\to\infty$, immediately follows from the expressions for the norm in MW_p^l derived in Chapter 3. Consider the case $lp \leq n$, $p>1$. Then $\|\zeta_r\gamma\|_{MW_p^l} \to 0$ as $r \to \infty$ because of (4.3.3/3). Therefore it suffices to approximate the multiplier γ with support in $B_{r/2}$ for a fixed r by functions from D.

Let γ_ρ be a mollification of γ with non-negative kernel K and radius ρ. We introduce the operators

$$T_* = \gamma \sum_j \varphi^{(j)} P^{(j)}, \qquad T_*^{(\rho)} = \gamma_\rho \sum_j \varphi^{(j)} P^{(j)}$$

(here we retain the same notation as in the definition of the operator T_* in the proof of Theorem 4.2.1). Obviously,

$$\|\gamma - \gamma_\rho\|_{MW_p^l} \leq \|(\gamma-\gamma_\rho) - (T_* - T_*^{(\rho)})\|_{W_p^l \to W_p^l} + \|T_* - T_*^{(\rho)}\|_{W_p^l \to W_p^l}.$$

For $lp<n$ as well as for $lp=n$, $0<l<1$, we have

$$\|(\gamma-\gamma_\rho) - (T_* - T_*^{(\rho)})\|_{W_p^l \to W_p^l}$$
$$\leq c\left(\sup_{\{e : d(e) \leq \delta\}} \frac{\|D_{p,l}(\gamma-\gamma_\rho); e\|_{L_p}}{[\operatorname{cap}(e, W_p^l)]^{1/p}} + \|\gamma-\gamma_\rho\|_{L_\infty}\right) \tag{7}$$

(see the proof of Theorem 4.2.2, where the restriction $l<m$ is insignificant, and the proof of Theorem 4.3.2/2). The right-hand side of (7) does not exceed

$$c\left(\sup_{\{e : d(e) \leq \delta\}} \left(\frac{\|D_{p,l}\gamma; e\|_{L_p}}{[\operatorname{cap}(e, W_p^l)]^{1/p}} + \frac{\|D_{p,l}\gamma_\rho; e\|_{L_p}}{[\operatorname{cap}(e, W_p^l)]^{1/p}} \right) + \|\gamma-\gamma_\rho\|_{L_\infty}\right).$$

Replacing here c by $2c$ we can omit the second summand because of the estimate

$$\|D_{p,l}\gamma_\rho; e\|_{L_p} \leq \int \rho^{-n} K(\xi/\rho)\, \|D_{p,l}\gamma; e_\xi\|_{L_p}\, d\xi$$

where $e_\xi = \{x : x+\xi \in e\}$. Further, we note that

$$\|T_* - T_*^{(\rho)}\|_{W_p^l \to W_p^l} \leq c(\delta, r)\, \|\gamma - \gamma_\rho; B_r\|_{W_p^l}.$$

Consequently,

$$\overline{\lim_{\rho\to 0}}\, \|\gamma-\gamma_\rho\|_{MW_p^l} \leq 2c \sup_{\{e:d(e)\leq\delta\}} \frac{\|D_{p,l}\gamma; e\|_{L_p}}{[\mathrm{cap}\,(e, W_p^l)]^{1/p}}$$

and it remains to make use of (2).

For $lp=n$, $l>1$, the proof follows the same lines provided that (7) is changed to the estimate

$$\|(\gamma-\gamma_\rho)-(T_*-T_*^{(\rho)})\|_{W_p^l\to W_p^l} \leq c\left(\sup_{\{e:d(e)\leq\delta\}} \frac{\|D_{p,l}(\gamma-\gamma_\rho); e\|_{L_p}}{[(\mathrm{cap}\,(e, W_p^l)]^{1/p}}\right.$$
$$\left. + \sup_{\{e:d(e)\leq\delta\}} \frac{\|\nabla(\gamma-\gamma_\rho); e\|_{L_p}}{[\mathrm{cap}\,(e, W_p^1)]^{1/p}} + \|\gamma-\gamma_\rho\|_{L_\infty}\right)$$

(see the proof of Theorem 4.3.3). □

4.4 The Essential Norm of a Differential Operator

In this section we obtain bounds for the essential norm of a differential operator (cf. 3.7).

Let $P(x, D_x)$ be the differential operator of order k, defined by (3.7/1), let P_0 be its principal homogeneous part, and let $\mathrm{ess}\,\|P\|_{W_p^h\to W_p^{h-k}}$ be the essential norm of the mapping $P: W_p^h \to W_p^{h-k}$, $h\geq k$.

Lemma 1. For all $\theta\in S^{n-1}$, $\|P_0(\cdot, \theta)\|_{L_\infty} \leq \mathrm{ess}\,\|P\|_{W_p^h\to W_p^{h-k}}$.

The proof is quite similar to that of Theorem 4.3.1, so we only outline it.

Let η, φ_k and Q_k be the same functions and the cube as in the proof of Theorem 4.3.1. We put

$$v_\xi(y) = |\xi|^{-h}\,\eta(y) \exp\left(i \sum_{j=1}^{n} [\xi_j] y_j\right),$$

where $\xi\in R^n\setminus\{0\}$, $[\xi_j]$ is the entire ξ_j. By the same arguments as in the proof of Theorem 4.3.1 we show that in the first place

$$\lim_{|\xi|\to\infty} \|v_\xi\|_{W_p^h} = A_h\,\|\eta\|_{L_p}, \qquad A_h = \mathrm{const}>0,$$

in the second place

$$\lim_{|\xi|\to\infty} \|\varphi_k P v_\xi\|_{W_p^{h-k}} = A_h\,\|\varphi_k P_0(\cdot, \xi\,|\xi|^{-1})\eta\|_{L_p}$$

and in the third place $\varphi_k T v_\xi \to 0$ as $|\xi|\to\infty$ in $\mathring{W}_p^h(Q_k)$. Here T is a compact operator in W_p^h.

Then

$$\overline{\lim_{|\xi|\to\infty}} \|\varphi_k P v_\xi\|_{W_p^{h-k}} = \overline{\lim_{|\xi|\to\infty}} \|\varphi_k (P-T) v_\xi\|_{W_p^{h-k}}$$

and by Lemma 4.1/4, for some $\sigma > 0$,

$$\overline{\lim_{|\xi|\to\infty}} \|\varphi_k P v_\xi\|_{W_p^{h-k}} \leq (1+O(k^{-\sigma})) \overline{\lim_{|\xi|\to\infty}} \|(P-T) v_\xi\|_{W_p^{h-k}}$$

$$\leq (1+O(k^{-\sigma})) \overline{\lim_{|\xi|\to\infty}} \|v_\xi\|_{W_p^h} (\text{ess}\, \|P\|_{W_p^h \to W_p^{h-k}} + \varepsilon).$$

Consequently,

$$\|\varphi_k P_0(\cdot, \xi\, |\xi|^{-1})\eta\|_{L_p} \leq (1+O(k^{-\sigma}))\, |\eta|_{L_p}\, \text{ess}\, |P\|_{W_p^h \to W_p^{h-k}}$$

and finally

$$\|P_0(\cdot, \xi\, |\xi|^{-1}\, \eta)\|_{L_p} \leq \|\eta\|_{L_p}\, \text{ess}\, \|P\|_{W_p^h \to W_p^{h-k}}. \quad \square$$

Lemma 2. (i) *The following estimate is valid:*

$$\text{ess}\, \|P\|_{W_p^h \to W_p^{h-k}} \leq c \sum_{|\alpha|\leq k} \text{ess}\, \|a_\alpha\|_{M(W_p^{h-|\alpha|} \to W_p^{h-k})}. \tag{1}$$

(ii) *If $p=1$ or $p(h-k)>n$, $p>1$ and P maps continuously W_p^h into W_p^{h-k}, then*

$$\text{ess}\, \|P\|_{W_p^h \to W_p^{h-k}} \sim \sum_{|\alpha|\leq k} \text{ess}\, \|a_\alpha\|_{M(W_p^{h-|\alpha|} \to W_p^{h-k})}. \tag{2}$$

Proof. (i) Let $\varepsilon > 0$ and let T_α be a compact operator mapping $W_p^{h-|\alpha|}$ into W_p^{h-k} and such that

$$\|a_\alpha - T_\alpha\|_{W_p^{h-|\alpha|} \to W_p^{h-k}} \leq \text{ess}\, \|a_\alpha\|_{M(W_p^{h-|\alpha|} \to W_p^{h-k})} + \varepsilon.$$

Since the operator $T = \sum_{|\alpha|\leq k} T_\alpha D^\alpha : W_p^h \to W_p^{h-k}$ is compact, then (1) follows.

(ii) We begin with the case $p=1$, $h-k \leq n$. Let $\varepsilon > 0$ and let T be a compact operator such that

$$\|P - T\|_{W_1^h \to W_1^{h-k}} \leq \text{ess}\, \|P\|_{W_1^h \to W_1^{h-k}} + \varepsilon.$$

Further, let $\eta_{\delta,x}$ and ζ_r be the 'truncating' functions introduced at the beginning of 4.1. Duplicating the proof of the first part of Theorem 4.2.1, with obvious changes, we obtain the estimate

$$\|\eta_{\delta,x} P\|_{W_1^h \to W_1^{h-k}} + \|\zeta_r P\|_{W_1^h \to W_1^{h-k}} \leq c(\text{ess}\, \|P\|_{W_1^h \to W_1^{h-k}} + \varepsilon) \tag{3}$$

which, by virtue of (3.7/3), is equivalent to

$$\sum_{|\alpha|\le k} (\|\eta_{\delta,x} a_\alpha\|_{M(W_1^{h-|\alpha|}\to W_1^{h-k})} + \|\zeta_r a_\alpha\|_{M(W_1^{h-|\alpha|}\to W_1^{h-k})})$$
$$\le c(\text{ess}\,\|P\|_{W_1^h\to W_1^{h-k}} + \varepsilon).$$

It remains to use Theorem 4.2.1.

Next let $p(h-k)>n$, $p\ge 1$. According to Lemma 3.7/1, $a_\alpha \in M(W_p^{h-|\alpha|}\to W_p^{h-k})$. The inequality

$$\|\zeta_r P\|_{W_p^h\to W_p^{h-k}} \le c(\text{ess}\,\|P\|_{W_p^h\to W_p^{h-k}} + \varepsilon),$$

where ε is small enough and r is sufficiently large, can be obtained in the same way as (3) above. The latter estimate and (3.7/3) yield

$$\sum_{|\alpha|\le k} \|\zeta_r a_\alpha\|_{M(W_p^{h-|\alpha|}\to W_p^{h-k})} \le c(\text{ess}\,\|P\|_{W_p^h\to W_p^{h-k}} + \varepsilon).$$

Therefore

$$\varlimsup_{|x|\to\infty} \sum_{|\alpha|\le k} \|a_\alpha; B_1(x)\|_{W_p^{h-k}} \le c\ \text{ess}\,\|P\|_{W_p^h\to W_p^{h-k}} \tag{4}$$

and, by Theorem 4.2.5/1,

$$\sum_{|\alpha|<k} \text{ess}\,\|a_\alpha\|_{M(W_p^{h-|\alpha|}\to W_p^{h-k})} \le c\ \text{ess}\,\|P\|_{W_p^h\to W_p^{h-k}}. \tag{5}$$

Applying the first part of the theorem to the operator $P-P_0$ and using (5), we arrive at

$$\text{ess}\,\|P_0\|_{W_p^h\to W_p^{h-k}} \le c\ \text{ess}\,\|P\|_{W_p^h\to W_p^{h-k}}$$

which, together with Lemma 1, shows that

$$\sum_{|\alpha|=k} \|a_\alpha\|_{L_\infty} \le c\ \text{ess}\,\|P\|_{W_p^h\to W_p^{h-k}}. \tag{6}$$

Taking into account Theorems 4.3.4/1, 4.3.4/3 and using (4), (6), we find

$$\sum_{|\alpha|=k} \text{ess}\,\|a_\alpha\|_{MW_p^{h-k}} \le c\ \text{ess}\,\|P\|_{W_p^h\to W_p^{h-k}}.$$

This and (5) imply the statement of the second part of the lemma. □

Consider now the matrix operator P defined by (3.7/4). Let ess $\|P\|$ denote the essential norm of the mapping (3.7/5).

Theorem. (i) *The following estimate is valid:*

$$\text{ess}\,\|P\| \le c \sum_{k=1}^{N} \sum_{j=1}^{M} \sum_{|\alpha|\le s_j+t_k} \text{ess}\,\|a_{jk}^{(\alpha)}\|_{M(W_p^{t_k+h-|\alpha|}\to W_p^{h-s_j})}.$$

(ii) *If* $p=1$ *or* $p(h-s_j)>n$, $j=1,\dots,M$, $p>1$ *and the mapping* (3.7/5) *is continuous, then*

$$\operatorname{ess} \|P\| \sim \sum_{k=1}^{N} \sum_{j=1}^{M} \sum_{|\alpha| \leqslant s_j + t_k} \operatorname{ess} \|a_{jk}^{(\alpha)}\|_{M(W_p^{t_k+h-|\alpha|} \to W_p^{h-s_j})}.$$

The proof obviously may be reduced to the scalar case considered in Lemma 2 (cf. the proof of Theorem 3.7/1).

It is clear that the last theorem can be immediately applied to the study of conditions for normal solvability in Sobolev spaces of elliptic systems with coefficients in multiplier spaces.

4.5 Singular Integral Operators with Symbols in Spaces of Multipliers

We conclude this chapter with a consideration of the application of multipliers in Sobolev spaces to the theory of singular integral operators. We show that the spaces MW_p^l and $\mathring{M}W_p^l$ are useful in the construction of the calculus of these operators acting in W_p^l, $1<p<\infty$, $l=0, 1, \ldots$.

First we cite basic definitions of the singular integrals theory (see Mikhlin [1], Mikhlin and Prößdorf [1], Stein [1]).

Let α be a bounded measurable function defined on $R^n \times B_1$ orthogonal to unity on ∂B_1, and let $\alpha_0^{(0)} \in L_\infty(R^n)$. The singular integral operator is defined by

$$(Au)(x) = \alpha_0^{(0)}(x)u(x) + \int_{R^n} \frac{\alpha(x, \theta)}{r^n} u(y)\, dy, \qquad x \in R^n, \tag{1}$$

where $r=|y-x|$, $\theta=(y-x)/r$ and the integral is interpreted in the sense of the Cauchy principal value. We express α as a series in spherical functions

$$\alpha(x, \theta) = \sum_{m=1}^{\infty} \sum_{k=1}^{k_m} \alpha_m^{(k)}(x) Y_m^{(k)}(\theta), \tag{2}$$

where k_m is the number of spherical functions $Y_m^{(k)}$ of order m. Then (1) and (2) imply the formal expansion

$$(Au)(x) = \alpha_0^{(0)}(x)u(x) + \sum_{m=1}^{\infty} \sum_{k=1}^{k_m} \alpha_m^{(k)}(x) \int_{R^n} \frac{Y_m^{(k)}(\theta)}{r^n} u(y)\, dy. \tag{3}$$

We put $(S_0^{(0)}u)(x) = u(x)$, $k_0 = 0$ and

$$(S_m^{(k)}u)(x) = \int_{R^n} \frac{Y_m^{(k)}(\theta)}{r^n} u(y)\, dy.$$

It is known that $S_m^{(k)} = \mu_m F^{-1} Y_m^{(k)} F$, where F is the Fourier transform in

R^n, $\mu_0 = 1$ and

$$\mu_m = i^{-m}\pi^{n/2}\frac{\Gamma(m/2)}{\Gamma((n+m)/2)}, \qquad |\mu_m| \sim m^{-n/2} \tag{4}$$

for $m \geqslant 1$. So the operator A can be written in the form

$$(Au)(x) = F^{-1}_{\xi\to x}[a(x, \xi/|\xi|)(Fu)(\xi)],$$

where a is the function defined by

$$a(x, \theta) = \sum_{m=0}^{\infty}\sum_{k=1}^{k_m} \mu_m \alpha_m^{(k)}(x) Y_m^{(k)}(\theta) \tag{5}$$

and is called the symbol of the singular integral operator.

Next we introduce the space $C^\infty(MW_p^l, \partial B_1)$ of infinitely differential functions defined on the sphere ∂B_1 with the range in MW_p^l. The space $C^\infty(\mathring{M}W_p^l, \partial B_1)$ is defined likewise.

Lemma 1. *If $a \in C^\infty(MW_p^l, \partial B_1)$ then the singular integral operator A with the symbol a is continuous in W_p^l and can be expressed in the form of the series $\sum_{m=0}^{\infty}\sum_{k=1}^{k_m} \alpha_m^{(k)} S_m^{(k)}$ which converges in the operator norm in $W_p^l \to W_p^l$.*

Proof. From (4), (5), and the definition of the space $C^\infty(MW_p^l, \partial B_1)$ it follows that for any positive integer N

$$\|\alpha_m^{(k)}; R^n\|_{MW_p^l} \leqslant C_N m^{-N}, \qquad m \geqslant 1.$$

It remains to make use of the fact that the singular convolution operator $S_m^{(k)}$ is continuous in W_p^l and its norm increases not faster than a certain degree of m as $m \to \infty$. □

Henceforth A, B, C are singular integral operators in R^n with symbols $a(x, \theta)$, $b(x, \theta)$, $c(x, \theta)$, where $x \in R^n$, $\theta \in \partial B_1$.

Theorem 1. *Let AB be a singular operator with the symbol ab and let $A \circ B$ be the composition of operators.*

If $a \in C^\infty(MW_p^l, \partial B_1)$ and there exists a function $b_\infty \in C^\infty(\partial B_1)$ such that $b - b_\infty \in C^\infty(\mathring{M}W_p^l, \partial B_1)$, then the operator $AB - A \circ B$ is compact in W_p^l.

Proof. By virtue of Lemma 1, it is sufficient to consider the operators A and B expressed in the form of finite sums

$$\sum_{m,k} \alpha_m^{(k)} S_m^{(k)}, \qquad \sum_{q,r} \beta_q^{(r)} S_q^{(r)}.$$

It is clear that

$$A \circ B = F^{-1}\Big(\sum_{m,k,q,r} \mu_m \alpha_m^{(k)} \mu_q \beta_q^{(r)} Y_m^{(k)} Y_q^{(r)}\Big) F$$

$$= \sum_{m,k,q,r} \alpha_m^{(k)} \beta_q^{(r)} \mu_m F^{-1} Y_m^{(k)} F \mu_q F^{-1} Y_q^{(r)} F = \sum_{m,k,q,r} \alpha_m^{(k)} \beta_q^{(r)} S_m^{(k)} S_q^{(r)}.$$

On the other hand,

$$AB = \sum_{m,k,q,r} \alpha_m^{(k)} S_m^{(k)} \beta_q^{(r)} S_q^{(r)}$$

$$= \sum_{m,k,q,r} \alpha_m^{(k)} \beta_q^{(r)} S_m^{(k)} S_q^{(r)} + \sum_{m,k,q,r} \alpha_m^{(k)} [S_m^{(k)}, \beta_q^{(r)}] S_q^{(r)},$$

where $[X, Y] = XY - YX$. Therefore

$$AB - A \circ B = \sum_{m,k,q,r} \alpha_m^{(k)} [S_m^{(k)}, \beta_q^{(r)}] S_q^{(r)} \tag{6}$$

and it remains to show that the commutator $[S_m^{(k)}, \beta_q^{(r)}]$ is compact in W_p^l. This assertion is contained in the following:

Lemma 2. *Let $\gamma \in \mathring{M}W_p^l$ and let A be a singular integral operator with the symbol $a(\theta)$, $a \in C^\infty(\partial B_1)$. Then the commutator $[\gamma, A]$ is compact in W_p^l.*

Proof. Let $\{\gamma_j\}$ be a sequence of functions, $\gamma_j \in D$, and let γ_j converge to γ in MW_p^l. Then the operators $(\gamma - \gamma_j)A$, $A(\gamma - \gamma_j)$ tend to zero in the operator norm in W_p^l. The compactness of the mapping $[\gamma_j, A]$ in W_p^l is well known (and can be easily verified). □

The following theorem contains the conditions for the operator $AB - A \circ B$ to be of order -1 in W_p^l (cf. Kohn, Nirenberg [1]).

Theorem 2. *If $a \in C^\infty(MW_p^{l+1}, \partial B_1)$ and $\nabla_x b \in C^\infty(MW_p^l, \partial B_1)$, then the operator $AB - A \circ B$ maps W_p^l continuously into W_p^{l+1}. Here AB is a singular operator with the symbol ab and $A \circ B$ is the composition of operators.*

This assertion follows from (6) and the next lemma.

Lemma 3. *Let function γ satisfy the Lipschitz condition and let $\nabla\gamma \in MW_p^l$. Further, let A be a singular integral operator with the symbol $a(\xi)$, $a \in C^\infty(\partial B_1)$. Then the commutator $[\gamma, A]$ satisfies the inequality*

$$\|[\gamma, A]\|_{W_p^l \to W_p^{l+1}} \le c\, \|\nabla\gamma; R^n\|_{MW_p^l}.$$

Proof. For $l=0$ the assertion is the known result due to Calderon [2]. Let the lemma be proved for all $l=0,1,\ldots,k-1$. Then, for all $u\in W_p^{k+1}$,

$$\|[\gamma, A]u\|_{W_p^{k+1}} \leqslant \sum_{j=1}^{n} \left\|\frac{\partial}{\partial x_j}[\gamma, A]u\right\|_{W_p^k} + \|[\gamma, A]u\|_{W_p^k}. \tag{7}$$

Since $MW_p^k \subset MW_p^{k-1}$, the last summand in (7) is estimated by induction hypothesis. Since $(\partial/\partial x_j)[\gamma, A] = (\partial\gamma/\partial x_j)A - A(\partial\gamma/\partial x_j) + [\gamma, A](\partial/\partial x_j)$, then

$$\left\|\frac{\partial}{\partial x_j}[\gamma, A]u\right\|_{W_p^k} \leqslant 2\,\|A\|_{W_p^k\to W_p^k} \left\|\frac{\partial\gamma}{\partial x_j}\right\|_{MW_p^k} \|u\|_{W_p^k} + \left\|[\gamma, A]\frac{\partial u}{\partial x_j}\right\|_{W_p^k}.$$

Applying the induction hypothesis to the last norm, we complete the proof. □

In conclusion, we formulate some corollaries on the regularization of a singular integral operator which follow from Theorems 1 and 2.

Corollary 1. Let there exist a function $a_\infty \in C^\infty(\partial B_1)$ such that $a - a_\infty \in C^\infty(\mathring{M}W_p^l, \partial B_1)$. Further, let $c = 1/a \in L_\infty(R^n \times \partial B_1)$. Then $c \in C^\infty(MW_p^l, \partial B_1)$ and $c - c_\infty \in C^\infty(\mathring{M}W_p^l, \partial B_1)$, where $c_\infty = 1/\alpha_\infty$. Moreover, operators $A \circ C - I$ and $C \circ A - I$ are compact in W_p^l.

Corollary 2. Let $a \in L_\infty(R^n \times \partial B_1)$ and let $\nabla_x a \in C^\infty(MW_p^l, \partial B_1)$. Further, let $c = 1/a \in L_\infty(R^n \times \partial B_1)$. Then $\nabla_x c \in C^\infty(MW_p^l, \partial B_1)$ and operators $A \circ C - I$, $C \circ A - I$ map W_p^l continuously into W_p^{l+1}.

Obviously, the condition of infinite differentiability of symbols on ∂B_1 everywhere can be changed by the condition of their sufficient smoothness.

5

Traces and Extensions of Multipliers

In the first section of this chapter we show that $MW_p^l(R^n)$ is the space of traces on R^n of multipliers in weighted Sobolev spaces on R^{n+m} in the same way as $W_p^l(R^n)$ is the space of traces of functions which belong to weighted Sobolev spaces on R^{n+m}. In 5.1.5 we give applications of these results to the first boundary value problem for an elliptic operator in R_+^{n+1}. Here we prove the unique solvability of the problem in a space of multipliers on R_+^{n+1} provided that the boundary Dirichlet data belong to certain classes of multipliers on R^n. In 5.2 we study the traces of functions in $MW_p^l(R^{n+m})$ on R^n. We show that the space of traces coincides with $MW_p^{l-m/p}(R^n)$ if $\{l-m/p\}>0$.

5.1 $MW_p^l(R^n)$ as the Space of Traces of Multipliers in Weighted Sobolev Spaces

5.1.1 Preliminaries

Let $R^{n+m}=\{z=(x, y): x\in R^n, y\in R^m\}$, $m>0$. For $U\in C_0^\infty(R^{n+m})$ we introduce the norm

$$\langle U\rangle_{k,p,\beta}=\left(\int_{R^{n+m}} |y|^{p\beta}\, |\nabla_{k,z} U|^p \, dz\right)^{1/p}.$$

One can easily see that for $k>r$, $\beta>r-m/p$,

$$\langle U\rangle_{k-r,p,\beta-r}\leq c\langle U\rangle_{k,p,\beta}.$$

Let $W_{p,\beta}^k(R^{n+m})$ denote the completion of the space $C_0^\infty(R^{n+m})$ with respect to the norm $\langle U\rangle_{k,p,\beta}+\|U; R^{n+m}\|_{L_p}$.

The following known assertion (see Lizorkin [1], Uspenskii [1]) gives the characteristics of traces on R^n of functions in $W_{p,\beta}^k(R^{n+m})$.

Lemma 1. (i) *Let U be an arbitrary function contained in $W_{p,\beta}^k(R^{n+m})$, and let $l=k-\beta-m/p$ be the fractional positive number,*

$l<k$. Then for almost all $x\in R^n$ there exists the limit $u(x)=\lim_{|y|\to 0} U(x,y)$. The function u belongs to the space $W_p^l(R^n)$ and the estimate

$$\|u; R^n\|_{W_p^l} \leq c\,\|U; R^{n+m}\|_{W^k_{p,\beta}} \tag{1}$$

is valid.

(ii) *Let l be a fractional positive number. There exists a linear extension operator $E: W_p^l(R^n)\ni u \to U \in W^k_{p,\beta}(R^{n+m})$, where $k>l$, $\beta=k-l-m/p$.*

Lemma 2. *Let r be a fractional number and let ω be an m-tuple multi-index, $|\omega|<r$. Further, let*

$$R_\omega(h,x)=D^\omega\gamma(x+h)-\sum_{|\nu|<r-|\omega|} D^{\nu+\omega}\gamma(x)\frac{h^\nu}{\nu!}. \tag{2}$$

Then

$$\left(\int_{R^n} |h|^{p(|\omega|-r)-n}\,|R_\omega(h,x)|^p\,dh\right)^{1/p} \leq c(D_{p,r}\gamma)(x). \tag{3}$$

Proof. From the identity

$$R_\omega(h,x)=([r]-|\omega|)\int_0^1 \sum_{|\nu|=[r]-|\omega|}\frac{h^\nu}{\nu!}$$
$$\times[(D^{\nu+\omega}\gamma)(x+th)-D^{\nu+\omega}\gamma(x)](1-t)^{[r]-|\omega|-1}\,dt$$

and the Minkowski inequality it follows that the left-hand side of (3) does not exceed the value

$$c\int_0^1\left(\int_{R^n} |h|^{-p\{r\}-n}\sum_{|\alpha|=[r]}|(D^\alpha\gamma)(x+th)-(D^\alpha\gamma)(x)|^p\,dh\right)^{1/p} dt$$

which is equivalent to $(D_{p,r}\gamma)(x)$. □

5.1.2 A Lemma on the Extension Operator

Following Stein [1], we introduce an extension operator that maps functions defined on R^n into functions on R^{n+m} by $(T\gamma)(x,y)=\int\zeta(t)\gamma(x+|y|\,t)\,dt$, where $\zeta\in C^\infty(R^n)\cap L(R^n)$, $\int\zeta(x)\,dx=1$.

Lemma. *Let $\{r\}>0$, $p\in[1,\infty)$ and let q be an integer, $q>r$. Further, let*

$$\int x^\alpha\zeta(x)\,dx=0, \qquad 0<|\alpha|\leq[r] \tag{1}$$

and

$$K_q^r = \int (1+|x|)^r \sum_{j=0}^{q} \sup_{\partial B_{|x|}} |\nabla_{j,x}\zeta| \, (1+|x|)^j \, dx. \tag{2}$$

The following estimate is valid:

$$\left(\int_{R^m} |y|^{p(q-r)-m} |\nabla_{q,z}(T\gamma)|^p \, dy \right)^{1/p} \leqslant c K_q^r (D_{p,r}\gamma)(x). \tag{3}$$

Proof. Let τ, σ, ρ, ω be m-tuple multi-indices such that $|\tau|+|\sigma|=q$, $\rho=0$, $\omega=\tau$ if $|\tau|\leqslant r$ and $\rho=\tau-\omega$, ω is an arbitrary multi-index with $|\omega|=[r]$ and $\omega<\tau$ if $|\tau|>r$. We have

$$D_x^\tau D_y^\sigma \int \zeta(t)\gamma(x+|y|\, t)\, dt = D_x^\rho D_y^\sigma \int \zeta(t) D_x^\omega \gamma(x+|y|\, t)\, dt$$
$$= D_y^\sigma \left(|y|^{-n-|\rho|} \int (D^\rho \zeta)\left(\frac{\xi - x}{|y|} \right) D^\omega \gamma(\xi)\, d\xi \right)$$
$$= D_y^\sigma \left(|y|^{-n-|\rho|} \int (D^\rho \zeta)\left(\frac{\xi - x}{|y|} \right) R_\omega(\xi - x, x)\, d\xi \right),$$

where R_ω is the function defined by (5.1.1/10). Here we used the identity

$$D_y^\sigma \left(|y|^{-n-|\rho|} \int (D^\rho \zeta)\left(\frac{\xi - x}{|y|} \right) (\xi - x)^\nu \, d\xi \right)$$
$$= D_y^\sigma \left(|y|^{|\nu|-|\rho|} \int D^\rho \zeta(\xi) \xi^\nu \, d\xi \right) = 0. \tag{4}$$

It is clear that

$$\left| D_y^\sigma \left(|y|^{-n-|\rho|} (D^\rho \zeta)\left(\frac{\xi - x}{|y|} \right) R_\omega(\xi - x, x) \right) \right|$$
$$\leqslant |y|^{r-|\rho|-|\sigma|-|\omega|} \varphi\left(\frac{\xi - x}{|y|} \right) \frac{|R_\omega(\xi - x, x)|}{|\xi - x|^{r-|\omega|+n}},$$

φ being a function for which the estimate

$$\varphi(\xi) \leqslant c\, |\xi|^{r-|\omega|+n} \sum_{i=0}^{|\sigma|} |\nabla_{i+|\rho|}\zeta(\xi)|\, (|\xi|^i + 1) \tag{5}$$

holds. Since $|\rho|+|\omega|=|\tau|$, $|\tau|+|\sigma|=q$, then

$$\int_{R^m} |y|^{p(q-r)-m} |D_x^\tau D_y^\sigma (T\gamma)|^p \, dy$$
$$\leqslant c \int_{R^m} \left(\int_{R^n} \varphi\left(\frac{\xi - x}{|y|} \right) \frac{|R_\omega(\xi - x, x)|}{|\xi - x|^{r-|\omega|+n}} \, d\xi \right)^p \frac{dy}{|y|^m}.$$

By introducing spherical coordinates, we write the right-hand side in the form

$$c\int_0^\infty \frac{d\lambda}{\lambda}\left(\int_0^\infty\int_{\partial B_1}\varphi\left(\frac{t\theta}{\lambda}\right)\frac{|R_\omega(t\theta, x)|}{t^{r-|\omega|}}\frac{dt}{t}\,d\theta\right)^p.$$

This value is not more than

$$c\int_0^\infty \frac{d\lambda}{\lambda}\left(\int_0^\infty Q\left(\frac{t}{\lambda}\right)g(t)\frac{dt}{t}\right)^p \leq c\left(\int_0^\infty Q(t)\frac{dt}{t}\right)^p\int_0^\infty g(t)^p\frac{dt}{t},$$

where $Q(t)=\sup_{|\theta|=1}\varphi(t\theta)$, $g(t)=t^{|\omega|-r}\int_{\partial B_1}|R_\omega(t\theta, x)|\,d\theta$. This and (5) imply

$$\int_{R^m}|y|^{p(q-r)-m}|D_x^\tau D_y^\sigma(T\gamma)|^p\,dy$$

$$\leq c\left(\int_{R^n}|\xi|^{r-|\omega|}\sum_{i=0}^{|\sigma|}(|\xi|+1)^i\sup_{\partial B_{|\xi|}}|\nabla_{i+|\rho|}\zeta(\xi)|\,d\xi\right)^p$$

$$\times\int_{R^n}|h|^{p(|\omega|-r)-n}|R_\omega(h, x)|^p\,dh.$$

Reference to Lemma 5.1.1/2 completes the proof. □

5.1.3 The Main Theorem

The following theorem shows that $MW_p^l(R^n)$ coincides with the space of traces on R^n of functions in $MW_{p,\beta}^k(R^{n+m})$, i.e. in the space of multipliers in $W_{p,\beta}^k(R^{n+m})$.

Theorem. (i) *Let $\{l\}>0$, $\Gamma\in MW_{p,\beta}^k(R^{n+m})$ with $\beta=k-l-m/p$ and $k>l$. Further, let γ be the trace of Γ on R^n, its existence being proved by $\Gamma\in W_{p,\beta,\mathrm{loc}}^k(R^{n+m})$ and Lemma 5.1.1/1. Then $\gamma\in MW_p^l(R^n)$ and the estimate*

$$\|\gamma; R^n\|_{MW_p^l}\leq c\,\|\Gamma; R^{n+m}\|_{MW_{p,\beta}^k} \tag{1}$$

is valid.

(ii) *Let $\{l\}>0$, $s=0, 1, \ldots,$ and k be an integer, $k>l$. Further, let $\nabla_s\gamma\in MW_p^l(R^n)$ and let $T\gamma$ be the extension of γ to R^{n+m} defined in 5.1.2 with ζ subjected to the conditions*

$$\int(1+|x|)^{l+s}\sum_{j=0}^{k+s}\sup_{\partial B_{|x|}}|\nabla_{j,x}\zeta|\,(1+|x|)^j\,dx=K<\infty, \tag{2}$$

$$\int x^\alpha\zeta(x)\,dx=0, \qquad 0<|\alpha|\leq[l]+s. \tag{3}$$

Then $\nabla_{s,z}(T\gamma)\in MW^k_{p,\beta}(R^{n+m})$, $\beta=k-l-m/p$, *and*

$$\|\nabla_{s,z}(T\gamma); R^{n+m}\|_{MW^k_{p,\beta}} \leq cK\,\|\nabla_{s,x}\gamma; R^n\|_{MW^l_p}. \tag{4}$$

Proof. (i) Let $U\in W^k_{p,\beta}(R^{n+m})$ and let $U(x,0)=u(x)$. We have

$$\|\gamma u; R^n\|_{W^l_p}\leq c\,\|\Gamma U; R^{n+m}\|_{W^k_{p,\beta}} \leq c\,\|\Gamma; R^{n+m}\|_{MW^k_{p,\beta}}\,\|U; R^{n+m}\|_{W^k_{p,\beta}}.$$

Using the second part of Lemma 4.1.1 and the arbitrariness of U, we arrive at (1).

(ii) Let χ be any m-tuple multi-index with $|\chi|=s$. It is clear that

$$\langle UD^\chi_y(T\gamma)\rangle_{k,p,\beta}\leq c \sum_{|\varepsilon|+|\mu|+|\nu|=k} \langle D^\nu_z U D^\mu_x D^{\chi+\varepsilon}_y(T\gamma)\rangle_{0,p,\beta}. \tag{5}$$

For $|\mu|<s$ we have

$$D^\mu_x D^{\chi+\varepsilon}_y(T\gamma)(z)=D^{\chi+\varepsilon}_y\,|y|^{-n}\int \zeta\Big(\frac{\xi-x}{|y|}\Big) \times\Big[D^\mu_\xi\gamma(\xi)-\sum_{|\varkappa|\leq s-|\mu|-1}\frac{(\xi-x)^\varkappa}{\varkappa!}\,D^{\varkappa+\mu}\gamma(x)\Big]\,d\xi \tag{6}$$

and for $|\mu|\geq s$,

$$D^\mu_x D^{\chi+\varepsilon}_y(T\gamma)(z)=D^{\mu_1}_x D^{\chi+\varepsilon}_y\Big(|y|^{-n}\int\zeta\Big(\frac{\xi-x}{|y|}\Big)D^{\mu_2}_\xi\gamma(\xi)\,d\xi\Big),$$

where $\mu=\mu_1+\mu_2$, $|\mu_1|>0$, $|\mu_2|=s$. So in both cases we have

$$|D^\mu_x D^{\chi+\varepsilon}_y(T\gamma)(z)|\leq cK\,\|\nabla_s\gamma; R^n\|_{L_\infty}\,|y|^{-|\mu|-|\varepsilon|}. \tag{7}$$

Hence for $|\nu|>l$ we obtain

$$\begin{aligned}\langle D^\nu_z U D^\mu_x D^{\chi+\varepsilon}_y(T\gamma)\rangle_{0,p,\beta} &\leq cK\,\|\nabla_s\gamma; R^n\|_{L_\infty}\langle U\rangle_{|\nu|,p,|\nu|-l-m/p}\\ &\leq cK\,\|\nabla_s\gamma; R^n\|_{L_\infty}\langle U\rangle_{k,p,\beta}.\end{aligned} \tag{8}$$

Now let $|\nu|<l$. It is clear that

$$\begin{aligned}&\langle D^\nu_z U D^\mu_x D^{\chi+\varepsilon}_y(T\gamma)\rangle_{0,p,\beta}\\ &\quad\leq\langle R_\nu D^\mu_z D^{\chi+\varepsilon}_y(T\gamma)\rangle_{k,p,\beta}+\sum_{j=0}^{[l]-|\nu|}\Big(\int_{R^{n+m}}|y|^{p(k-l-j)-m}\\ &\qquad\times|D^\mu_z D^{\chi+\varepsilon}_y(T\gamma)(z)|^p\,|(\nabla_{j,y}D^\nu_z U)(x,0)|^p\,dz\Big)^{1/p},\end{aligned} \tag{9}$$

where

$$\begin{aligned}R_\nu(z)&=D^\nu_z U(z)-\sum_{|\tau|\leq[l]-|\nu|}(D^\tau_y D^\nu_z U)(x,0)\,\frac{y^\tau}{\tau!}\\ &=([l]-|\nu|-1)\sum_{|\tau|=[l]-|\nu|+1}\frac{y^\tau}{\tau!}\int_0^1(D^\tau_y D^\nu_z U)(x,ty)(1-t)^{[l]-|\nu|}\,dt.\end{aligned}$$

By (7) and Minkowski's inequality,

$$\begin{aligned}
&\langle R_\nu D_z^\mu D_y^{\varkappa+\varepsilon}(T\gamma)\rangle_{k,p,\beta} \\
&\leq cK\,\|\nabla_s\gamma; R^n\|_{L_\infty}\left(\int |y|^{p(1-\{l\})-m}\left(\int_0^1 |(\nabla_{[l]+1,z}U)(x,ty)|\,dt\right)^p dz\right)^{1/p} \\
&\leq cK\,\|\nabla_s\gamma; R^n\|_{L_\infty}\langle U\rangle_{[l]+1,p,1-\{l\}-m/p} \\
&\leq cK\,\|\nabla_s\gamma; R^n\|_{L_\infty}\langle U\rangle_{k,p,\beta}.
\end{aligned}$$

By Lemma 5.1.2 with $q=|\mu|+s+|\varepsilon|$, $r=l+s-j-|\nu|$, the sum on the right in (9) does not exceed

$$cK\sum_{j=0}^{[l]-|\nu|}\left(\int_{R^n} |(\nabla_{j,y}D_z^\nu U)(x,0)|^p\,[(D_{p,l-j-|\nu|}\nabla_s\gamma)(x)]^p\,dx\right)^{1/p}. \tag{10}$$

For $p>1$ this, by virtue of (3.2.1/5), is majorized by

$$cK\sum_{j=0}^{[l]-|\nu|}\sup_e\left(\frac{\int_e |D_{p,l-j-|\nu|}\nabla_s\gamma|^p\,dx}{\operatorname{cap}(e, W_p^{l-j-|\nu|})}\right)^{1/p}\|(\nabla_{j+\nu,z}U)(\cdot,0); R^n\|_{W_p^{l-j-|\nu|}}$$

which, by Theorem 3.2.7/2, is not more than

$$cK\,\|\nabla_s\gamma; R^n\|_{MW_p^l}\,\|U; R^{n+m}\|_{W_{p,\beta}^k}. \tag{11}$$

The same estimate results in the case $p=1$ from Lemma 3.4.2 and Theorem 3.4.2. So, for $|\nu|<l$,

$$\langle D_z^\nu U D_x^\mu D_y^{\varkappa+\varepsilon}(T\gamma)\rangle_{0,p,\beta}\leq cK\,\|\nabla_s\gamma; R^n\|_{MW_p^l}\,\|U; R^{n+m}\|_{W_{p,\beta}^k}.$$

It follows by insertion of this estimate and (8) into (5) that

$$\langle UD_y^\varkappa(T\gamma)\rangle_{k,p,\beta}\leq cK\,\|\nabla_s\gamma; R^n\|_{MW_p^l}\,\|U; R^{n+m}\|_{W_{p,\beta}^k}.$$

According to (7) with $|\mu|=|\varepsilon|=0$, we have $|D_y^\varkappa(T\gamma)|\leq cK\,\|\nabla_s\gamma; R^n\|_{L_\infty}$. Therefore

$$\|UD_y^\varkappa(T\gamma); R^{n+m}\|_{L_p}\leq cK\,\|\nabla_s\gamma; R^n\|_{L_\infty}\,\|U; R^{n+m}\|_{L_p}. \quad \square$$

5.1.4 Extension of Multipliers from the Hyperplane to the Half-Space

An assertion analogous to Theorem 5.1.3 is also valid for the space of multipliers $MW_{p,\beta}^k(R_+^{n+1})$, where $R_+^{n+1}=\{z=(x,y): x\in R^n, y>0\}$ and $W_{p,\beta}^k(R_+^{n+1})$ is the completion of $C_0^\infty(\overline{R_+^{n+1}})$ with respect to the norm

$$\left(\int_{R_+^{n+1}} y^{p\beta}\,|\nabla_{k,z}U|^p\,dz\right)^{1/p}+\|U; R_+^{n+1}\|_{L_p}.$$

If we put $m=1$ and replace R^m by $R_+^1=\{y: y>0\}$ in the statement of Lemma 5.1.2, then it remains valid without the condition (5.1.2/1) as well. We need to verify only that

$$D_y^\sigma(y^{|\nu|-|\rho|})\int D^\rho\zeta(\xi)\xi^\nu\,d\xi=0. \tag{1}$$

For $|\nu|<|\rho|$ after integration by parts we obtain that the integral in (1) is equal to zero. In the case $|\nu|\geqslant|\rho|$ the function $D_y^\sigma(y^{|\nu|-|\rho|})$ vanishes identically, since $\sigma>\sigma+[r]-q=[r]-|\tau|=[r]-|\omega|-|\rho|\geqslant|\nu|-|\rho|$.

The condition (5.1.3/3) was used only in (5.1.3/6) in the proof of Theorem 5.1.3. Since $|\chi|+|\varepsilon|=s+|\varepsilon|>s-|\mu|-1\geqslant|\chi|$, then for $m=1$ and $y\in R_+^1$ the equality (5.1.3/6) remains valid without the condition (5.1.3/3). Therefore we have the following assertion:

Theorem. (i) *Let $\{l\}>0$, $\Gamma\in MW_{p,\beta}^k(R_+^{n+1})$, $\beta=k-l-1/p$, $k>l$ and $\gamma(x)=\Gamma(x,0)$. Then $\gamma\in MW_p^l(R^n)$ and*

$$\|\gamma;R^n\|_{MW_p^l}\leqslant c\,\|\Gamma;R_+^{n+1}\|_{MW_{p,\beta}^k}.$$

(ii) *Let $\{l\}>0$ and $\nabla_s\gamma\in MW_p^l(R^n)$. Further, let $T\gamma$ be the extension of γ to R_+^{n+1} defined in 5.1.2, where the function ζ is subjected to the condition (5.1.3/2) only. Then $\nabla_s(T\gamma)\in MW_{p,\beta}^k(R_+^{n+1})$, $k>l$, $\beta=k-l-1/p$, and the estimate*

$$\|\nabla_s(T\gamma);R_+^{n+1}\|_{MW_{p,\beta}^k}\leqslant cK\,\|\nabla_s\gamma;R^n\|_{MW_p^l}$$

holds.

5.1.5 Application to the First Boundary Value Problem in a Half-Space

In the half-space R_+^{n+1} let us consider the Dirichlet problem

$$L(D)U=0 \quad \text{for} \quad y\geqslant 0,$$

$$\partial^j U/\partial y^j=\varphi_j \quad \text{for} \quad y=0, \qquad j=0,\ldots,m-1,$$

where L is a homogeneous differential elliptic operator of order $2m$ with constant coefficients.

Theorem. *Let $\nabla_{m-1-j}\varphi_j\in MW_p^l(R^n)$, where $0<l<1$, $1\leqslant p<\infty$. Then there exists one and only one solution of the Dirichlet problem such that $\nabla_{m-1}U\in MW_{p,k-l-1/p}^k(R_+^{n+1})$, $k\geqslant1$. This solution satisfies the estimate*

$$\|\nabla_{m-1}U;R_+^{n+1}\|_{MW_{p,k-l-1/p}^k}\leqslant K\sum_{j=0}^{m-1}\|\nabla_{m-1-j}\varphi_j;R^n\|_{MW_p^l},$$

where K is a constant which depends on L and n, p, m, k, l.

Proof. If $U\in MW_{p,k-l-1/p}^k(R_+^{n+1})$ is a solution of the homogeneous problem, then $\|\nabla_{m-1}U; R_+^{n+1}\|_{L_\infty} < \infty$ and hence $U=0$ (see, for instance, Agmon, Douglis, Nirenberg [1]).

It is known (see Agmon, Douglis, Nirenberg [1], Ch. I, §2) that the existence of the solutions follows from the assumption $\nabla_{m-1-j}\varphi_j \in L_\infty(R^n)$, $j=0,1,\dots,m-1$. The solution satisfies the equality

$$D_x^\alpha \frac{\partial^i}{\partial y^i} U(x,y) = \sum_{j=0}^{m-1} \sum_{|\beta|=m-1-j} \int_{R^n} K_{i,j,\beta}(x-\xi, y) D_\xi^\beta \varphi_j(\xi)\, d\xi,$$

where $0 \leq i \leq m-1$, α is any multi-index of order $m-1-i$ and $K_{i,j,\beta}(z)$ are positive homogeneous functions of order $-n$, smooth in $R^{n+1}\backslash\{0\}$ and such that $K_{i,j,\beta}(x,0)=0$ for $x\neq 0$. These conditions imply

$$(|x|^2+y^2)^{1/2}\, |\nabla_x K_{i,j,\beta}(x,y)| + |K_{i,j,\beta}(x,y)| \leq cy(|x|^2+y^2)^{-(n+1)/2}$$

which shows that the function $\zeta(x) = K_{i,j,\beta}(x,1)$ satisfies (5.1.3/2) for $s=0$, $0<l<1$. It remains to use Theorem 5.1.4. □

5.2 Traces of Functions in $MW_p^l(R^{n+m})$ on R^n

In this section we show that the space of restrictions of functions in $MW_p^l(R^{n+m})$ ($lp>n$, $l-m/p$ is fractional) to R^n coincides with $MW_p^{l-m/p}(R^n)$.

5.2.1 Auxiliary Assertions

We shall use the extension operator T defined in 5.1.2. We assume that the conditions (5.1.3/2) and (5.1.3/3) for $k=[l]$, $s=1$ are valid.

Lemma. *Let $\sigma \in (0, l]$. Then*

$$\left(\int_{2|\eta|<|y|} |\nabla_{[\sigma],y}(T\gamma)(x, y+\eta) - \nabla_{[\sigma],y}(T\gamma)(x,y)|^p\, |\eta|^{-m-p\{\sigma\}}\, d\eta\right)^{1/p}$$
$$\leq cK\, |y|^{-\sigma}\, \|\gamma; R^n\|_{L_\infty}, \tag{1}$$

$$\left(\int_{R^n} |\nabla_{[\sigma],x}(T\gamma)(x+h, y) - \nabla_{[\sigma],x}(T\gamma)(x,y)|^p\, |h|^{-n-p\{\sigma\}}\, dh\right)^{1/p}$$
$$\leq cK\, |y|^{-\sigma}\, \|\gamma; R^n\|_{L_\infty}. \tag{2}$$

Proof. By A_y and B_x we denote the left-hand sides in (1) and (2).

Obviously,

$$A_y \leq \|\gamma; R^n\|_{L_\infty} \left\{ \int_{2|\eta|<|y|} \left(\int_{R^n} \left| \nabla_{[\sigma],y} \left[\zeta\left(\frac{\xi-x}{|y+\eta|}\right) |y+\eta|^{-n} - \zeta\left(\frac{\xi-x}{|y|}\right) |y|^{-n} \right] \right| d\xi \right)^p |\eta|^{-m-p\{\sigma\}} d\eta \right\}$$

$$\leq \|\gamma; R^n\|_{L_\infty} \left\{ \int_{2|\eta|<|y|} \frac{|\eta|^p \, d\eta}{|\eta|^{m+p\{\sigma\}}} \times \left(\int_0^1 dz \int_{R^n} |\varphi_{\xi-x}^{(\sigma+1)}[|y| + z(|y+\eta| - |y|)]| \, d\xi \right)^p \right\}^{1/p},$$

where

$$\varphi_{\xi-x}^{(l)}(t) = \frac{\partial^{[l]}}{\partial t^{[l]}} \left(t^{-n} \zeta\left(\frac{\xi-x}{t}\right)\right).$$

Therefore

$$A_y \leq cK \|\gamma; R^n\|_{L_\infty} \times \left\{ \int_{2|\eta|<|y|} |\eta|^{p(1-\{\sigma\})-m} \, d\eta \left(\int_0^1 (|y| + z(|y+\eta| - |y|))^{-[\sigma]-1} \, dz \right)^p \right\}^{1/p}$$

$$\leq cK \|\gamma; R^n\|_{L_\infty} |y|^{-\sigma}.$$

The inequality (1) is proved. By making the change of variables $\xi - x = |y| \, \Xi$, $h = |y| \, H$, we obtain

$$B_x \leq |y|^{-\sigma} \left\{ \int_{R^n} \left(\int_{R^n} |\nabla_{[\sigma],\Xi}[\zeta(\Xi + H) - \zeta(\Xi)]| \, d\Xi \right)^p |H|^{-n-p\{\sigma\}} \, dH \right\}^{1/p}.$$

We divide the integral over H into two integrals, the first of which is over the ball B_1. We have

$$\int_{B_1} \leq \int_{B_1} \left(\int_{R^n} \left| \sum_{i=1}^n H_i \int_0^1 \frac{\partial}{\partial \Xi_i} \nabla_{[\sigma],\Xi} \zeta(\Xi + zH) \, dz \right| d\Xi \right)^p |H|^{-n-p\{\sigma\}} \, dH$$

$$\leq \left(\int_{R^n} |\nabla_{[\sigma]+1} \zeta(\Xi)| \, d\Xi \right)^p \int_{B_1} |H|^{-n+(1-\{\sigma\})p} \, dH \leq cK^p.$$

Finally,

$$\int_{R^n \setminus B_1} \leq \int_{R^n \setminus B_1} \left(\int_{R^n} (|\nabla_{[\sigma]} \zeta(\Xi + H)| + |\nabla_{[\sigma]} \zeta(\Xi)|) \, d\Xi \right)^p |H|^{-n-p\{\sigma\}} \, dH$$

$$= 2^p \left(\int_{R^n} |\nabla_{[\sigma]} \zeta(\Xi)| \, d\Xi \right)^p \int_{R^n \setminus B_1} |H|^{-n-p\{\sigma\}} \, dH \leq cK^p. \quad \square$$

Let d_j denote the number of all derivatives of order j with respect to variables $y_1, \dots, y_m$ and let $[W_p^\sigma(R^n)]^{d_j}$ be the Cartesian product of d_j copies of the space $W_p^\sigma(R^n)$. It is known that there exists an extension operator E defined on vector-functions $(\boldsymbol{\varphi}_0, \boldsymbol{\varphi}_1, \dots, \boldsymbol{\varphi}_{[l-m/p]})$, where $\boldsymbol{\varphi}_j$ is the d_j-tuple vector-function. This operator maps into the space of scalar functions and has the following properties:

(i) E is a continuous operator:

$$\prod_{j=0}^{[l-m/p]} [W_p^{[l-j-m/p]}(R^n)]^{d_j} \to W_{p,k-l}^k(R^{n+m}). \tag{3}$$

(ii) The relation $(\nabla_j E\boldsymbol{\varphi})(x, 0) = \boldsymbol{\varphi}_j(x)$ with $j = 0, 1, \dots, [l-m/p]$ holds.

In what follows we use the fact that the space $W_{p,1-\{l\}}^{[l]+1}(R^{n+m})$ is imbedded into $W_p^l(R^{n+m})$ (see Uspenskii [1]) as well as the generalized Hardy inequality

$$\int_{R^m} |y|^{-pl}\, |V|^p\, dy \le c\, \|V; R^m\|_{W_p^l}^p, \tag{4}$$

where V is a function in $W_p^l(R^m)$ subjected to conditions

$$(\nabla_j V)(x, 0) = 0, \qquad j = 0, 1, \dots, [l-m/p]. \tag{5}$$

Moreover, we shall apply the following equivalent norm in $W_p^l(R^{n+m})$:

$$\|U; R^{n+m}\|_{W_p^l}$$
$$\sim \left(\int_{R^n} dx \int_{R^m} dy \int_{R^m} |\nabla_{[l],y} U(x, y+\eta) - \nabla_{[l],y} U(x, y)|^p\, |\eta|^{-m-p\{l\}}\, d\eta\right)^{1/p}$$
$$+ \left(\int_{R^m} dy \int_{R^n} dx \int_{R^n} |\nabla_{[l],x} U(x+h, y) - \nabla_{[l],x} U(x, y)|^p\, |h|^{-n-p\{l\}}\, dh\right)^{1/p}$$
$$+ \|U; R^{n+m}\|_{L_p}. \tag{6}$$

5.2.2 The Main Result

Theorem. (i) *Let $lp > n$, $1 \le p < \infty$ and $l - m/p$ be fractional. Further, let $\Gamma \in MW^l(R^{n+m})$ and $\gamma(x) = \Gamma(x, 0)$. Then $\gamma \in MW_p^{l-m/p}(R^n)$ and*

$$\|\gamma; R^n\|_{MW_p^{l-m/p}} \le c\, \|\Gamma; R^{n+m}\|_{MW_p^l}. \tag{1}$$

(ii) *Let the kernel ζ satisfy* (5.1.3/2) *and* (5.1.3/3). *If $\gamma \in MW_p^{l-m/p}(R^n)$ with $lp > m$, $1 \le p < \infty$ and fractional $l - m/p$, then $T\gamma \in MW_p^l(R^{n+m})$ and*

$$\|T\gamma; R^{n+m}\|_{MW_p^l} \le cK\, \|\gamma; R^n\|_{MW_p^{l-m/p}}. \tag{2}$$

Proof. (i) Let $U \in W_p^l(R^{n+m})$, $U(x, 0) = u(x)$. We have

$$\|\gamma u; R^n\|_{W_p^{l-m/p}} \le c\, \|\Gamma U; R^{n+m}\|_{W_p^l} \le c\, \|\Gamma; R^{n+m}\|_{MW_p^l}\, \|U; R^{n+m}\|_{W_p^l}$$

which implies (1).

(ii) It is sufficient to suppose l to be fractional, since for integer l the result is contained in Theorem 5.1.3.

Let $U \in W_p^l(R^{n+m})$ and

$$\varphi(x) = (U(x, 0), (\nabla_y U)(x, 0), \ldots, (\nabla_{[l-m/p],y} U)(x, 0)).$$

We introduce the function $V = U - E\varphi$ where E is the extension operator which was considered in 5.2.1. Then

$$\|U T\gamma; R^{n+m}\|_{W_p^l} \leq \|(T\gamma)E\varphi; R^{n+m}\|_{W_p^l} + \|VT\gamma; R^{n+m}\|_{W_p^l}.$$

Since $W_{p,1-\{l\}}^{[l]+1}(R^{n+m}) \subset W_p^l(R^{n+m})$, the first summand on the right does not exceed

$$c\,\|(T\gamma)E\varphi; R^{n+m}\|_{W_{p,1-\{l\}}^{[l]+1}}$$

which, by Theorem 5.1.3, is not more than

$$cK\,\|\gamma; R^n\|_{MW_p^{l-m/p}}\,\|E\varphi; R^{n+m}\|_{W_{p,1-\{l\}}^{[l]+1}}.$$

Since E performs the continuous mapping (5.2.1/3), then

$$\|(T\gamma)E\varphi; R^{n+m}\|_{W_p^l}$$
$$\leq cK\,\|\gamma; R^n\|_{MW_p^{l-m/p}} \sum_{j=0}^{[l-m/p]} \|(\nabla_{j,y} U)(\cdot, 0); R^n\|_{W_p^{l-m/p-j}}.$$

Consequently,

$$\|(T\gamma)E\varphi; R^{n+m}\|_{W_p^l} \leq cK\,\|\gamma; R^n\|_{MW_p^{l-m/p}}\,\|U; R^{n+m}\|_{W_p^l}. \tag{3}$$

Let us prove the inequality

$$\|VT\gamma; R^{n+m}\|_{W_p^l} \leq cK\,\|\gamma; R^n\|_{L_\infty}\,\|V; R^{n+m}\|_{W_p^l}. \tag{4}$$

It is easily seen that

$$\Big(\int_{R^n} dx \int_{R^m} dy \int_{2|\eta|>|y|} |(\nabla_{[l],y}(VT\gamma))(x, y+\eta)$$
$$- (\nabla_{[l],y}(VT\gamma))(x, y)|^p\, |\eta|^{-m-p\{l\}}\, d\eta\Big)^{1/p}$$
$$\leq c\Big(\int_{R^{n+m}} |\nabla_{[l],y}(VT\gamma)|^p\, |y|^{-p\{l\}}\, dz\Big)^{1/p} \tag{5}$$

which, by (5.1.3/7), where $s = 0$, $\mu = 0$, does not exceed

$$c\sum_{j=0}^{[l]} \Big(\int_{R^{n+m}} |\nabla_{[l]-j,y} T\gamma|^p\, |\nabla_{j,y} V|^p\, |y|^{-p\{l\}}\, dz\Big)^{1/p}$$
$$\leq cK\,\|\gamma; R^n\|_{L_\infty} \sum_{j=0}^{[l]} \Big(\int_{R^{n+m}} |\nabla_{j,y} V|^p\, |y|^{(j-l)p}\, dz\Big)^{1/p}.$$

This and (5.2.1/4) show that the left-hand side of (5) is not more than $cK\,\|\gamma;R^n\|_{L_\infty}\|V;R^{n+m}\|_{W_p^l}$. The expression

$$\Big(\int_{R^n}dx\int_{R^m}dy\int_{2|\eta|<|y|}|\nabla_{[l],y}[(VT\gamma)(x,y+\eta)-(VT\gamma)(x,y)]|^p\,|\eta|^{-m-p\{l\}}\,d\eta\Big)^{1/p}$$

is majorized by

$$c\sum_{j=0}^{[l]}\Big(\int_{R^n}dx\int_{R^m}dy\int_{2|\eta|<|y|}|(\nabla_{[l]-j,y}T\gamma)(x,y+\eta)|^p\times|\nabla_j(V(x,y+\eta)-V(x,y))|^p\,|\eta|^{-m-p\{l\}}\,d\eta\Big)^{1/p}$$
$$+c\sum_{j=0}^{[l]}\Big(\int_{R^n}dx\int_{R^m}dy\,|(\nabla_{j,y}V)(x,y)|^p\times\int_{2|\eta|<|y|}|\nabla_{[l]-j,y}(T\gamma(x,y+\eta)-T\gamma(x,y))|^p\,|\eta|^{-m-p\{l\}}\,d\eta\Big)^{1/p}. \quad (6)$$

Since $|(\nabla_{[l]-j,y}T\gamma)(x,y+\eta)|\le cK\,\|\gamma;R^n\|_{L_\infty}|y|^{j-[l]}$, the first sum does not exceed

$$cK\,\|\gamma;R^n\|_{L_\infty}\sum_{j=0}^{[l]-1}\Big(\int_{R^n}dx\int_{R^m}|y|^{p(j-[l])}\int_{2|\eta|<|y|}|\eta|^{-m+p(1-\{l\})}\times\Big(\int_0^1|\nabla_{j+1,y}V(x,y+t\eta)|\,dt\Big)^p d\eta\Big)^{1/p}+cK\,\|\gamma;R^n\|_{L_\infty}\|V;R^{n+m}\|_{W_p^l}.$$

Here we have used the relation (5.2.1/6). Further, we have

$$\int_{R^m}|y|^{(j-[l])p}\int_{2|\eta|<|y|}|\eta|^{-m+p(1-\{l\})}\Big(\int_0^1|\nabla_{j+1,y}V(x,y+t\eta)|\,dt\Big)^p d\eta\,dy$$
$$\le c\int_0^1dt\int_{R^m}|\eta|^{-m+p(1-\{l\})}\int_{2|\eta|<|y|}|y|^{(j-[l])p}\,|\nabla_{j+1,y}V(x,y+t\eta)|^p\,dy\,d\eta$$
$$\le c\int_{R^m}|\eta|^{-m+p(1-\{l\})}\int_{|\chi|>|\eta|}|\chi|^{(j-[l])p}\,|\nabla_{j+1,x}V(x,\chi)|^p\,d\chi\,d\eta$$
$$=c\int_{R^m}|\chi|^{p(j+1-l)}\,|\nabla_{j+1,x}V(x,\chi)|^p\,d\chi$$

which, according to (5.2.1/4), does not exceed $c\,\|V(x,\cdot);R^m\|^p_{W_p^l}$ for a.a. $x\in R^n$. Thus, the first sum in (6) is not more than $cK\,\|V;R^{n+m}\|_{W_p^l}$. Using

(5.2.1/1), we find that the second sum in (6) has the majorant

$$cK\,\|\gamma; R^n\|_{L_\infty} \sum_{j=0}^{[l]} \Big(\int_{R^{n+m}} |(\nabla_{j,y} V)(z)|^p\, |y|^{(j-l)p}\, dz\Big)^{1/p}$$

which, by virtue of (5.2.1/4), is not more than $cK\,\|\gamma; R^n\|_{L_\infty} \|V; R^{n+m}\|_{W_p^l}$.

To obtain a bound for

$$\Big(\int_{R^m} dy \int_{R^n} dx \int_{R^n} |\nabla_{[l],x}[(VT\gamma)(x+h, y)$$
$$-(VT\gamma)(x, y)]|^p\, |h|^{-n-p\{l\}}\, dh\Big)^{1/p} \tag{7}$$

it suffices to estimate the integrals

$$\int_{R^m} dy \int_{R^n} |\nabla_{j,x} V(x, y)|^p$$
$$\times \int_{R^n} |\nabla_{[l]-j,x}[(T\gamma)(x+h, y)-(T\gamma)(x, y)]|^p\, |h|^{-n-p\{l\}}\, dh\, dx,$$

$$\int_{R^m} dy \int_{R^n} |[\nabla_{[l]-j,x}(T\gamma)](x, y)|^p$$
$$\times \int_{R^n} |\nabla_{j,x}(V(x+h, y)-V(x, y))|^p\, |h|^{-n-p\{l\}}\, dh\, dx.$$

The first is estimated by Lemma 5.2.1 and inequality (5.2.1/4). It does not exceed

$$c(K\,\|\gamma; R^n\|_{L_\infty} \|V; R^{n+m}\|_{W_p^l})^p.$$

The second integral is majorized by

$$cK^p\,\|\gamma; R^n\|_{L_\infty}^p \int_{R^n} dx \int_{R^n} |h|^{-n-p\{l\}} \int_{R^m} |y|^{p(j-[l])}$$
$$\times |\nabla_{j,x}(V(x+h, y)-V(x, y))|^p\, dy\, dh$$
$$\leqslant cK^p\,\|\gamma; R^n\|_{L_\infty}^p \int_{R^n} dx \int_{R^n} |h|^{-n-p\{l\}}$$
$$\times \int_{R^m} |\nabla_{[l],z}(V(x+h, y)-V(x, y))|^p\, dy\, dh$$
$$\leqslant c(K\,\|\gamma; R^n\|_{L_\infty} \|V; R^{n+m}\|_{W_p^l})^p.$$

Here we have used (5.2.1/4) and (5.2.1/6). So inequality (4) is proved.

By putting $\gamma = 1$, $T\gamma = 1$ in (3) we obtain the estimate

$$\|E\varphi; R^{n+m}\|_{W_p^l} \leqslant c\,\|U; R^{n+m}\|_{W_p^l}$$

which, together with (4) and the equality $V = U - E\varphi$, shows

$$\|VT\gamma; R^{n+m}\|_{W_p^l} \leqslant cK\,\|\gamma; R^n\|_{L_\infty} \|U; R^{n+m}\|_{W_p^l}. \quad \square$$

6

Multipliers in a Pair of Sobolev Spaces in a Domain

In this chapter we study multipliers in pairs of Sobolev spaces in a domain. Section 6.1 concerns the special Lipschitz domain G, i.e. $G = \{(x, y) : x \in R^n, y > \varphi(x)\}$, where φ is a function satisfying the Lipschitz condition. We find necessary and sufficient conditions for a function to belong to the space $M(W_p^m(G) \to W_p^l(G))$. In 6.2 we show that the extension operator due to Stein [1] maps continuously

$$M(W_p^m(G) \to W_p^l(G)) \to M(W_p^m(R^n) \to W_p^l(R^n)).$$

Analogous results for the space $M(W_p^m(\Omega) \to W_p^l(\Omega))$, where Ω is a bounded domain of the Lipschitz class $C^{0,1}$, are obtained in 6.3. A description of the space $ML_p^1(\Omega)$, where $L_p^1(\Omega) = \{u \in L_{p,\mathrm{loc}}(\Omega) : \nabla u \in L_p(\Omega)\}$ and Ω is an arbitrary bounded domain, is given. We show that, in general, the restriction to Ω of a multiplier in $W_p^1(R^n)$ is not a multiplier in $W_p^1(\Omega)$.

Further, in 6.4 we study the influence of a change of variables upon Sobolev spaces. Here we give the characteristics of classes of mappings ((p, l)-diffeomorphisms) which preserve the space W_p^l, as well as classes of non-smooth manifolds on which the space W_p^l is correctly defined. These characteristics are given in terms of multipliers. In conclusion, a change of variables is considered as an operator in the pair of Sobolev spaces $W_p^m(V) \to W_p^l(U)$.

In 6.5 we prove a modification of the implicit function theorem with multipliers in its statement. We consider a function u in a special Lipschitz domain G. We assume that $\operatorname{grad} u \in MW_p^{l-1}(G)$, $l \geq 2$, that u vanishes on ∂G, i.e. for $y = \varphi(x)$, and that $\partial u / \partial y$ is separated from zero on ∂G. We show that $\operatorname{grad} \varphi \in MW_p^{l-1-1/p}(R^{n-1})$.

In 6.6 we study the solvability of the Dirichlet problem for an elliptic second order equation in the space $MW_2^1(\Omega)$. Finally, in 6.7 we give the description of the space $M(\mathring{W}_p^m(\Omega) \to W_p^l(\Omega))$.

The results of the present chapter are mainly taken from the authors' papers [6], [9].

6.1 Multipliers in a Special Lipschitz Domain

6.1.1 Special Lipschitz Domains

Let $z=(x, y)$, where $x \in R^{n-1}$, $y \in R^1$. By a special Lipschitz domain we mean $G=\{z \in R^n : x \in R^{n-1}, y>\varphi(x)\}$, where φ is a function satisfying the Lipschitz conditions $|\varphi(x_1)-\varphi(x_2)| \leqslant L\,|x_1-x_2|$.

In Stein [1] (§ 3, Ch. 6) it is shown that there exists a function $z \to \delta^*(z)$ with properties:

(i) $\delta^* \in C^\infty(R^n \backslash \partial G)$ and, for any multi-index α,

$$|D^\alpha \delta^*| \leqslant c_\alpha (\delta^*)^{1-|\alpha|},$$

where c_α are constants depending on L.

(ii) For all $z \in R^n \backslash G$,

$$2[\varphi(x)-y] \leqslant \delta^*(z) \leqslant a[\varphi(x)-y], \tag{1}$$

where $a=\text{const}>2$.

We introduce the operator $\mathfrak{E}$ which performs the extension to the whole of R^n of a function f defined on G. Namely, if $z \in R^n \backslash \bar{G}$, then we put

$$(\mathfrak{E}f)(z)=\int_1^2 f(x, y+\lambda\delta^*(z))\psi(\lambda)\,d\lambda, \tag{2}$$

where ψ is a function in $C([1, 2])$ such that

$$\int_1^2 \psi(\lambda)\,d\lambda=1, \qquad \int_1^2 \lambda^k \psi(\lambda)\,d\lambda=0, \qquad k=1, 2, \ldots, l. \tag{3}$$

The operator $\mathfrak{E}$ maps $W_p^l(G)$ continuously into $W_p^l(R^n)$ (see Stein [1]).

Let K be a non-negative function in $C_0^\infty(B_1)$ with support in the cone $\{z : y>2L\,|x|\}$. With function f defined on R^n we associate its mollification with radius h,

$$[\mathfrak{R}(h)f](z)=\int_{R^n} f(z+h\zeta)K(\zeta)\,d\zeta. \tag{4}$$

It is clear that if $z \in G$ then $[\mathfrak{R}(h)f](z)$ depends only on values of f in G.

6.1.2 Auxiliary Assertions

Here, as in 6.1.1, G is a special Lipschitz domain in R^n.

Lemma 1. *Let w be a measurable non-negative function defined on G and let $0 \leqslant l<m$, $1<p<\infty$. The best constant C in the inequality*

$$\int_G (|\nabla_l u(z)|^p+|u(z)|^p)w(z)\,dz \leqslant C\,\|u; G\|^p_{W_p^m}, \qquad u \in W_p^m(G), \tag{1}$$

is equivalent to

$$\mathcal{L}=\sup\frac{\int_e w(z)\,dz}{\operatorname{cap}(e, W_p^{m-l}(R^n))}, \tag{2}$$

where the supremum is taken over all compact subsets e of the domain G.

Proof. We extend w by zero to the exterior of G. Then (1) implies the same inequality with G replaced by R^n. Now the desired lower estimate for the constant immediately follows from Lemma 1.3.1/3.

Let us obtain the upper bound for C. By w_ε we denote a function which coincides with w on the set $\{z\in G:\operatorname{dist}(z,\partial G)>\varepsilon, w(z)<1/\varepsilon\}$, $\varepsilon>0$, and vanishes elsewhere. By virtue of Lemma 1.3.1/3 for all $v\in C_0^\infty(R^n)$,

$$\int_{R^n}(|\nabla_l v|^p+|v|^p)w_\varepsilon\,dz\leqslant c\sup_E\frac{\int_E w_\varepsilon(z)\,dz}{\operatorname{cap}(E, W_p^{m-l}(R^n))}\|v; R^n\|_{W_p^m}^p, \tag{3}$$

where the supremum is taken over all compact subsets of R^n. From the definition of the function w_ε and the monotonicity of the capacity, it follows that this supremum does not exceed (2). Let $u\in W_p^m(G)$. Then $\mathfrak{C}u\in W_p^m(R^n)$. Approximating $\mathfrak{C}u$ by $C_0^\infty(R^n)$-functions in the metrics of the space $W_p^m(R^n)$, we obtain from (3)

$$\int_{R^n}(|\nabla_l\mathfrak{C}u|^p+|\mathfrak{C}u|^p)w_\varepsilon\,dz\leqslant c\mathcal{L}\,\|\mathfrak{C}u; R^n\|_{W_p^m}^p.$$

Since $\mathfrak{C}u=u$ in G, $w_\varepsilon=0$ in $R^n\backslash G$ and the operator $\mathfrak{C}: W_p^m(G)\to W_p^m(R^n)$ is continuous,

$$\int_{R^n}(|\nabla_l u|^p+|u|^p)w_\varepsilon\,dz\leqslant c\mathcal{L}\,\|u; G\|_{W_p^m}^p.$$

Passing to the limit as $\varepsilon\to 0$ in the left-hand side, we complete the proof. □

From Lemma 1 and the definition of capacity immediately follows:

Corollary. *The following relation is valid:*

$$\|\gamma; G\|_{M(W_p^m\to L_p)}\sim\sup_{e\subset G}\frac{\|\gamma; e\|_{L_p}}{[\operatorname{cap}(e, W_p^m(R^n))]^{1/p}}.$$

From the existence of the operator $\mathfrak{C}$ and the interpolation property of Sobolev spaces in R^n it follows that the spaces $W_p^k(G)$ have the same interpolation property. In particular,

$$\|\gamma; G\|_{M(W_p^{m-j}\to W_p^{l-j})}\leqslant c\,\|\gamma; G\|_{M(W_p^m\to W_p^l)}^{(l-j)/l}\,\|\gamma; G\|_{M(W_p^{m-l}\to L_p)}^{j/l}. \tag{4}$$

We introduce some notation. Let γ be a function defined on G whose generalized derivatives of order k are locally summable with degree p. We put

$$f_k(\gamma; e)=\frac{\|\nabla_k\gamma; e\|_{L_p}}{[\mathrm{cap}\,(e, W_p^{m-l+k}(R^n))]^{1/p}}, \tag{5}$$

where $m\geq l$, $0\leq k\leq l$ and e is a compact subset of $\bar{G}$. Further, let

$$s_k(\gamma)=\sup_{e\subset G} f_k(\gamma; e), \tag{6}$$

where the supremum is taken over all compact subsets of G of positive n-dimensional measure. If $m=l$, then $s_0(\gamma)=\|\gamma, G\|_{L_\infty}$.

We note that the value $s_k(\gamma)$ does not change if in its definition we replace e by any compact subset of $\bar{G}$ of positive n-dimensional measure. In fact, for any $\varepsilon>0$ there exists a compactum $E\subset\bar{G}$ such that

$$\sup_{e\subset\bar{G}} f_k(\gamma; e)\leq(1+\varepsilon)f_k(\gamma; E). \tag{7}$$

Let $E_\delta=\{z\in E: y\geq\varphi(x)+\delta\}$, $\delta>0$. It is clear that

$$f_k(\gamma; e)\leq\frac{\|\nabla_k\gamma; E\|_{L_p}}{[\mathrm{cap}\,(E_\delta, W_p^{m-l+k}(R^n))]^{1/p}}.$$

Since, for small δ,

$$\|\nabla_k\gamma; E\|_{L_p}\leq(1+\varepsilon)\,\|\nabla_k\gamma; E_\delta\|_{L_p},$$

then

$$f_k(\gamma; E)\leq(1+\varepsilon)f_k(\gamma; E_\delta)$$

which, together with (7), yields

$$\sup_{e\subset\bar{G}} f_k(\gamma; e)\leq(1+\varepsilon)^2\sup_{e\subset G} f_k(\gamma; e).$$

It remains to make use of the arbitrariness of ε.

Lemma 2. *Let $\mathfrak{R}(h)\gamma$ be the mollification of γ, specified by* (6.1.1/4). *Then*

$$\|\mathfrak{R}(h)\gamma; G\|_{M(W_p^m\to W_p^l)}\leq\|\gamma; G\|_{M(W_p^m\to W_p^l)}\leq\varliminf_{h\to0}\|\mathfrak{R}(h)\gamma; G\|_{M(W_p^m\to W_p^l)}, \tag{8}$$

$$\varliminf_{h\to0} s_k(\mathfrak{R}(h)\gamma)\geq s_k(\gamma)\geq cs_k(\mathfrak{R}(h)\gamma). \tag{9}$$

The proof of (8) is the same as that of Lemma 1.3.1/1. Inequality (9) follows from (8) with $l=0$, and the corollary.

6.1.3 Description of the Space of Multipliers

Theorem 1. *Let m and l be integers, $m \geqslant l \geqslant 0$, $p \in (1, \infty)$. The space $M(W_p^m(G) \to W_p^l(G))$ consists of functions γ which are locally summable with degree p, along with their generalized derivatives up to the order l, and such that $s_l(\gamma) + s_0(\gamma) < \infty$. The relation*

$$\|\gamma; G\|_{M(W_p^m \to W_p^l)} \sim s_l(\gamma) + s_0(\gamma)$$

holds.

The proof is similar to that of Theorem 1.3.2/2 except that we should use Lemmas 6.1.2/1 and 6.1.2/2 as well as interpolation inequality (6.1.2/4).

From Theorem 1 and the inequality

$$\|\nabla_j \gamma; G\|_{M(W_p^{m-l+j} \to L_p)} \leqslant c \, \|\gamma; G\|_{M(W_p^m \to W_p^l)}^{j/l} \, \|\gamma; G\|_{M(W_p^{m-l} \to L_p)}^{1-j/l},$$

where $j = 1, \ldots, l-1$ (cf. (1.3.2/11), it follows that

$$\|\gamma; G\|_{M(W_p^m \to W_p^l)} \sim \sum_{j=0}^{l} s_j(\gamma). \tag{1}$$

Next we turn to the study of the space $M(W_1^m(G) \to W_1^l(G))$.

Lemma 1. *Let G be a special Lipschitz domain and let w be a measurable function defined on G. Then the best constant C in*

$$\|wu; G\|_{L_1} \leqslant C \, \|u; G\|_{W_1^m}, \qquad u \in W_1^m(G), \tag{2}$$

is equivalent to

$$N = \sup_{z \in R^n, \rho \in (0,1)} \rho^{m-n} \, \|w; B_\rho(z) \cap G\|_{L_1}.$$

Proof. We extend w by zero to the exterior of G. Then we obtain from (2) the same inequality with G replaced by R^n. Now the desired lower bound for the constant C follows from Theorem 1.3.5/1. The same theorem yields

$$\|wv; R^n\|_{L_1} \leqslant cN \, \|v; R^n\|_{W_1^m}$$

for all $v \in W_1^m(R^n)$. By minimizing the right-hand side over all extensions of $u \in W_1^m(G)$, we arrive at (2) with the constant cN. □

Remark 1. Obviously, by replacing the condition $\rho \in (0, 1)$ in the definition of N by $\rho \in (0, C)$, where C is an arbitrary positive constant, we obtain an equivalent value. The same is true for the change of $z \in R^n$ to $z \in G$.

Theorem 2. *Let G be a special Lipschitz domain. The space $M(W_1^m(G)\to W_1^l(G))$ consists of functions γ which are locally summable in $\bar{G}$ together with their generalized derivatives of order l and such that*

$$\|\nabla_l\gamma; B_\rho(z)\cap G\|_{L_1}+\rho^{-l}\,\|\gamma; B_\rho(z)\cap G\|_{L_1}\leq \text{const}\,\rho^{-m+n}$$

for all $z\in R^n$, $\rho\in(0,1)$.

The following relation is valid:

$$\|\gamma; G\|_{M(W_1^m\to W_1^l)}\sim \sup_{z\in R^n,\rho\in(0,1)}\rho^{m-n}\sum_{j=0}^{l}\rho^{j-l}\,\|\nabla_j\gamma; B_\rho(z)\cap G\|_{L_1}.$$

(The same relation obviously holds if we consider only $z\in G$ in the right-hand side).

Proof. Let us insert the function $\zeta\to u(\zeta)=\Phi((\zeta-z)/\rho)$, where $z\in R^n$, $\Phi\in C_0^\infty(B_2)$, $\Phi=1$ on B_1, and $\rho\in(0,1)$, into the inequality

$$\|\gamma u; G\|_{W_1^l}\leq\|\gamma; G\|_{M(W_1^m\to W_1^l)}\,\|u; G\|_{W_1^m}. \tag{3}$$

Since, for $j=0,1,\ldots,l-1$,

$$\rho^{j-l}\,\|\nabla_j(\gamma u); B_{2\rho}(z)\cap G\|_{L_1}\leq c\,\|\nabla_l(\gamma u); B_{2\rho}(z)\cap G\|_{L_1},$$

then from (3) follows

$$\rho^{j-l}\,\|\nabla_j\gamma; B_\rho(z)\cap G\|_{L_1}\leq\|\gamma; G\|_{M(W_1^m\to W_1^l)}\rho^{-m+n}.$$

Thus the required lower estimate for the norm in $M(W_1^m(G)\to W_1^l(G))$ is obtained.

The upper estimate follows from the obvious inequality

$$\|\gamma u; G\|_{W_1^l}\leq c\sum_{0\leq k+j\leq l}\|\,|\nabla_j u|\,|\nabla_k\gamma|; G\|_{L_1}$$

and the above lemma. □

The following theorem shows that the description of the space $M(W_p^m(G)\to W_p^l(G))$ is especially simple provided that $mp>n$, $p>1$ or $m\geq n$, $p=1$. This theorem can be derived from Theorems 1, 2 but we present a direct proof.

Theorem 3. *Let G be a special Lipschitz domain and either $mp>n$, $p>1$, or $m\geq n$, $p=1$.*

The space $M(W_p^m(G)\to W_p^l(G))$ consists of functions γ which are locally summable with degree p in $\bar{G}$ together with their generalized derivatives of order l and such that $\|\gamma; B_1(z)\cap G\|_{W_p^l}\leq$const for any $z\in R^n$. Moreover,

$$\|\gamma; G\|_{M(W_p^m\to W_p^l)}\sim\sup_{z\in R^n}\|\gamma; B_1(z)\cap G\|_{W_p^l}.$$

(The same relation is obviously true if we consider only $z \in G$ on the right above).

Proof. By putting the function $\zeta \to u(\zeta) = \varphi(\zeta - z)$, where $z \in R^n$, $\varphi \in C_0^\infty(B_2)$, $\varphi = 1$ on B_1, into

$$\|\gamma u; G\|_{W_p^l} \leq \|\gamma; G\|_{M(W_p^m \to W_p^l)} \|u; G\|_{W_p^m},$$

we obtain

$$\|\gamma; B_1 \cap G\|_{W_p^l} \leq c\, \|\gamma; G\|_{M(W_p^m \to W_p^l)}.$$

By $\{B^{(j)}\}_{j \geq 1}$ we denote a covering of R^n by unit balls with a finite multiplicity which depends only on n. We have

$$\|\gamma u; B^{(j)} \cap G\|_{W_p^l} \leq c \sum_{i=0}^{l} \|\, |\nabla_i u|\, |\nabla_{l-i}\gamma|; B^{(j)} \cap G\|_{L_p}$$

$$\leq c \sum_{i=0}^{l} \|\nabla_i u; B^{(j)} \cap G\|_{L_{q_i}} \|\nabla_{l-i}\gamma; B^{(j)} \cap G\|_{L_{q_i p/(q_i - p)}},$$

where $q_i = pn/[n - p(m-i)]$ provided that $n > p(m-i)$, $q_i = \infty$ provided that $n < p(m-i)$ and q_i is an arbitrary positive number in the case $n = p(m-i)$. According to the Sobolev imbedding theorem,

$$\|\nabla_i u; B^{(j)} \cap G\|_{L_{q_i}} \leq c\, \|u; B^{(j)} \cap G\|_{W_p^m},$$

$$\|\nabla_{l-i}\gamma; B^{(j)} \cap G\|_{L_{q_i p/(q_i - p)}} \leq c\, \|\gamma; B^{(j)} \cap G\|_{W_p^l}.$$

Consequently,

$$\|\gamma u; B^{(j)} \cap G\|_{W_p^l}^p \leq c\, \|\gamma; B^{(j)} \cap G\|_{W_p^l}^p \|u; B^{(j)} \cap G\|_{W_p^m}^p. \tag{4}$$

By summing over j and applying the inequality $\sum a_j^\alpha \leq (\sum a_j)^\alpha$, where $a_j \geq 0$, $\alpha \geq 1$, we complete the proof. □

6.2 Extension of Multipliers to the Complement of a Special Lipschitz Domain

Theorem. *Let* $\gamma \in M(W_p^m(G) \to W_p^l(G))$, $1 \leq p < \infty$. *Then* $\mathfrak{E}\gamma \in M(W_p^m(R^n) \to W_p^l(R^n))$ *and*

$$\|\mathfrak{E}\gamma; R^n\|_{M(W_p^m \to W_p^l)} \leq c\, \|\gamma; G\|_{M(W_p^m \to W_p^l)}. \tag{1}$$

Proof. Let $u \in C_0^\infty(R^n)$. We have

$$\|u\mathfrak{E}\gamma; R^n\|_{W_p^l} \leq \|u\mathfrak{E}\gamma; R^n \backslash G\|_{L_p} + \|\nabla_l(u\mathfrak{E}\gamma); R^n \backslash G\|_{L_p} + \|u\gamma; G\|_{W_p^l}. \tag{2}$$

Let us estimate the first summand on the right above. It is clear that

$$|(\mathfrak{C}\gamma)(z)| \leq c \int_1^2 |\gamma(x, y+\lambda\delta^*(z))| \, d\lambda = c(\delta^*)^{-1} \int_{\delta^*}^{2\delta^*} |\gamma(x, y+s)| \, ds.$$

By using property (ii) of δ^*, we obtain for $z \in R^n \setminus G$

$$|(\mathfrak{C}\gamma)(z)| \leq c(\varphi(x)-y)^{-1} \int_{2(\varphi(x)-y)}^{a(\varphi(x)-y)} |\gamma(x, y+s)| \, ds$$

$$= c \int_2^a |\gamma(x, y+t(\varphi(x)-y))| \, dt. \tag{3}$$

This and the Minkowski inequality imply

$$\|u\mathfrak{C}\gamma; R^n \setminus G\|_{L_p} \leq c \Big[\int_{R^n \setminus G} |u(z)|^p \Big(\int_2^a |\gamma(x, y+t(\varphi(x)-y))| \, dt \Big)^p dz \Big]^{1/p}$$

$$\leq c \int_2^a \Big(\int_{R^n \setminus G} |u(z)\gamma(x, y+t(\varphi(x)-y))|^p \, dz \Big)^{1/p} dt.$$

Let $p>1$. By making use of Lemma 6.1.2/1, we have

$$\|u\mathfrak{C}\gamma; R^n \setminus G\|_{L_p}$$

$$\leq C \sup_{2<t<a} \sup_{E \subset R^n \setminus G} \frac{[\int_E |\gamma(x, y+t(\varphi(x)-y))|^p \, dz]^{1/p}}{[\operatorname{cap}(E, W_p^{m-l}(R^n))]^{1/p}} \|u; R^n \setminus G\|_{W_p^{m-l}}. \tag{4}$$

By $e(t)$ we denote the image of a compactum E under the mapping $z \to \zeta = (\xi, \eta)$, where $\xi = x$, $\eta = y + t(\varphi(x)-y)$. Since $\eta \geq y + 2(\varphi(x)-y) > \varphi(x) = \varphi(\xi)$ for $y < \varphi(x)$, this mapping maps $R^n \setminus G$ into G. It is clear that $z \to \zeta$ is Lipschitzian uniformly with respect to t and that the inverse mapping has the same property. From this and Definition (2.1.2/3) of capacity it follows that

$$\operatorname{cap}(E, W_p^{m-l}(R^n)) \sim \operatorname{cap}(e(t), W_p^{m-l}(R^n)).$$

Therefore the right-hand side of (4) does not exceed

$$c \sup_{2<t<a} \sup_{e(t) \subset G} f_0(\gamma; e(t)) \|u; R^n \setminus G\|_{W_p^{m-l}}.$$

Thus, for $p>1$,

$$\|u\mathfrak{C}\gamma; R^n \setminus G\|_{L_p} \leq c s_0(\gamma) \|u; R^n \setminus G\|_{W_p^{m-l}}. \tag{5}$$

In the case $p=1$, using Lemma 6.1.3/1 in place of Lemma 6.1.2/1, we get the following analogue of inequality (4):

$$\|u\mathfrak{C}\gamma; R^n\setminus G\|_{L_1} \leqslant c \sup_{2<t<a} \sup_{\sigma\in R^n\setminus G, \rho\in(0,1)} \rho^{m-l-n} \int_{B_\rho(\sigma)\setminus G} |\gamma(x, y+t(\varphi(x)-y))|\, dz \times \|u; R^n\setminus G\|_{W_p^{m-l}}.$$

Applying the properties of the mapping $z\to\zeta$ that we have just used above, we arrive at

$$\|u\mathfrak{C}\gamma; R^n\setminus G\|_{L_1} \leqslant c \sup_{\sigma\in G, \rho\in(0,1)} \rho^{m-l-n} \int_{B_{c\rho}(\sigma)\cap G} |\gamma(z)|\, dz\, \|u; R^n\setminus G\|_{W_1^{m-l}}. \tag{6}$$

Clearly, we may assume the last integral to be over $B_\rho(\sigma)\cap G$ by means of an appropriate change of the coefficient before the supremum. For simplicity of notation in the case $p=1$ we shall denote by $s_j(\gamma)$ the value

$$\sup_{\sigma\in G, \rho\in(0,1)} \rho^{m-l+j-n} \int_{B_\rho(\sigma)\cap G} |\nabla_j\gamma(z)|\, dz$$

for the rest of the proof.

Obviously, the second summand in the right-hand side of (2) is not more than

$$c\sum_{j=0}^{l} \|\, |\nabla_{l-j}u|\, |\nabla_j(\mathfrak{C}\gamma)|; R^n\setminus G\|_{L_p}. \tag{7}$$

By (5) and (6) the summand corresponding to $j=0$ does not exceed $cs_0(\gamma)\,\|u; R^n\setminus G\|_{W_p^m}$.

Next we estimate the other summands in (7). Applying the operator $D_x^\varkappa D_y^k$, $|\varkappa|+k=j\geqslant 1$, to $\mathfrak{C}\gamma$ and formally differentiating under the integral, we obtain the linear combination of expressions of the form

$$\int_1^2 (D_x^\rho D_y^r\gamma)(x, \varphi(\lambda, z)) \prod_{\nu=1}^{r} D_x^{\alpha_\nu} D_y^{a_\nu}[\varphi(\lambda, z)]\psi(\lambda)\, d\lambda, \tag{8}$$

where $0\leqslant\rho\leqslant\varkappa$, $|\rho|+r\leqslant|\varkappa|+k$, $\sum_{\nu=1}^r \alpha_\nu=\varkappa-\rho$, $\sum_{\nu=1}^r a_\nu=k$, and $\varphi(\lambda, z)=y+\lambda\delta^*(z)$. We take for each of derivatives $(D_x^\rho D_y^r\gamma)(x, \varphi(\lambda, z))$, with $|\rho|+r<|\varkappa|+k$, a Taylor series expansion

$$\begin{aligned}(D_x^\rho D_y^r\gamma)(x, y+\lambda\delta^*)(z)) = &\sum_{j=0}^{|\varkappa|+k-|\rho|-r-1} (j!)^{-1}[(\lambda-1)\delta^*(z)]^j \\ &\times(D_x^\rho D_y^{r+j}\gamma)(x, y+\delta^*(z)) + \frac{1}{(|\varkappa|+k-|\rho|-r)!} \\ &\times\int_{\delta^*(z)}^{\lambda\delta^*(z)} [\lambda\delta^*(z)-t]^{|\varkappa|+k-|\rho|-r-1} \\ &\times(D_x^\rho D_y^{|\varkappa|+k-|\rho|}\gamma)(x, y+t)\, dt.\end{aligned} \tag{9}$$

By virtue of (6.1.1/3),

$$\int_1^2 (\lambda-1)^j \prod_{\nu=1}^{r} D_x^{\alpha_\nu} D_y^{a_\nu}[y+\lambda\delta^*(z)]\psi(\lambda)\, d\lambda = 0.$$

The modulus of the integral on the right in (9) is majorized by

$$c[\delta^*(z)]^{|\varkappa|+k-|\rho|-r-1} \int_{\delta^*(z)}^{\lambda\delta^*(z)} |(D_x^{\rho} D_y^{|\varkappa|+k-|\rho|}\gamma)(x, y+t)|\, dt.$$

Moreover, by the inequality $|D_z^{\tau}\delta^*| \leqslant c_{|\tau|}(\delta^*)^{1-|\tau|}$ we have

$$\left|\prod_{\nu=1}^{r} D_x^{\alpha_\nu} D_y^{a_\nu}[\varphi(\lambda, z)]\right| \leqslant c \prod_{\nu=1}^{r} (\delta^*(z))^{1-|\alpha_\nu|-a_\nu} = c(\delta^*(z))^{r+|\rho|-|\varkappa|-k}. \quad (10)$$

Consequently, for $|\rho|+r<|\varkappa|+k$ the modulus of the integral (8) does not exceed

$$c(\delta^*(z))^{-1} \int_1^2 d\lambda \int_{\delta^*(z)}^{\lambda\delta^*(z)} |\nabla_{|\varkappa|+k}\gamma(x, y+t)|\, dt$$
$$\leqslant c(\delta^*)^{-1} \int_{\delta^*}^{2\delta^*} |(\nabla_{|\varkappa|+k}\gamma)(x, y+s)|\, ds. \quad (11)$$

In the case $|\rho|+r=|\varkappa|+k$ it follows directly from (10) that the integral (8) is majorized by (11) also.

Thus, for $j \geqslant 1$,

$$|\nabla_j(\mathfrak{C}\gamma)(z)| \leqslant c \int_2^a |\nabla_j\gamma(x, y+t(\varphi(x)-y))|\, dt$$

(cf. (3)). Hence, duplicating the arguments used in the proof of (5) and (6) and applying (3), we find

$$\|\, |\nabla_{l-j}u|\, |\nabla_j(\mathfrak{C}\gamma)|; R^n\backslash G\|_{L_p} \leqslant cs_j(\gamma)\, \|u; R^n\backslash G\|_{W_p^m}.$$

Thus the second summand in (2) is not more than

$$c \sum_{j=0}^{l} s_j(\gamma)\, \|u; R^n\backslash G\|_{W_p^m}$$

which, together with (2), (5) and (6), gives

$$\|u\mathfrak{C}\gamma; R^n\|_{W_p^l} \leqslant c \sum_{j=0}^{l} s_j(\gamma)\, \|u; R^n\backslash G\|_{W_p^m} + \|\gamma; G\|_{M(W_p^m \to W_p^l)}\, \|u; G\|_{W_p^m}.$$

Reference to (6.1.3/1) completes the proof. □

6.3 Multipliers in a Bounded Domain

6.3.1 Domains of the Class $C^{0,1}$

We say that a bounded domain Ω belongs to the class $C^{k,1}$, $k=0, 1, \ldots$, if for any point of its boundary $\partial\Omega$ there exists a neighbourhood in which $\partial\Omega$ admits (in a Cartesian coordinate system) the representation $y=\varphi(x)$, where φ is a function whose derivatives of order k satisfy the Lipschitz condition.

Let $\Omega \in C^{0,1}$ and let ε be a small positive number. We construct a covering of the set $\Gamma_\varepsilon = \{z \in R^n : \text{dist}(z, \partial\Omega) \leq \varepsilon\}$ by domains $U_1, \ldots, U_N$ with the following property: for any $i=1, \ldots, N$ there exists a special Lipschitz domain G_i such that $U_i \cap \Omega = U_i \cap G_i$. We add the set $U_0 = \Omega \backslash \Gamma_{\varepsilon/2}$ to the collection $\{U_i\}_{i\geq 1}$.

By $\{\varphi_i\}_{i=0}^N$ we denote a set of non-negative functions with the following properties: (i) $\varphi_i \in C_0^\infty(U_i)$; (ii) $\sum_{0\leq i\leq N} \varphi_i^2 = 1$ on Ω. To define the extension operator from Ω to R^n we note that for $i=1, \ldots, N$ there exists a linear continuous operator $\mathfrak{C}_i : W_p^l(U_i \cap \Omega) \to W_p^l(R^n)$, $p \geq 1$. Let $\mathfrak{C}_0$ denote the operator of extension by zero to the exterior of the set U_0. Obviously, the operator

$$\mathfrak{C} = \sum_{0\leq i\leq N} \varphi_i \mathfrak{C}_i \varphi_i \tag{1}$$

performs the extension of a function defined on Ω to R^n. We introduce the operator $\mathfrak{R}(h)$, $h>0$, by

$$\mathfrak{R}(h)\gamma = \sum_{0\leq i\leq N} \varphi_i \mathfrak{R}_i(h)(\varphi_i \gamma), \tag{2}$$

where $\mathfrak{R}_i(h)$ is the mollification operator defined by (6.1.1/4) for a special domain G_i.

6.3.2 Auxiliary Assertions

Lemmas 6.1.2/1 and 6.1.3/1 remain valid for any bounded domain of the class $C^{0,1}$ except that in their proof by $\mathfrak{C}$ we mean the operator specified by (6.3.1/1).

To transfer Theorem 6.1.3/1 to bounded domains of the class $C^{0,1}$ we need the following two lemmas.

Lemma 1. The following estimate is valid:

$$\sup_{e\subset\Omega} \frac{\|\nabla_k \gamma; e\|_{L_p}}{[\text{cap}(e, W_p^m(R^n))]^{1/p}}$$

$$\leq c \sup_{z\in\Omega, \rho\in(0,1)} \rho^{m-n/p}(\|\nabla_l \gamma; B_\rho(z)\|_{L_p} + \rho^{-l}\|\gamma; B_\rho(z)\|_{L_p}), \tag{1}$$

where $p>1$, $k=0, 1, \ldots, l-1$.

Proof. If $mp > n$, then the capacity $\text{cap}(e, W_p^m(R^n))$ of any compact set e in Ω is bounded from above and is separated from zero. So the left-hand side of (1) is equivalent to the norm $\|\nabla_k \gamma; \Omega\|_{L_p}$. Further, $\|\nabla_k \gamma; \Omega\|_{L_p} \leq c(\|\nabla_l \gamma; \Omega\|_{L_p} + \|\gamma; \Omega\|_{L_p})$ which immediately implies (1).

For $mp \leq n$ the left-hand side of (1) does not exceed

$$c \sup_{e \subset \Omega} \frac{\|\nabla_k \gamma; e\|_{L_q}}{[\text{cap}(e, W_p^s(R^n))]^{1/p}}, \tag{2}$$

where $s < m$, $q > p$ and the numbers s and q are sufficiently close to m and p respectively. From the inequality

$$\left(\int_\Omega |u|^q \, d\mu\right)^{p/q} \leq c \sup_{z \in \Omega, \rho \in (0,1)} \rho^{ps-n}[\mu(B_\rho(z) \cap \Omega)]^{p/q} \|u; R^n\|^p_{W_p^s},$$

where μ is a measure in Ω and $u \in C_0^\infty(R^n)$ (see Lemma 1.3.6/1), it follows that (2) is not more than

$$c \sup_{z \in \Omega, \rho \in (0,1)} \rho^{s-n/p} \|\nabla_k \gamma; B_\rho(z)\|_{L_q}.$$

Since q is close to p, then

$$\rho^{k-l+n(1/p-1/q)} \|\nabla_k \gamma; B_\rho(z)\|_{L_q} \leq c(\|\nabla_l \gamma; B_\rho(z)\|_{L_p} + \rho^{-l} \|\gamma; B_\rho(z)\|_{L_p}).$$

Consequently (2) is majorized by

$$c \sup_{z \in \Omega, \rho \in (0,1)} \rho^\mu (\rho^{m-n/p} \|\nabla_l \gamma; B_\rho(z)\|_{L_p} + \rho^{m-l-n/p} \|\gamma; B_\rho(z)\|_{L_p}),$$

where $\mu = l - k + n(1/q - 1/p) + s - m$. Since $\mu > 0$, the result follows. □

Lemma 2. The following inequalities are valid:

$$c\,\|\mathfrak{R}(h)\gamma; \Omega\|_{M(W_p^m \to W_p^l)} \leq \|\gamma; \Omega\|_{M(W_p^m \to W_p^l)} \leq \varliminf_{h \to 0} \|\mathfrak{R}(h)\gamma; \Omega\|_{M(W_p^m \to W_p^l)}, \tag{3}$$

$$\varliminf_{h \to 0} s_k(\mathfrak{R}(h)\gamma) \geq s_k(\gamma), \tag{4}$$

$$s_k(\mathfrak{R}(h)\gamma) \leq c[s_k(\gamma) + s_0(\gamma)], \tag{5}$$

where $s_k(\gamma)$ is the same as in (6.1.2/6) with G replaced by Ω.

Proof. It is clear that

$$\|\mathfrak{R}(h)\gamma; \Omega\|_{M(W_p^m \to W_p^l)} \leq c \sum_{0 \leq i \leq N} \|\mathfrak{R}_i(h)(\varphi_i \gamma); G_i\|_{M(W_p^m \to W_p^l)}.$$

This and Lemma 6.1.2/2 imply

$$\|\mathfrak{R}(h)\gamma; \Omega\|_{M(W_p^m \to W_p^l)} \leq c \sum_{0 \leq i \leq N} \|\varphi_i\gamma; G_i\|_{M(W_p^m \to W_p^l)}$$

$$= c \sum_{0 \leq i \leq N} \|\varphi_i\gamma; \Omega\|_{M(W_p^m \to W_p^l)}.$$

Since $\varphi \in C_0^\infty(R^n)$, the left inequality in (3) is proved.

The right inequality in (3) follows via the chain

$$\|u\gamma; \Omega\|_{W_p^l} = \lim_{h \to +0} \|u\mathfrak{R}(h)\gamma; \Omega\|_{W_p^l} \leq \underline{\lim}_{h \to 0} \|\mathfrak{R}(h)\gamma; \Omega\|_{M(W_p^m \to W_p^l)} \|u; \Omega\|_{W_p^m}.$$

Obviously, for any compactum $E \subset \Omega$,

$$\underline{\lim}_{h \to 0} s_k(\mathfrak{R}(h)\gamma) \geq \frac{\lim_{h \to 0} (\int_E |\nabla_k(\mathfrak{R}(h)\gamma)|^p \, dx)^{1/p}}{[\text{cap}(E, W_p^{m-l+k}(R^n))]^{1/p}}. \tag{6}$$

Since $\varphi_i\mathfrak{R}_i(h)(\varphi_i\gamma) \to \varphi_i^2\gamma$ in $W_p^k(\Omega)$ and $\sum_i \varphi_i^2 = 1$, the right-hand side of (6) is equal to $f_k(\gamma; E)$. Thus, (4) is proved.

Next we turn to estimate (5). Since $\varphi \in C_0^\infty(R^n)$, then, for small enough h,

$$\int_E |\nabla_k[\varphi_i\mathfrak{R}_i(h)(\varphi_i\gamma)]|^p \, dx \leq c \sum_{0 \leq j \leq k} \int_{E \cap \bar{G}_i} |\nabla_j[\mathfrak{R}_i(h)(\varphi_i\gamma)]|^p \, dx.$$

By virtue of Lemma 6.1.2/2,

$$\sup_{e \subset \bar{G}_i} \frac{\int_e |\nabla_j[\mathfrak{R}_i(h)(\varphi_i\gamma)]|^p \, dx}{\text{cap}(e, W_p^{m-l+k}(R^n))} \leq \sup_{e \subset \bar{G}_i} \frac{\int_e |\nabla_j(\varphi_i\gamma)|^p \, dx}{\text{cap}(e, W_p^{m-l+k}(R^n))}.$$

Therefore

$$s_k(\mathfrak{R}(h)\gamma) \leq c \sum_{0 \leq i \leq N} \sum_{0 \leq j \leq k} \sup_{e \subset \bar{G}_i} \frac{\int_e |\nabla_j\gamma|^p \, dx}{\text{cap}(e, W_p^{m-l+k}(R^n))}.$$

It remains to use Lemma 1, noting beforehand that the right-hand side of (1) does not exceed $c(s_l(\gamma) + s_0(\gamma))$. □

6.3.3 Description of Spaces of Multipliers in a Bounded Domain of the Class $C^{0,1}$

Theorem 1. *Let m, l be integers, $m \geq l \geq 0$, $p \in (1, \infty)$. The space $M(W_p^m(\Omega) \to W_p^l(\Omega))$ consists of functions $\gamma \in W_p^l(\Omega)$ such that*

$$\sup_{e \subset \Omega} \left(\frac{\|\nabla_l\gamma; e\|_{L_p}}{[\text{cap}(e, W_p^m(R^n))]^{1/p}} + \frac{\|\gamma; e\|_{L_p}}{[\text{cap}(e, W_p^{m-l}(R^n))]^{1/p}} \right) < \infty. \tag{1}$$

The following inequalities are valid:

$$c\sum_{j=0}^{l}\sup_{e\subset\Omega}\frac{\|\nabla_j\gamma; e\|_{L_p}}{[\operatorname{cap}(e, W_p^{m-l+j}(R^n))]^{1/p}}\leqslant\|\gamma;\Omega\|_{M(W_p^m\to W_p^l)}$$

$$\leqslant c\left(\sup_{e\subset\Omega}\frac{\|\nabla_l\gamma; e\|_{L_p}}{[\operatorname{cap}(e, W_p^m(R^n))]^{1/p}}+\sup_{e\subset\Omega}\frac{\|\gamma; e\|_{L_p}}{[\operatorname{cap}(e, W_p^{m-l}(R^n))]^{1/p}}\right). \quad (2)$$

(Note that, by Theorem 2.6, the last summands in (1) and (2) can be replaced by $\|\gamma;\Omega\|_{L_1}$.)

The proof follows the same lines as that of Theorem 1.3.2/2 except that, in place of the usual mollification operator, we must use the operator $\mathfrak{R}(h)$ specified by (6.3.1/2) and Lemma 6.3.2/2.

The following assertion is proved in the same way as Theorem 6.1.3/2.

Theorem 2. *Let m, l be integers, $m\geqslant l\geqslant 0$. The space $M(W_1^m(\Omega)\to W_1^l(\Omega))$ consists of functions $\gamma\in W_1^l(\Omega)$ such that*

$$\|\nabla_l\gamma; B_\rho(z)\cap\Omega\|_{L_1}+\rho^{-l}\|\gamma; B_\rho(z)\cap\Omega\|_{L_1}\leqslant\text{const}\,\rho^{n-m}$$

for all $z\in\Omega$ and $\rho\in(0,1)$. The following relation holds:

$$\|\gamma;\Omega\|_{M(W_1^m\to W_1^l)}\sim\sup_{z\in\Omega,\rho\in(0,1)}\rho^{m-n}\sum_{j=0}^{l}\rho^{j-l}\|\nabla_j\gamma; B_\rho(z)\cap\Omega\|_{L_1}.$$

From Theorem 6.1.3/3 immediately follows:

Theorem 3. *Let $mp>n$, $p\in(1,\infty)$ or $m\geqslant n$, $p=1$. The space $M(W_p^m(\Omega)\to W_p^l(\Omega))$ coincides with the space $W_p^l(\Omega)$.*

The corollary of Theorem 6.2 and formula (6.3.1/1) is:

Theorem 4. *Let $\mathfrak{E}$ be the extension operator defined by (6.3.1/1) and let $\gamma\in(W_p^m(\Omega)\to W_p^l(\Omega))$, $p\geqslant 1$. Then $\mathfrak{E}\gamma\in M(W_p^m(R^n)\to W_p^l(R^n))$ and*

$$\|\mathfrak{E}\gamma; R^n\|_{M(W_p^m\to W_p^l)}\leqslant c\,\|\gamma;\Omega\|_{M(W_p^m\to W_p^l)}.$$

6.3.4 The Essential Norm and Compact Multipliers in a Bounded Domain of the Class $C^{0,1}$

Let Ω be a bounded domain of the class $C^{0,1}$. As in the case of the whole space R^n, we associate with any element $\gamma\in M(W_p^m(\Omega)\to W_p^l(\Omega))$ the essential norm

$$\operatorname{ess}\|\gamma;\Omega\|_{M(W_p^m\to W_p^l)}=\inf_{\{T\}}\|\gamma-T;\Omega\|_{W_p^m\to W_p^l}$$

where $\{T\}$ is the collection of all compact linear operators: $W_p^m(\Omega) \to W_p^l(\Omega)$.

The derivation of two-sided estimates for the essential norm in $M(W_p^m(\Omega) \to W_p^l(\Omega))$ needs no new argument beyond those given in Chapter 4 and is even simpler, since here m and l are integers and the region Ω is bounded. The role of the operator T_* used in Chapter 4 is played by the mapping T_*, defined as follows:

$$(T_* u)(x) = \gamma(x) \sum_j \varphi^{(j)}(x) P^{(j)}(u; x)$$

(cf. the proof of the second part of Theorem 4.2.1/1). Here we use the following notation: $\{\varphi^{(j)}\}$ is a smooth finite partition of unity subordinate to the covering of $\bar{\Omega}$ by open balls $K_\delta^{(j)}$ with radius δ and with centres $x_j \in \Omega$; $P^{(j)}$ are polynomials of the form

$$\sum_{|\beta| \leq m-1} \left(\frac{x - x_j}{\delta}\right)^\beta \delta^{-n} \int_{K_\delta^{(j)} \cap \Omega} \psi_\beta \left(\frac{y - y_j}{\delta}\right) u(y)\, dy$$

where $\psi_\beta \in C_0^\infty(B_1)$.

In the same way as in Chapter 4, majorants for ess $\|\gamma; \Omega\|_{W_p^m \to W_p^l}$ can be obtained from upper bounds for the norms $\|(\gamma - T_*)u; \Omega\|_{W_p^l}$ which are collected in the next assertion (cf. Remark 4.2.5/2).

Lemma. *Let $\gamma \in M(W_p^m(\Omega) \to W_p^l(\Omega))$ where m and l are integers, $m \geq l > 0$.*

(i) *If $p > 1$ and $mp < n$, then*

$$\|\gamma - T_*; \Omega\|_{W_p^m \to W_p^l} \leq \sup_{\{e \subset \Omega : d(e) \leq \delta\}} \left(\frac{\|\nabla_l \gamma; e\|_{L_p}}{[\text{cap}\,(e, W_p^m)]^{1/p}} + \frac{\|\gamma; e\|_{L_p}}{[\text{cap}\,(e, W_p^{m-l})]^{1/p}} \right).$$

In particular,

$$\|\gamma - T_*; \Omega\|_{W_p^l \to W_p^l} \leq c \left(\sup_{\{e \subset \Omega : d(e) \leq \delta\}} \frac{\|\nabla_l \gamma; e\|_{L_p}}{[\text{cap}\,(e, W_p^l)]^{1/p}} + \|\gamma; \Omega\|_{L_\infty} \right).$$

If $p > 1$, $mp = n$, then one should replace δ by $\delta^{1/2}$ in the right-hand sides of these inequalities.

(ii) *If $m \leq n$, then*

$$\|\gamma - T_*; \Omega\|_{W_1^m \to W_1^l} \leq c\delta^{m-n} \sup_{x \in \Omega} (\|\nabla_l \gamma; B_\delta(x) \cap \Omega\|_{L_1} + \delta^{-l} \|\gamma; B_\delta(x) \cap \Omega\|_{L_1}).$$

In particular,

$$\|\gamma - T_*; \Omega\|_{W_1^l \to W_1^l} \leq c \left(\delta^{l-n} \sup_{x \in \Omega} \|\nabla_l \gamma; B_\delta(x) \cap \Omega\|_{L_1} + \|\gamma; \Omega\|_{L_\infty} \right).$$

Now we state a theorem on two-sided estimates for the essential norm.

Theorem 1. *Let $\gamma \in M(W_p^m(\Omega) \to W_p^l(\Omega))$, where m and l are integers, $m \geq l \geq 0$.*

(i) *If $p>1$ and $mp \leq n$, then*

$$\operatorname{ess} \|\gamma; \Omega\|_{M(W_p^m \to W_p^l)} \sim \lim_{\delta \to 0} \sup_{\{e \subset \Omega : d(e) \leq \delta\}} \left(\frac{\|\gamma; e\|_{L_p}}{[\operatorname{cap}(e, W_p^{m-l})]^{1/p}} + \frac{\|\nabla_l \gamma; e\|_{L_p}}{[\operatorname{cap}(e, W_p^m)]^{1/p}} \right).$$

In particular,

$$\operatorname{ess} \|\gamma; \Omega\|_{MW_p^l} \sim \|\gamma; \Omega\|_{L_\infty} + \lim_{\delta \to 0} \sup_{\{e \subset \Omega : d(e) \leq \delta\}} \frac{\|\nabla_l \gamma; e\|_{L_p}}{[\operatorname{cap}(e, W_p^l)]^{1/p}}.$$

(ii) *If $m<n$, then*

$$\operatorname{ess} \|\gamma; \Omega\|_{M(W_1^m \to W_1^l)} \sim \overline{\lim_{\delta \to 0}}\, \delta^{m-n} \sup_{x \in \Omega} (\delta^{-l} \|\gamma; B_\delta(x) \cap \Omega\|_{L_1} + \|\nabla_l \gamma; B_\delta(x) \cap \Omega\|_{L_1}).$$

In particular,

$$\operatorname{ess} \|\gamma; \Omega\|_{MW_1^l} \sim \|\gamma; \Omega\|_{L_\infty} + \overline{\lim_{\delta \to 0}}\, \delta^{l-n} \sup_{x \in \Omega} \|\nabla_l \gamma; B_\delta(x) \cap \Omega\|_{L_1}.$$

(iii) *If $mp>n$, $p \in (1, \infty)$ or $m \geq n$, $p=1$, then*

$$\operatorname{ess} \|\gamma; \Omega\|_{M(W_p^m \to W_p^l)} = 0 \quad \text{for } m>l$$

and

$$\operatorname{ess} \|\gamma; \Omega\|_{MW_p^l} \sim \|\gamma; \Omega\|_{L_\infty} \quad \text{for} \quad m=l.$$

This immediately implies:

Proposition 1. *The function $\gamma \in M(W_p^m(\Omega) \to W_p^l(\Omega))$, $m>l$, belongs to the space $\mathring{M}(W_p^m(\Omega) \to W_p^l(\Omega))$ of compact multipliers if and only if*

$$\lim_{\delta \to 0} \sup_{\{e \subset \Omega : d(e) \leq \delta\}} \left(\frac{\|\gamma; e\|_{L_p}}{[\operatorname{cap}(e, W_p^{m-l})]^{1/p}} + \frac{\|\nabla_l \gamma; e\|_{L_p}}{[\operatorname{cap}(e, W_p^m)]^{1/p}} \right) = 0$$

for $p \in (1, \infty)$ and $mp \leq n$;

$$\lim_{\delta \to 0} \delta^{m-n} \sup_{x \in \Omega} (\delta^{-l} \|\gamma; B_\delta(x) \cap \Omega\|_{L_1} + \|\nabla_l \gamma; B_\delta(x) \cap \Omega\|_{L_1}) = 0$$

for $m<n$. Finally, $\mathring{M}(W_p^m(\Omega) \to W_p^l(\Omega)) = W_p^l(\Omega)$ if $mp>n$, $p \in (1, \infty)$ or $m \geq n$, $p=1$.

Similarly to Theorem 4.2.6/3, one can obtain the following description of the space of compact multipliers.

Proposition 2. $\mathring{M}(W_p^m(\Omega) \to W_p^l(\Omega))$, $m > l$, *is the completion of* $C^\infty(\bar{\Omega})$ *with respect to the norm in* $M(W_p^m(\Omega) \to W_p^l(\Omega))$.

In correspondence to this assertion we denote by $\mathring{M}W_p^l(\Omega)$ the completion of $C^\infty(\bar{\Omega})$ with respect to the norm of $MW_p^l(\Omega)$.

The following proposition is an analogue of Theorem 4.3.5/2.

Proposition 3. *The function* γ *belongs to* $\mathring{M}W_p^l(\Omega)$ *if and only if* $\gamma \in C(\bar{\Omega})$ *and one of the following conditions is fulfilled:*

(i) *If* $pl \leq n$, $p > 1$, *then*

$$\sup_{\{e \subset \Omega : d(e) \leq \delta\}} \frac{\|\nabla_l \gamma; e\|_{L_p}}{[\text{cap}\,(e, W_p^l)]^{1/p}} = o(1) \quad \text{as} \quad \delta \to 0. \tag{1}$$

(ii) *If* $l < n$, *then*

$$\delta^{l-n} \sup_{x \in \Omega} \|\nabla_l \gamma; B_\delta(x)\|_{L_1} = o(1) \quad \text{as} \quad \delta \to 0. \tag{2}$$

(iii) *If* $pl > n$, $p > 1$ *or* $l \geq n$, $p = 1$, *then* $\mathring{M}W_p^l(\Omega) = W_p^l(\Omega)$.

To conclude this section we consider the connection of the essential norm with the constant K in

$$\|\gamma u; \Omega\|_{W_p^l} \leq K \|u; \Omega\|_{W_p^m} + C(\gamma) \|u; \Omega\|_{L_p}. \tag{3}$$

Theorem 2. *Let* $\gamma \in M(W_p^m(\Omega) \to W_p^l(\Omega))$, $m \geq l$, *and let* $\inf K$ *be the infimum of numbers* K *for which there exists a constant* $C(\gamma)$ *such that* (3) *is valid for all* $u \in W_p^m(\Omega)$. *Then*

$$\inf K \leq \text{ess}\, \|\gamma; \Omega\|_{M(W_p^m \to W_p^l)} \leq c \inf K \tag{4}$$

where $c = c(\Omega, n, p, l, m)$.

Proof. We insert $u - \sum_j \varphi^{(j)} P^{(j)}$ in place of u in (3). Then

$$\|\gamma u - T_* u; \Omega\|_{W_p^l} \leq K \Big\|u - \sum_j \varphi^{(j)} P^{(j)}; \Omega\Big\|_{W_p^m} + C(\gamma) \Big\|u - \sum_j \varphi^{(j)} P^{(j)}; \Omega\Big\|_{L_p}.$$

We have

$$\Big\|u - \sum_j \varphi^{(j)} P^{(j)}; \Omega\Big\|_{W_p^m}^p = \Big\|\sum_j \varphi^{(j)}(u - P^{(j)}); \Omega\Big\|_{W_p^m}^p$$
$$\leq c \sum_j \sum_{k=0}^m \delta^{-kp} \|u - P^{(j)}; K_\delta^{(j)} \cap \Omega\|_{W_p^{m-k}}^p.$$

Since Ω is Lipschitz, then, for some $c \geq 1$,

$$\|u - P^{(j)}; K_\delta^{(j)} \cap \Omega\|_{W_p^{m-k}} \leq c\delta^k \|u; K_{c\delta}^{(j)} \cap \Omega\|_{W_p^m}.$$

Therefore

$$\|(\gamma - T_*)u; \Omega\|_{W_p^l} \leq c(K + C(\gamma)\delta^m)\, \|u; \Omega\|_{W_p^m}$$

and so

$$\text{ess}\, \|\gamma; \Omega\|_{M(W_p^m \to W_p^l)} \leq c(K + C(\gamma)\delta^m)$$

for any small enough δ. The right estimate in (4) follows.

We now turn to the left estimate in (4). According to the definition of the essential norm, for some compact operator $T: W_p^m(\Omega) \to W_p^l(\Omega)$ and for all $u \in W_p^m(\Omega)$ we have

$$\|\gamma u; \Omega\|_{W_p^l} \leq (\text{ess}\, \|\gamma; \Omega\|_{W_p^m \to W_p^l} + \varepsilon)\, \|u; \Omega\|_{W_p^m} + \|Tu; \Omega\|_{W_p^l}.$$

We need to show that for any $\varepsilon > 0$ one can find a constant C_ε such that

$$\|Tu; \Omega\|_{W_p^l} \leq \varepsilon\, \|u; \Omega\|_{W_p^m} + C_\varepsilon\, \|u; \Omega\|_{L_p}. \tag{5}$$

We suppose the contrary. Then for some $\varepsilon > 0$ there exist a sequence $\{u_j\}$, $\|u_j; \Omega\|_{W_p^m} = 1$ and a number sequence $\{k_j\}$, $k_j \to +\infty$ so that

$$\|Tu_j; \Omega\|_{W_p^l} > \varepsilon + k_j\, \|u_j; \Omega\|_{L_p}. \tag{6}$$

Since the operator $T: W_p^m(\Omega) \to W_p^l(\Omega)$ is bounded and the sequence $\{u_j\}$ has the unit norm in $W_p^m(\Omega)$ then by (6) we have $u_j \to 0$ in $L_p(\Omega)$. From $\{u_j\}$ we select a subsequence weakly converging in $W_p^m(\Omega)$ for which we retain the notation $\{u_j\}$. Let v designate its weak limit. Then, for any $g \in L_{p'}(\Omega)$, $p + p' = pp'$, we have $\int_\Omega g u_j\, dx \to \int_\Omega g v\, dx$ and so $v = 0$ because $u_j \to 0$ in $L_p(\Omega)$. Since T transforms a sequence weakly converging in $W_p^m(\Omega)$ into a sequence strongly converging in $W_p^l(\Omega)$ then $\|Tu_j; \Omega\|_{W_p^l} \to 0$ contrary to (6). □

Proposition 4. *If $m = l$ and $\gamma \in \mathring{M}(W_p^l(\Omega))$ then* $\inf K = \|\gamma; \Omega\|_{L_\infty}$.

Proof. Let γ_1 be a function in $C^\infty(\bar{\Omega})$ such that $\|\gamma - \gamma_1; \Omega\|_{MW_p^l} < \varepsilon$. Clearly,

$$\|\gamma_1 u; \Omega\|_{W_p^l} \leq \|\gamma_1; \Omega\|_{L_\infty} \|u; \Omega\|_{W_p^l} + c \sum_{j=1}^{l} \|\, |\nabla_j \gamma_1|\, |\nabla_{l-j} u|; \Omega\|_{L_p}.$$

Hence

$$\|\gamma u; \Omega\|_{W_p^l} \leq (\|\gamma; \Omega\|_{L_\infty} + 2\varepsilon)\, \|u; \Omega\|_{W_p^l} + c\, \|\gamma_1; \Omega\|_{C^l}\, \|u; \Omega\|_{W_p^{l-1}}$$

and, since $\|u; \Omega\|_{W_p^{l-1}} \leq \varepsilon\, \|u; \Omega\|_{W_p^l} + c(\varepsilon)\, \|u; \Omega\|_{L_p}$, then $\inf K \leq \|\gamma; \Omega\|_{L_\infty}$.

Let us estimate $\inf K$ from below. Insert $u_\delta(x) = \delta^{l-n/p}\eta((x-y)/\delta)$ into (3) where $m = l$. We use the notation $y \in \bar{\Omega}$, $\delta > 0$, $\eta \in C_0^\infty(B_1)$, $\eta(0) = 1$. By definition of $\mathring{M}W_p^l(\Omega)$ the function γ can be assumed to be smooth on

$\bar{\Omega}$. One can easily see that

$$\|\gamma u_\delta; \Omega\|_{W_p^l} = |\gamma(y)| \, \|u_\delta; \Omega\|_{W_p^l} + o(1)$$

and $\|u_\delta; \Omega\|_{L_p} = o(1)$ as $\delta \to 0$.

This, together with (3) and the inequality $\underline{\lim}_{\delta\to 0} \|u_\delta; \Omega\|_{W_p^l} > 0$, gives $|\gamma(y)| \leq K$. □

We note that the inequality

$$\|\gamma u; \Omega\|_{W_p^l} \leq (\|\gamma; \Omega\|_{L_\infty} + \varepsilon) \, \|u; \Omega\|_{W_p^l} + C(\gamma; \varepsilon) \, \|u; \Omega\|_{L_p}$$

with arbitrarily small $\varepsilon > 0$ is used in the L_p-theory of elliptic boundary value problems (see, for example, Triebel [4], 5.3.4).

In conclusion, we state an obvious corollary of Theorem 2 and Proposition 4.

Corollary. *If* $\gamma \in \mathring{M}(W_p^l(\Omega))$ *then*

$$\|\gamma; \Omega\|_{L_\infty} \leq \text{ess} \, \|\gamma; \Omega\|_{MW_p^l} \leq c \, \|\gamma; \Omega\|_{L_\infty}. \tag{7}$$

Similarly to Theorem 4.3.1, one can prove that the left estimate in (7) holds for all $\gamma \in MW_p^l(\Omega)$.

6.3.5 The Space $ML_p^1(\Omega)$ for an Arbitrary Bounded Domain

Let Ω be a domain in R^n with compact closure. By $L_p^1(\Omega)$ we denote the space of functions in $L_{p,\text{loc}}(\Omega)$ with the first generalized derivatives summable in Ω with degree p, $p \in [1, \infty)$. We supply $L_p^1(\Omega)$ with the norm

$$\|u; \Omega\|_{L_p^1} = \|\nabla u; \Omega\|_{L_p} + \|u; \omega\|_{L_p},$$

where ω is a non-empty domain contained in Ω along with its closure. One can easily check that the change of ω leads to an equivalent norm.

Obviously, if Ω is a domain of the class $C^{0,1}$, then $L_p^1(\Omega) = W_p^1(\Omega)$. Comparing Theorems 1.3.2/2, 1.3.5/1 with Theorems 6.3.3/1, 6.3.3/2 and using Theorem 6.3.3/4, we obtain that for Lipschitz domains the space $ML_p^1(\Omega)$ coincides with the space of restrictions to Ω of multipliers in $W_p^1(R^n)$. The following example shows that this fails for arbitrary domains.

Example. Let Ω be the union of rectangles

$$A_m = \{x : 2^{1-m} - \delta_m < x_1 < 2^{1-m}, 2/3 < x_2 < 1\},$$

$$B_m = \{x : 2^{1-m} - \varepsilon_m < x_1 < 2^{1-m}, 1/3 \leq x_2 \leq 2/3\},$$

$$C = \{x : 0 < x_1 < 1, 0 < x_2 < 1/3\},$$

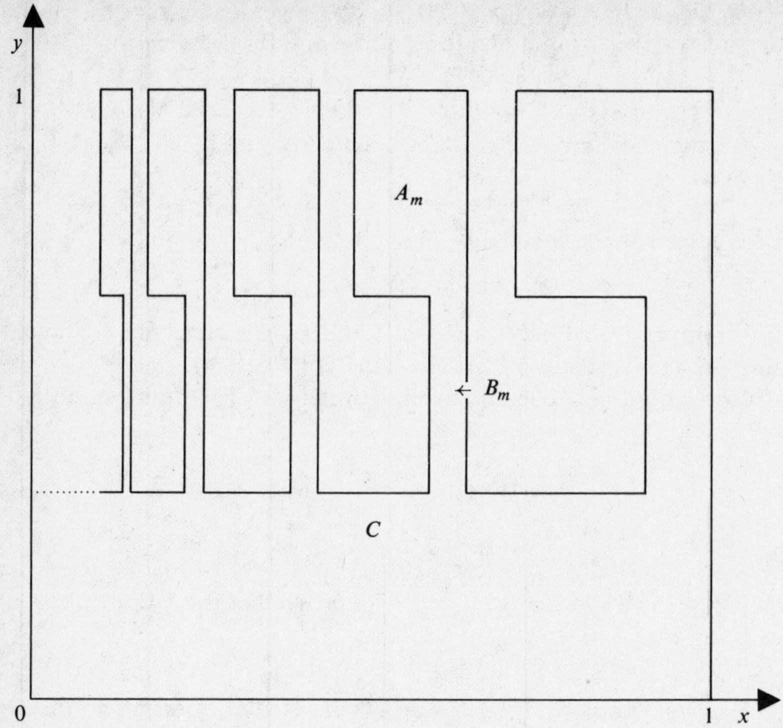

Figure 1

where $\delta_m = 2^{-m-1}$, $\varepsilon_m = 2^{-(m+1)\beta}$, $\beta \geqslant 1$, $m = 1, 2, \ldots$ (see Figure 1). This domain was proposed in 1933 by Nikodym [1] as an example of failure of the Poincaré inequality (see also Maz'ya [11]).

We shall show that the function $\gamma(x) = x_1^\lambda$ is a multiplier in $L_p^1(\Omega)$ if and only if $\lambda \geqslant (\beta + p - 1)/p$. It is clear that this function is a multiplier in $W_{p,\mathrm{loc}}^1(R^n)$ even for $\lambda > (p-1)/p$. So in the case $(p-1)/p < \lambda < (\beta + p - 1)/p$ the restriction to Ω of the multiplier x_1^λ in $W_{p,\mathrm{loc}}^1(R^n)$ is not a multiplier in $L_p^1(\Omega)$.

The necessity of the condition $\lambda \geqslant (\beta + p - 1)/p$ can be checked very easily. Namely, let u be a continuous function equal to $(2^{m\beta} m^{-2})^{1/p}$ in A_m, to zero in C and let u be linear in B_m. Clearly,

$$\|\nabla u; \Omega\|_{L_p}^p = \sum_{m=1}^{\infty} 2^{m\beta} m^{-2} \operatorname{mes}_2 B_m < \infty,$$

$$\|\nabla(\gamma u); \Omega\|_{L_p}^p \geqslant c \sum_{m=1}^{\infty} 2^{m\beta} m^{-2} 2^{-(\lambda-1)pm} \operatorname{mes}_2 A_m.$$

The last series diverges if $\lambda < (\beta + p - 1)/p$.

Now let $\lambda \geq (\beta + p - 1)/p$. By B_m^+ and B_m^- we denote that the rectangle B_m is respectively raised and lowered by one-third. Obviously,

$$\delta_m^{-1} \int_{A_m} |u|^p \, dx - \varepsilon_m^{-1} \int_{B_m^+} |u|^p \, dx \leq c \delta_m^{p-1} \int_{A_m} |\partial u / \partial x_2|^p \, dx.$$

Moreover,

$$\frac{1}{2} \int_{B_m \cup B_m^+} |u|^p \, dx - \int_{B_m^-} |u|^p \, dx \leq c \int_{B_m \cup B_m^+ \cup B_m^-} |\partial u / \partial x_1|^p \, dx.$$

Therefore

$$\int_{A_m \cup B_m} |u|^p \, dx \leq c \frac{\delta_m}{\varepsilon_m} \left(\int_{A_m \cup B_m \cup B_m^-} |\nabla u|^p \, dx + \int_{B_m^-} |u|^p \, dx \right). \tag{1}$$

Obviously,

$$\|u \nabla \gamma; \Omega\|_{L_p}^p \leq c \left(\int_C |u|^p \, dx + \sum_{m=1}^{\infty} \delta_m^{(\lambda-1)p} \int_{A_m \cup B_m} |u|^p \, dx \right).$$

Taking into account (1), we obtain that the sum on the right does not exceed

$$c \sum_{m=1}^{\infty} \delta_m^{(\lambda-1)p+1-\beta} \left(\int_{A_m \cup B_m \cup B_m^-} |\nabla u|^p \, dx + \int_{B_m^-} |u|^p \, dx \right)$$
$$\leq c \sup_m \delta_m^{(\lambda-1)p+1-\beta} \left(\int_\Omega |\nabla u|^p \, dx + \int_C |u|^p \, dx \right).$$

Consequently,

$$\|\nabla(\gamma u); \Omega\|_{L_p} \leq c (\|\nabla u; \Omega\|_{L_p} + \|u; C\|_{L_p}),$$

i.e. $\gamma \in ML_p^1(\Omega)$.

It can easily be shown that $x_1^\lambda \in MW_p^1(\Omega)$ if and only if $\lambda \geq 1$. Therefore for $(p-1)/p < \lambda < 1$ the restriction to Ω of $x_1^\lambda \in MW_{p,\mathrm{loc}}^1$ does not belong to $MW_p^1(\Omega)$.

We describe below the space of multipliers in $L_p^1(\Omega)$, where Ω is an arbitrary bounded domain. This description is obtained as a corollary of theorems due to Maz'ya [11] on necessary and sufficient conditions for validity of imbeddings of spaces of functions with first derivatives in $L_p(\Omega)$.

In what follows, by g and G we denote the so-called admissible subsets of Ω, i.e. bounded sets such that $\Omega \cap \partial g$ and $\Omega \cap \partial G$ are manifolds of the class C^∞. Further, let $\operatorname{clos}_\Omega g$ be the closure of g with respect to Ω. We also need the following function of the pair of sets g and G:

$$p\text{-cap}_\Omega(g, G) = \inf \{ \|\nabla u; \Omega\|_{L_p}^p : u \in C^\infty(\Omega),\ u = 0 \text{ on } G,\ u = 1 \text{ on } g \}.$$

Theorem 1 (Maz'ya [11]). *Let u be an arbitrary function in $C^\infty(\Omega)$, $u=0$ on G.*

(i) *For $p>1$, the inequality*

$$\|\gamma u;\Omega\|_{L_p}\leqslant C\,\|\nabla u;\Omega\|_{L_p} \tag{2}$$

holds if and only if $\int_g |\gamma|^p\,dx\leqslant \text{const}\; p\text{-cap}_\Omega(g,G)$, where g is any admissible set with $\text{clos}_\Omega\, g\subset\Omega\backslash\text{clos}_\Omega\, G$.

The best constant C in (2) for $p>1$ satisfies the inequality

$$\frac{(p-1)^{p-1}}{p^p}C^p\leqslant\sup_g\frac{\|\gamma;g\|^p_{L_p}}{p\text{-cap}_\Omega(g,G)}\leqslant C^p.$$

(ii) *For $p=1$, inequality (2) holds if and only if $\int_g|\gamma|\,dx\leqslant$ const $s(\Omega\cap\partial g)$ where g is any admissible set with $\text{clos}_\Omega\, g\subset\Omega\backslash\text{clos}_\Omega\, G$ and s is the $(n-1)$-dimensional area.*

The best constant C in (2) for $p=1$ is defined by the equality $C=\sup_g(\|\gamma;g\|_{L_1}/s(\Omega\cap\partial g))$.

From Theorem 1 there easily follows a description of the space $ML^1_p(\Omega)$.

Theorem 2. The function γ belongs to $ML^1_p(\Omega)$ if and only if $\gamma\in L_\infty(\Omega)\cap MW^1_{p,\text{loc}}(\Omega)$ and, for some admissible set G with $\bar G\subset\Omega$,

$$\sup\frac{\|\nabla\gamma;g\|^p_{L_p}}{p\text{-cap}_\Omega(g,G)}<\infty\quad\textit{for}\quad p>1,$$

$$\sup\frac{\|\nabla\gamma;g\|_{L_1}}{s(\Omega\cap\partial g)}<\infty\qquad\textit{for}\quad p=1,$$

where the supremums are over all admissible sets g with $\text{clos}_\Omega\, g\subset\Omega\backslash\bar G$.

Proof. The necessity of the condition $\gamma\in MW^1_{p,\text{loc}}(\Omega)$ is obvious and the necessity of the boundedness of γ follows from the inequality

$$\|\gamma^N u;\Omega\|^{1/N}_{L^1_p}\leqslant\|\gamma;\Omega\|_{ML^1_p}\|u;\Omega\|_{L^1_p},\qquad N=1,2,\ldots.$$

The other assertions follow from Theorem 1 and the estimate

$$\|u\nabla\Gamma-\nabla(\Gamma u);\Omega\|_{L_p}\leqslant\|\gamma;\Omega\|_{L_\infty}\|\nabla u;\Omega\|_{L_p},$$

where $\Gamma=\eta\gamma$, $\eta\in C^\infty(R^n)$, $\eta=0$ on G, $\eta=1$ in a neighbourhood of $\partial\Omega$, $0\leqslant\eta\leqslant 1$. □

Remark. By Theorem 6.3.3/1 for $p>1$ the condition $\gamma\in MW^1_{p,\text{loc}}(\Omega)$

can be replaced by

$$\sup \frac{\|\nabla\gamma; g\|_{L_p}^p}{\operatorname{cap}(g, W_p^1(R^n))} < \infty.$$

Here the supremum is over all admissible sets g placed at an arbitrary fixed positive distance from $\partial\Omega$.

According to Theorem 6.3.3/2, $\gamma \in MW_{1,\text{loc}}^1(\Omega)$ if and only if

$$\sup r^{1-n} \|\nabla\gamma; B_r(x)\|_{L_1} < \infty,$$

where the supremum is taken over all balls $B_r(x)$, $x \in \Omega$, placed at an arbitrary fixed positive distance from $\partial\Omega$.

6.4 Change of Variables in Norms of Sobolev Spaces

In this section we introduce and study certain classes of differentiable mappings which are considered as operators in pairs of Sobolev spaces. In 6.4.1, using the spaces of multipliers, we define the so-called (p, l)-diffeomorphisms. We show that these mappings preserve the space W_p^l. With the help of (p, l)-diffeomorphisms we introduce in 6.4.2 the class of (p, l)-manifolds on which the latter space is defined correctly. Finally, Subsection 6.4.3 concerns mappings of the class $T_p^{m,l}$, i.e. the mappings $U \to V$ which generate a continuous operator $W_p^m(V) \to W_p^l(U)$, where $p \geq 1$, m and l are integers, U and V are open sets in R^n.

6.4.1 (p, l)-*Diffeomorphisms*

Let U be an open set in R^n. In this subsection we consider the space $W_p^l(U)$ not only for $l = 0, 1, \ldots$ but also for fractional $l > 0$. In the latter case,

$$\|u; U\|_{W_p^l} = \|u; U\|_{W_p^{[l]}} + \sum_{j=0}^{[l]} \left(\int_U \int_U |\nabla_j u(x) - \nabla_j u(y)|^p \, |x-y|^{-n-p\{l\}} \, dx \, dy \right)^{1/p}.$$

Together with U we consider an open set $V \subset R^n$ and introduce a Lipschitz mapping $\varkappa = (\varkappa_1, \ldots, \varkappa_n) : U \to V$ such that the determinant $\det \varkappa'$ retains its sign and is separated from zero. If elements of the Jacobi matrix $\varkappa'$ belong to the space of multipliers $MW_p^{l-1}(U)$, $p \geq 1$, $l \geq 1$, then, by definition, the mapping $\varkappa$ is a (p, l)-diffeomorphism.

In what follows, $\|\varkappa'; U\|_{MW_p^{l-1}}$ is the sum of norms of elements of the matrix $\varkappa'$ in the space $MW_p^{l-1}(U)$.

The present subsection concerns certain properties of (p, l)-diffeomorphisms.

Lemma 1. *Let $u \in W_p^l(V)$, $l \geq 1$, and let $\varkappa : U \to V$ be a (p, l)-diffeomorphism. Then $u \circ \varkappa \in W_p^l(U)$ and*

$$\|u \circ \varkappa; U\|_{W_p^l} \leq c\, \|u; V\|_{W_p^l}. \tag{1}$$

Proof. In order to be precise, let $\det \varkappa' > 0$. We put $\lambda = \inf \det \varkappa'$. Obviously,

$$\|u \circ \varkappa; U\|_{W_p^1} \leq \|(\varkappa')^*(\nabla u) \circ \varkappa; U\|_{L_p} + \|u \circ \varkappa; U\|_{L_p}$$
$$\leq \lambda^{-1/p} (\|\varkappa'; U\|_{L_\infty} \|\nabla u; V\|_{L_p} + \|u; V\|_{L_p}).$$

Moreover, if $\{l\} > 0$, then

$$\|u \circ \varkappa; U\|_{W_p^{\{l\}}} = \left(\int_U \int_U |u(\varkappa(x)) - u(\varkappa(y))|^p\, |x - y|^{-n - p\{l\}}\, dx\, dy \right)^{1/p}$$
$$+ \|u \circ \varkappa; U\|_{L_p} \leq (\lambda^{-2/p}\, \|\varkappa'; U\|_{L_\infty}^{n/p + \{l\}} + \lambda^{-1/p})\, \|u; U\|_{W_p^{\{l\}}}.$$

Suppose that inequality (1) is valid for $[l] = 1, \ldots, k-1$. Then, for $[l] = k$,

$$\|u \circ \varkappa; U\|_{W_p^l} = \|\nabla(u \circ \varkappa); U\|_{W_p^{l-1}} + \|u \circ \varkappa; U\|_{L_p}$$
$$= \|(\varkappa')^*(\nabla u) \circ \varkappa; U\|_{W_p^{l-1}} + \|u \circ \varkappa; U\|_{L_p}$$
$$\leq \|\varkappa'; U\|_{MW_p^{l-1}}\, \|(\nabla u) \circ \varkappa; U\|_{W_p^{l-1}} + \|u \circ \varkappa; U\|_{L_p}.$$

and it remains to use the induction hypothesis. □

Lemma 2. *If $\varkappa$ is a (p, l)-diffeomorphism, then $\varkappa^{-1}$ is also a (p, l)-diffeomorphism.*

Proof. For $l = 1$ the statement is contained in the definition of a (p, l)-diffeomorphism.

Let $1 < l < 2$ and let u be an arbitrary scalar function in $W_p^{l-1}(V)$. We have

$$\|u(\varkappa^{-1})'; V\|_{W_p^{l-1}} = \|[(u \circ \varkappa)(\varkappa')^{-1}] \circ \varkappa^{-1}; V\|_{W_p^{l-1}}. \tag{2}$$

Since $\varkappa$ is a bi-Lipschitz mapping, then the last norm does not exceed $c\, \|(u \circ \varkappa)(\varkappa')^{-1}; U\|_{W_p^{l-1}}$. Consequently,

$$\|u(\varkappa^{-1})'; V\|_{W_p^{l-1}} \leq c\, \|(\varkappa')^{-1}; U\|_{MW_p^{l-1}}\, \|u \circ \varkappa; U\|_{W_p^{l-1}}.$$

It remains to use Lemma 1 and the condition $\varkappa' \in MW_p^{l-1}(U)$.

Suppose that the lemma is proved for $[l] = 1, \ldots, k-1$. Let $[l] = k$. The condition $\varkappa' \in MW_p^{l-1}(U)$ implies $\varkappa' \in MW_p^{l-2}(U)$ and hence, by induction hypothesis, $(\varkappa^{-1})' \in MW_p^{l-2}(V)$. This and Lemma 1 yield $(u \circ \varkappa)(\varkappa')^{-1} \circ \varkappa^{-1} \in W_p^{l-1}(V)$, if $(u \circ \varkappa)(\varkappa')^{-1} \in W_p^{l-1}(U)$. Here, as before, u is a scalar function in $W_p^{l-1}(V)$. The latter inclusion holds, since $(\varkappa')^{-1} \in MW_p^{l-1}(U)$ and, by Lemma 1, the function $u \circ \varkappa$ belongs to the space $W_p^{l-1}(U)$. □

Lemma 3. *Let $\gamma \in MW_p^l(V)$, $l \geqslant 1$, and let $\varkappa$ be a (p, l)-diffeomorphism. Then $\gamma \circ \varkappa \in MW_p^l(U)$ and*

$$\|\gamma \circ \varkappa; U\|_{MW_p^l} \leqslant c\,\|\gamma; V\|_{MW_p^l}. \tag{3}$$

Proof. By virtue of Lemma 1, for all $u \in W_p^{l-1}(U)$,

$$\begin{aligned}\|(\gamma \circ \varkappa)u; U\|_{W_p^l} &= \|[(u \circ \varkappa^{-1})\gamma] \circ \varkappa; U\|_{W_p^l}\\ &\leqslant c\,\|\gamma; V\|_{MW_p^l}\|u \circ \varkappa^{-1}; V\|_{W_p^l}.\end{aligned} \tag{4}$$

Since, according to Lemma 2, $\varkappa^{-1}$ is a (p, l)-diffeomorphism, Lemma 1 implies the estimate $\|u \circ \varkappa^{-1}; V\|_{W_p^l} \leqslant c\,\|u; U\|_{W_p^l}$, which, being inserted into (4), completes the proof. □

Lemma 4. *Let U, V and W be open sets in R^n and let $\varkappa_1: U \to V$, $\varkappa_2: V \to W$ be (p, l)-diffeomorphisms. Then their composition $\varkappa_2 \circ \varkappa_1: U \to W$ is also a (p, l)-diffeomorphism.*

Proof. The matrix $(\varkappa_2 \circ \varkappa_1)'$ is equal to the product of matrices $(\varkappa_2' \circ \varkappa_1)$ and $\varkappa_1'$. Therefore

$$\|(\varkappa_2 \circ \varkappa_1)'; U\|_{MW_p^{l-1}} \leqslant \|\varkappa_2' \circ \varkappa_1; U\|_{MW_p^{l-1}}\|\varkappa_1'; U\|_{MW_p^{l-1}}.$$

We estimate the first factor using (3), with l changed to $l-1$, γ changed to $\varkappa_2'$, and $\varkappa$ to $\varkappa_1$. □

Let

$$P(z, D_z)u = \sum_{|\alpha| \leqslant k} p_\alpha(z) D_z^\alpha u \tag{5}$$

be a differential operator on U and let $\varkappa$ be a (p, l)-diffeomorphism $U \to V$, $l \geqslant k$. Further, let Q be a differential operator on V, defined by the equality $Q(u \circ \varkappa^{-1}) = (Pu) \circ \varkappa^{-1}$. By virtue of Lemmas 1 and 2, the operator Q maps $W_p^l(V)$ continuously into $W_p^{l-k}(V)$ if and only if the operator P maps $W_p^l(U)$ continuously into $W_p^{l-k}(U)$.

By $O_{p,\mathrm{loc}}^{l,k}(U)$ we denote the class of operators of form (5) such that $p_\alpha \in M(W_{p,\mathrm{loc}}^{l-|\alpha|}(U) \to W_{p,\mathrm{loc}}^{l-k}(U))$ for any multi-index α with $|\alpha| \leqslant k$.

Lemma 5. *The operator P belongs to the class $O_{p,\mathrm{loc}}^{l,k}(U)$ if and only if $Q \in O_{p,\mathrm{loc}}^{l,k}(V)$.*

Proof. Let $\zeta = \varkappa(z)$. We have

$$D^\alpha[v(\varkappa(z))] = \sum_{1 \leqslant |\beta| \leqslant |\alpha|} (D^\beta v)(\varkappa(z)) \sum_s c_s \prod_{i=1}^n \prod_j D^{s_{ij}} \varkappa_i(z), \tag{6}$$

where the sum is taken over all multi-indices $s=(s_{ij})$ such that

$$\sum_{i,j} s_{ij}=\alpha, \qquad |s_{ij}|\geq 1, \sum_{i,j}(|s_{ij}|-1)=|\alpha|-|\beta|. \tag{7}$$

Let

$$Q(\zeta, D_\zeta)=\sum_{|\beta|\leq k} q_\beta(\zeta)D_\zeta^\beta.$$

By virtue of (6),

$$q_\beta=\sum_{|\beta|\leq|\alpha|\leq k}(p_\alpha\circ\varkappa^{-1})\sum_s c_s\prod_{i=1}^n\prod_j(D^{s_{ij}}\varkappa_i)\circ\varkappa^{-1}. \tag{8}$$

Since $\nabla\varkappa_i\in MW_{p,\mathrm{loc}}^{l-1}(U)\subset MW_{p,\mathrm{loc}}^{l-r}(U)$, $1\leq r\leq l$,

$$D^{s_{ij}}\varkappa_i\in M(W_{p,\mathrm{loc}}^{l-r}(U)\to W_{p,\mathrm{loc}}^{l-r-|s_{ij}|+1}(U)).$$

Therefore

$$\prod_{i,j}D^{s_{ij}}\varkappa_i\in M(W_{p,\mathrm{loc}}^{l-|\beta|}(U)\to W_{p,\mathrm{loc}}^{l-|\beta|-\Sigma_{i,j}(|s_{ij}|-1)}(U))$$

or, what is the same,

$$\prod_{i,j}D^{s_{ij}}\varkappa_i\in M(W_{p,\mathrm{loc}}^{l-|\beta|}(U)\to W_{p,\mathrm{loc}}^{l-|\alpha|}(U)).$$

It remains to use the condition $p_\alpha\in M(W_{p,\mathrm{loc}}^{l-|\alpha|}(U)\to W_{p,\mathrm{loc}}^{l-k}(U))$ together with Lemmas 1 and 2. □

By Lemma 3.7/1, the condition $P\in O_{p,\mathrm{loc}}^{l,k}(U)$ is sufficient for the operator P to map $W_{p,\mathrm{loc}}^{l}(U)$ into $W_{p,\mathrm{loc}}^{l-k}(U)$. The inclusion $P\in O_{p,\mathrm{loc}}^{l,k}(U)$ is also necessary for $p=1$ or $p(l-k)>n$ (see Lemma 3.7/1).

6.4.2. (*p*, *l*)-*Manifolds*

In terms of (p, l)-diffeomorphisms we can define in a standard manner (see, for instance, de Rham [1], Hörmander [1]) a class of non-smooth n-dimensional manifolds on which Sobolev spaces can be naturally considered.

We recall that a topological space $\mathfrak{M}$ is called an n-dimensional manifold if there exists a collection of homeomorphisms $\{\varphi\}$ of open sets $U_\varphi\in\mathfrak{M}$ onto open subsets of R^n with $\mathfrak{M}=\cup U_\varphi$.

The pair (φ, U_φ) is called a map (or the coordinate system) and the set of maps is called an atlas.

We say that two maps (φ, U_φ) and (ψ, U_ψ) have (p, l)-overlapping, if the mapping $\varphi\psi^{-1}:\varphi(U_\varphi\cap U_\psi)\to\psi(U_\varphi\cap U_\psi)$ is a (p, l)-diffeomorphism. By Lemma 6.4.1/2, the same is true for the inverse mapping.

If any two maps have (p, l)-overlapping then we have a (p, l)-atlas. The maximal (p, l)-atlas on $\mathfrak{M}$ is called a (p, l)-structure. By a (p, l)-manifold we mean a manifold with a (p, l)-structure.

Since $MW_p^{l-1}(R^n) = W_{p,\text{unif}}^{l-1}(R^n)$ for $p(l-1) > n$, then for these values of p and l the structure on a (p, l)-manifold belongs to the class C^1, whereas for $p(l-1) \leq n$ such manifolds are Lipschitzian.

For functions defined on a (p, l)-manifold Ω we introduce the space $W_{p,\text{loc}}^l(\Omega)$: namely, $u \in W_{p,\text{loc}}^l(\Omega)$ if the function $u \circ \varphi^{-1}$ belongs to the space $W_{p,\text{loc}}^l(\varphi(U_\varphi))$ for each map (φ, U_φ).

With the help of Lemmas 6.4.1/1 and 6.4.1/2 we can prove, in a standard way (see Hörmander [1], Theorem 2.6.2), that to define the space $W_{p,\text{loc}}^l(\Omega)$ it suffices to use only one arbitrary (p, l)-atlas; i.e., the following is valid:

Theorem. *If a function u defined on a (p, l)-manifold Ω is such that $u \circ \varphi^{-1} \in W_{p,\text{loc}}^l(\varphi(U_\varphi))$ for any map of some atlas, then $u \in W_{p,\text{loc}}^l(\Omega)$. If $\eta_\varphi \in C_0^\infty(\varphi(U_\varphi))$ and the open sets*

$$V_\varphi = \{x \in U_\varphi : \eta_\varphi(\varphi(x)) \neq 0\}$$

cover Ω, then to define the topology in the space $W_{p,\text{loc}}^l(\Omega)$ it suffices to introduce semi-norms $u \to \|\eta_\varphi(u \circ \varphi^{-1})\|_{W_p^l}$.

If a manifold Ω is compact, then the topology in it can be induced by the norm $\sum_\varphi \|\eta_\varphi(u \circ \varphi^{-1})\|_{W_p^l}$, where the sum is taken over all maps of a certain atlas.

Let Ω be a (p, l)-manifold. We define a differential operator of order k, $k \leq l$, as a linear mapping P of the space $W_{p,\text{loc}}^l(\Omega)$ into the space $W_{p,\text{loc}}^{l-k}(\Omega)$, provided that for any map (φ, U_φ) there exists a differential operator P_φ in the class $O_{p,\text{loc}}^{l,k}(\varphi(U_\varphi))$ such that $(Pu) \circ \varphi^{-1} = P_\varphi(u \circ \varphi^{-1})$ on $\varphi(U_\varphi)$.

By virtue of Lemma 6.4.1/5, it suffices to restrict oneself to maps of certain atlas.

Replacing the space R^n by the closed half-space $\overline{R_+^n} = \{\zeta \in R^n : \zeta_n \geq 0\}$ in the definition of a (p, l)-manifold $\mathfrak{M}$, we obtain the definition of a (p, l)-manifold $\mathfrak{M}$ with the boundary $\partial\mathfrak{M}$.

Let l be an integer, $l \geq 2$, and let $\mathfrak{M}$ be a (p, l)-manifold. If $p(l-1) \leq n$ we additionally assume that the (p, l)-structure on $\mathfrak{M}$ belongs to the class C^1. Then the implicit function theorem which will be proved in 6.5.2 implies that the (p, l)-structure on $\mathfrak{M}$ induces the $(p, l-1/p)$-structure on $\partial\mathfrak{M}$.

In 6.5.1 we show that the closure of the special Lipschitz domain $G = \{z = (x, y) : x \in R^{n-1}, y > \varphi(x)\}$ belonging to $M_p^{l-1/p}$ is a (p, l)-manifold with a boundary.

The classes of manifolds we have introduced are convenient for an extension of L_p-theory of elliptic boundary value problems. Without being concerned with full generality, we shall show in Chapter 7 that this is true for the boundary value problems in a bounded subdomain of R^n which belongs to $M_p^{l-1/p}$.

6.4.3 Differentiable Mappings of One Sobolev Space into Another

Let U and V be domains in R^n. We say that the mapping $\varkappa : U \to V$ belongs to the class $T_p^{m,l}$ if $u \circ \varkappa \in W_p^l(U)$ for any $u \in W_p^m(V)$ and

$$\|u \circ \varkappa; U\|_{W_p^l} \leqslant c\, \|u; V\|_{W_p^m}. \tag{1}$$

We limit consideration to integers m and l, $m \geqslant l \geqslant 1$. For $m = l$ we shall write T_p^l instead of $T_p^{l,l}$.

In this subsection we give sufficient and, for some values of p, m, l, necessary and sufficient conditions for a mapping to belong to the class $T_p^{m,l}$. In particular, for $m = l$ we obtain a wider set of mappings than the class of (p, l)-diffeomorphisms.

In what follows, $\varkappa = (\varkappa_1, \ldots, \varkappa_n)$ is a one-to-one mapping with $\varkappa' \in W_1^{l-1}(U)$ such that $\det \varkappa'$ does not change its sign and

$$\int_U u(\varkappa(z))\, |\det \varkappa'(z)|\, dz = \int_V u(\zeta)\, d\zeta \tag{2}$$

for any $u \in L_1(V)$. (For sufficient conditions under which (2) is valid see, for instance, Vodopianov, Goldstein, Rešetnjak [1].)

Since $\varkappa$ is a mapping of the class $W_1^l(U)$, then for any $u \in C^l(V)$ and any multi-index α with $|\alpha| \leqslant l$,

$$D^\alpha[u(\varkappa(z))] = \sum_{1 \leqslant |\beta| \leqslant |\alpha|} \varphi_\beta^\alpha(z)(D^\beta u)(\varkappa(z)) \tag{3}$$

a.e. in U. Here and henceforth,

$$\varphi_\beta^\alpha = \sum_s c_s \prod_{i=1}^n \prod_j D^{s_{ij}} \varkappa_i,$$

where the sum is taken over all collections of multi-indices $s = (s_{ij})$ governed by conditions (6.4.1/7). We note that another and more explicit expression for functions φ_β^α is presented by Fraenkel [1].

Proposition 1. *If* $1/|\det(\varkappa' \circ \varkappa^{-1})|^{1/p} \in M(W_p^m(V) \to L_p(V))$ *and* $(\varphi_\beta^\alpha/|\det \varkappa'|^{1/p}) \circ \varkappa^{-1} \in M(W_p^{m-|\beta|}(V) \to L_p(V))$ *for all multi-indices* α *and* β *with* $l \geqslant |\alpha| \geqslant |\beta| \geqslant 1$, *then the mapping* $\varkappa$ *belongs to the class* $T_p^{m,l}$.

The proof of inequality (1) for any $u \in C^l(V) \cap W_p^m(V)$ follows directly

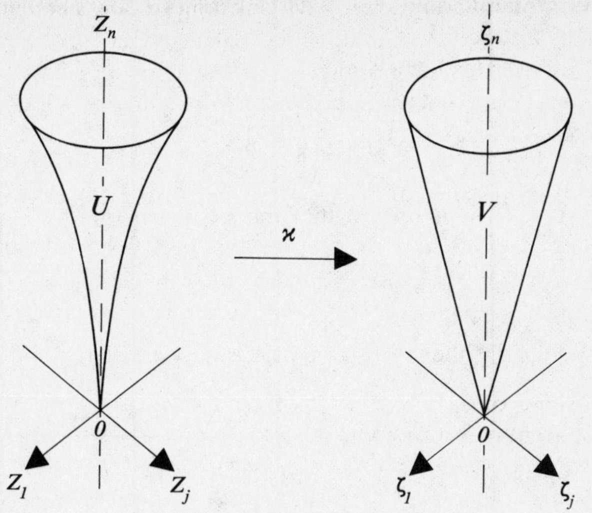

Figure 2

from (2) and (3). The additional assumption $u \in C^l(V)$ can be removed by approximation.

We give an example which shows that the conditions of Proposition 1 are exact.

Example. We consider the domain $U=\{z: z_1^2+\cdots+z_{n-1}^2<z_n^{2\gamma}, 0<z_n<1\}$, $\gamma>0$, and the mapping $\varkappa: z \to \zeta$ with $\zeta_i = z_i$, $1 \leqslant i \leqslant n-1$, and $\zeta_n = z_n^\gamma$. It is clear that $\varkappa$ transforms U into the cone $V=\{\zeta: \zeta_1^2+\cdots+\zeta_{n-1}^2<\zeta_n^2, 0<\zeta_n<1\}$. (See Figure 2.) We show that $\varkappa \in T_p^{m,l}$ if and only if either

$$p(m-1)<n,\ \gamma \geqslant \frac{pl-1}{pm-1} \quad \text{or} \quad p(m-1)\geqslant n,\ \gamma>\frac{pl-1}{p+n-1}. \tag{4}$$

Let $u(\zeta)=\zeta_n^\sigma$, where $\sigma=1$ in the case $p(m-1)\geqslant n$ and σ is a fractional number in $(m-n/p, (pl-1+\gamma-\gamma n)/\gamma p]$ for $p(m-1)<n$. Clearly, $u \in W_p^m(V)$. On the other hand, $u(\varkappa(z))=z_n^{\gamma\sigma}$ and therefore

$$\|\nabla_l(u \circ \varkappa^{-1}); U\|_{L_p}^p \geqslant c \int_0^1 z_n^{p(\gamma\sigma-l)} \int_0^{z_n^\gamma} r^{n-2}\, dr\, dz_n = \infty.$$

Thus, conditions (4) are necessary.

Now we turn to the proof of sufficiency. By straightforward computation we have $\varphi_\beta^\alpha(z)=\text{const}\, z_n^{|\beta|\gamma-|\alpha|}$ and $\det \varkappa'(z)=\gamma z_n^{\gamma-1}$. Consequently, $(\varphi_\beta^\alpha \circ \varkappa^{-1})(\zeta)=\text{const}\, \zeta_n^{|\beta|-|\alpha|/\gamma}$ and $\det(\varkappa' \circ \varkappa^{-1})(\zeta)=\gamma z_n^{1-1/\gamma}$. To check the conditions of Proposition 1 it suffices to verify the inequalities

$$\int_V |\zeta|^{(1-\gamma)/\gamma} |w|^p \, d\zeta \leqslant c \, \|w; V\|^p_{W^m_p},$$

$$\int_V |\zeta|^{p(|\beta|-|\alpha|/\gamma)+(1-\gamma)/\gamma} |w|^p \, d\zeta \leqslant c \, \|w; V\|^p_{W^{m-|\beta|}_p}$$

for any $w \in W^m_p(V)$. It is known that the Hardy inequality

$$\int_V \frac{|w|^p}{|\zeta|^{pk}} \, d\zeta \leqslant c \, \|w; V\|^p_{W^s_p}, \qquad pk < n,$$

holds if and only if either $ps \geqslant n$ or $ps < n$ and $k \leqslant s$. It remains to note that

$$(1-\gamma)/\gamma > -n, \qquad (1-\gamma)/\gamma p \geqslant -m$$

and that, by virtue of (4),

$$p(|\beta| - |\alpha|/\gamma) + (1-\gamma)/\gamma \geqslant p(1 - l/\gamma) + (1-\gamma)/\gamma > -n,$$

$$|\beta| - |\alpha|/\gamma + (1-\gamma)/\gamma p \geqslant |\beta| - m \quad \text{for} \quad p(m-1) < n. \quad \square$$

Remark. For $|\beta| = |\alpha| = l = m$, one of the conditions of Proposition 1 takes the form

$$\varphi^\alpha_\beta / |\det \varkappa'|^{1/p} \in L_\infty(U).$$

We have

$$\frac{\partial^l}{\partial z_{i_1} \cdots \partial z_{i_l}} u(\varkappa(z)) = \sum_{k_1, \ldots, k_l} \left(\frac{\partial^l u}{\partial \zeta_{k_1} \cdots \partial \zeta_{k_l}} \right)(\varkappa(z)) \prod_{\nu=1}^{l} \partial \varkappa_{k_\nu} / \partial z_{i_\nu} + \ldots, \tag{5}$$

where the terms with differentiation with respect to ζ of order less than l are omitted. So the condition mentioned above is equivalent to

$$\frac{\partial \varkappa_{k_1}}{\partial z_{i_1}} \cdots \frac{\partial \varkappa_{k_l}}{\partial z_{i_l}} \bigg/ |\det \varkappa'|^{1/p} \in L_\infty(U). \tag{6}$$

Here $k_1, \ldots, k_l, i_1, \ldots, i_l$ are numbers with values $1, \ldots, n$. The inclusion (6) can be rewritten in the form

$$|\varkappa'|^{pl} / |\det \varkappa'| \leqslant C, \tag{7}$$

where $C = \text{const}$ and $|\varkappa'| = (\sum^n_{i,j=1} (\partial \varkappa_i / \partial z_i)^2)^{1/2}$.

According to the Hadamard inequality,

$$|\det \varkappa'| \leqslant \prod_{i=1}^n \left(\sum_{j=1}^n \left(\frac{\partial \varkappa_i}{\partial z_j} \right)^2 \right)^{1/2}.$$

So $|\det \varkappa'| \leq n^{-n/2} |\varkappa'|^n$. Consequently, for $pl > n$, $|\varkappa'| \leq (Cn^{-n/2})^{1/(pl-n)}$, i.e. $\varkappa$ is a Lipschitz mapping. In the case $pl < n$ we obtain

$$|\det \varkappa'| \geq (n^{pl/2} C^{-1})^{n/(n-pl)}. \tag{8}$$

Suppose that the mapping $\varkappa$ satisfies (7) together with $\varkappa^{-1}$. Then $\varkappa$ is Lipschitz for $pl < n$ also. Indeed, by changing $\varkappa$ to $\varkappa^{-1}$ in (8), we get

$$|\det \varkappa'| \leq (n^{pl/2} C^{-1})^{n/(pl-n)}.$$

This estimate and (7) yield

$$|\varkappa'| \leq n^{n/2(pl-n)} C^{(2n-pl)/(n-pl)pl}.$$

Thus, for $pl \neq n$, the mapping $\varkappa$ which belongs to T_p^l together with $\varkappa^{-1}$ is bi-Lipschitz.

The class of mappings which perform the isomorphism $W_p^1(U) \approx W_p^1(V)$ is studied by Lewis [1], Gehring [1], Vodopianov, Goldstein [2], where it is shown, in particular, that for $p \geq n$ such mappings are bi-Lipschitz.

For $pl = n$, condition (7) means that $\varkappa$ is a mapping with bounded distortion (see Rešetnjak [1]). Mappings subjected to (7) with $p = l = 1$ are called subareal since they either decrease the area of $(n-1)$-dimensional surfaces or increase it with a finite coefficient (see Maz'ya [3], [11]).

Proposition 2. *Inequality (7) is necessary for $\varkappa \in T_p^l$. It is equivalent to $\varkappa \in T_p^1$. (Hence, by interpolation, $T_p^l \subset T_p^k$, $k = 1, \ldots, l-1$.)*

Proof. We put

$$u(\zeta) = \eta(\zeta) |\lambda|^{-l} \exp(i(\lambda, \zeta)),$$

where $\eta \in C_0^\infty(V)$, $\lambda \in \mathbb{C}^n$, into (1). Applying the Cauchy formula

$$D_\lambda^\gamma P(0; z) = \gamma! (2\pi i)^{-n} \int_{|\lambda_1|=1} \cdots \int_{|\lambda_n|=1} P(\lambda; z) \lambda^{-\gamma} \frac{d\lambda_1}{\lambda_1} \cdots \frac{d\lambda_n}{\lambda_n}$$

to the polynomial

$$\lambda \to P(\lambda; z) = |\lambda|^l D^\alpha[u(\varkappa(z))],$$

we find that its coefficients belong to $L_p(U)$. Therefore we may pass to the limit as $|\lambda| \to \infty$ in (1). As a result, for any unit vector $\theta = (\theta_1, \ldots, \theta_n)$ we obtain

$$\left\| \eta \, |\det(\varkappa' \circ \varkappa^{-1})|^{-1/p} \sum_{|\gamma|=l} \theta^\gamma D^\gamma P(0; \cdot) \circ \varkappa^{-1}; V \right\|_{L_p} \leq c \, \|\eta; V\|_{L_p}$$

which, by virtue of (5), can be written as follows:

$$\left\| \eta \, |\det(\varkappa' \circ \varkappa^{-1})|^{-1/p} \sum_{k_1,\dots,k_l} \theta_{k_1} \cdots \theta_{k_l} \left(\frac{\partial \varkappa_{k_1}}{\partial z_{i_1}} \cdots \frac{\partial \varkappa_{k_l}}{\partial z_{i_l}} \right) \circ \varkappa^{-1}; V \right\|_{L_p}$$

$$\leqslant c \, \|\eta; V\|_{L_p}.$$

Since η is arbitrary, we obtain that the functions

$$|\det(\varkappa' \circ \varkappa^{-1})|^{-1/p} \frac{\partial(\theta, \varkappa)}{\partial z_{i_1}} \cdots \frac{\partial(\theta, \varkappa)}{\partial z_{i_l}}$$

are bounded; then condition (7) holds. □

The next assertion on conditions for a mapping to belong to $T_p^{m,l}$ follows directly from Proposition 1 and Theorem 6.3.3/1.

Proposition 3. *Let V be a bounded domain of the class $C^{0,1}$ and let $p \in (1, \infty)$. If, for any compact set $e \subset \bar{V}$,*

$$\mathrm{mes}_n \, \varkappa^{-1}(e) \leqslant c \, \mathrm{cap}(e, W_p^m)$$

and for all multi-indices α, β with $l \geqslant |\alpha| \geqslant |\beta| \geqslant 1$,

$$\sup_{e \subset \bar{V}} \frac{\|\varphi_\beta^\alpha; \varkappa^{-1}(e)\|_{L_p}}{[\mathrm{cap}(e, W_p^{m-|\beta|})]^{1/p}} < \infty,$$

then $\varkappa \in T_p^{m,l}$.

Now we present two propositions on necessary and sufficient conditions for a mapping to belong to the class $T_p^{m,l}$.

Proposition 4. *Let V be a bounded domain of the class $C^{0,1}$. The mapping $\varkappa$ belongs to $T_1^{m,l}$ if and only if for any ball $B_r(\zeta)$ with $\zeta \in \bar{V}$, $r \in (0, 1)$,*

$$\mathrm{mes}_n \, \varkappa^{-1}(B_r(\zeta) \cap V) \leqslant c r^{n-m}$$

and for all multi-indices α, β with $l \geqslant |\alpha| \geqslant |\beta| \geqslant 1$,

$$\sup_{\zeta \in \bar{V}, r \in (0,1)} r^{m-|\beta|-n} \|\varphi_\beta^\alpha; \varkappa^{-1}(B_r(\zeta) \cap V)\|_{L_1} < \infty.$$

Proof. The sufficiency is an immediate corollary of Proposition 1 and Theorem 6.3.3/2.

On the other hand, the inclusion $\varkappa \in T_1^{m,l}$ is equivalent to

$$\|u/\det(\varkappa' \circ \varkappa^{-1}); V\|_{L_1} + \sum_{1 \leqslant |\alpha| \leqslant l} \left\| \sum_{1 \leqslant |\beta| \leqslant |\alpha|} \left(\frac{\varphi_\beta^\alpha}{\det \varkappa'} \right) \circ \varkappa^{-1} D^\beta u; V \right\|_{L_1}$$

$$\leqslant c \, \|u; V\|_{W_1^m}.$$

It remains to note that the proof of Theorem 3.7/1 relating R^n suits a Lipschitz domain as well. □

The following two assertions can be proved with the same arguments.

Proposition 5. *Let V be a bounded domain of the class $C^{0,1}$ and let $(m-l)p>n$, $p\in(1,\infty)$. The mapping $\varkappa$ is an element of the class $T_p^{m,l}$ if and only if $\text{mes}_n U<\infty$ and $\varphi_\beta^\alpha\in L_p(U)$ for all multi-indices α, β with $l\geqslant|\alpha|\geqslant|\beta|\geqslant 1$.*

Proposition 6. *Let V be a bounded domain of the class $C^{0,1}$ and let $p>n$. The mapping $\varkappa$ belongs to T_p^l if and only if inequality (7) is valid and $\varphi_\beta^\alpha\in L_p(U)$ for $l\geqslant|\alpha|\geqslant|\beta|\geqslant 1$.*

For instance, for $p>n$ the inclusion $\varkappa\in T_p^2$ is equivalent to the conditions

$$|\varkappa'|^{2p}/|\det \varkappa'|\in L_\infty(U) \quad \text{and} \quad \nabla\varkappa'\in L_p(U).$$

Similarly, $\varkappa\in T_p^3$ for $p>n$ if and only if

$$|\varkappa'|^{3p}/|\det \varkappa'|\in L_\infty(U), \qquad \nabla_2\varkappa'\in L_p(U)$$

and

$$\sum_{1\leqslant\rho,\sigma,\tau\leqslant n}\frac{\partial\varkappa_r}{\partial z_\rho}\frac{\partial^2\varkappa_s}{\partial z_\sigma\,\partial z_\tau}\in L_p(U), \qquad r,s=1,\ldots,n.$$

If both U and V are bounded and belong to $C^{0,1}$, then the conditions of Proposition 5 can be simplified. Namely, the following assertion holds.

Proposition 7. *Let U and V be bounded domains in $C^{0,1}$ and $p>1$, $(m-l)p>n$. The mapping $\varkappa$ belongs to $T_p^{m,l}$ if and only if $\varkappa\in W_p^l(U)$.*

Proof. Since

$$\{\varphi_\beta^\alpha\}_{|\alpha|=l}^{|\beta|=1}=\{D^\alpha\varkappa_i\}_{|\alpha|=l}^{1\leqslant i\leqslant n},$$

$$\{\varphi_\beta^\alpha\}_{|\alpha|=l}^{|\beta|=l}=\left\{\prod_{\nu=1}^{l}\partial\varkappa_{k_\nu}/\partial z_{i_\nu}\right\}_{1\leqslant k_\nu\leqslant n}^{1\leqslant i_\nu\leqslant n},$$

then the necessity follows from Proposition 5.

By definition of φ_β^α we obtain

$$\|\varphi_\beta^\alpha; U\|_{L_p}\leqslant c\sum_s\prod_{i=1}^n\prod_j\|D^{s_{ij}}\varkappa_i; U\|_{L_{p|\alpha|/|s_{ij}|}}$$

which, together with the Gagliardo [1]–Nirenberg [1] inequality

$$\|D^{s_{ij}}\varkappa_i; U\|_{L_{p|\alpha|/|s_{ij}|}} \leqslant c\,\|\varkappa_i; U\|_{W_p^{|\alpha|}}^{|s_{ij}|/|\alpha|}\,\|\varkappa_i; U\|_{L_\infty}^{1-|s_{ij}|/|\alpha|},$$

completes the proof. □

6.5 Implicit Function Theorems

6.5.1 Special Lipschitz Domains of the Class $M_p^{l-1/p}$

Let $p \geqslant 1$ and l be an integer, $l > 1$. We say that the special Lipschitz domain $G = \{(x, y): x \in R^{n-1},\ y > \varphi(x)\}$ belongs to the class $M_p^{l-1/p}$ if $\nabla\varphi \in MW_p^{l-1-1/p}(R^{n-1})$. By definition, $M_p^{1-1/p} = C^{0,1}$.

An arbitrary function $\nu \in C_0^\infty(R^{n-1})$ generates the operator $\gamma \to T\gamma$ defined by

$$(T\gamma)(\zeta) = \int_{R^{n-1}} \nu(\tau)\gamma(\tau\eta + \xi)\,d\tau, \quad \zeta = (\xi, \eta),\ \xi \in R^{n-1},\ \eta > 0. \tag{1}$$

According to Theorem 5.1.4, T is a continuous mapping of $MW_p^{k-1/p}(R^{n-1})$ into $MW_p^k(R_+^n)$, $k = 1, 2, \ldots$.

Let, in addition, $\nu \geqslant 0$, $\nu(\tau) = 0$ for $|\tau| > 1$ and $\int_{R^{n-1}} \nu(\tau)\,d\tau = 1$. Then $\Phi = T\varphi$ is an extension of φ onto R_+^n.

Since

$$\nabla_\xi \Phi(\xi, \eta) = \int_{R^{n-1}} \nu(\tau)(\nabla_\xi\varphi)(\tau\eta + \xi)\,d\tau,$$

$$\frac{\partial\Phi}{\partial\eta}(\xi, \eta) = \int_{R^{n-1}} \nu(\tau)\tau(\nabla\varphi)(\tau\eta + \xi)\,d\tau,$$

then, by Theorem 5.1.4, $\nabla\Phi \in MW_p^{l-1}(R_+^n)$ and the estimate

$$\|\nabla\Phi; R_+^n\|_{MW_p^{l-1}} \leqslant c\,\|\nabla\varphi; R^{n-1}\|_{MW_p^{l-1-1/p}} \tag{2}$$

holds.

In the case $l = 1$ the left-hand side is majorized by $c\,\|\nabla\varphi; R^{n-1}\|_{L_\infty}$.

In what follows, we do not distinguish the exceptional case $l = 1$ whenever the norm $\|\cdot\|_{MW_p^{l-1-1/p}}$ occurs. So $\|\cdot\|_{MW_p^{-1/p}}$ denotes the norm in L_∞ although literally this is not correct.

Let L be the Lipschitz constant of a function φ and let K be an arbitrary constant satisfying the inequality $K > L$.

We introduce the mapping

$$R_+^n \ni (\xi, \eta) \xrightarrow{\ \lambda\ } (x, y) \in G \tag{3}$$

by equalities

$$x = \xi, \qquad y = K\eta + \Phi(\xi, \eta). \tag{4}$$

Lemma 1. *The mapping λ is (p, l)-diffeomorphism.*

Proof. We have

$$\lambda' = \begin{pmatrix} & & 0 \\ & I & \vdots \\ & & 0 \\ \dfrac{\partial\Phi}{\partial\xi_1}, \ldots, & \dfrac{\partial\Phi}{\partial\xi_{n-1}}, & K + \dfrac{\partial\Phi}{\partial\eta} \end{pmatrix},$$

where I is the unit $(n-1)\times(n-1)$-matrix and $\det\lambda' = K + \partial\Phi/\partial\eta$. By virtue of (1), $|\partial\Phi/\partial\eta| \leq L$, therefore $\det\lambda' \geq K - L \geq 0$. This and the equality $\Phi(\xi, 0) = \varphi(\xi)$ imply that λ is a one-to-one bi-Lipschitz mapping of the half-space R^n_+ onto G. Inequality (2) shows that elements of the matrix λ' belong to the space $MW^{l-1}_p(R^n_+)$. □

By Lemma 6.4.2 the reverse to λ mapping $\varkappa$ is (p, l)-diffeomorphism of G onto R^n_+. The mapping $(x, y) \xrightarrow{\varkappa} (\xi, \eta)$ is given by $\xi = x$, $\eta = u(x, y)$, where u is the unique solution of the equation

$$y = Ku + \Phi(x, u). \tag{5}$$

The restrictions of mappings λ and $\varkappa$ to $R^{n-1} = \partial R^n_+$ and ∂G will be also denoted by λ and $\varkappa$ respectively.

Let G be a special Lipschitz domain. By $W^{l-1/p}_p(\partial G)$ we denote the totality of traces on ∂G of functions in $W^l_p(G)$. In a similar way we define the space $W^{l-1/p}_p(\Gamma)$, where Γ is a subset of the boundary ∂G. The trace of a function f on the boundary of the domain will be denoted by tr.

Since G can be mapped onto R^n_+ by a (p, l)-diffeomorphism, then by Theorem 5.1.4 and Lemma 6.4.1/3 the space $MW^{l-1/p}_p(\partial G)$ is the space of traces of functions in $MW^l_p(G)$.

6.5.2 Implicit Function Theorems

Theorem 1. *Let G be a special Lipschitz domain $\{z = (x, y) : x \in R^{n-1}, y > \varphi(x)\}$ and let u be a function defined on G and satisfying the conditions*: (i) $\operatorname{grad} u \in MW^{l-1}_p(G)$, *where l is an integer, $l \geq 2$,* (ii) $\operatorname{tr} u = 0$, (iii) $\inf \operatorname{tr}(\partial u/\partial y) > 0$. *Then*

$$\operatorname{grad}\varphi \in MW^{l-1-1/p}_p(R^{n-1}).$$

Proof. Since $\operatorname{grad} u \in MW^{l-1}_p(G)$, then $u \in W^{l-1}_{p,\mathrm{loc}}(\bar{G})$ and, in particular, $\partial u/\partial y \in W^1_{p,\mathrm{loc}}(\bar{G})$. We introduce the bi-Lipschitz mapping $G \ni z \xrightarrow{\varkappa} \zeta = (\xi, \eta) \in R^n_+$ by $\xi = x$, $\eta = y - \varphi(x)$. Further, let $v(\zeta) = u(\xi, \eta + \varphi(\xi))$, where $\eta \geq 0$, $\xi \in R^{n-1}$. Obviously,

$$v_{\xi_i} \circ \varkappa_0 = u_{x_i} + \varphi_{x_i} u_y, \qquad v_\eta \circ \varkappa_0 = u_y \quad \text{and} \quad \det \varkappa_0' = 1.$$

Therefore $v_{\xi_i \eta} \circ \varkappa_0 = u_{x_i y} + \varphi_{x_i} u_{yy}$ and consequently $v_{\xi_i \eta} \in L_{p,\text{loc}}(\overline{R^n_+})$. Then, because $v(\xi, +0) = 0$ for a.e. $\xi \in R^{n-1}$ and $v \in W^1_{p,\text{loc}}(\overline{R^n_+})$, it follows that

$$v(\xi, \eta) = \int_0^\eta v_t(\xi, t)\, dt.$$

By differentiating both parts with respect to ξ_i and applying Hölder's inequality, we find

$$\int_{|\xi|<r} |v_{\xi_i}(\xi, \eta)|^p \, d\xi \leq \eta^{p-1} \int_{|\xi|<r} \int_0^\eta |v_{\xi_i t}(\xi, t)|^p \, dt\, d\xi$$

with any positive r. So $v_{\xi_i}(\xi, +0) = 0$ for a.e. ξ or, which is the same,

$$\lim_{y \to \varphi(x)+0} (u_{x_i} + \varphi_{x_i} u_y) = 0 \quad \text{for} \quad \text{a.e. } x \in R^{n-1}.$$

This gives

$$\operatorname{grad} \varphi(x) = \frac{-\operatorname{grad}_x u(x, \varphi(x)+0)}{u_y(x, \varphi(x)+0)} \quad \text{for} \quad \text{a.e. } x \in R^{n-1}. \tag{1}$$

Since $\operatorname{grad} u \in MW^1_p(G)$ and G is a special Lipschitz domain, then $\operatorname{tr} u_{x_i}$, $\operatorname{tr} u_y \in MW^{1-1/p}_p(\partial G)$ or, which is equivalent, the functions

$$x \to u_{x_i}(x, \varphi(x)+0), \qquad x \to u_y(x, \varphi(x)+0)$$

belong to $MW^{1-1/p}_p(R^{n-1})$. From $\inf u_y(x, \varphi(x)+0) > 0$ we conclude that the function

$$x \to u_{x_i}(x, \varphi(x)+0)/u_y(x, \varphi(x)+0)$$

belongs to $MW^{1-1/p}_p(R^{n-1})$. (The inclusion $\gamma^{-1} \in MW^l_p$, provided that $\gamma^{-1} \in L_\infty$ and $\gamma \in MW^l_p$, is verified in the same way as Corollary 2.4/2). So $\varphi_{x_i} \in MW^{1-1/p}_p(R^{n-1})$.

Suppose now that $\varphi_{x_i} \in MW^{k-1-1/p}_p(R^{n-1})$, where k is an integer, $2 \leq k < l$. Then the mapping $\lambda: R^n_+ \to G$ introduced in 6.5.1 is a (p, k)-diffeomorphism. This, together with the inclusion u_{x_i}, $u_y \in MW^k_p(G)$, implies $u_{x_i} \circ \lambda$, $u_y \circ \lambda \in MW^k_p(R^n_+)$. Thus $\operatorname{tr}(u_{x_i} \circ \lambda)$, $\operatorname{tr}(u_y \circ \lambda) \in MW^{k-1/p}_p(R^{n-1})$ and, since $\operatorname{tr}(u_y \circ \lambda)$ is separated from zero, the function $\operatorname{tr}(u_{x_i} \circ \lambda)/\operatorname{tr}(u_y \circ \lambda)$ belongs to the space $MW^{k-1/p}_p(R^{n-1})$. It remains to note that (1) can be rewritten in the form $\operatorname{grad} \varphi = -\operatorname{tr}(\operatorname{grad}_x u \circ \lambda)/\operatorname{tr}(u_y \circ \lambda)$. □

Next we prove the local variant of Theorem 1.

Theorem 2. *Let G be a special Lipschitz domain, let ω be a $(n-1)$-dimensional domain and let U be the cylinder $\{z : y \in R^1,\ x \in \omega\}$. Further, let u be a function defined on $G \cap U$ with properties*: (i) grad $u \in MW^{l-1}_{p,\mathrm{loc}}(U \cap \bar{G})$, *where l is an integer, $l \geq 2$*; (ii) tr $u = 0$ *on $U \cap \partial G$*; (iii) *the function* tr $(\partial u/\partial y)$ *is separated from zero on any compact subset of $U \cap \partial G$. Then* grad $\varphi \in MW^{l-1-1/p}_{p,\mathrm{loc}}(\omega)$.

Proof. By repeating the beginning of the proof of Theorem 1, we arrive at (1) for almost all $x \in \omega$. The second part of the proof is also retained except that in place of spaces $W^s_p(G)$, $W^{s-1/p}_p(\partial G)$, $W^s_p(R^n_+)$, $W^{s-1/p}_p(R^{n-1})$ we must use the spaces $W^s_{p,\mathrm{loc}}(U \cap G)$, $W^{s-1/p}_{p,\mathrm{loc}}(U \cap \partial G)$, $W^s_{p,\mathrm{loc}}(\varkappa(U \cap \bar{G}))$, $W^{s-1/p}_{p,\mathrm{loc}}(\omega)$, where $\varkappa = \lambda^{-1}$. □

Remark 1. Theorems 1 and 2 are exact in the following sense. If grad $\varphi \in MW^{l-1-1/p}_p(R^{n-1})$, then there exists a function u defined on G with properties (i)–(iii). The role of such a function can be played by a solution of the equation (6.5.1/5). In fact, (6.5.1/5) implies $u(x, \varphi(x)) = 0$ and $\partial u/\partial y = (K + \partial\Phi/\partial u)^{-1} \geq (K+L)^{-1}$. Since $\varkappa$ is the (p, l)-diffeomorphism, then grad $u \in MW^{l-1}_p(G)$.

Remark 2. To conclude this section we formulate the implicit mapping theorem analogues to Theorem 1 which can be proved in the same way.

Theorem 3. *Let l and s be integers, $n > s > n-(l-1)p \geq 0$, $z = (x, y)$, $x \in R^s$, $y \in R^{n-1}$ and let u, φ be the mappings $R^n \to R^{n-s}$ and $R^s \to R^{n-s}$ respectively with the properties*:

(i) $u_z \in MW^{l-1}_p(R^n)$;
(ii) tr $u = 0$, *where* tr *is the restriction to the surface* $\{z : x \in R^s,\ y = \varphi(x)\}$;
(iii) *there exists the inverse matrix* $(\mathrm{tr}\, u_y)^{-1}$ *and its norm is uniformly bounded on the surface* $\{z : x \in R^s,\ y = \varphi(x)\}$.

Then $\varphi_x \in MW^{l-1-(n-s)/p}_p(R^s)$.

In the same way as for Theorem 2, one can state the local variant of Theorem 3.

6.6 The Dirichlet Problem for the Second-Order Elliptic Equation in the Space of Multipliers

It is useful to compare the present section with 5.1.5, where we proved the solvability in the space of multipliers of the first boundary value problem for an arbitrary order elliptic equation with constant coefficients in R_+^{n+1}.

By Ω we denote a bounded open subset of R^n. Let

$$Lu = -\sum_{i,j=1}^{n} \frac{\partial}{\partial x_i}\left(a_{ij}(x)\frac{\partial u}{\partial x_j}\right).$$

Suppose that the coefficients a_{ij} are real measurable bounded functions on Ω and that the matrix $\|a_{ij}\|$ is symmetric and uniformly positive definite.

We consider the Dirichlet problem $Lu=0$ in Ω, $u-g\in \mathring{W}_2^1(\Omega)$, where $g\in W_2^1(\Omega)$. This problem is uniquely solvable.

Theorem. *If* $g\in MW_2^1(\Omega)$, *then* $u\in MW_2^1(\Omega)$. *Moreover,* $u-g\in M(W_2^1(\Omega)\to \mathring{W}_2^1(\Omega))$ *and*

$$\|u;\Omega\|_{MW_2^1}\leqslant c\,\|g;\Omega\|_{MW_2^1}. \tag{1}$$

Proof. By the maximum principle for generalized solutions of $Lu=0$ we have

$$\|u;\Omega\|_{L_\infty}\leqslant \|g;\Omega\|_{L_\infty}. \tag{2}$$

So the function $\gamma=u-g$ belongs to $\mathring{W}_2^1(\Omega)\cap L_\infty(\Omega)$. The definition of a generalized solution yields for any $\varphi\in \mathring{W}_2^1(\Omega)$

$$\int_\Omega a_{ij}\frac{\partial\varphi}{\partial x_i}\frac{\partial\gamma}{\partial x_j}\,dx = -\int_\Omega a_{ij}\frac{\partial\varphi}{\partial x_i}\frac{\partial g}{\partial x_j}\,dx. \tag{3}$$

Let v be an arbitrary function in $W_2^1(\Omega)\cap L_\infty(\Omega)$. It is easily seen that γv and γv^2 belong to $\mathring{W}_2^1(\Omega)$. Set $\varphi=\gamma v^2$ in (3). Then

$$\int_\Omega a_{ij}\frac{\partial(\gamma v)}{\partial x_i}\frac{\partial(\gamma v)}{\partial x_j}\,dx = \int_\Omega a_{ij}\frac{\partial v}{\partial x_i}\frac{\partial v}{\partial x_j}\gamma^2\,dx - \int_\Omega a_{ij}\frac{\partial(\gamma v)}{\partial x_i}v\frac{\partial g}{\partial x_j}\,dx - \int_\Omega \gamma v a_{ij}\frac{\partial v}{\partial x_i}\frac{\partial g}{\partial x_j}\,dx.$$

Consequently,

$$c\,\|\nabla(\gamma v);\Omega\|_{L_2}^2\leqslant \|\gamma;\Omega\|_{L_\infty}^2\|\nabla v;\Omega\|_{L_2}^2 + \|\nabla(\gamma v);\Omega\|_{L_2}\|v\nabla g;\Omega\|_{L_2} + \|\gamma;\Omega\|_{L_\infty}\|\nabla v;\Omega\|_{L_2}\|v\nabla g;\Omega\|_{L_2}. \tag{4}$$

Clearly,

$$\|v\nabla g;\Omega\|_{L_2}\leqslant \|vg;\Omega\|_{W_2^1} + \|g;\Omega\|_{L_\infty}\|v;\Omega\|_{W_2^1}.$$

Since $\|g;\Omega\|_{L_\infty}\leqslant \|g;\Omega\|_{MW_2^1}$, then

$$\|v\nabla g;\Omega\|_{L_2}\leqslant 2\,\|g;\Omega\|_{MW_2^1}\|v;\Omega\|_{W_2^1}.$$

This, together with (2), (4), implies

$$c\,\|\nabla(\gamma v); \Omega\|_{L_2}^2 \leq 8\,\|g; \Omega\|_{MW_2^1}^2 \|v; \Omega\|_{W_2^1}^2 + 2\,\|\nabla(\gamma v); \Omega\|_{L_2} \|g; \Omega\|_{MW_2^1} \|v; \Omega\|_{W_2^1}.$$

So $\|\gamma v; \Omega\|_{W_2^1} \leq c\,\|g; \Omega\|_{MW_2^1} \|v; \Omega\|_{W_2^1}$ or, which is the same,

$$\|uv; \Omega\|_{W_2^1} \leq c\,\|g; \Omega\|_{MW_2^1} \|v; \Omega\|_{W_2^1}. \tag{5}$$

Since $W_2^1(\Omega) \cap L_\infty(\Omega)$ is dense in $W_2^1(\Omega)$ and $\gamma v \in \mathring{W}_2^1(\Omega)$ for all $v \in W_2^1(\Omega) \cap L_\infty(\Omega)$, then for any $v \in W_2^1(\Omega)$ we have (5) and $\gamma v \in \mathring{W}_2^1(\Omega)$. □

Remark 1. Let Ω be a bounded domain in R^n with $\partial\Omega \in C^{0,1}$. By Theorem 5.1.4, $MW_2^{1/2}(R^{n-1})$ is the space of traces on R^{n-1} of functions in $MW_2^1(R_+^n)$. This clearly implies that $\varphi \in MW_2^{1/2}(\partial\Omega)$ admits the extension $g \in MW_2^1(\Omega)$ for which

$$\|g; \Omega\|_{MW_2^1} \sim \|\varphi; \partial\Omega\|_{MW_2^{1/2}}.$$

This, together with the theorem, proves the unique solvability of the Dirichlet problem

$$Lu = 0 \quad \text{in} \quad \Omega, \qquad u\,|_{\partial\Omega} = \varphi \in MW_2^{1/2}(\partial\Omega)$$

in $MW_2^1(\Omega)$.

Remark 2. Let Ω be a bounded domain in R^n with $\partial\Omega \in C^{0,1}$. In the theorem we used the space $M(W_p^m(\Omega) \to \mathring{W}_p^l(\Omega))$. Let us show that

$$M(W_p^m(\Omega) \to \mathring{W}_p^l(\Omega)) = \mathring{W}_p^l(\Omega) \cap M(W_p^m(\Omega) \to W_p^l(\Omega)).$$

We denote the left-hand side above by A and the right-hand side by B. Since $1 \in W_p^m(\Omega)$, then $A \subset B$. Let $u \in W_p^m(\Omega)$, $\gamma \in B$ and $u_\nu \in C^\infty(\bar{\Omega})$, $u_\nu \to u$ in $W_p^m(\Omega)$ Then $\gamma u_\nu \in \mathring{W}_p^l(\Omega)$ and

$$\|\gamma u - \gamma u_\nu; \Omega\|_{W_p^l} \leq \|\gamma; \Omega\|_{M(W_p^m \to W_p^l)} \|u - u_\nu; \Omega\|_{W_p^m} = o(1).$$

Consequently $\gamma u \in \mathring{W}_p^l(\Omega)$, that is, $\gamma \in A$.

In comparison with $M(W_p^m(\Omega) \to \mathring{W}_p^l(\Omega))$ the description of $M(\mathring{W}_p^m(\Omega) \to W_p^l(\Omega))$ is more interesting. We give it in the next section.

6.7 The Space $M(\mathring{W}_p^m(\Omega) \to W_p^l(\Omega))$

6.7.1 Auxiliary Results

Let Ω be a domain in R^n and let $p \in (1, \infty)$. We define the capacity of a compact set $e \subset \Omega$ by

$$\operatorname{cap}(e, \mathring{W}_p^l(\Omega)) = \inf\{\|u; \Omega\|_{W_p^l}^p : u \in C_0^\infty(\Omega),\ u \geq 1 \text{ on } e\}.$$

The definition of the capacity cap $(e, \mathring{L}_p^l(\Omega))$, where $\mathring{L}_p^l(\Omega)$ is the completion of $C_0^\infty(\Omega)$ with respect to the norm $\|\nabla_l u; \Omega\|_{L_p}$, is analogous.

Obviously, if $e_1 \subset e_2$ and $\Omega_1 \supset \Omega_2$, then

$$\text{cap}\,(e_1, \mathring{W}_p^l(\Omega_1)) \leqslant \text{cap}\,(e_2, \mathring{W}_p^l(\Omega_2)).$$

The same property of monotonicity has the capacity cap $(e, \mathring{L}_p^l(R^n))$. It is also clear that the capacity cap $(e, \mathring{L}_p^l(R^n))$ acquires the factor d^{n-pl} under the similarity transform with coefficient d. The capacity cap $(e, \mathring{L}_p^l(R^n))$ vanishes for any compact set e provided that $n \leqslant lp$, $p > 1$ or $n < l$, $p = 1$.

The Sobolev theorem on the imbedding of $W_p^l(R^n)$ into $L_\infty(R^n)$ for $lp > n$, $p > 1$, implies that the capacity cap $(e, \mathring{W}_p^l(R^n))$ is separated from zero.

We present some other known properties of capacity:

(ii) Let $lp < n$ and let e be a compact subset of the ball B_ρ. Then

$$\text{cap}\,(e, \mathring{L}_p^l(B_\rho)) \leqslant c\, \text{cap}\,(e, \mathring{L}_p^l(R^n)),$$

where c does not depend on ρ.

(ii) For all compact subsets e of the ball B_1,

$$\text{cap}\,(e, \mathring{W}_p^l(B_2)) \sim \text{cap}\,(e, \mathring{W}_p^l(R^n)).$$

We recall certain properties of capacity that were used in preceding chapters:

(iii) If $\rho \leqslant 1$, then

$$\text{cap}\,(\bar{B}_\rho, W_p^l(R^n)) \sim \begin{cases} \rho^{n-pl} & \text{if either} \quad n > pl,\ p > 1 \quad \text{or} \quad n \geqslant l,\ p = 1; \\ (\log 2/\rho)^{1-p} & \text{if} \quad n = pl,\ p > 1. \end{cases}$$

(iv) If $\rho > 1$, then cap $(\bar{B}_\rho, W_p^l(R^n)) \sim \rho^n$

(v) If $n > pl$, then cap $(e, W_p^l(R^n)) \geqslant c(\text{mes}_n\, e)^{(n-pl)/n}$.

(vi) If $n = pl$ and $d(e) \leqslant 1$, then

$$\text{cap}\,(e, W_p^l(R^n)) \geqslant c(\log(2^n/\text{mes}_n\, e))^{1-p}.$$

To reveal the dependence of certain constants upon the diameter of the domain, we introduce the following new norm in $W_p^k(\Omega)$:

$$|||u; \Omega|||_{W_p^k} = \sum_{j=0}^{k} d^{j-k}\, \|\nabla_j u; \Omega\|_{L_p}, \tag{1}$$

where d is the diameter of Ω (see 3.1.1).

The norm in $M(W_p^m(\Omega) \to W_p^l(\Omega))$, generated by the norm (1), will be denoted by

$$|||\gamma; \Omega|||_{M(W_p^m \to W_p^l)}. \tag{2}$$

In the following theorem we present the norms equivalent to the norm

(2), the equivalence being understood in the sense that their ratios are bounded and separated from zero by constants independent of d.

Theorem. *Let Ω be a domain with $\partial\Omega \in C^{0,1}$ and diameter $d<\infty$.*
(i) *If $p\in(1,\infty)$, then*

$$\|\gamma;\Omega\|_{M(W_p^m\to W_p^l)}\sim \sup_{e\subset\Omega}\frac{\|\nabla_l\gamma;e\|_{L_p}}{[\operatorname{cap}(e,\mathring{L}_p^m(B_{ad}))]^{1/p}}+\sup_{e\subset\Omega}\frac{\|\gamma;e\|_{L_p}}{[\operatorname{cap}(e,\mathring{L}_p^{m-l}(B_{ad}))]^{1/p}},$$

where $a>1$ and B_{ad} is a ball with centre $O\in\bar\Omega$. In the case $mp<n$ we can change B_{ad} for R^n. If $m=l$, then the second summand is equal to $\|\gamma;\Omega\|_{L_\infty}$.
(ii) *If either $pm>n$, $p>1$, or $m\geqslant n$, $p=1$, then the relation*

$$\|\gamma;\Omega\|_{M(W_p^m\to W_p^l)}\sim d^{m-n/p}\|\gamma;\Omega\|_{W_p^l} \tag{3}$$

holds.
(iii) *If $m<n$, then*

$$\|\gamma;\Omega\|_{M(W_1^m\to W_1^l)}\sim \sup_{\substack{x\in\Omega,\\ 2\rho<\operatorname{dist}(x,\partial\Omega)}}\rho^{m-n}(\|\nabla_l\gamma;B_\rho(x)\|_{L_1}+\rho^{-l}\|\gamma;B_\rho(x)\|_{L_1}).$$

Proof. For $d=1$ the assertions formulated above are contained in Theorems 6.3.3/1–6.3.3/3. (To obtain (i) one must use in addition properties (i), (ii) of the capacity.) The transition from $d=1$ to $d\in(0,\infty)$ is performed by the similarity transformation. □

6.7.2 Description of the Space $M(\mathring{W}_p^m(\Omega)\to W_p^l(\Omega))$

Let Ω be a domain in R^n with compact closure and $\partial\Omega\in C^{0,1}$. In the next theorem, Q_j are cubes with edge-length d_j constituting a Whitney covering of Ω (see Stein [1] § 1, ch. 6). Furthermore, let Q_j^* be a cube homothetic to Q_j with edge-length $9d_j/8$. Cubes Q_j^* form a covering of Ω with a finite multiplicity which depends only on n (see Stein [1]).

Theorem. *Let $1<p<\infty$ and $mp<n$. The relation*

$$\|\gamma;\Omega\|_{M(\mathring{W}_p^m\to W_p^l)}\sim\sup_j\sup_{e\subset Q_j}\left(\frac{\|\nabla_l\gamma;e\|_{L_p}}{[\operatorname{cap}(e,\mathring{L}_p^m(Q_j^*))]^{1/p}}+\frac{\|\gamma;e\|_{L_p}}{[\operatorname{cap}(e,\mathring{L}_p^{m-l}(Q_j^*))]^{1/p}}\right) \tag{1}$$

holds.

Proof. By $r(x)$ we denote the regularized distance from $x\in\Omega$ to $\partial\Omega$

(see Stein [1], § 2, ch. 6). From the Hardy inequality

$$\|r^{-m+j}\nabla_j u; \Omega\|_{L_p} \leq c\, \|u; \Omega\|_{\mathring{W}_p^m}, \qquad j = 0, 1, \ldots, m-1 \tag{2}$$

it follows that

$$\|u; \Omega\|_{\mathring{W}_p^m}^p \sim \sum_j \|\|u; Q_j\|\|_{W_p^m}^p \sim \sum_j \|\|u; Q_j^*\|\|_{W_p^m}^p \tag{3}$$

where $\|\|\cdot\|\|$ is the norm defined by (6.7.1/1). By (3) we have an equivalent norm in $M(\mathring{W}_p^m(\Omega) \to W_p^l(\Omega))$ described in terms of the space $M(W_p^m(Q_j) \to W_p^l(Q_j))$. Namely,

$$\|\gamma; \Omega\|_{M(\mathring{W}_p^m \to W_p^l)} \sim \sup_j \|\|\gamma; Q_j\|\|_{M(W_p^m \to W_p^l)}. \tag{4}$$

In fact,

$$\|\gamma u; \Omega\|_{W_p^l}^p \leq \sum_j \|\|\gamma u; Q_j\|\|_{W_p^l}^p \leq \sup_j \|\|\gamma; Q_j\|\|_{M(W_p^m \to W_p^l)}^p \sum_j \|\|u; Q_j\|\|_{W_p^m}^p$$

which, together with (3), gives the required upper bound for the norm of γ in $M(\mathring{W}_p^m(\Omega) \to W_p^l(\Omega))$. On the other hand, let $u \in W_p^m(Q_j)$ and let v be an extension of u onto Q_j^* such that

$$\|\|v; Q_j^*\|\|_{W_p^m} \leq c\, \|u; Q_j\|_{W_p^m}. \tag{5}$$

By φ we denote a function in $C_0^\infty(Q_j^*)$ equal to 1 on Q_j and such that $\nabla_k \varphi = o(d_j^{-k})$. We have

$$\begin{aligned} \|\|\gamma u; Q_j\|\|_{W_p^l} &\leq \|\|\gamma\varphi v; Q_j^*\|\|_{W_p^l} \leq c\, \|\gamma\varphi v; \Omega\|_{W_p^l} \\ &\leq c\, \|\gamma; \Omega\|_{M(\mathring{W}_p^m \to W_p^l)} \|\varphi v; \Omega\|_{W_p^m} \\ &\leq c\, \|\gamma; \Omega\|_{M(\mathring{W}_p^m \to W_p^l)} \|\|v; Q_j^*\|\|_{W_p^m}. \end{aligned}$$

By this and (5) we find the lower bound for the norm of γ in $M(\mathring{W}_p^m(\Omega) \to W_p^l(\Omega))$. Relation (4) is proved.

It remains to use Theorem 6.7.1. The proof is complete. □

Corollary 1. *Let $1 < p < \infty$. The relation*

$$\|\gamma; \Omega\|_{M(\mathring{W}_p^m \to W_p^l)} \sim \sup_{e \subset \Omega} \left(\frac{\|\nabla_l \gamma; e\|_{L_p}}{[\operatorname{cap}(e, \mathring{W}_p^m(\Omega))]^{1/p}} + \frac{\|\gamma; e\|_{L_p}}{[\operatorname{cap}(e, \mathring{W}_p^{m-l}(\Omega))]^{1/p}} \right) \tag{6}$$

holds. It remains true if we add the restriction $d(e) \leq \delta(e)$, where $d(e)$ is the diameter of e and $\delta(e)$ is the distance from e to $\partial\Omega$ on the right.

Proof. From (3) there follows the inequality

$$\operatorname{cap}(e, \mathring{W}_p^m(\Omega)) \geq c \sum_j \operatorname{cap}(e \cap Q_j; \mathring{L}_p^m(Q_j^*)), \qquad e \subset \Omega,$$

which leads to the required lower bound for the norm of γ in $M(\mathring{W}_p^m(\Omega) \to W_p^l(\Omega))$. The upper bound follows via the chain of inequalities

$$c \operatorname{cap}(e, \mathring{L}_p^m(Q_j^*)) \geqslant \operatorname{cap}(e, \mathring{W}_p^m(Q_j^*)) \geqslant \operatorname{cap}(e, \mathring{W}_p^m(\Omega)). \quad \square$$

Corollary 2. *Let $pm > n$, $p > 1$ or $m \geqslant n$, $p = 1$. Then*

$$\|\gamma; \Omega\|_{M(\mathring{W}_p^m \to W_p^l)} \sim \sup_j d_j^{m-n/p} |||\gamma; Q_j|||_{W_p^l}.$$

Corollary 3. *The following relation holds:*

$$\|\gamma; \Omega\|_{M(\mathring{W}_1^m \to W_1^l)} \sim \sup_{\substack{x \in \Omega, \\ 2\rho < \delta(x)}} \rho^{m-n} (\|\nabla_l \gamma; B_\rho(x)\|_{L_1} + \rho^{-l} \|\gamma; B_\rho(x)\|_{L_1}).$$

Corollaries 2 and 3 immediately follow from (4) and Theorem 6.7.1.

6.7.3 More on (p, l)-Diffeomorphisms

Let U and V be bounded domains in R^n and let $\varkappa$ be a mapping $U \to V$ such that $\det \varkappa'$ is positive and is separated from zero. As in 6.4, we say that $\varkappa$ is the (p, l)-diffeomorphism, $l \geqslant 1$, if elements of the Jacobi matrix $\varkappa'$ belong to the space $MW_p^{l-1}(U)$.

In this subsection we present some properties of (p, l)-diffeomorphisms related to the norm $|||\cdot|||$ which will be used in the next chapter.

(i) Let $u \in W_p^l(V)$ and let $\varkappa : U \to V$ be a (p, l)-diffeomorphism. Then $u \circ \varkappa \in W_p^l(U)$ and

$$|||u \circ \varkappa; U|||_{W_p^l} \leqslant c\, |||u; V|||_{W_p^l}.$$

(ii) If $\varkappa$ is a (p, l)-diffeomorphism, then $\varkappa^{-1}$ is also a (p, l)-diffeomorphism, i.e.

$$|||(\varkappa^{-1})'; U|||_{MW_p^{l-1}} \leqslant c.$$

(iii) Let $\gamma \in MW_p^l(V)$ and let $\varkappa$ be a (p, l)-diffeomorphism. Then $\gamma \circ \varkappa \in MW_p^l(U)$ and

$$|||\gamma \circ \varkappa; U|||_{MW_p^l} \leqslant c\, |||\gamma; V|||_{MW_p^l}.$$

The constants c in (i)–(iii) depend on $\inf \det \varkappa'$, p, l, n and the norm $|||\varkappa'; U|||_{MW_p^{l-1}}$. An analogous statement relates the following property:

(iv) Let U, V and W be open sets in R^n and let $\varkappa_1 : U \to V$, $\varkappa_2 : V \to W$ be (p, l)-diffeomorphisms. Then

$$|||(\varkappa_2 \circ \varkappa_1)'; U|||_{MW_p^{l-1}} \leqslant c.$$

Assertions (i)–(iv) are proved in 6.4 for the usual norm in W_p^l. The transition to the norm (6.7.1/1) does not change the proof.

7

Regularity of the Boundary in L_p-Theory of Elliptic Boundary Value Problems

7.1 Description of the Results

The purpose of this chapter is to give applications of the theory of multipliers, developed earlier, to elliptic boundary value problems in domains with 'non-regular' boundaries.

We consider an operator $\{P, \operatorname{tr} P_1, \ldots, \operatorname{tr} P_h\}$ of the general elliptic boundary value problem with smooth coefficients in a bounded domain $\Omega \subset R^n$. We assume that $\operatorname{ord} P = 2h \leq l$, $\operatorname{ord} P_j = k_j < l$, $1 < p < \infty$, and denote the trace operator on the boundary $\partial\Omega$ by tr.

It is well known that the mapping

$$\{P; \operatorname{tr} P_j\}: W_p^l(\Omega) \to W_p^{l-2h}(\Omega) \times \prod_{j=1}^{h} W_p^{l-k_j-1/p}(\partial\Omega) \tag{1}$$

is Fredholm, i.e. it has a finite index and a closed range, provided that the boundary is sufficiently smooth. In particular, for all $u \in W_p^l$ the *a priori* estimate

$$\|u; \Omega\|_{W_p^l} \leq c\left(\|Pu; \Omega\|_{W_p^{l-2h}} + \sum_{j=1}^{h} \|\operatorname{tr} P_j u; \partial\Omega\|_{W_p^{l-k_j-1/p}} + \|u; \Omega\|_{L_1}\right) \tag{2}$$

holds; the last norm on the right can be omitted in the case of a unique solution (see Agmon, Douglis, Nirenberg [1], Hörmander [1] *et al.*).

The analytical background to these fundamental assertions of 'elliptic L_p-theory' is the study of the boundary value problem with constant coefficients in R_+^n and subsequent localization of the original problem, with the help of a partition of unity together with a local mapping of the domain onto half-space. The smoothness of the coefficients, and hence that of the solution of the obtained boundary value problem in R_+^n, depends on the smoothness of the surface $\partial\Omega$. It is well known that the above-mentioned properties of the operator (1) fail where the boundary has singularities.

The effect of the singularities of the boundary on the solvability of elliptic problems in $W_p^l(\Omega)$ has been studied in many papers which can be divided provisionally into two groups. In some of them (Kondrat'ev [1], [2], Veržbinskii, Maz'ya [1], [2], Maz'ya, Plamenevskii [1]–[3] *et al.*) such singularities of the boundary as conical points, edges, polyhedral angles etc. are considered. In the other group of papers (Maz'ya [6], Kondrat'ev and Eidelman [1] *et al.*), in which singularities are not localized, attention is focused on the smoothness conditions for the boundary that guarantee the validity of various estimates for solutions. For example, in the paper of Kondrat'ev and Eidelman [1] the operator of the general elliptic boundary value problem defined in $W_2^l(\Omega)$ is studied and the following condition, exact in a certain sense, for its Fredholm solvability is given. The surface $\partial\Omega$ is locally defined by an equation $y=\varphi(x)$ and the modulus of continuity $\omega_{l-1}(t)$ of the vector-function $\nabla_{l-1}\varphi$ satisfies the condition $\int[\omega_{l-1}(t)/t]^2\,dt<\infty$.

The results of the present chapter also belong to this second group. The boundary of the domain is characterized in terms of spaces of multipliers for $p(l-1)\leqslant n$ or in terms of Sobolev spaces for $p(l-1)>n$. We use the usual procedure of localization of the boundary value problem. The novel aspect is the application of properties of multipliers—in particular, the theorems on their traces on the boundary. This makes less stringent the conditions for the domain Ω that ensure the main results of elliptic L_p-theory.

Sections 7.2 and 7.4 contain auxiliary results. In 7.3 we show that the mapping (1) is Fredholm in the case $p(l-1)\leqslant n$ provided that the domain Ω satisfies the condition $N_p^{l-1/p}$. For each point of the boundary there exists a neighbourhood U and a function φ such that $U\cap\Omega=\{(x,y)\in U: x\in R^{n-1}, y>\varphi(x)\}$ and

$$\|\nabla\varphi; R^{n-1}\|_{MW_p^{l-1-1/p}}\leqslant\delta.$$

Here δ is a small constant and MW_p^s is the space of multipliers in W_p^s for $s>0$ and the space L_∞ for $s\leqslant 0$.

For $p(l-1)>n$ the mapping $\{P;\operatorname{tr} P_j\}$ is Fredholm provided that the domain Ω belongs to the class $W_p^{l-1/p}$.

In 7.5 we consider specifically the first boundary value problem for a strongly elliptic operator P in divergence form. We study two variants of this problem which differ in the description of the boundary data. In the first formulation we require a solution $u\in W_p^l(\Omega)$ of the equation $Pu=f\in W_p^{l-2h}(\Omega)$, $l\geqslant h$, satisfying the condition $u-g\in W_p^l(\Omega)\cap \mathring{W}_p^h(\Omega)$, where g is a given function in $W_p^l(\Omega)$. It is shown that this problem has a unique solution if Ω satisfies the condition $N_p^{l-h+1/p'}$ for $p(l-h)\leqslant n$ or Ω belongs to the class $W_p^{l-h+1/p'}$ for $p(l-h)>n$. In the second, stronger formulation, the boundary data are prescribed by means of the differential operators P_j, $1\leqslant j\leqslant h$. Such a problem is solvable for $h>1$ if Ω belongs to the class

$M_p^{l-1/p}$ and if the Lipschitz constants of functions φ that locally define $\partial\Omega$ are small. In the case $p(l-1)>n$, this condition is equivalent to $\Omega\in W_p^{l-1/p}$.

The inclusion $\Omega\in W_p^{l-1/p}$ for $p(l-1)>n$ is not only sufficient but also necessary for solvability of the Dirichlet problem in the second formulation (see 7.6).

In 7.7 we give an equivalent formulation of $N_p^{l-1/p}$ in terms of the capacity and obtain some simpler conditions ensuring $N_p^{l-1/p}$. For instance, if the norm $\|\nabla\varphi; R^{n-1}\|_{L_\infty}$ is small and φ is a function contained in the Besov space $B_{q,p}^{l-1/p}(R^{n-1})$ with $q\in[p(n-1)/(p(l-1)-1),\infty]$ for $p(l-1)<n$ and $q\in(p,\infty]$ for $p(l-1)=n$, then Ω satisfies the condition $N_p^{l-1/p}$. Putting $q=\infty$ we obtain that $N_p^{l-1/p}$ follows from the convergence of the integral $\int_0 [\omega_{l-1}(t)/t]^p\,dt$, where ω_{l-1} is the modulus of continuity of the vector-function $\nabla_{l-1}\varphi$.

By virtue of the imbedding $B_{\infty,p}^{l-1/p}\subset W_p^{l-1/p}$, the last condition is also sufficient for $\Omega\in W_p^{l-1/p}$. Using this fact, one can immediately derive the following assertions, generalizing the basic results of the paper by Kondrat'ev and Eidelman [1], from our theorems.

The inclusion $\omega_{l-1}(t)/t\in L_p(0,1)$ provides the Fredholm solvability of the operator $\{P, \operatorname{tr} P_j\}$ as well as the unique solvability of the Dirichlet problem in the second formulation. Moreover, the unique solvability in $W_p^l(\Omega)$ of the Dirichlet problem in the first formulation is obtained provided that $\omega_{l-h}(t)/t\in L_p(0,1)$.

In section 7.7 we note that even these, the roughest of our sufficient conditions, are precise in a sense.

The proof of one theorem in 7.7 which contains a local formulation of $N_p^{l-1/p}$ is carried out in 7.8.

7.2 Change of Variables in Differential Operators

Consider the domain $G=\{z=(x,y)\in R^n : x\in R^{n-1},\ y>\varphi(x)\}$, where φ is a function satisfying the Lipschitz condition $|\varphi(x_1)-\varphi(x_2)|\leq L\,|x_1-x_2|$. The following assertion characterizes coefficients of a differential operator under a change of variables.

Proposition. *Let G be a special Lipschitz domain and let λ be an arbitrary (p,l)-diffeomorphism $R_+^n\to G$, $\varkappa=\lambda^{-1}$. Further, let*

$$R(z,D_z)=\sum_{0\leq|\alpha|\leq h} a_\alpha(z)D_z^\alpha,\qquad z\in G,$$

and

$$S(\zeta,D_\zeta)=\sum_{0\leq|\beta|\leq h} b_\beta(\zeta)D_\zeta^\beta,\qquad \zeta\in R_+^n,$$

be differential operators in G and R^n_+ such that

$$Sv = [R(v \circ \varkappa)] \circ \lambda. \tag{1}$$

If $a_\alpha \in M(W_p^{l-|\alpha|}(G) \to W_p^{l-h}(G))$ for all multi-indices α, then $b_\beta \in M(W_p^{l-|\beta|}(R^n_+) \to W_p^{l-h}(R^n_+))$ and

$$\|b_\beta; R^n_+\|_{M(W_p^{l-|\beta|} \to W_p^{l-h})} \leq c \sum_{|\beta| \leq |\alpha| \leq h} \|a_\alpha; G\|_{M(W_p^{l-|\alpha|} \to W_p^{l-h})}. \tag{2}$$

The proof follows the same lines as that of Lemma 6.4.1/5.

We note that, according to (6.4.1/8), the equality

$$b_\beta = \sum_{|\beta| \leq |\alpha| \leq h} (a_\alpha \circ \lambda) \sum c_s \prod_{i,j} (D_z^{s_{ij}} \varkappa_i) \circ \lambda \tag{3}$$

holds.

Lemma 1. *Let G denote a special Lipschitz domain and let λ be an arbitrary (p, l)-diffeomorphism $R^n_+ \to G$. Further, let R be a homogeneous differential operator of order h with constant coefficients and let S be an operator defined by (1). Then*

$$\|S - R; R^n_+\|_{M(W_p^l \to W_p^{l-h})} \leq c \, \|I - \lambda'; R^n_+\|_{MW_p^{l-1}}, \tag{4}$$

where c is a continuous function of the norm of λ' in $MW_p^{l-1}(R^n_+)$. (Here and henceforth by the norm of a matrix we mean the sum of norms of its elements.)

Proof. We introduce the notations $\varkappa = \lambda^{-1}$ and $a = \|I - \lambda'; R^n_+\|_{MW_p^{l-1}}$. Let then $S_1(\zeta, D_\zeta)$ denote the principal homogeneous part of the operator S. Since $S_1(\zeta, \rho) = S((\varkappa')^* \rho) \circ \lambda$ for any vector $\rho \in R^n$, then each coefficient of S_1 differs from the corresponding coefficient of R by $O(a)$ in the norm of $MW_p^{l-h}(R^n_+)$. Hence, $\|S_1 - R; R^n_+\|_{M(W_p^l \to W_p^{l-h})} \leq ca$.

Consider the coefficients of the operator S which multiply the derivatives of order $|\beta| < h$. Let formula (3) relate the coefficients a_α and b_β of operators R and S. Since R is homogeneous, then $|\alpha| = h$ in (3). Hence by (2) every summand in (3) with $|\beta| < h$ contains at least one factor $D^{s_{ij}} \varkappa_i(z)$ for which $|s_{ij}| > 1$. Noting that such a factor is equal to $D^{s_{ij}}[\varkappa_i(z) - z_i]$, we obtain

$$\|b_\beta; R^n_+\|_{M(W_p^{l-|\beta|} \to W_p^{l-h})} \leq c \, \|I - \varkappa'; G\|_{MW_p^{l-1}} \leq ca$$

(see the proof of Lemma 6.4.1/5). Therefore, $\|S - S_1; R^n_+\|_{M(W_p^l \to W_p^{l-h})} \leq ca$. $\square$

Duplicating the proof of Lemma 1 with obvious changes and using the properties of (p, l)-diffeomorphisms in 6.4.1, we obtain the following local variant of Lemma 1.

Lemma 2. Let all the conditions of the proposition be satisfied. Then, for each $v \in W_p^l(R_+^n)$ *with support in* $B_r \cap \overline{R_+^n}$,

$$\|(S-R)v; R_+^n\|_{W_p^{l-h}} \leq c\, \|I-\lambda'; B_r \cap R_+^n\|_{MW_p^{l-1}} \|v; R_+^n\|_{W_p^l}, \tag{5}$$

where c is a constant independent of $r \in (0, 1)$.

For $p(l-1) > n$ it follows from (6.7.1/3) that (5) is equivalent to

$$\|(S-R)v; R_+^n\|_{W_p^{l-h}} \leq cr^{l-1-n/p} \|I-\lambda'; B_r \cap R_+^n\|_{W_p^{l-1}} \|v; R_+^n\|_{W_p^l}. \tag{6}$$

7.3 The Fredholm Property of the Elliptic Boundary Value Problem

7.3.1 *Bounded Domains of the Classes $M_p^{l-1/p}$, $W_p^{l-1/p}$ and the Condition $N_p^{l-1/p}$*

Let Ω be a bounded domain of the class $C^{0,1}$. We introduce the class $M_p^{l-1/p}$ $(l = 2, 3, \ldots)$ of domains Ω, satisfying the following condition. For each point of the boundary $\partial\Omega$ there exists a neighbourhood in which $\partial\Omega$ is specified (in a certain Cartesian coordinate system) by a function φ such that $\nabla\varphi \in MW_p^{l-1-1/p}(R^{n-1})$. Moreover, by definition, $M_p^{1-1/p} = C^{0,1}$.

We say that Ω belongs to the class $W_p^{l-1/p}$ if $\partial\Omega$ can be locally specified by a function $\varphi \in W_p^{l-1/p}(R^{n-1})$. Since $MW_p^{l-1-1/p}(R^{n-1}) \subset W_{p,\text{loc}}^{l-1-1/p}(R^{n-1})$, $l \geq 2$, and $C^{0,1}(R^{n-1}) \subset W_{p,\text{loc}}^{1-1/p}(R^{n-1})$, then any bounded domain from $M_p^{l-1/p}$ belongs to $W_p^{l-1/p}$.

According to Proposition 3.2.8 for $p(l-1) > n$ we have

$$\|\nabla\varphi; R^{n-1}\|_{MW_p^{l-1-1/p}} \sim \sup_{x \in R^{n-1}} \|\nabla\varphi; B_1(x)\|_{W_p^{l-1-1/p}}.$$

Therefore the classes $M_p^{l-1/p}$ and $W_p^{l-1/p}$ coincide for $p(l-1) > n$.

For a bounded domain Ω of the class $C^{0,1}$ we denote by $W_p^{l-1/p}(\partial\Omega)$ the space of traces on $\partial\Omega$ of functions in $W_p^l(\Omega)$. Taking into account the analogous fact for special Lipschitz domains of the class $M_p^{l-1/p}$ (see 6.5.1), we obtain that $MW_p^{l-1/p}(\partial\Omega)$ is the space of traces of functions in $MW_p^l(\Omega)$.

Let $P, P_1, \ldots, P_h$ be differential operators in $\bar{\Omega}$ of orders $2h, k_1, \ldots, k_h$, respectively, where $2h \leq l$, $k_j < l$. Suppose that the coefficients of the operators P and P_j belong to $C^{l-2h}(\bar{\Omega})$ and $C^{l-k_j}(\bar{\Omega})$, respectively. (This restriction can be removed by the use of spaces of multipliers, but we do not want to complicate the formulations.) We assume that the operators P, tr $P_1, \ldots,$ tr P_h form an elliptic boundary value problem in every point $O \in \partial\Omega$ with respect to the hyperplane $y = 0$ and that P is an elliptic operator in Ω.

In our further exposition the following additional condition upon Ω will play an important role:

The condition $N_p^{l-1/p}$. For each point $O \in \partial\Omega$ there exists a neighbourhood U and a special Lipschitz domain $G = \{z = (x, y) : x \in R^{n-1}, y > \varphi(x)\}$ such that $U \cap \Omega = U \cap G$ and

$$\|\nabla\varphi; R^{n-1}\|_{MW_p^{l-1-1/p}} \leq \delta.$$

Here $p(l-1) \leq n$ and δ is a constant which depends on the coefficients of the principal homogeneous parts of $P, P_1, \ldots, P_h$ calculated at point O in the coordinate system (x, y). For $l = 1$ the role of the last inequality is played by $\|\nabla\varphi; R^{n-1}\|_{L_\infty} \leq \delta$.

Obviously, the domains satisfying the condition $N_p^{l-1/p}$ belong to the class $M_p^{l-1/p}$ and, therefore, to the class $W_p^{l-1/p}$. In 7.7 we give an equivalent formulation of $N_p^{l-1/p}$ and discuss sufficient conditions for it.

7.3.2 A priori L_p-Estimate for Solutions and Other Properties of the Elliptic Boundary Value Problem

In the next two theorems we consider separately the cases $p(l-1) \leq n$ and $p(l-1) > n$.

Theorem 1. *If* $p(l-1) \leq n$, $1 < p < \infty$, *and if* Ω *satisfies the condition* $N_p^{l-1/p}$, *then* (7.1/2) *holds for any* $u \in W_p^l(\Omega)$.

Proof. We retain the notation used in the formulation of the condition $N_p^{l-1/p}$. Let U be a closed ball with a small radius, $\sigma \in C_0^\infty(U)$, and R and R_j be the principal homogeneous parts of operators P and P_j with 'frozen' coefficients at point O. Clearly,

$$\|(P-R)(\sigma u); U \cap \Omega\|_{W_p^{l-2h}} \leq \varepsilon \|\sigma u; U \cap \Omega\|_{W_p^l} + c \|\sigma u; U \cap \Omega\|_{W_p^{l-1}}, \quad (1)$$

where ε is a small positive number (the required smallness is defined by the coefficients of the operators $R, R_1, \ldots, R_h$). The analogous estimate is valid for the norm of $(P_j - R_j)(\sigma u)$ in $W_p^{l-k_j}(U \cap \Omega)$.

By $N_p^{l-1/p}$ the Lipschitz constant of φ is small, so we can put $K = 1$ in the definition (6.5.1/4) of the mapping $\lambda : R_+^n \to G$. Then, from (6.5.1/2), we obtain that λ' differs from the identity matrix by $O(\delta)$ in the norm of $MW_p^{l-1}(R_+^n)$. It is known (see, for instance, Agmon, Douglis, Nirenberg [1], Triebel [4]) that for all $v \in W_p^l(R_+^n)$ with supports in $B_1 \cap \overline{R_+^n}$,

$$\|v; R_+^n\|_{W_p^l} \leq c\left(\|Rv; R_+^n\|_{W_p^{l-2h}} + \sum_{j=1}^{h} \|\text{tr}\, R_j v; R^{n-1}\|_{W_p^{l-k_j-1/p}}\right).$$

Since δ is small, we can change here R to R_j and S to S_j (see Lemma

7.2/2). From the estimate obtained after this change, it follows that

$$\|\sigma u; U\cap\Omega\|_{W_p^l}\leq c\Big(\|R(\sigma u); U\cap\Omega\|_{W_p^{l-2h}} + \sum_{j=1}^{h}\|\mathrm{tr}\, R_j(\sigma u); U\cap\partial\Omega\|_{W_p^{l-kj-1/p}}\Big).$$

This and (1) entail

$$\|\sigma u;\Omega\|_{W_p^l}\leq c\Big(\|P(\sigma u);\Omega\|_{W_p^{l-2h}}+\sum_{j=1}^{h}\|\mathrm{tr}\, P_j(\sigma u);\partial\Omega\|_{W_p^{l-kj-1/p}}+\|\sigma u;\Omega\|_{W_p^{l-1}}\Big).$$

Summing over all sufficiently small neighbourhoods U that generate a covering of $\bar{\Omega}$, we come to

$$\|u;\Omega\|_{W_p^l}\leq c\Big(\|Pu;\Omega\|_{W_p^{l-2h}}+\sum_{j=1}^{h}\|\mathrm{tr}\, P_j u;\partial\Omega\|_{W_p^{l-k_j-1/p}}+\|u;\Omega\|_{W_p^{l-1}}\Big).$$

It remains to use the known inequality

$$\|u;\Omega\|_{W_p^{l-1}}\leq\varepsilon\,\|u;\Omega\|_{W_p^l}+c(\varepsilon)\,\|u;\Omega\|_{L_1},$$

where ε is any positive number. Estimate (7.1/2) is proved □

Theorem 2. If $p(l-1)>n$, $1<p<\infty$, and if $\Omega\in W_p^{l-1/p}$, then the conclusion of Theorem 1 holds.

Proof. From the condition $\Omega\subset W_p^{l-1/p}$ and the Sobolev embedding theorem, it follows that $\Omega\in C^1$. We place the origin at the point $O\in\partial\Omega$ and direct the axis Oy along the interior normal to $\partial\Omega$. Let U be the neighbourhood of O in the definition of the class $W_p^{l-1/p}$, i.e. $U\cap\Omega=U\cap G$, where $G=\{z: x\in R^{n-1},\ y>\varphi(x)\}$ and $\varphi\in W_p^{l-1/p}(R^{n-1})$. Let ε be a small positive number, which will be specified later, and let $B_\rho=\{z\in R^n: |z|<\rho\}$.

We choose a small number ρ such that $\|\nabla\varphi; B_\rho\cap R^{n-1}\|_{L_\infty}<\varepsilon$ and $\overline{B_{2\rho}}\subset U$. Let $\tau\in C_0^\infty(B_2)$, $\tau=1$ on B_1, and $\tau_\rho(z)=\tau(z/\rho)$. We introduce the function $\varphi^*=\varphi\tau_\rho$ on R^{n-1} and note that $\|\nabla\varphi^*; R^{n-1}\|_{L_\infty}<c\varepsilon$. Define also the extension Φ of φ^* onto R_+^n by $\Phi=T\varphi^*$. According to (6.5.1/2), where φ is changed by φ^*,

$$\|\nabla\Phi; R_+^n\|_{L_\infty}\leq c\varepsilon. \tag{2}$$

Since

$$\|\varphi^*; R^{n-1}\|_{W_p^{l-1/p}}\leq c(\rho)\,\|\varphi; R^{n-1}\|_{W_p^{l-1/p}},$$

then

$$\|\Phi; R_+^n\|_{W_p^l}\leq c(\rho)\,\|\varphi; R^{n-1}\|_{W_p^{l-1/p}}.$$

Now let r be a small positive number such that $r<\rho$ and

$$r^{l-1-n/p}\|\Phi; R_+^n\|_{W_p^l}<\varepsilon. \tag{3}$$

From the inequality

$$\|\nabla\Phi; B_r\cap R_+^n\|_{W_p^{l-1}}\leq c(\|\nabla_l\Phi; B_r\cap R_+^n\|_{L_p}+r^{1-l+n/p}\|\nabla\Phi; B_r\cap R_+^n\|_{L_\infty})$$

and the estimates (2), (3) it follows that

$$r^{l-1-n/p}\|\nabla\Phi; B_r\cap R_+^n\|_{W_p^{l-1}}\leq c\varepsilon.$$

According to (6.7.1/3), this means

$$\|\nabla\Phi; B_r\cap R_+^n\|_{MW_p^{l-1}}\leq c\varepsilon.$$

Using the function Φ, we define the mapping λ by (6.5.1/4) with $K=1$. By the last inequality,

$$\|I-\lambda'; B_r\cap R_+^n\|_{MW_p^{l-1}}\leq c\varepsilon.$$

Now it suffices to duplicate the arguments we have already used in Theorem 1, except that in place of the ball U we have B_r and instead of (7.2/5) we apply the estimate (7.2/6). □

The following assertion can be proved in a standard way from the *a priori* estimate (7.1/2) (see, for instance, Hörmander [1], §10.5; Triebel [4], 5.4.3).

Proposition 1. *Let the domain Ω satisfy the condition of either Theorem 1 or Theorem 2.*

(i) *If the kernel of the operator* (7.1/1) *is trivial, then the norm* $\|u;\Omega\|_{L_1}$ *in* (7.1/2) *can be omitted.*

(ii) *The kernel of the operator* (7.1/1) *is finite-dimensional.*

(iii) *The range of the operator* (7.1/1) *is closed.*

Proof. (i) Suppose that the assertion is not true. Then there exists a sequence of functions $\{v_m\}_{m\geq 1}$ in $W_p^l(\Omega)$ such that

$$\|v_m;\Omega\|_{W_p^l}=1, \tag{4}$$

$$\|Pv_m;\Omega\|_{W_p^{l-2h}}+\sum_{j=1}^{h}\|\mathrm{tr}\, P_jv_m;\partial\Omega\|_{W_p^{l-k_j-1/p}}\to 0. \tag{5}$$

We can select a subsequence of $\{v_m\}$, also denoted by $\{v_m\}$, which weakly converges in $W_p^l(\Omega)$ to a function $v\in W_p^l(\Omega)$. Since the imbedding operator $W_p^l(\Omega)\to L_1(\Omega)$ is compact, then $v_m\to v$ in $L_1(\Omega)$. Substituting the function v_m-v_k into (7.1/2) we obtain that $v_m\to v$ in $W_p^l(\Omega)$. This and (5) imply that $v\in\ker\{P;\mathrm{tr}\, P_j\}$, i.e. $v=0$. This contradicts (4).

(ii) From (7.1/2) it follows that, for all $v \in \ker\{P; \operatorname{tr} P_j\}$,

$$\|v; \Omega\|_{W_p^l} \leq c\,\|v; \Omega\|_{L_1}.$$

Therefore a unit sphere in $\ker\{P; \operatorname{tr} P_j\}$ considered as a subspace of $W_p^l(\Omega)$ is compact and the dimension of the kernel is finite.

(iii) Since $\dim\ker\{P; \operatorname{tr} P_j\} < \infty$, there exists a projection operator Π which projects parallel to $\ker\{P; \operatorname{tr} P_j\}$.

Duplicating the arguments used in part (i) of the present proof, we obtain

$$\|v; \Omega\|_{W_p^l} \leq c\left(\|Pv; \Omega\|_{W_p^{l-2h}} + \sum_{j=1}^{h} \|\operatorname{tr} P_j v; \partial\Omega\|_{W_p^{l-k_j-1/p}}\right)$$

for all $v \in \Pi W_p^l(\Omega)$, which implies that the range of the operator $\{P; \operatorname{tr} P_j\}$ is closed. □

Next we derive the local *a priori* estimate for solutions of the elliptic boundary value problem (cf. Agmon, Douglis, Nirenberg [1]).

Proposition 2. *Let the domain Ω satisfy the condition of either Theorem 1 or Theorem 2. Further, let U, V be open subsets of R^n, $U \subset V$, and $u \in W_p^l(V \cap \Omega)$. Then*

$$\|u; U\cap\Omega\|_{W_p^l} \leq c\left(\|Pu; V\cap\Omega\|_{W_p^{l-2h}} + \sum_{j=1}^{h} \|\operatorname{tr} P_j u; V\cap\partial\Omega\|_{W_p^{l-k_j-1/p}} + \|u; V\cap\Omega\|_{L_1}\right). \tag{6}$$

Proof. Let U and V be concentric balls with radii r and ρ, $r<\rho$. Further, let either $\bar{V} \subset \Omega$ or the centre of the balls be placed on $\partial\Omega$. Clearly, it suffices to prove the proposition under this additional assumption. We introduce the sets $C_0 = U \cap \Omega$, $C_k = \{x \in \Omega : \delta_k < \rho - |x| < \delta_{k-1}\}$ where $k = 1, 2, \ldots$ and $\delta_k = (\rho - r)2^{-k}$. Let $D_0 = C_0 \cup C_1$, $D_k = C_{k-1} \cup C_k \cup C_{k+1}$, $k = 1, 2, \ldots$. We construct a C^∞-partition of unity $\{\sigma_k\}_{k\geq 0}$ subordinate to the covering $\{D_k\}_{k\geq 0}$ of $V \cap \Omega$ such that $|D^\alpha \sigma_k| = O(\delta_k^{-|\alpha|})$ for any multi-index α.

By applying the *a priori* estimate (7.1/2) to $\sigma_k v$, we obtain

$$\|\sigma_k v; \Omega\|_{W_p^l} \leq c\left(\|P(\sigma_k v); \Omega\|_{W_p^{l-2h}} + \sum_{j=1}^{h} \|\operatorname{tr} P_j(\sigma_k v); \partial\Omega\|_{W_p^{l-k_j-1/p}} + \|\sigma_k v; \Omega\|_{L_1}\right),$$

which yields

$$\|\sigma_k v; \Omega\|_{W_p^l} \leq c\Big(\|\sigma_k Pv; \Omega\|_{W_p^{l-2h}} + \sum_{j=1}^{h} \|\sigma_k \operatorname{tr} P_j v; \partial\Omega\|_{W_p^{l-k_j-1/p}} + \delta_k^{-l}\|v; D_k\|_{W_p^{l-1}}\Big).$$

Let M be a sufficiently large positive number. We have

$$\sum_{k=0}^{\infty} \delta_k^M \|v; C_k\|_{W_p^l} \leq c\Big(\sum_{k=0}^{\infty} \delta_k^M \|\sigma_k Pv; \Omega\|_{W_p^{l-2h}} + \sum_{k=0}^{\infty} \delta_k^M \sum_{j=1}^{h} \|\sigma_k \operatorname{tr} P_j v; \partial\Omega\|_{W_p^{l-k_j-1/p}} + \sum_{k=0}^{\infty} \delta_k^{M-l}\|v; D_k\|_{W_p^{l-1}}\Big)$$

$$\leq c\Big(\|Pv; V\cap\Omega\|_{W_p^{l-2h}} + \sum_{j=1}^{h} \|\operatorname{tr} P_j v; V\cap\partial\Omega\|_{W_p^{l-k_j-1/p}} + \sum_{k=0}^{\infty} \delta_k^{M-l}\|v; D_k\|_{W_p^{l-1}}\Big).$$

Note that for any $\varepsilon > 0$ and for some positive N

$$\|v; D_k\|_{W_p^{l-1}} \leq \varepsilon\, \delta_k^l \|v; D_k\|_{W_p^l} + c(\varepsilon)\, \delta_k^{-N} \|v; D_k\|_{L_1}.$$

Consequently,

$$\sum_{k=0}^{\infty} \delta_k^M \|v; C_k\|_{W_p^l} \leq c\Big(\|Pv; V\cap\Omega\|_{W_p^{l-2h}} + \sum_{j=1}^{h} \|\operatorname{tr} P_j v; V\cap\partial\Omega\|_{W_p^{l-k_j-1/p}} + c_1(\varepsilon)\|v; V\cap\Omega\|_{L_1} + \varepsilon \sum_{k=0}^{\infty} \delta_k^M \|v; D_k\|_{W_p^l}\Big).$$

Clearly, the last sum can be omitted by changing c. □

Using the same properties of P, P_j, λ, $\varkappa$ as those used in the proof of Theorems 1 and 2, one can establish the following assertion by the known method (see, for instance, Eskin [1], ch. 6, §22).

Proposition 3. *Let the domain Ω satisfy the condition of either Theorem 1 or Theorem 2. There exists a linear bounded operator*

$$R: W_p^{l-2h}(\Omega) \times \prod_{j=1}^{h} W_p^{l-k_j-1/p}(\partial\Omega) \to W_p^l(\Omega)$$

such that $\{P; \operatorname{tr} P_j\}R = I + K$. Here I and K are the identity and compact operators respectively.

An immediate corollary of Propositions 1 and 3 is:

Theorem 3. *Let Ω satisfy $N_p^{l-1/p}$ for $p(l-1)\leq n$ and belong to $W_p^{l-1/p}$ for $p(l-1)>n$. Then the operator (7.1/1) is Fredholm, that is, its null-space is finite-dimensional and its range is closed and has a finite codimension.*

In the following sections we consider the Dirichlet problem in more detail.

7.4 Auxiliary Assertions

7.4.1 Some Properties of the Operator T

In this subsection, T is the operator defined by (6.5.1/1).

Lemma. *Let α be an n-tuple multi-index and let k, r be non-negative integers, $k\geq|\alpha|-r\geq 0$. Then the operator*

$$M(W_p^{k-1/p}(R^{n-1}))\ni\gamma\to\eta^r(D^\alpha T\gamma)(\zeta)\in M(W_p^k(R_+^n)\to W_p^{k-|\alpha|+r}(R_+^n))$$

is continuous.

Proof. Clearly,

$$\eta^r(D^\alpha T\gamma)(\zeta)=D^\beta\sum_{0\leq|\nu|\leq r}c_\nu\eta^{|\nu|}(D^\nu T\gamma)(\zeta),$$

where β is a multi-index of order $|\alpha|-r$, $c_\nu=\text{const}$. The operator $\gamma\xrightarrow{T_\nu}\eta^{|\nu|}(D^\nu T\gamma)(\zeta)$ has the same form as T and therefore it maps $M(W_p^{k-1/p}(R^{n-1}))$ into $M(W_p^k(R_+^n))$. Hence, by Lemma 3.8.2/2, the continuity of the operator

$$M(W_p^{k-1/p}(R^{n-1}))\ni\gamma\to D^\beta T_\nu\gamma\in M(W_p^k(R_+^n)\to W_p^{k-|\alpha|+r}(R_+^n))$$

follows. □

From this lemma immediately follows:

Corollary. *Let G be a special Lipschitz domain, let α be a positive n-tuple multi-index and let r be a non-negative integer, $l\geq|\alpha|-r>0$. Then the function $\zeta\to\eta^r(D^\alpha\Phi)(\zeta)$ belongs to the space $M(W_p^{l-1}(R_+^n)\to W_p^{l-|\alpha|+r}(R_+^n))$ and*

$$\|\eta^rD^\alpha\Phi;R_+^n\|_{M(W_p^{l-1}\to W_p^{l-|\alpha|+r})}\leq c\,\|\nabla\varphi;R^{n-1}\|_{MW_p^{l-1-1/p}}.$$

7.4.2 Properties of the Mappings λ and $\varkappa$

Let G be a special Lipschitz domain and let λ be the mapping (6.5.1/3) defined by equalities (6.5.1/4). As in 6.5.1, by $\varkappa$ we denote the inverse mapping to λ. We shall assume that $L<1$ and $K=1$.

From Corollary 7.4.1 immediately follows:

Corollary. *Let α be a multi-index. Let r be a non-negative integer, $l\geqslant|\alpha|-r+1>0$ and $\lambda(\zeta)=\{\lambda_1(\zeta),\dots,\lambda_n(\zeta)\}$. Then the function $\zeta\to\eta^r(D^\alpha\lambda_i')(\zeta)$ belongs to the space $M(W_p^{l-1}(R_+^n)\to W_p^{l-1-|\alpha|+r}(R_+^n))$ and*

$$\|\eta^r D^\alpha(\lambda'-I);\, R_+^n\|_{M(W_p^{l-1}\to W_p^{l-|\alpha|+r})}\leqslant c\,\|\nabla\varphi;\, R^{n-1}\|_{MW_p^{l-1-1/p}}.$$

The similar assertion on the mapping $\varkappa$ needs a separate proof.

Lemma. *Let α be a multi-index, let r be a non-negative integer, $l\geqslant|\alpha|-r+1>0$ and $\varkappa(z)=\{\varkappa_1(z),\dots,\varkappa_n(z)\}$. Then the function $z\to(\eta^r D^\alpha\varkappa_i')(z)$ belongs to the space $M(W_p^{l-1}(G)\to W_p^{l-1-|\alpha|+r}(G))$ and*

$$\|\eta^r D^\alpha(\varkappa'-I);\, G\|_{M(W_p^{l-1}\to W^{l-|\alpha|+r-1})}\leqslant c\,\|\nabla\varphi;\, R^{n-1}\|_{MW_p^{l-1-1/p}}.$$

Proof. For $|\alpha|=0$ the result follows from the definition of the (p,l)-diffeomorphism. Suppose that the lemma is proved for $|\alpha|<N$. Let $|\alpha|=N$, $r<N$. For any multi-index δ of order $N-1$,

$$(D^\delta\varkappa')(z)=D^\delta[\lambda'(\varkappa(z))]^{-1}$$
$$=\sum_{1\leqslant|\beta|\leqslant|\delta|}[D^\beta(\lambda')^{-1}](\varkappa(z))\sum c_s\prod_{i=1}^{n}\prod_j D^{s_{ij}}\varkappa_i(z),$$

where the summation is taken over all collections of multi-indices $s=(s_{ij})$ such that $\sum s_{ij}=\delta$, $|s_{ij}|\geqslant1$, $\sum(|s_{ij}|-1)=|\delta|-|\beta|$. Therefore the expression $(\eta^r D^\delta\varkappa')(z)$ is the sum of the products of two factors

$$\pi_1(z)=c_\beta[\eta^{|\beta|}D^\beta(\lambda')^{-1}](\varkappa(z)),$$
$$\pi_2(z)=\prod_{i=1}^{n}\prod_j(\eta^{r-|\beta|}D^{s_{ij}}\varkappa_i)(z).$$

The corollary implies that the function $\zeta\to\eta^{|\beta|}D^\beta(\lambda')^{-1}$ belongs to the space $MW_p^{l-1}(R_+^n)$ and, since $\varkappa$ is a (p,l)-diffeomorphism, then $\pi_1\in MW_p^{l-1}(G)$.

We introduce positive integers σ_{ij} such that $\sigma_{ij}\leqslant|s_{ij}|$, $\sum(\sigma_{ij}-1)=r-|\beta|$. Then

$$\pi_2(z)=\prod_{i=1}^{n}\prod_j(\eta^{\sigma_{ij}-1}D^{s_{ij}}\varkappa_i)(z).$$

Since $|\beta|\geq 1$, we have $|s_{ij}|\leq N-1$ and by the induction hypothesis the function $z\to(\eta^{\sigma_{ij}-1}D^{s_{ij}}\varkappa_i)(z)$ belongs to $M(W_p^q(G)\to W_p^{q-1-|s_{ij}|+\sigma_{ij}}(G))$, $q=|s_{ij}|-\sigma_{ij},\ldots,l-1$. Hence

$$\pi_2\in M(W_p^{l-1}(G)\to W_p^{l-2-\Sigma(|s_{ij}|-\sigma_{ij})}(G))=M(W_p^{l-1}(G)\to W_p^{l-2-|\delta|+r}(G)).$$

Taking into account that $|\alpha|=1+|\delta|$ and $\pi_1\in MW_p^{l-1}(G)$, we obtain $\pi_1\pi_2\in M(W_p^{l-1}(G)\to W_p^{l-1-|\alpha|-r}(G))$. □

Since the space W_p^l is invariant under (p, l)-diffeomorphisms, a function u on ∂G belongs to $W_p^{l-1/p}(\partial G)$ if and only if $u\circ\text{tr}\,\lambda\in W_p^{l-1/p}(R^{n-1})$. We put

$$\|u;\partial G\|_{W_p^{l-1/p}}=\|u\circ\text{tr}\,\lambda;R^{n-1}\|_{W_p^{l-1/p}}.$$

Taking here an arbitrary (p, l)-diffeomorphism $R^n_+\to G$ instead of λ, one obtains the equivalent norm (see Lemmas 6.4.1/1, 6.4.1/2, and 6.4.1/4).

7.4.3 Invariance of the Space $W_p^l\cap \mathring{W}_p^h$ under a Change of Variables

In this subsection we present auxiliary assertions which will be used later in the study of conditions for solvability of the Dirichlet problem in $W_p^l(\Omega)$.

Let, as before, $G=\{z=(x, y): x\in R^{n-1}, y>\varphi(x)\}$ and let G belong to the class $M_p^{l-h+1/p'}$, where l and h are integers, $l\geq h\geq 1$. In other words, $\nabla\varphi\in MW_p^{l-h-1/p}(R^{n-1})$ if $l>h$ and $\varphi\in C^{0,1}(R^{n-1})$ if $l=h$. Let λ be the mapping $R^n_+\ni(\xi,\eta)\to(x, y)\in G$, defined by (6.5.1/4), and let $\varkappa=\lambda^{-1}$.

Lemma 1. *Let $v\in(\mathring{W}_p^k\cap W_p^{t+k})(G)$, $0\leq t\leq l-h$, and let $G\subset M_p^{l-h+1/p'}$. Then*

$$\|\eta^{-k}v;G\|_{W_p^t}\leq c\,\|v;G\|_{W_p^{t+k}}. \tag{1}$$

Proof. Since $\eta(z)\sim y-\varphi(x)$, then (1) with $t=0$ follows from the Hardy inequality,

$$\int_{\varphi(x)}^{\infty}|v(x, y)|^p\frac{dy}{(y-\varphi(x))^{pk}}\leq c\int_{\varphi(x)}^{\infty}\left|\frac{\partial^k v}{\partial y^k}(x, y)\right|^p dy \tag{2}$$

a.e. in R^{n-1}. Let the lemma be proved for all $t<T$ and $k>K$. We have

$$\|\eta^{-K}v;G\|_{W_p^T}\leq\|\nabla(\eta^{-K}v);G\|_{W_p^{T-1}}+\|\eta^{-K}v;G\|_{L_p}.$$

The second summand in the right-hand side is estimated by (2) and the first one is not more than

$$\|\eta^{-K}\,\nabla v;G\|_{W_p^{T-1}}+K\,\|\eta^{-K-1}v\,\nabla\eta;G\|_{W_p^{T-1}}. \tag{3}$$

Since $G\subset M_p^{l-h+1/p'}$, then $\nabla\eta\in MW_p^{l-h}(G)\subset MW_p^{T-1}(G)$. So the sum (3) is bounded by

$$\|\eta^{-K}\nabla v; G\|_{W_p^{T-1}} + c\,\|\eta^{-K-1}v; G\|_{W_p^{T-1}}.$$

Using the induction hypothesis, we complete the proof. □

Lemma 2. *For all $u \in W_p^l(G) \cap \mathring{W}_p^h(G)$, the following inequality holds:*

$$\|u\circ\lambda; R_+^n\|_{W_p^l} \leqslant c\,\|u; G\|_{W_p^l}. \tag{4}$$

Proof. We have

$$\int_{R_+^n} |D_\zeta^\alpha[u(\lambda(\zeta))]|^p\, d\zeta$$

$$\leqslant c \sum_{1\leqslant|\beta|\leqslant l} \int_{R_+^n} \Big|(D^\beta u)(\lambda(\zeta)) \sum_s c_s \prod_{i,j} D_\zeta^{s_{ij}}\lambda_i(\zeta)\Big|^p d\zeta,$$

where α is an arbitrary positive multi-index of order l, $s = (s_{ij})$ is the set of multi-indices satisfying (6.4.1/7). Hence,

$$\int_{R_+^n} |D_\zeta^\alpha[u(\lambda(\zeta))]|^p\, d\zeta$$

$$\leqslant c \sum_{1\leqslant|\beta|\leqslant l} \int_G \Big|(D^\beta u)(z) \sum_s c_s \prod_{i,j} (D^{s_{ij}}\lambda_i)(\varkappa(z))\Big|^p dz\, \|\varkappa'; G\|_{L_\infty}^p. \tag{5}$$

Let $|\beta| \geqslant h$ or, what is the same, $l-|\beta| \leqslant l-h$. Since $\nabla\lambda_i \in MW_p^k(R_+^n)$ for $k \leqslant l-h$, then $D^{s_{ij}}\lambda_i \in M(W_p^k(R_+^n) \to W_p^{k-|s_{ij}|+1}(R_+^n))$ for $|s_{ij}|-1 \leqslant k \leqslant l-h$. Consequently,

$$\prod_{i,j} D^{s_{ij}}\lambda_i \in M(W_p^{\sum_{i,j}(|s_{ij}|-1)}(R_+^n) \to L_p(R_+^n)) = M(W_p^{l-|\beta|}(R_+^n) \to L_p(R_+^n)).$$

Taking into account that $\varkappa$ is a $(p, l-h)$-diffeomorphism, we obtain

$$\prod_{i,j} D^{s_{ij}}\lambda_i \circ \varkappa \in M(W_p^{l-|\beta|}(G) \to L_p(G)).$$

Therefore the summands on the right in (5) which correspond to multi-indices β of order $|\beta| \geqslant h$ are bounded by $c\,\|u; G\|_{W_p^l}^p$.

Now suppose $|\beta| \leqslant h-1$. By (1) the function $z \to \eta(z)^{|\beta|-h}(D^\beta u)(z)$ belongs to the space $W_p^{l-h}(G)$. We denote by σ_{ij} the integers that satisfy the conditions $1 \leqslant \sigma_{ij} \leqslant |s_{ij}|$, $\sum_{i,j}(\sigma_{ij}-1) = h-|\beta|$. Such integers exist, since $\sum_{i,j}(|s_{ij}|-1) = l-|\beta|$ and $l \geqslant h$. By Corollary 7.4.2, the function $\zeta \to \eta^{\sigma_{ij}-1}(D^{s_{ij}}\lambda_i)(\zeta)$ belongs to $M(W_p^k(R_+^n) \to W_p^{k-|s_{ij}|+\sigma_{ij}}(R_+^n))$. Taking into account the identity

$$\prod_{i,j} (D^{s_{ij}}\lambda_i)(\zeta) = \eta^{|\beta|-h} \prod_{i,j} \eta^{\sigma_{ij}-1}(D^{s_{ij}}\lambda_i)(\zeta),$$

we observe that the function $\zeta \to \eta^{h-|\beta|}\prod(D^{s_{ij}}\lambda_i)(\zeta)$ belongs to $M(W_p^{l-h}(R_+^n) \to L_p(R_+^n))$. Since $\varkappa$ is a $(p, l-h)$-diffeomorphism, the function $z \to \eta(z)^{h-|\beta|}\prod(D^{s_{ij}}\lambda_i)(\varkappa(z))$ is an element of $M(W_p^{l-h}(G) \to L_p(G))$. Therefore the summands with $|\beta| \leqslant h-1$ on the right-hand side of (5) are bounded by $c\,\|u; G\|_{W_p^l}^p$. Because λ is a bi-Lipschitz mapping, we have $\|u; G\|_{L_p} \sim \|u \circ \lambda; R_+^n\|_{L_p}$. This and (5) give the estimate (4). □

Now we prove an analogous assertion relating to the mapping $\varkappa$.

Lemma 3. *For each $v \in W_p^l(R_+^n) \cap \mathring{W}_p^h(R_+^n)$, the following inequality holds:*

$$\|v \circ \varkappa; G\|_{W_p^l} \leqslant c\,\|v; R_+^n\|_{W_p^l}. \tag{6}$$

Proof. It is sufficient to derive the estimate $\int_G |D_z^\alpha[v(\varkappa(z))]|^p\,dz \leqslant c\,\|v; R_+^n\|_{W_p^l}^p$. The left-hand side is dominated by

$$c \sum_{1 \leqslant |\beta| \leqslant l} \int_{R_+^n} \left|(D_\zeta^\beta v)(\zeta) \sum_s c_s \prod_{i,j} (D^{s_{ij}}\varkappa_i)(\lambda(\zeta))\right|^p d\zeta\, \|\lambda'; R_+^n\|_{L_\infty}^p \tag{7}$$

(cf. (5)). Let $|\beta| \geqslant h$. Repeating the same arguments as in Lemma 2 with the change of R_+^n to G, λ to $\varkappa$, u to v and *vice versa*, we obtain that the summands in (7) with $|\beta| \geqslant h$ are bounded by $c\,\|v; R_+^n\|_{W_p^l}^p$.

Now let $|\beta| \leqslant h-1$. Since $v \in W_p^l(R_+^n) \cap \mathring{W}_p^h(R_+^n)$, then $\eta^{h-|\beta|}(D^\beta v)(\zeta)$ belongs to $W_p^{l-h}(R_+^n)$ and its norm is dominated by $c\,\|v; R_+^n\|_{W_p^l}$. According to Lemma 7.4.2, $\eta^{|s_{ij}|-1}(D^{s_{ij}}\varkappa_i)(\lambda(\zeta))$ is a multiplier in $W_p^{l-h}(R_+^n)$. Hence $\eta^{|\beta|-h}\prod(D^{s_{ij}}\varkappa_i)(\lambda(\zeta))$ is a multiplier in the same space. □

7.4.4 The Space W_p^{-k} for a Special Lipschitz Domain

We assume G to be a special Lipschitz domain. In other words, put $l = h$ in the conjectures of the preceding subsection. We retain the notation of 7.4.3.

We introduce the space $W_p^{-k}(G)$ of linear functionals on $\mathring{W}_{p'}^k(G)$, where $p + p' = pp'$, $k = 0, 1, \dots$.

One can immediately check the continuity of the operator $D^\alpha : W_p^s(G) \to W_p^{s-|\alpha|}(G)$ for any $s = 0, \pm 1, \dots$.

By Lemmas 7.4.3/2 and 7.4.3/3, the mapping λ performs the isomorphism $\mathring{W}_{p'}^k(G) \approx \mathring{W}_{p'}^k(R_+^n)$. Therefore λ maps $W_p^{-k}(G)$ onto $W_p^{-k}(R_+^n)$ isomorphically.

The following assertion, which is proved in a standard way, gives one of the possible realizations of $W_p^{-k}(G)$.

Proposition. *Any linear functional on $\mathring{W}_{p'}^k(G)$ can be identified with a* *distribution* $f \in D'(G)$ *of the form*

$$f(z) = \sum_{|\alpha| \leq k} D^\alpha f_\alpha(z), \tag{1}$$

where f_α *is a function such that* $\eta^{k-|\alpha|} f_\alpha \in L_p(G)$.

The norm of this functional is equivalent to the norm

$$\|f\| = \inf \left\| \left(\sum_{|\alpha| \leq k} \eta^{2(k-|\alpha|)} f_\alpha^2 \right)^{1/2}; G \right\|_{L_p},$$

the infimum being taken over all collections $\{f_\alpha\}_{|\alpha| \leq k}$ *in* (1).

Proof. By Lemma 7.4.1/1, the space $\mathring{W}^k_{p'}(G)$ can be supplied with the norm

$$\left\| \left(\sum_{|\alpha| \leq k} \eta^{2(k-|\alpha|)} (D^\alpha u)^2 \right)^{1/2}; G \right\|_{L_{p'}}.$$

Therefore the right-hand side of (1) is the linear functional on $\mathring{W}^k_{p'}(G)$ and

$$\|f\| \leq \left\| \left(\sum_{|\alpha| \leq k} \eta^{2(k-|\alpha|)} f_\alpha^2 \right)^{1/2}; G \right\|_{L_p}.$$

To express an arbitrary linear functional on $\mathring{W}^k_{p'}$ in the form (1), we consider the space $\mathbf{L}_{p'}(G)$ of vectors $\mathbf{v} = \{v_\alpha\}_{|\alpha| \leq k}$ with components in $L_{p'}(G)$ endowed with the norm $\|(\sum_{|\alpha| \leq k} v_\alpha^2)^{1/2}; G\|_{L_{p'}}$. Further, let

$$\Lambda_k = \{(-1)^{|\alpha|} \eta^{|\alpha|-k} D^\alpha\}_{|\alpha| \leq k}.$$

Since the space $\mathring{W}^k_{p'}(G)$ is complete, the range of the operator $\Lambda_k : \mathring{W}^k_{p'}(G) \to \mathbf{L}_{p'}(G)$ is a closed subspace of $\mathbf{L}_{p'}(G)$. Let $u \in \mathring{W}^k_{p'}(G)$ and let $f(u)$ be the value of the functional $f \in W_p^{-k}(G)$ on u. We define the functional Φ by $\Phi(\mathbf{v}) = f(u)$ on the set of vectors $\mathbf{v}$ which can be expressed in the form $\Lambda_k u$. Then $\|\Phi\| = \|f\|$ and, by the Hahn–Banach theorem, Φ can be extended to a linear functional on $\mathbf{L}_{p'}(G)$ with the same norm. Consequently,

$$\Phi(\Lambda_k u) = \sum_{|\alpha| \leq k} \int_G g_\alpha (-1)^{|\alpha|} \eta^{|\alpha|-k} D^\alpha u \, dz,$$

where $g_\alpha \in L_p(G)$. It remains to put $f_\alpha = \eta^{|\alpha|-k} g_\alpha$. □

Lemma. *Let* $v \in \mathring{W}^{t+k}_p(G)$, *where* $t < 0$, k *is a non-negative integer,* $t + k \geq 0$. *Then inequality* (7.4.3/1) *is valid.*

Proof. We have

$$\|\eta^{-k}v; G\|_{W_p^t} = \sup_{w \in \mathring{W}_{p'}^{-t}} \frac{(\eta^{-k}v, w)}{\|w\|_{W_{p'}^{-t}}}.$$

By Lemma 7.4.1,

$$(\eta^{-k}v, w) \leqslant \|\eta^{-k-t}v; G\|_{L_p} \|\eta^t w\|_{L_{p'}}$$
$$\leqslant c\,\|v; G\|_{W_p^{t+k}} \|w\|_{W_{p'}^{-t}}. \quad \square$$

Corollary. Let $\eta^k v \in W_p^{t+k}(G)$, where t and k are the same numbers as in the lemma. Then $\|v; G\|_{W_p^t} \leqslant c\,\|\eta^k v; G\|_{W_p^{t+k}}$.

Proof. Let $w \in \mathring{W}_{p'}^{-t}$. We have

$$(v, w) \leqslant \|\eta^k v; G\|_{W_p^{t+k}} \|\eta^{-k} w; G\|_{W_{p'}^{-t-k}}.$$

Using the lemma with p and t replaced by p' and $-t-k$, we obtain

$$\|\eta^{-k} w; G\|_{W_{p'}^{-t-k}} \leqslant c\,\|w; G\|_{W_{p'}^{-t}}. \quad \square$$

7.4.5 Auxiliary Assertions on Differential Operators in Divergence Form

Lemma 1. Let $G \in M_p^{l-h+1/p'}$, $l \geqslant h$, and

$$Pu = \sum_{|\alpha|,|\beta| \leqslant h} (-1)^{|\alpha|} D^\alpha (a_{\alpha\beta}(z) D^\beta u). \tag{1}$$

If $\eta^{h-|\beta|} a_{\alpha\beta} \in M(W_p^{l-h}(G) \to W_p^{l-2h+|\alpha|}(G))$ for $l-2h+|\alpha| \geqslant 0$ and $\eta^{2h-|\alpha|-|\beta|} a_{\alpha\beta} \in MW_p^{l-h}(G)$ for $l-2h+|\alpha|<0$, then P is a continuous operator $(W_p^l \cap \mathring{W}_p^h)(G) \to W_p^{l-2h}(G)$ and its norm does not exceed cA, where

$$A = \sum_{|\beta| \leqslant h} \Big(\sum_{|\alpha| \geqslant 2h-l} \|\eta^{h-|\beta|} a_{\alpha\beta}; G\|_{M(W_p^{l-h} \to W_p^{l-2h+|\alpha|})}$$
$$+ \sum_{|\alpha|<2h-l} \|\eta^{2h-|\alpha|-|\beta|} a_{\alpha\beta}; G\|_{MW_p^{l-h}} \Big).$$

Proof. Let $u \in (W_p^l \cap \mathring{W}_p^h)(G)$. According to Lemma 7.4.3/1, $\eta^{|\beta|-h} D^\beta u \in W_p^{l-h}(G)$. Consequently, $a_{\alpha\beta} D^\beta u \in W_p^{l-2h+|\alpha|}(G)$ and $D^\alpha(a_{\alpha\beta} D^\beta u) \in W_p^{l-2h}(G)$ for $|\alpha| \geqslant 2h-l$.

Now let $|\alpha|<2h-l$. By virtue of Corollary 7.4.4, we have

$$\|D^\alpha(a_{\alpha\beta} D^\beta u); G\|_{W_p^{l-2h}}$$
$$\leqslant \|a_{\alpha\beta} D^\beta u; G\|_{W_p^{l-2h+|\alpha|}} \leqslant c\,\|\eta^{h-|\alpha|} a_{\alpha\beta} D^\beta u; G\|_{W_p^{l-h}}$$
$$\leqslant c\,\|\eta^{2h-|\alpha|-|\beta|} a_{\alpha\beta}; G\|_{MW_p^{l-h}} \|\eta^{|\beta|-h} D^\beta u; G\|_{W_p^{l-h}}.$$

Using inequality (7.4.3/1) to estimate the latter norm, we arrive at

$$\|D^\alpha(a_{\alpha\beta}D^\beta u); G\|_{W_p^{l-2h}} \leqslant cA\,\|u; G\|_{W_p^l}. \quad \square$$

Henceforth in this subsection, R is a differential operator of order $2h$ with constant coefficients of the form

$$R(D) = \sum_{|\alpha|,|\beta|=h} (-1)^{|\alpha|} D^\alpha(a_{\alpha\beta}D^\beta)$$

and S is the operator defined by (7.2/1).

We retain the notation as well as the assumptions imposed on the domain G in Subsection 7.4.3.

Lemma 2. *For all $v \in (W_p^l \cap \mathring{W}_p^h)(R_+^n)$, $l \geqslant h$,*

$$\|(S-R)v; R_+^n\|_{W_p^{l-2h}} \leqslant c\,\|I-\lambda'; R_+^n\|_{MW_p^{l-h}}\,\|v; R_+^n\|_{W_p^l},$$

where c is a continuous function of the norm of λ' in $MW_p^{l-h}(R_+^n)$ which is independent of v.

Proof. Changing variables in the bilinear form $(R\varphi, \psi)$, where $\varphi, \psi \in D(G)$, we obtain

$$S(\zeta, D_\zeta) = \frac{1}{\det \lambda'(\zeta)} \sum_{|\mu|,|\delta|\leqslant h} (-1)^{|\mu|} D_\zeta^\mu[\det \lambda'(\zeta) f_{\mu\delta}(\zeta) D_\zeta^\delta],$$

where

$$f_{\mu\delta}(\zeta) = \sum_{|\alpha|=|\beta|=h} (a_{\alpha\beta}\circ\lambda) \sum_{\{s\}} c_s \prod_{i,j} (D^{s_{ij}}\varkappa_i)(\lambda(\zeta)) \sum_{\{t\}} c_t \prod_{i,j} (D^{t_{ij}}\varkappa_i)(\lambda(\zeta)).$$

By $\{s\}$ and $\{t\}$ we denote the collections of multi-indices s_{ij} and t_{ij} such that

$$\sum_{i,j} s_{ij} = \alpha, \qquad |s_{ij}| \geqslant 1, \qquad \sum_{i,j} (|s_{ij}|-1) = h - |\mu|,$$

$$\sum_{i,j} t_{ij} = \beta, \qquad |t_{ij}| \geqslant 1, \qquad \sum_{i,j} (|t_{ij}|-1) = h - |\delta|.$$

Since $uD^\mu v = \sum_{\mu\geqslant\gamma>0} c_{\mu\gamma} D^\gamma(vD^{\mu-\gamma}u)$, then

$$S(\zeta, D_\zeta) = \sum_{|\gamma|,|\delta|\leqslant h} (-1)^{|\gamma|} D_\zeta^\gamma(b_{\gamma\delta}(\zeta) D_\zeta^\delta)$$

where

$$b_{\gamma\delta}(\zeta) = \sum_{\mu\geqslant\gamma>0} (-1)^{|\mu|-|\gamma|} c_{\mu\gamma} \left[D_\zeta^{\mu-\gamma}\left(\frac{1}{\det \lambda'(\zeta)}\right)\right] \det \lambda'(\zeta) f_{\mu\delta}(\zeta).$$

In particular, if $|\gamma|=|\delta|=h$, then

$$b_{\gamma\delta}(\zeta)=\sum_{|\alpha|=|\beta|=h} a_{\alpha\beta}P_{\alpha\beta\gamma\delta}(\varkappa'\circ\lambda),$$

where $P_{\alpha\beta\gamma\delta}$ is a polynomial of elements of the matrix $\varkappa'$ and $P_{\alpha\beta\gamma\delta}(I)=1$ if $\alpha=\gamma$, $\beta=\delta$ and $P_{\alpha\beta\gamma\delta}(I)=0$ if $\alpha\neq\gamma$ or $\beta\neq\delta$.

Let

$$S_0(\zeta, D_\zeta)=\sum_{|\gamma|=|\delta|=h}(-1)^h D_\zeta^\gamma(b_{\gamma\delta}(\zeta)D_\zeta^\delta), \qquad S_1=S-S_0.$$

It is clear that

$$\begin{aligned}\|(S_0-R)v; R_+^n\|_{W_p^{l-2h}} &\leq \sum_{\alpha,\beta,\gamma,\delta}\|(P_{\alpha\beta\gamma\delta}(\varkappa'\circ\lambda)-P_{\alpha\beta\gamma\delta}(I))D_\zeta^\delta v; R_+^n\|_{W_p^{l-h}}\\ &\leq c\,\|I-\lambda'; R_+^n\|_{MW_p^{l-h}}\,\|v; R_+^n\|_{W_p^l}\\ &\leq c\,\|\nabla\varphi; R^{n-1}\|_{MW_p^{l-h+1/p'}}\,\|v; R_+^n\|_{W_p^l}.\end{aligned}$$

Next we derive an analogous estimate for the norm $\|(S-S_0)v; R_+^n\|_{W_p^{l-2h}}$. According to Lemma 1, it suffices to prove the following inequalities:

$$\|\eta^{h-|\delta|}b_{\gamma\delta}; R_+^n\|_{M(W_p^{l-h}\to W_p^{l-2h+|\gamma|})}\leq c\,\|\nabla\varphi; R^{n-1}\|_{MW_p^{l-h+1/p'}} \tag{2}$$

for $|\gamma|\leq h$, $|\delta|\leq h$, $|\gamma|+|\delta|<2h$, $l-2h+|\gamma|\geq 0$,

$$\|\eta^{2h-|\gamma|-|\delta|}b_{\gamma\delta}; R_+^n\|_{MW_p^{l-h}}\leq c\,\|\nabla\varphi; R^{n-1}\|_{MW_p^{l-h+1/p'}} \tag{3}$$

for $|\gamma|\leq h$, $|\delta|\leq h$, $|\gamma|+|\delta|<2h$, $l-2h+|\gamma|<0$.

By Corollary 7.4.2,

$$D_\zeta^{\mu-\gamma}(1/\det\lambda'(\zeta))\in M(W_p^{l-h}(R_+^n)\to W_p^{l-h-|\mu|+|\gamma|}(R_+^n)) \tag{4}$$

and, for $\mu>\gamma$,

$$\|D_\zeta^{\mu-\gamma}(1/\det\lambda'(\zeta)); R_+^n\|_{M(W_p^{l-h}\to W_p^{l-h-|\mu|+|\gamma|})}\leq c\,\|\nabla\varphi; R^{n-1}\|_{MW_p^{l-h+1/p'}}. \tag{5}$$

We prove that

$$\eta^{h-|\delta|}f_{\mu\delta}\in M(W_p^{l-h-|\mu|+|\gamma|}(R_+^n)\to W_p^{l-2h+|\gamma|}(R_+^n)). \tag{6}$$

Applying Corollary 7.4.2 once more, we obtain

$$\eta^{h-|\delta|}\prod_{i,j}(D^{t_{ij}}\varkappa_i)\circ\lambda\in MW_p^{l-h}(R_+^n),$$

$$\begin{aligned}\prod_{i,j}(D^{s_{ij}}\varkappa_i)\circ\lambda &\in M(W_p^{l-h}(R_+^n)\to W_p^{l-2h+|\mu|}(R_+^n))\\ &\subset M(W_p^{l-h-|\mu|+|\gamma|}(R_+^n)\to W_p^{l-2h+|\gamma|}(R_+^n))\end{aligned} \tag{7}$$

and therefore inclusion (6) holds. Since, for $\mu=\gamma$, at least one of the exponents t_{ij}, s_{ij} is more than unity, then

$$\|\eta^{h-|\delta|}f_{\gamma\delta};R_+^n\|_{M(W_p^{l-h}\to W_p^{l-2h+|\gamma|})}\leq c\,\|\nabla\varphi;R^{n-1}\|_{MW_p^{l-h+1/p'}}. \tag{8}$$

Now (2) immediately follows from (4)–(6), (8).

Next we turn to the proof of (3). By virtue of Corollary 7.4.2, the inclusion (7) is valid and moreover

$$\eta^{|\mu|-|\gamma|}D_\zeta^{\mu-\gamma}(1/\det\lambda'(\zeta))\in MW_p^{l-h}(R_+^n),$$

$$\eta^{h-|\mu|}\prod_{i,j}(D^{s_{ij}}\varkappa_i)\circ\lambda\in MW_p^{l-h}(R_+^n).$$

To obtain (3) it remains to note that either we always have $\mu>\gamma$ or one of exponents t_{ij}, s_{ij} is more than unity, and to apply Corollary 7.4.2 once more. □

With minor modifications to the above proof and with the use of properties of (p, l)-diffeomorphisms presented in 6.4.1, we arrive at the following local variant of Lemma 2:

Lemma 3. For all $v\in(W_p^l\cap\mathring{W}_p^h)(R_+^n)$, $l\geq h$, with supports in $B_r\cap\overline{R_+^n}$, the inequality

$$\|(S-R)v;R_+^n\|_{W_p^{l-2h}}\leq c\,\|\|I-\lambda';B_r\cap R_+^n\|\|_{MW_p^{l-h}}\,\|v;R_+^n\|_{W_p^l} \tag{9}$$

is valid. For $p(l-h)>n$ it follows from (6.7.1/3) that (9) is equivalent to

$$\|(S-R)v;R_+^n\|_{W_p^{l-2h}}\leq cr^{l-h-n/p}\,\|\|I-\lambda';B_r\cap R_+^n\|\|_{W_p^{l-h}}\,\|v;R_+^n\|_{W_p^l}. \tag{10}$$

7.5 Solvability of the Dirichlet Problem in $W_p^l(\Omega)$

7.5.1 Generalized Formulation of the Dirichlet Problem

Let Ω be an open subset of R^n and let P be the operator (7.4.5/1), where $a_{\alpha\beta}\in C^{l-h}(\bar{\Omega})$, $l\geq h$. Further, let for all $u\in C_0^\infty(\Omega)$ the Gårding inequality

$$\operatorname{Re}\int_\Omega\sum_{|\alpha|=|\beta|=h}a_{\alpha\beta}D^\alpha u\overline{D^\beta u}\,dz\geq c\,\|u;\Omega\|^2_{W_2^h} \tag{1}$$

hold.

We say that $u\in W_p^l(\Omega)$ is a solution of the Dirichlet problem in $W_p^l(\Omega)$ if

$$Pu=f,\qquad u-g\in W_p^l(\Omega)\cap\mathring{W}_p^h(\Omega), \tag{2}$$

where f and g are given functions in spaces $W_p^{l-2h}(\Omega)$ and $W_p^l(\Omega)$ respectively.

By $W_p^{-k}(\Omega)$, $k=1,2,\ldots$, we mean the space of linear continuous functionals on $\mathring{W}_{p'}^k(\Omega)$.

7.5.2 An a priori Estimate for the Solution of the Generalized Dirichlet Problem

Following the proof of Theorem 7.3.2/1 and using Lemma 7.4.5/2 in place of Lemma 7.2.1/1, we arrive at:

Theorem 1. *If $p(l-h)\leq n$, $1<p<\infty$, and if Ω satisfies the condition $N_p^{l-h+1/p'}$, then*

$$\|u;\Omega\|_{W_p^l}\leq c(\|Pu;\Omega\|_{W_p^{l-2h}}+\|u;\Omega\|_{L_1}) \tag{1}$$

for all $u\in(W_p^l\cap\mathring{W}_p^h)(\Omega)$.

Duplicating the proof of Theorem 7.3.2/2 and using estimate (7.4.5/10) instead of (7.2.1/6), we obtain:

Theorem 2. *If $p(l-h)>n$, $1<p<\infty$, and $\Omega\in W_p^{l-h+1/p'}$, then Theorem 1 is valid.*

Next we state two corollaries of (1) which are analogous to Propositions 7.3.2/1 and 7.3.2/2.

Proposition 1. *Let Ω satisfy the conditions of either Theorem 1 or Theorem 2.*

(i) *If the kernel of the operator*

$$P:(W_p^l\cap\mathring{W}_p^h)(\Omega)\to W_p^{l-2h}(\Omega) \tag{2}$$

is trivial, then the norm $\|u;\Omega\|_{L_1}$ in (1) can be omitted.

(ii) *The kernel of the operator (2) is finite-dimensional and the range of this operator is closed.*

Proposition 2. *Let Ω satisfy the conditions of either Theorem 1 or Theorem 2. Further, let U and V be open subsets of R^n, $\bar{U}\subset V$ and $u\in(W_p^l\cap\mathring{W}_p^h)(\Omega)$. Then*

$$\|u;U\cap\Omega\|_{W_p^l}\leq c(\|Pu;V\cap\Omega\|_{W_p^{l-2h}}+\|u;V\cap\Omega\|_{L_1}).$$

7.5.3 Solvability of the Generalized Dirichlet Problem

Let the Gårding inequality

$$\operatorname{Re}(Pu,u)\geq c\,\|u;\Omega\|^2_{W_2^h} \tag{1}$$

hold for all $u \in C_0^\infty(\Omega)$. Then it is well known that the equation $Pu = f$ with $f \in W_2^{-h}(\Omega)$ is uniquely solvable in $\mathring{W}_2^h(\Omega)$. (One can easily see that (7.5.1/1) follows from (1).)

Theorem. *Let Ω satisfy the condition $N_p^{l-h+1/p'}$ for $p(l-h) \leq n$ and let Ω belong to the class $W_p^{l-h+1/p'}$ for $p(l-h) > n$.*

(i) *If $f \in W_p^{l-2h}(\Omega) \cap W_2^{-h}(\Omega)$, $g \in W_p^l(\Omega) \cap W_2^h(\Omega)$, $1 < p < \infty$, and if $u \in W_2^h(\Omega)$ is such that $Pu = f$, $u - g \in \mathring{W}_2^h(\Omega)$, then $u \in W_p^l(\Omega)$ and $u - g \in \mathring{W}_p^h(\Omega)$.*

(ii) *The problem (7.5.1/2) has one and only one solution, $u \in W_p^l(\Omega)$.*

Proof. It is sufficient to assume $g = 0$.

(i) Let at first $p(l-h) \leq n$. We put $\varphi_\varepsilon(x) = \varepsilon + \Phi(x, \varepsilon)$, where Φ is an extension of φ defined by $\Phi = T\varphi$ (cf. 6.5.1). We introduce the domain $G_\varepsilon = \{z = (x, y) : x \in R^{n-1}, \ y > \varphi_\varepsilon(x)\}$. Since $1 + \partial\Phi/\partial\eta > 0$, then $\{G_\varepsilon\}$ is a monotonous family; $\overline{G_\varepsilon} \subset G$, $G_\varepsilon \to G$ as $\varepsilon \to +0$. By the first part of Theorem 5.1.4, we have

$$\|\nabla \varphi_\varepsilon ; R^{n-1}\|_{MW_p^{l-h+1/p'}} \leq c \, \|\nabla \Phi; R_+^n\|_{MW_p^{l-h+1}}.$$

This and inequality (6.5.1/2) imply

$$\|\nabla \varphi_\varepsilon ; R^{n-1}\|_{MW_p^{l-h+1/p'}} \leq c \, \|\nabla \varphi ; R^{n-1}\|_{MW_p^{l-h+1/p'}}, \tag{2}$$

with constant c independent of ε. Let $\Omega_\varepsilon = \Omega \backslash (\bar{U} \backslash G_\varepsilon)$ and let u_ε be a solution of $Pu = f$ in $\mathring{W}_2^h(\Omega_\varepsilon)$. It is known (see, for instance, Nečas [1], 6.6, Ch. 3), that $u_\varepsilon \to u$ in $\mathring{W}_2^h(\Omega)$. We denote by U_1 an open set such that $\overline{U_1} \subset U$. Since $\varphi_\varepsilon \in C^\infty(R^{n-1})$, then, by the known theorem on the regularity of weak solutions of elliptic boundary value problems near a smooth part of a boundary, we have $u_\varepsilon \in W_p^l(U_1 \cap \Omega_\varepsilon)$. This and Proposition 7.5.2/2 imply

$$\|u_\varepsilon ; U_2 \cap \Omega_\varepsilon\|_{W_p^l} \leq c(\|f; U_1 \cap \Omega_\varepsilon\|_{W_p^{l-2h}} + \|u_\varepsilon ; U_1 \cap \Omega_\varepsilon\|_{L_1}),$$

where U_2 is an open set, $\overline{U_2} \subset U_1$, and c does not depend on ε. So the left-hand side is uniformly bounded with respect to ε. Now, if we fix a domain ω such that $\bar{\omega} \subset \Omega$, then the upper limit $\overline{\lim}_{\varepsilon \to +0} \|u_\varepsilon ; \omega\|_{W_p^l}$ is bounded by a constant independent of ω. From $\{u_\varepsilon\}$ we select a sequence which is weakly convergent in $W_p^l(\omega)$. This sequence converges in $W_2^h(\omega)$, hence its weak limit in $W_p^l(\omega)$ coincides with u. Therefore, $u \in W_p^l(\omega)$, and the $W_p^l(\omega)$-norm of u is uniformly bounded with respect to ω. So $u \in W_p^l(\Omega)$. The identity of spaces $W_p^h(\Omega) \cap \mathring{W}_2^h(\Omega)$ and $\mathring{W}_p^h(\Omega)$ for domains Ω of the class $C^{0,1}$ is known.

The case $p(l-1) > n$ can be treated in the same way, with (2) replaced by

$$\|\varphi_\varepsilon ; R^{n-1}\|_{W_p^{l-h+1/p'}} \leq c \, \|\varphi ; R^{n-1}\|_{W^{l-h+1/p'}}.$$

(ii) For $p \geq 2$, the assertion follows from the unique solvability of the problem in $\mathring{W}_2^h(\Omega)$, together with the first part of the theorem.

Consider the case $p<2$. We denote by P^t the operator formally conjugate to P. The coefficients of P^t belong to $C^{l-2h}(\bar{\Omega})$ and Gårding's inequality (1) holds for P^t, too. Recall that Ω satisfies $N_p^{1/p'}$, i.e. the Lipschitz constants of functions φ that locally specify $\partial\Omega$ are small. Hence the equation $P^t v = F \in W_{p'}^{-h}(\Omega)$, with $v \in \mathring{W}_{p'}^h(\Omega)$, is uniquely solvable in $\mathring{W}_{p'}^h(\Omega)$. Let u be a solution of the homogeneous problem (7.5.1/2) and let $\{v_m\}_{m \geq 1}$ be a sequence of functions in $C_0^\infty(\Omega)$, which converges to v in $\mathring{W}_{p'}^h(\Omega)$. Then

$$0 = \lim \sum_{|\alpha|,|\beta| \leq h} (a_{\alpha\beta} D^\alpha u, D^\beta v_m) = \sum_{|\alpha|,|\beta| \leq h} (a_{\alpha\beta} D^\alpha u, D^\beta v) = (u, F),$$

and the uniqueness of the solution of problem (7.5.1/2) follows.

Let $f_m \in C^l(\bar{\Omega})$, $m = 1, 2, \ldots$, $f_m \to f$ in $W_p^{l-2h}(\Omega)$. We denote by u_m a function in $\mathring{W}_2^h(\Omega)$ satisfying $Pu_m = f_m$. According to the first part of the theorem, $u_m \in W_p^l(\Omega) \cap \mathring{W}_p^h(\Omega)$. By the first part of Proposition 7.5.2/1,

$$\|u_m - u_k; \Omega\|_{W_p^l} \leq c\, \|f_m - f_k; \Omega\|_{W_p^{l-2h}}.$$

So $\{u_m\}$ converges in $W_p^l(\Omega) \cap \mathring{W}_p^h(\Omega)$ and its limit satisfies $Pu = f$. □

7.5.4 *The Dirichlet Problem in Terms of Traces*

The first boundary value problem (7.5.1/2) is not a particular case of the general boundary value problem formulated in 7.3.1. In the present subsection we study the Dirichlet problem in another formulation which is analogous to that considered in 7.3.1.

Let P be the elliptic operator (7.4.5/1) with coefficients $a_{\alpha\beta}$ in $C^{l-h}(\bar{\Omega})$, $l \geq h$, for which inequality (7.5.3/1) is valid. We assume Ω to be of the class $C^{0,1}$.

We introduce a sufficiently small finite open covering $\{U\}$ of $\bar{\Omega}$ and a corresponding partition of unity $\{\zeta_U\}$. Let $P_{jU} = \partial^{j-1}/\partial y^{j-1}$ $(j = 1, \ldots, h)$ if $U \cap \partial\Omega \neq \emptyset$, and $P_{jU} = 0$ if $U \cap \partial\Omega = \emptyset$. The Dirichlet boundary conditions will be prescribed by operators $P_j = \sum_U \zeta_U P_{jU}$.

The new formulation of the Dirichlet problem is the following. We seek for a function $u \in W_p^l(\Omega)$ such that

$$Pu = f \quad \text{in} \quad \Omega, \qquad \operatorname{tr} P_j u = f_j \quad \text{on} \quad \partial\Omega, \qquad j = 1, \ldots, h, \tag{1}$$

where f and f_j are given functions in spaces $W_p^{l-2h}(\Omega)$ and $W_p^{l-j+1/p'}(\partial\Omega)$ respectively.

It is clear that any solution of the problem (7.5.1/2) is a solution of (1) with $f_j = \operatorname{tr} P_j g$. The following lemma shows that the opposite statement holds provided that $\Omega \in M_p^{l-1/p}$.

Lemma. *Let $G=\{z=(x,y):x\in R^{n-1}, y>\varphi(x)\}$ be a domain of the class $M_p^{l-1/p}$ and let $f_1,\ldots,f_h$ be arbitrary functions in $W_p^{l-1/p}(\partial G),\ldots,W_p^{l-h+1/p'}(\partial G)$. Then there exists a function $g\in W_p^l(G)$ such that $\operatorname{tr}(\partial^{j-1}g/\partial y^{j-1})=f_j$, $j=1,\ldots,h$.*

Proof. We use the notation Φ, λ, $\varkappa$, introduced in 6.5.1. We have

$$[(\partial/\partial y)^{j-1}g]\circ\lambda=[(K+\partial\Phi/\partial\eta)^{-1}(\partial/\partial\eta)]^{j-1}(g\circ\lambda).$$

Since $\nabla\Phi\in MW_p^l(R_+^n)$,

$$[(\partial/\partial y)^{j-1}g]\circ\lambda=\sum_{\nu=1}^{j}a_{\nu j}(\partial/\partial\eta)^{\nu-1}(g\circ\lambda),\qquad j=1,\ldots,h, \tag{2}$$

where $a_{\nu j}\in M(W_p^{l-\nu+1}(R_+^n)\to W_p^{l-j+1}(R_+^n))$, $a_{jj}=(K+\partial\Phi/\partial\eta)^{1-j}$. We note that (2) is a triangular algebraic system with respect to $(\partial/\partial\eta)^{\nu-1}(u\circ\lambda)$, so that

$$(\partial/\partial\eta)^{\nu-1}(g\circ\lambda)=\sum_{j}^{\nu}b_{j\nu}[(\partial/\partial y)^{j-1}g]\circ\lambda,\qquad \nu=1,\ldots,h,$$

where $b_{j\nu}\in M(W_p^{l-j+1}(R_+^n)\to W_p^{l-\nu+1}(R_+^n))$. Taking into account that $\operatorname{tr} b_{j\nu}\in M(W_p^{l-j+1/p'}(R^{n-1})\to W_p^{l-\nu+1/p'}(R^{n-1}))$, we obtain $(\operatorname{tr} b_{j\nu})f_j\circ\lambda\in W_p^{l-\nu+1/p'}(R^{n-1})$. Therefore there exists a function $H\in W_p^l(R_+^n)$ such that

$$\operatorname{tr}(\partial/\partial\eta)^{\nu-1}H=\sum_{\nu=1}^{j}(\operatorname{tr} b_{j\nu})f_j\circ\lambda.$$

Setting $g=H\circ\varkappa$, we complete the proof. □

Since both formulations (7.5.1/2) and (1) of the Dirichlet problem are equivalent for domains of the class $M_p^{l-1/p}$, then, from Theorem 7.5.3, we obtain:

Theorem. *Let any of the following conditions hold:*
(α) $h=1$, $p(l-1)\le n$; Ω satisfies $N_p^{l-1/p}$;
(β) $h=1$, $p(l-1)>n$; $\Omega\in W_p^{l-1/p}$;
(γ) $h>1$, $\Omega\in M_p^{l-1/p}$ and $\partial\Omega$ is locally defined by equations of the form $y=\varphi(x)$, where φ is a function with a small Lipschitz constant (for $p(l-1)>n$, this is equivalent to $\Omega\in W_p^{l-1/p}$).
Then the operator

$$\{P;P_j\}:W_p^l(\Omega)\to W_p^{l-2h}(\Omega)\times\prod_{j=1}^{h}W_p^{l-j+1/p'}(\partial\Omega)$$

is an isomorphism.

Proof. For $h=1$ the assertion follows from Theorem 7.5.3. Let $h>1$.

According to (3.2.1/1),

$$\|\nabla\varphi; R^{n-1}\|_{MW_p^{l-h-1/p}} \leqslant c\,\|\nabla\varphi; R^{n-1}\|^{\alpha}_{MW_p^{l-1-1/p}}\,\|\nabla\varphi; R^{n-1}\|^{1-\alpha}_{L_\infty}$$

with $\alpha=(p(l-h)-1)/(p(l-1)-1)$. Consequently, from (γ) it follows that Ω satisfies $N_p^{l-h+1/p'}$ if $p(l-h)\leqslant n$, and $\Omega\in W_p^{l-h+1/p'}$ if $p(l-h)>n$. By Theorem 7.5.3 the proof is complete. □

Thus, by changing the formulation of the Dirichlet problem, we have obtained its solvability in $W_p^l(\Omega)$ under stricter assumptions on Ω (compare the last theorem with Theorem 7.5.3). The exception is the second-order operator P, i.e., $h=1$, when admissible classes of domains coincide for both formulations.

In the following section we discuss the necessity of conditions of the last theorem.

7.6 Necessity of Assumptions on the Domain

7.6.1 Counterexample for the Domain of the Class $M_2^{3/2}\cap C^1$ which does not Satisfy the Condition $N_2^{3/2}$

In this subsection we give an example which shows that for $p(l-1)\leqslant n$ and for $h=1$ the condition $N_p^{l-1/p}$ in part (α) of Theorem 7.5.4 cannot be changed by the assumption that a domain belongs to the class $M_p^{l-1/p}\cap C^{l-1}$. To be precise, here we construct the domain $\Omega\in M_2^{3/2}\cap C^1$ for which the problem

$$-\Delta u=f \quad \text{in} \quad \Omega, \qquad \operatorname{tr} u=0 \quad \text{on} \quad \partial\Omega \tag{1}$$

is solvable in $W_2^2(\Omega)$ for not all $f\in L_2(\Omega)$. This means that smallness of the norm $\|\nabla\varphi; R^{n-1}\|_{MW_2^{1/2}}$ in the condition $N_2^{3/2}$ is essential for solvability of the problem (1) in $W_2^2(\Omega)$.

Let, in a neighbourhood of O, the domain Ω be specified by the inequality $y>-C\varphi(x)$, where C is a positive constant and $\varphi(x)=\eta(x,0)\,|x_1|/\log(1/|x_1|)$. Here and henceforth η is a function in $C_0^\infty(B_{1/2})$, $\eta=1$ on the ball $B_{1/4}$.

We introduce the domain

$$\omega=\{\xi: x_1+ix_2: |\xi|<1/2,\ x_2>-C\,|x_1|/\log(1/|x_1|)\}$$

and by $\zeta(\xi)$ we denote the conformal mapping of ω onto the half-disc $\{\zeta: \operatorname{Im}\zeta>0, |\zeta|<1\}$ with the fixed point $\xi=0$. Let $\xi=i\rho\exp(i\theta)$ and let $\omega=\{\xi: \rho<1/2, |\theta|<\pi/2+\varphi(\rho)\}$. One can easily check that

$$\varphi(\rho)=C(\log 1/\rho)^{-1}+O((\log 1/\rho)^{-3}).$$

According to an asymptotic formula due to Warschawski [1],

$$\operatorname{Im} \zeta(\xi) = c \exp\left(-\pi \int_\rho^{1/2} \frac{dr}{r(\pi + 2\varphi(r))}\right)\left(\cos \frac{\pi\theta}{\pi + 2\varphi(\rho)} + o(1)\right)$$
$$= c\rho(\log 1/\rho)^{2C/\pi}\left(\cos \frac{\pi\theta}{\pi + 2\varphi(\rho)} + o(1)\right).$$

It is clear that, for $C \geqslant \pi/4$,

$$\int_\omega \frac{(\operatorname{Im} \zeta(\xi))^2}{\rho^4(\log \rho)^2} dx_1\, dx_2 = \infty. \tag{2}$$

Next we need the following:

Lemma. *If $h \in W_2^2(\omega) \cap \mathring{W}_2^1(\omega)$, then*

$$\int_\omega \frac{h^2\, dx_1\, dx_2}{\rho^4(\log \rho)^2} \leqslant c\, \|h; \omega\|^2_{W_2^2}. \tag{3}$$

Proof. First we show that, for any $g \in W_2^1(\omega)$,

$$\int_\omega g^2 \frac{dx_1\, dx_2}{\rho^2(\log \rho)^2} \leqslant c\, \|g; \omega\|^2_{W_2^1}. \tag{4}$$

Clearly, to prove (4) it suffices to assume that ω is the half-disc $\{\xi : \rho < 1/2, |\theta| < \pi/2\}$. After integration of the Hardy inequality

$$\int_0^{1/2} g^2 \frac{d\rho}{\rho(\log \rho)^2} \leqslant 4 \int_0^{1/2} \left(\frac{\partial g}{\partial \rho}\right)^2 \rho\, d\rho$$

with respect to variable θ, we arrive at (4). Putting $g = \partial h/\partial x_1$, $g = \partial h/\partial x_2$ into (4), we obtain

$$\int_\omega (\nabla h)^2 \frac{dx_1\, dx_2}{\rho^2(\log \rho)^2} \leqslant c\, \|h; \omega\|^2_{W_2^2}.$$

Let $C_\rho = \{\xi : |\xi| = \rho\}$. Since $h = 0$ on $\partial\omega$, then, for almost all $\rho > 0$,

$$\int_{\omega \cap C_\rho} h^2\, d\theta \leqslant c \int_{\omega \cap C_\rho} (\partial h/\partial \theta)^2\, d\theta \leqslant c\rho^2 \int_{\omega \cap C_\rho} (\nabla h)^2\, d\theta,$$

which leads to

$$\int_\omega h^2 \frac{dx_1\, dx_2}{\rho^4(\log \rho)^2} \leqslant c \int_\omega (\nabla h)^2 \frac{dx_1\, dx_2}{\rho^2(\log \rho)^2}. \quad \square$$

From (2) and (3) it follows that the function u defined on Ω by $u(z) = \eta(2z) \operatorname{Im} \zeta(x_1 + ix_2)$ does not belong to $W_2^2(\Omega)$. On the other hand,

u is contained in $\mathring{W}_2^1(\Omega)$ and satisfies the equation $-\Delta u=f$, $f\in L_2(\Omega)$. Consequently, the boundary value problem (1) is solvable in $W_2^2(\Omega)$ if and only if $C<\pi/4$.

Obviously, the domain Ω is of the class C^1. We show that this domain belongs to $M_2^{3/2}$, i.e. $\nabla\varphi\in MW_2^{1/2}(R^{n-1})$. With this aim in view we verify that the gradient of the function ψ defined as

$$\psi(z)=\eta(z)r\log r,\quad \text{where}\quad r=(x_1^2+y^2)^{1/2},$$

belongs to the space $MW_2^1(R^n)$ (cf. Theorem 5.1.3). Clearly, $\nabla\psi\in L_\infty(R^n)$ and it remains to prove that $\nabla_2\psi\in M(W_2^1(R^n)\to L_2(R^n))$. In fact, for all $u\in W_2^1(R^n)$,

$$\|u\,\nabla_2\psi;R^n\|_{L_2}^2\leqslant c\int_{B_{1/2}}\left|\frac{u}{r\log r}\right|^2 dz\leqslant c\,\|u;B_1\|_{W_p^1}^2.$$

Thus, $\Omega\in M_2^{3/2}$.

7.6.2 Necessary Conditions for the Solvability of the Dirichlet Problem

The next assertion, which immediately follows from the implicit function Theorem 6.5.2/2, shows that the condition $\Omega\in W_p^{l-1/p}$ with $p(l-1)>n$ is necessary for solvability of the Dirichlet problem (7.5.4/1) in $W_p^l(\Omega)$ for an operator P of higher than second order.

Theorem 1. *Let Ω be a bounded domain of the class $C^{0,1}$. If there exists a solution $u\in W_p^l(\Omega)$ $(p(l-1)>n$, l is an integer, $l\geqslant 2h$, $1<p<\infty$ and $h>1)$ of the problem*

$$Pu=0\ \text{ in }\ \Omega,\qquad \operatorname{tr} u=0,\qquad \operatorname{tr} P_2u=1,\qquad \operatorname{tr} P_ju=0,\qquad j=3,\dots,h,\qquad (1)$$

then $\Omega\in W_p^{l-1/p}$.

Under the additional assumption $\Omega\in C^{l-2,1}$, we can prove the necessity of $\Omega\in W_p^{l-1/p}$ for $p(l-1)\leqslant n$.

Theorem 2. *Let Ω be of the class $C^{l-2,1}$ and let there exist a solution $u\in W_p^l(\Omega)$ $(p(l-1)\leqslant n$, l is an integer, $l\geqslant 2h$, $1<p<\infty$, $h>1)$ of the problem (1). Then $\Omega\in W_p^{l-1/p}$.*

Proof. We use the same notation as in the formulation of Theorem 6.5.2/2. Since $\operatorname{grad}_x u$, $u_y\in W_{p,\mathrm{loc}}^{l-1}(U\cap\bar G)$ and $\varphi\in C^{l-2,1}(R^{n-1})$, then $\operatorname{grad}_x u\circ\lambda$, $u\circ\lambda\in W_{p,\mathrm{loc}}^{l-1}(\varkappa(U\cap\bar G))$. Therefore $\operatorname{tr}(\operatorname{grad}_x u\circ\lambda)$ and $\operatorname{tr}(u_y\circ\lambda)\in W_{p,\mathrm{loc}}^{l-1-1/p}(\omega)$. Now we note that Ω is of the class C^h and so, by the known coercive estimate for solutions of the elliptic boundary value problem in the variational form (cf. Agmon, Douglis, Nirenberg [1]), we have $u \in W_q^h(\Omega)$ for any $q<\infty$. In particular, grad $u \in C(\bar{\Omega})$. It is known that the space $(W_{p,\text{loc}}^{l-1-1/p} \cap L_\infty)(\omega)$ is a multiplication algebra. So the vector function $\text{grad}\, \varphi = \text{tr}\,(\text{grad}_x\, u \circ \lambda)/\text{tr}\,(u_y \circ \lambda)$ belongs to $W_{p,\text{loc}}^{l-1-1/p}(\omega)$. □

We shall consider briefly the case of the second-order operator P.

Theorem 3. Let l be an integer, $l \geq 2$, $1<p<\infty$, $h=1$, and $P(1) \leq 0$. Let Ω be a domain of the class C^1 and let the normal to $\partial\Omega$ satisfy the Diny condition. If, for a non-positive function $f \in C_0^\infty(\Omega)$, there exists a solution $u \in W_p^l(\Omega)$ of the problem

$$Pu=f \quad \text{in} \quad \Omega, \qquad \text{tr}\, u=0, \tag{2}$$

then $\Omega \in W_p^{l-1/p}$.

Proof. It is sufficient to note that the interior normal derivative in any point of $\partial\Omega$ is positive, and then to duplicate the proof of Theorem 2. □

7.7 Domains Satisfying the Condition $N_p^{l-1/p}$

Let, as in 7.5, 7.6, Ω be a bounded domain of $C^{0,1}$. We denote by O an arbitrary point of $\partial\Omega$ and introduce a neighbourhood U of O such that $\Omega \cap U = G \cap U$ with $G=\{(x, y): x \in R^{n-1},\ y>\varphi(x)\}$.

By Theorem 3.2.7/2, the norm of $\nabla\varphi$ in $MW_p^{l-1-1/p}(R^{n-1})$ is equivalent to

$$\sup_{e \subset R^{n-1}} (\|D_{p,l-1/p}\varphi; e\|_{L_p}/[\text{cap}\,(e, W_p^{l-1-1/p}(R^{n-1})]^{1/p}) + \|\nabla\varphi; R^{n-1}\|_{L_\infty}. \tag{1}$$

(Here we can restrict ourselves to compact sets e with $d(e) \leq 1$, cf. Remark 3.2.7/1.) So the condition $N_p^{l-1/p}$ means that (1) is sufficiently small.

The following assertion, of which the proof is postponed to Section 7.8, gives a local formulation of $N_p^{l-1/p}$ in terms of the capacity.

Theorem 1. Let $p(l-1) \leq n$. The condition $N_p^{l-1/p}$ is equivalent to the following one. For any point $O \in \partial\Omega$ there exists a neighbourhood U and a special Lipschitz domain $G=\{z=(x, y): x \in R^{n-1},\ y>\varphi(x)\}$ such that $U \cap \Omega = U \cap G$ and

$$\lim_{\varepsilon \to 0} \Big(\sup_{e \subset B_\varepsilon} (\|D_{l-1/p}(\varphi; B_\varepsilon); e\|_{L_p}/[\text{cap}\,(e, W_p^{l-1-1/p}(R^{n-1}))]^{1/p}$$

$$+ \|\nabla\varphi; B_\varepsilon\|_{L_\infty} \Big) \leq c\delta. \tag{2}$$

Here B_ε is a ball with centre at O and radius ε; c is a constant which depends on l, p, n; δ is a constant in the definition of $N_p^{l-1/p}$ and, finally,

$$D_{j-1/p}(\varphi; B_\varepsilon)(x) = \left(\int_{B_\varepsilon} |\nabla_{j-1}\varphi(x) - \nabla_{j-1}\varphi(y)|^p \, |x-y|^{-n+2-p} \, dy \right)^{1/p}.$$

Theorem 1 and properties (v), (vi) of the capacity formulated in 6.7.1 lead to:

Corollary 1. (i) *If $n > p(l-1)$ and*

$$\lim_{\varepsilon \to 0} \left(\sup_{e \subset B_\varepsilon} \frac{\|D_{l-1/p}(\varphi; B_\varepsilon); e\|_{L_p}}{(\text{mes}_{n-1} e)^{[n-p(l-1)]/(n-1)p}} + \|\nabla\varphi; B_\varepsilon\|_{L_\infty} \right) < c\delta,$$

then $N_p^{l-1/p}$ holds.

(ii) *If $n = p(l-1)$ and*

$$\lim_{\varepsilon \to 0} \left(\sup_{e \subset B_\varepsilon} \|D_{l-1/p}(\varphi; B_\varepsilon); e\|_{L_p} \, |\log(\text{mes}_{n-1} e)|^{(1-p)/p} + \|\nabla\varphi; B_\varepsilon\|_{L_\infty} \right) < c\delta,$$

then $N_p^{l-1/p}$ holds.

One can derive another test for $N_p^{l-1/p}$ in terms of the space $B_{q,p}^m$ (cf. Corollary 2 below).

We say that the domain Ω in $C^{0,1}$ belongs to $B_{q,p}^{l-1/p}$ ($l = 1, 2, \ldots$) if, for any point of $\partial\Omega$, there exists a neighbourhood in which $\partial\Omega$ is specified in Cartesian coordinates by a function φ satisfying

$$\int_{R^{n-1}} \left(\int_{R^{n-1}} |\nabla_{l-1}\varphi(x+h) - \nabla_{l-1}\varphi(x)|^q \, dx \right)^{p/q} |h|^{2-n-p} \, dh < \infty.$$

Corollary 2. *Let $p(l-1) \leq n$ and let Ω be a bounded Lipschitz domain of the class $B_{q,p}^{l-1/p}$ with $q \in [p(n-1)/(p(l-1)-1), \infty]$ if $p(l-1) < n$, and $q \in (p, \infty]$ if $p(l-1) = n$. Further, let $\partial\Omega$ be locally defined in Cartesian coordinates by $y = \varphi(x)$, where φ is a function with a Lipschitz constant less than $c\delta$. Then the condition $N_p^{l-1/p}$ holds.*

Proof. We have

$$\begin{aligned} \|D_{l-1/p}(\varphi; B_\varepsilon); e\|_{L_p}^p &\leq \int_{B_\varepsilon} |h|^{-n+2-p} \, dh \int_e |\nabla_{l-1}\varphi(x+h) - \nabla_{l-1}\varphi(x)|^p \, dx \\ &\leq (\text{mes}_{n-1} e)^{1-p/q} \int_{B_\varepsilon} |h|^{-n+2-p} \, dh \\ &\quad \times \left(\int_{B_\varepsilon} |\nabla_{l-1}\varphi(x+h) - \nabla_{l-1}\varphi(x)|^q \, dx \right)^{p/q}. \end{aligned}$$

Then the result follows from the first part of Corollary 1. □

Corollary 2 can be made sharper in the case $p(l-1)=n$, by virtue of the second part of Corollary 1, if one uses the Orlicz space $L_{t^p(\log_+ t)^{p-1}}$ instead of L_q, but we shall not go into this in detail.

Setting $q=\infty$ in Corollary 2, we obtain a simple sufficient condition for $N_p^{l-1/p}$ in terms of the modulus of continuity $\omega(t)$ of $\nabla_{l-1}\varphi$:

$$\int_0 (\omega(t)/t)^p \, dt<\infty. \tag{3}$$

Since $B_{\infty,p}^{l-1/p}\subset W_p^{l-1/p}$, then (3) is sufficient for $\Omega\in W_p^{l-1/p}$.

We show that (3) is in a sense the exact condition for solvability of the Dirichlet problems (7.6.2/1) and (7.6.2/2) in $W_p^l(\Omega)$.

We have shown in 3.3.2 that, for any increasing function $\omega\in C[0,1]$ satisfying inequalities (3.3.2/4) as well as the condition $\int_0^1 (\omega(t)/t)^p \, dt=\infty$, one can construct a function φ on R^{n-1} such that

(i) the modulus of continuity of $\nabla_{l-1}\varphi$ does not exceed $c\omega$ with $c=\text{const}$;

(ii) $\operatorname{supp}\varphi\subset Q_{2\pi}$, where $Q_d=\{x\in R^{n-1}:|x_i|<d\}$;

(iii) $\varphi\notin W_p^{l-1/p}(R^{n-1})$.

By Ω we denote a bounded domain in R^n such that

$$\Omega\cap\{z:x\in Q_{3\pi},\ |y|<1\}=\{z:x\in Q_{3\pi},\ \varphi(x)<y<1\}.$$

Further, we assume that $\partial\Omega$ is a surface of the class C^∞ in the exterior of the set $\{z:x\in Q_{2\pi},\ y=\varphi(x)\}$.

By virtue of Theorems 7.6.2/1–7.6.2/3, the problems (7.6.2/1), (7.6.2/2) in this Ω have no solutions in $W_p^l(\Omega)$.

To conclude this section we make two remarks:

Remark 1. Suppose that, for any point $O\in\partial\Omega$, there exists a neighbourhood U such that $U\cap\Omega$ is C^l-diffeomorphic to the domain $\{(x,y):y>\varphi(x_1,\ldots,x_{n-s})\}$, $2\leqslant s\leqslant n-1$, i.e. 'the dimensions of singularities of $\partial\Omega$ are not less than $s-1$'. Then all properties of domains that satisfy $N_p^{l-1/p}$ remain valid after the change of $n-1$ to $n-s$. This follows from the formulation of $N_p^{l-1/p}$, from Theorem 5.1.3 and from the obvious fact that, if ψ is defined in R^n and ψ depends on $n-s+1$ variables only, the norms

$$\|\psi;R^n\|_{MW_p^l},\qquad \|\psi;R^{n-s+1}\|_{MW_p^l}$$

are equivalent.

Remark 2. Clearly, the condition $\Omega\in W_p^{l-1/p}$ with $p(l-1)>n$ excludes edges and conic points. However, the condition $N_p^{l-1/p}$, with $p(l-1)<n$, admits such singularities. Consider, for example, the domain

$G=\{z:y>cr_s\}$, where $r_s^2=x_{s+1}^2+\cdots+x_{n-1}^2$, $c=\text{const}$. By Hardy's inequality, the first derivatives of the function $R^n\ni z\to(y^2+r_s^2)^{1/2}$ belong to $MW_p^{l-1}(R^n)$ if $p(l-1)<n-s$. So the first derivatives of the function $R^{n-1}\ni x\to r_s$ are elements of $MW_p^{l-1-1/p}(R^{n-1})$. The smallness of the constant c assures the condition $N_p^{l-1/p}$.

7.8 Proof of Theorem 7.1

7.8.1 A Local Form of the Condition $N_p^{l-1/p}$

Let η be an even function in $C_0^\infty(-1,1)$, $\eta=1$ on $(-1/2,1/2)$. We put

$$\eta_\varepsilon(z)=\begin{cases}\eta(|z|/\varepsilon), & \text{if}\quad p(l-1)<n,\\ \eta(\log\varepsilon/\log|z|), & \text{if}\quad p(l-1)=n.\end{cases}$$

Clearly, $\operatorname{supp}\eta_\varepsilon\subset B_\varepsilon^{(n)}$ and

$$|\nabla_j\eta_\varepsilon(z)|\leqslant\begin{cases}c\varepsilon^{-j}, & \text{if}\quad p(l-1)<n,\\ c\,|\log|z||^{-1}\,|z|^{-j}, & \text{if}\quad p(l-1)=n.\end{cases}\tag{1}$$

Lemma 1. Let $p(l-1)=n$. Then

$$\int_{B_\varepsilon}(|\nabla_j\eta_\varepsilon(x)-\nabla_j\eta_\varepsilon(y)|^p/|x-y|^{n-2+p})\,dy\leqslant c_j\,|\log|x||^{-p}\,|x|^{1-p-pj},\tag{2}$$

where $x\in B_\varepsilon$, $j=0,1,\ldots$.

Proof. Since

$$D^\alpha\eta_\varepsilon(x)=|x|^{-|\alpha|}\sum_{k=1}^{|\alpha|}\sigma_k(\log\varepsilon/\log|x|)(\log\varepsilon)^{-k},$$

where $\sigma_k\in C_0^\infty(-1,1)$, then

$$|\nabla_j\eta_\varepsilon(x)-\nabla_j\eta_\varepsilon(y)|\leqslant\begin{cases}c_j\,|\log\varepsilon|^{-1}\,|x-y|\,|x|^{-j-1}\\ \quad\text{if } |x|/2\leqslant|y|\leqslant 2\,|x|,\\ c_j\,|\log\varepsilon|^{-1}\,(\max\{|x|,|y|\})^{-j}\\ \quad\text{if } j>0 \text{ and } |y|<|x|/2 \text{ or } |x|<|y|/2,\\ c_j\,|\log\varepsilon|^{-1}\,|\log(|x|/|y|)|\\ \quad\text{if } j=0 \text{ and } |y|<|x|/2 \text{ or } |x|<|y|/2.\end{cases}$$

These inequalities imply

$$\int_{B_\varepsilon}(|\nabla_j\eta_\varepsilon(x)-\nabla_j\eta_\varepsilon(y)|^p/|x-y|^{n-2+p})\,dy\leqslant c_j\,|\log\varepsilon|^{-p}\,|x|^{1-p-pj},$$

which is equivalent to (2) for $x \in B_\varepsilon \backslash B_{\varepsilon^3}$. Let $x \in B_{\varepsilon^3}$. We have

$$\int_{B_\varepsilon \backslash B_\varepsilon^2} \frac{|\delta_{0,j} - \nabla_j \eta_\varepsilon(y)|^p}{|y|^{n-2+p}}\, dy \sim \int_{B_\varepsilon} \frac{|\nabla_j \eta_\varepsilon(x) - \nabla_j \eta_\varepsilon(y)|^p}{|x-y|^{n-2+p}}\, dy,$$

where $\delta_{0,j} = 1$ for $j = 0$ and $\delta_{0,j} = 0$ for $j > 0$. Consequently,

$$\int_{B_\varepsilon \backslash B_\varepsilon^2} (|\delta_{0,j} - \nabla_j \eta_\varepsilon(y)|^p / |y|^{n-2+p})\, dy \leq c\, |\log \varepsilon|^{-p}\, |x|^{1-p-pj}.$$

Setting $|x| = \varepsilon^3$ in the last inequality and observing that $t^{3(p+pj-1)}\, |\log t|^p$ increases near $t = 0$, we obtain (2). □

The aim of this subsection is the proof of the following assertion containing a local formulation of $N_p^{l-1/p}$. Hereafter, without loss of generality, we assume $\varphi(0) = 0$.

Lemma 2. *The condition $N_p^{l-1/p}$ can be written as*

$$\overline{\lim_{\varepsilon \to 0}}\, \|\nabla(\eta_\varepsilon \varphi); R^{n-1}\|_{MW_p^{l-1-1/p}} \leq c\delta, \tag{3}$$

where c is a constant which depends on l, p, n, and δ is a constant in the definition of $N_p^{l-1/p}$.

Proof. Clearly, (3) implies $N_p^{l-1/p}$. In order to obtain the converse, it is sufficient to derive the estimate

$$\|\nabla(\eta_\varepsilon \varphi); R^{n-1}\|_{MW_p^{l-1-1/p}} \leq c\, \|\nabla \varphi; R^{n-1}\|_{MW_p^{l-1-1/p}}. \tag{4}$$

Let Φ be an extension of φ, defined in 6.5.1. By Theorem 5.1.4,

$$|\nabla(\eta_\varepsilon \varphi); R^{n-1}\|_{MW_p^{l-1-1/p}} \leq c\, \|\nabla(\eta_\varepsilon \Phi); R_+^n\|_{MW_p^{l-1}}. \tag{5}$$

For any function $u \in W_p^{l-1}(R_+^n)$, we have

$$\|u\, \nabla(\eta_\varepsilon \Phi); R_+^n\|_{W_p^{l-1}} \leq c\Big(\sum_{j=0}^{l-1} \|\Phi\, |\nabla_{j+1} \eta_\varepsilon|\, |\nabla_{l-1-j} u|; R_+^n\|_{L_p} + \sum_{j=0}^{l-1} \sum_{k=0}^{j} \||\nabla_{j+1-k} \Phi|\, |\nabla_k \eta_\varepsilon|\, |\nabla_{l-1-j} u|; R_+^n\|_{L_p} \Big). \tag{6}$$

Let at first $p(l-1) < n$. The first sum in the right-hand side of (6) is dominated by

$$c\, \|\nabla \Phi; R_+^n\|_{L_\infty} \sum_{j=0}^{l-1} \|r^{-j}\, \nabla_{l-1-j} u; R_+^n\|_{L_p}$$

and the second one is not more than

$$\sum_{j=0}^{l-1} \sum_{k=0}^{j} \|\nabla_{j+1-k} \Phi; R_+^n\|_{M(W_p^{j-k} \to L_p)}\, \|r^{-k}\, \nabla_{l-1-j} u; R_+^n\|_{W^{j-k}}.$$

From Lemma 3.8.2/2 and the inclusion $MW_p^s(R_+^n) \subset MW_p^t(R_+^n)$, $s > t$, it follows that

$$\|\nabla_{j+1-k}\Phi; R_+^n\|_{M(W_p^{j-k} \to L_p)} \leqslant c\,\|\nabla\Phi; R_+^n\|_{MW_p^{j-k}} \leqslant c\,\|\nabla\Phi; R_+^n\|_{MW_p^{l-1}}. \tag{7}$$

Moreover, by Hardy's inequality

$$\|r^{-k}\,\nabla_{l-1-j}u; R_+^n\|_{W_p^{j-k}} \leqslant c\,\|u; R_+^n\|_{W_p^{l-1}}$$

with $p(l-1) < n$, $l-1 \geqslant j \geqslant k$, we obtain

$$\|\nabla(\eta_\varepsilon\Phi); R_+^n\|_{MW_p^{l-1}} \leqslant c\,\|\nabla\Phi; R_+^n\|_{MW_p^{l-1}}. \tag{8}$$

This, together with (6.5.1/2) and (5), implies (4) for $p(l-1) < n$.

Let now $p(l-1) = n$. The first sum in the right-hand side of (6) does not exceed

$$c\,\|\nabla\Phi; R_+^n\|_{L_\infty}\Big(|\log\varepsilon|^{-1}\sum_{j=0}^{l-2}\|r^{-j}\,\nabla_{l-1-j}u; R_+^n\|_{L_p}$$
$$+\|r^{1-l}(\log r)^{-1}u; B_{1/2}\cap R_+^n\|_{L_p}\Big),$$

and the second one is not more than

$$\sum_{j=0}^{l-1}\sum_{k=0}^{j}\|\nabla_{j+1-k}\Phi; R_+^n\|_{M(W_p^{j-k}\to L_p)}\,\|r^{-k}(\log r)^{-1}\eta_{1/2}\,\nabla_{l-1-j}u; R_+^n\|_{W_p^{j-k}}.$$

Using Hardy's inequality

$$\|r^{-k}(\log r)^{-1}v; R_+^n\|_{W_p^{j-k}} \leqslant c\,\|v; R_+^n\|_{W_p^j},$$

where v is a function with support in $B_{1/2}$, we obtain

$$\|r^{-k}(\log r)^{-1}\eta_{1/2}\,\nabla_{l-1-j}u; R_+^n\|_{W_p^{j-k}} \leqslant c\,\|u; R_+^n\|_{W_p^{l-1}},$$

which, together with (7), gives (8). This, together with (6.5.1/2) and (5), leads to (4) for $p(l-1) = n$. □

7.8.2 Estimate for s_1

Clearly, if (7.7/1) is dominated by $c\delta$, then (7.7/2) holds. So we must prove the sufficiency of (7.7/2). According to Remark 3.2.7/1, the condition (7.8.1/3) means that

$$\sup_e \frac{\|D_{p,l-1/p}(\eta_\varepsilon\varphi); e\|_{L_p}}{[\mathrm{cap}\,(e; W_p^{l-1-1/p}(R^{n-1})]^{1/p}} + \|\nabla(\eta_\varepsilon\varphi); B_\varepsilon\|_{L_\infty} < c\delta \tag{1}$$

for sufficiently small $\varepsilon > 0$. Hereafter, e is a compact in R^{n-1} with $d(e) < 1$. Since $\varphi(0) = 0$, then (7.8.1/1) implies $\|\nabla(\eta_\varepsilon\varphi); B_\varepsilon\|_{L_\infty} \leqslant c\,\|\nabla\varphi; B_\varepsilon\|_{L_\infty}$.

In (1) the first summand is majorized by

$$\sup_{e\subset R^{n-1}}\left(\frac{\|D_{p,l-1/p}(\eta_\varepsilon\varphi);e\backslash B_e\|_{L_p}}{[\operatorname{cap}(e\backslash B_\varepsilon,W_p^{l-1-1/p}(R^{n-1}))]^{1/p}}\right.$$
$$\left.+\frac{\|D_{p,l-1/p}(\eta_\varepsilon\varphi);e\cap B_\varepsilon\|_{L_p}}{[\operatorname{cap}(e\cup B_\varepsilon,W_p^{l-1-1/p}(R^{n-1}))]^{1/p}}\right) \tag{2}$$

(if $e\backslash B_\varepsilon$ or $e\cap B_\varepsilon$ has zero capacity, then the corresponding summand is equal to zero). Consequently, the supremum in (1) is not more than $s_1+s_2+s_3$, where

$$s_1=\sup_{e\subset R^{n-1}\backslash B_\varepsilon}\frac{\|D_{p,l-1/p}(\eta_\varepsilon\varphi);e\|_{L_p}}{[\operatorname{cap}(e,W_p^{l-1-1/p}(R^{n-1}))]^{1/p}},$$

$$s_2=\sup_{e\subset B_\varepsilon}\frac{(\int_e dx\int_{R^{n-1}\backslash B_e}|\nabla_{l-1}(\eta_\varepsilon\varphi)(x)-\nabla_{l-1}(\eta_\varepsilon\varphi)(y)|^p\,|x-y|^{-n+2-p}\,dy)^{1/p}}{[\operatorname{cap}(e,W_p^{l-1-1/p}(R^{n-1})]^{1/p}},$$

$$s_3=\sup_{e\subset B_\varepsilon}\frac{\|D_{l-1/p}(\eta_\varepsilon\varphi;B_\varepsilon);e\|_{L_p}}{[\operatorname{cap}(e,W_p^{l-1-1/p}(R^{n-1}))]^{1/p}}.$$

The aim of this subsection is to give an estimate for s_1.

Lemma. *If (7.7/2) holds, then $s_1\leqslant c\delta$ for sufficiently small ε.*

Proof. We have

$$s_1^p=\sup_{e\subset R^{n-1}\backslash B_\varepsilon}\frac{\int_{B_\varepsilon}|\nabla_{l-1}(\eta_\varepsilon\varphi)(y)|^p\,dy\int_e|x-y|^{2-n-p}\,dx}{\operatorname{cap}(e,W_p^{l-1-1/p}(R^{n-1}))}.$$

Let $q=(n-1)/(p(l-1)-1)$ if $p(l-1)<n$, and let $q\in[1,\infty)$ if $p(l-1)=n$. Since $y\in\operatorname{supp}\eta_\varepsilon\subset B_{\varepsilon/2}$, then

$$\int_e|x-y|^{2-n-p}\,dx\leqslant(\operatorname{mes}_{n-1}e)^{1-1/q}\left(\int_{R^{n-1}\backslash B_\varepsilon}|x-y|^{(2-n-p)q}\,dx\right)^{1/q}$$
$$\leqslant c(\operatorname{mes}_{n-1}e)^{1-1/q}\varepsilon^{2-n-p+(n-1)/q}. \tag{3}$$

We use the inequalities

$$\operatorname{cap}(e,W_p^{l-1-1/p}(R^{n-1}))\geqslant\begin{cases}c(\operatorname{mes}_{n-1}e)^{1-1/q} & \text{if}\quad p(l-1)<n,\\ c(\log(2^n/\operatorname{mes}_{n-1}e))^{1-p} & \text{if}\quad p(l-1)=n.\end{cases} \tag{4}$$

Then, for $p(l-1)<n$,

$$\int_e|x-y|^{2-n-p}\,dx\leqslant c\varepsilon^{p(l-2)+1-n}(\operatorname{mes}_{n-1}e)^{1-1/q}.$$

In the case $p(l-1)=n$, we have

$$(\log(2^n/\text{mes}_{n-1}\,e))^{p-1}\int_e |x-y|^{2-n-p}\,dx$$
$$\leqslant c(\text{mes}_{n-1}\,e)^{1-1/q}(\log(2^n/\text{mes}_{n-1}\,e))^{p-1}\varepsilon^{2-n-p+(n-1)/q}. \tag{5}$$

If $\text{mes}_{n-1}\,e \leqslant \varepsilon^{n-1}$, then the right-hand side is dominated by $c\varepsilon^{1-p}\,|\log\varepsilon|^{p-1}$. If $\text{mes}_{n-1}\,e > \varepsilon^{n-1}$, then, setting $q=1$ in (5), we obtain the same majorant $c\varepsilon^{1-p}\,|\log\varepsilon|^{p-1}$. Thus

$$s_1^p \leqslant \begin{cases} c\varepsilon^{p(l-2)+1-n}\displaystyle\int_{B_\varepsilon} |\nabla_{l-1}(\eta_\varepsilon\varphi)|^p\,dy & \text{if } p(l-1)<n,\\ c\varepsilon^{1-p}\,|\log\varepsilon|^{p-1}\displaystyle\int_{B_\varepsilon} |\nabla_{l-1}(\eta_\varepsilon\varphi)(y)|^p\,dy & \text{if } p(l-1)=n. \end{cases}$$

In the case $p(l-1)<n$, this implies

$$s_1^p \leqslant c\varepsilon^{p(l-2)+1-n}\left(\varepsilon^{(1-l)p}\int_{B_\varepsilon}|\varphi|^p\,dy+\sum_{j=0}^{l-2}\varepsilon^{-jp}\int_{B_\varepsilon}|\nabla_{l-1-j}\varphi|^p\,dy\right).$$

We introduce the notation $\langle v; B_\varepsilon\rangle_{p,l-1-1/p} = \|D_{l-1-1/p}(v; B_\varepsilon); B_\varepsilon\|_{L_p}$ and use the inequality

$$\int_{B_\varepsilon}|\nabla_{l-2-j}v|^p\,dy$$
$$\leqslant c\left(\varepsilon^{p(j+1)-1}\langle v; B_\varepsilon\rangle^p_{p,l-1-1/p}+\varepsilon^{p(j+2-l)}\int_{B_\varepsilon}|v|^p\,dy\right). \tag{6}$$

Then

$$s_1^p \leqslant c\left(\varepsilon^{-p+1-n}\int_{B_\varepsilon}|\varphi|^p\,dy+\varepsilon^{1-n}\int_{B_\varepsilon}|\nabla\varphi|^p\,dy+\varepsilon^{p(l-1)-n}\langle\varphi; B_\varepsilon\rangle^p_{p,l-1/p}\right)$$
$$\leqslant c(\varepsilon^{p(l-1)-n}\langle\varphi; B_\varepsilon\rangle^p_{p,l-1/p}+\|\nabla\varphi; B_\varepsilon\|^p_{L_\infty}),$$

which, together with (7.7/2), gives $s_1 \leqslant c\delta$.

Let now $p(l-1)=n$. In the case $l=2$, we have

$$s_1^p \leqslant c\varepsilon^{1-p}\,|\log\varepsilon|^{p-1}\,\|\nabla\varphi; B_\varepsilon\|^p_{L_\infty}\left(\int_{B_\varepsilon}|\eta_\varepsilon|^p\,dy+\int_{B_\varepsilon}|y|^p\,|\nabla\eta_\varepsilon|^p\,dy\right).$$

Since $\int_{B_\varepsilon}|\eta_\varepsilon|^p\,dy \leqslant c\int_{B_\varepsilon}|y|^p\,|\nabla\eta_\varepsilon|^p\,dy \leqslant c\varepsilon^{p-1}\,|\log\varepsilon|^{-p}$, then $s_1^p \leqslant c\,|\log\varepsilon|^{-1}\,\|\nabla\varphi; B_\varepsilon\|^p_{L_\infty}$.

Suppose $l>2$, $p(l-1)=n$. We write

$$s_1^p \leqslant c\varepsilon^{1-p}\,|\log\varepsilon|^{p-1}\Big(|\log\varepsilon|^{-p}\int_{B_\varepsilon}|y|^{(1-l)p}\,|\varphi|^p\,dy$$

$$+\sum_{j=1}^{l-2}|\log\varepsilon|^{-p}\int_{B_\varepsilon}|y|^{-jp}\,|\nabla_{l-1-j}\varphi|^p\,dy+\int_{B_\varepsilon}|\eta_\varepsilon\,\nabla_{l-1}\varphi|^p dy\Big). \tag{7}$$

The first summand in brackets does not exceed

$$c\varepsilon^{p-1}\,|\log\varepsilon|^{-p}\,\|\nabla\varphi;B_\varepsilon\|_{L_\infty}^p. \tag{8}$$

Using the inequality

$$\int_{B_\varepsilon}|y|^{-jp}\,|\nabla_{l-2-j}v|^p\,dy\leqslant c\Big(\int_{B_\varepsilon}|\nabla_{l-2}v|^p\,dy+\varepsilon^{p(2-l)}\int_{B_\varepsilon}|v|^p\,dy\Big)$$

with $v=\partial\varphi/\partial y_i$, we conclude that the sum with respect to j in (7) is majorized by

$$c\varepsilon^{p-1}\,|\log\varepsilon|^{-p}\Big(\varepsilon^{1-p}\int_{B_\varepsilon}|\nabla_{l-1}\varphi|^p\,dy+\|\nabla\varphi;B_\varepsilon\|_{L_\infty}^p\Big). \tag{9}$$

We apply (6) with $j=0$ to the vector-function $v=\nabla\varphi$. Then (9) is not more than

$$c\varepsilon^{p-1}\,|\log\varepsilon|^{-p}\,(\langle\varphi;B_\varepsilon\rangle_{p,l-1/p}^p+\|\nabla\varphi;B_\varepsilon\|_{L_\infty}^p). \tag{10}$$

Now we turn to the estimate for the last integral in (7). The inequality $\|w;B_\varepsilon\|_{L_p}\leqslant c\varepsilon^{1-1/p}\langle w;B_\varepsilon\rangle_{p,1-1/p}$ holds for all w defined on B_ε and vanishing outside $B_{(1-c)\varepsilon}$, $c\in(0,1)$. Hence,

$$\varepsilon^{1-p}\int_{B_\varepsilon}|\eta_\varepsilon\,\nabla_{l-1}\varphi|^p\,dy\leqslant c\langle\nabla_{l-1}(\eta_\varepsilon\varphi);B_\varepsilon\rangle_{p,1-1/p}^p$$

$$\leqslant c\Big(\langle\nabla_{l-1}\varphi;B_\varepsilon\rangle_{p,1-1/p}^p$$

$$+\int_{B_\varepsilon}|\nabla_{l-1}\varphi(x)|^p\,dx\int\frac{|\eta_\varepsilon(x)-\eta_\varepsilon(y)|^p}{|x-y|^{n-2+p}}\,dy\Big). \tag{11}$$

According to Lemma 7.8.1/1,

$$\int\frac{|\eta_\varepsilon(x)-\eta_\varepsilon(y)|^p}{|x-y|^{n-2+p}}\,dy\leqslant c\,|\log\varepsilon|^{-p}\,|x|^{1-p}.$$

Therefore,

$$\langle \nabla_{l-1}(\eta_\varepsilon \varphi); B_\varepsilon \rangle^p_{p,1-1/p} \leq c\left(\langle \nabla_{l-1}\varphi; B_\varepsilon \rangle^p_{p,1-1/p} + |\log \varepsilon|^{-p} \int_{B_\varepsilon} |\nabla_{l-1}\varphi(y)|^p \, |y|^{1-p} \, dy \right).$$

Since $p(l-1)=n$, $l>2$, then $p<n$ and Hardy's inequality

$$\int_{B_\varepsilon} |v|^p \, |y|^{1-p} \, dy \leq c\left(\langle v; B_\varepsilon \rangle^p_{p,1-1/p} + \varepsilon^{1-p} \int_{B_\varepsilon} |v|^p \, dy \right)$$

holds. Setting here $\nabla_{l-1}\varphi$ instead of v, we get

$$\langle \nabla_{l-1}(\eta_\varepsilon \varphi); B_\varepsilon \rangle^p_{p,1-1/p} \leq c\left(\langle \varphi; B_\varepsilon \rangle^p_{p,l-1/p} + \varepsilon^{1-p} \, |\log \varepsilon|^{-p} \int_{B_\varepsilon} |\nabla_{l-1}\varphi(y)|^p \, dy \right).$$

By (6), the last integral does not exceed $\varepsilon^{p-1}(\langle \varphi; B_\varepsilon \rangle^p_{p,l-1/p} + \|\nabla \varphi; B_\varepsilon\|^p_{L_\infty})$ and, thereby,

$$\langle \nabla_{l-1}(\eta_\varepsilon \varphi); B_\varepsilon \rangle^p_{p,1-1/p} \leq c(\langle \varphi; B_\varepsilon \rangle^p_{p,l-1/p} + |\log \varepsilon|^{-p} \, \|\nabla \varphi; B_\varepsilon\|^p_{L_\infty}). \tag{12}$$

Consequently

$$\int_{B_\varepsilon} |\eta_\varepsilon \nabla_{l-1}\varphi|^p \, dy \leq c\varepsilon^{p-1}(\langle \varphi; B_\varepsilon \rangle^p_{p,l-1/p} + |\log \varepsilon|^{-p} \, \|\nabla \varphi; B_\varepsilon\|^p_{L_\infty}). \tag{13}$$

Substituting (8), (10), and (13) into (7), we derive

$$s_1^p \leq c(|\log \varepsilon|^{p-1} \langle \varphi; B_\varepsilon \rangle^p_{p,l-1/p} + \|\nabla \varphi; B_\varepsilon\|^p_{L_\infty})$$

which, together with (7.7/2), gives $s_1 \leq c\delta$. □

7.8.3 Estimate for s_2

Lemma. *If (7.7/2) holds, then $s_2 \leq c\delta$ for sufficiently small ε.*

Proof. Clearly,

$$s_2^p = \sup_{e \subset B_\varepsilon} \frac{\int_e |\nabla_{l-1}(\eta_\varepsilon \varphi)|^p \, dx \int_{R^{n-1} \setminus B_\varepsilon} |x-y|^{2-n-p} \, dy}{\operatorname{cap}(e, W_p^{l-1-1/p}(R^{n-1}))}.$$

Let $p(l-1)<n$. By q we denote any number sufficiently close to p, $q>p$. We have

$$s_2^p \leq c \sup_{e \subset B_\varepsilon} \sum_{j=0}^{l-1} \varepsilon^{-(j+1)p+1} \frac{\int_e |\nabla_{l-1-j}\varphi|^p \, dx}{\operatorname{cap}(e, W_p^{l-1-1/p}(R^{n-1}))}$$
$$\leq c \sum_{j=0}^{l-1} \varepsilon^{-(j+1)p+1} \sup_{e \subset B_\varepsilon} (\operatorname{mes}_{n-1} e)^{1-p/q} \frac{(\int_e |\nabla_{l-1-j}\varphi|^q \, dx)^{p/q}}{\operatorname{cap}(e, W_p^{l-1-1/p}(R^{n-1}))}. \tag{1}$$

From the theorem of Adams stated in Lemma 1.4/1 and the equality $W_p^{l-1}(R^n)|_{R^{n-1}} = W_p^{l-1-1/p}(R^{n-1})$, it follows that

$$\left(\int_{R^{n-1}} |u|^q \, d\mu\right)^{p/q} \leq c \sup_{x\in B_\varepsilon, \rho\in(0,\varepsilon)} \frac{[\mu(B_\rho(x))]^{p/q}}{\operatorname{cap}(B_\rho, W_p^{l-1-1/p}(R^{n-1}))} \|u; R^{n-1}\|^p_{W_p^{l-1-1/p}},$$

where μ is a measure with supp $\mu \subset B_\varepsilon$. Therefore

$$\sup_{e\subset B_\varepsilon} \frac{[\mu(e)]^{p/q}}{\operatorname{cap}(e, W_p^{l-1-1/p}(R^{n-1}))} \leq c \sup_{x\in B_\varepsilon, \rho\in(0,\varepsilon)} \frac{[\mu(B_\rho(x))]^{p/q}}{\operatorname{cap}(B_\rho, W_p^{l-1-1/p}(R^{n-1}))}. \tag{2}$$

This and (1) imply

$$s_2^p \leq c \sum_{j=0}^{l-1} \varepsilon^{-(j+1)p+1+(n-1)(1-p/q)} \times \sup_{x\in B_\varepsilon, \rho\in(0,\varepsilon)} \left(\left(\int_{B_\rho(x)} |\nabla_{l-1-j}\varphi|^q \, dy\right)^{p/q} \Big/ \rho^{n-p(l-1)}\right).$$

Since $-(j+1)p+1+(n-1)(1-p/q) \leq (n-1)(1-p/q)+1-p<0$ for any q sufficiently close to p, then

$$s_2^p \leq c \sum_{j=0}^{l-2} \sup_{x\in B_\varepsilon, \rho\in(0,\varepsilon)} \left(\int_{B_\rho(x)} |\nabla_{l-1-j}\varphi|^q \, dy/\rho^{n-1-q(l-j-2)}\right)^{p/q} + c\,\|\nabla\varphi; B_\varepsilon\|^p_{L_\infty}.$$

We use the inequality

$$\rho^{l-j-2-(n-1)/q}\left(\int_{B_\rho} |\nabla_{l-2-j}v|^q \, dy\right)^{1/q} \leq c\rho^{l-1-1/p-(n-1)/p}\langle v; B_\rho\rangle_{p,l-1-1/p} + c\left(\rho^{1-n}\int_{B_\rho} |v|^p \, dy\right)^{1/p}, \tag{3}$$

$j=0,\ldots,l-2$, which follows by similarity transformation from the continuity of the embedding $W_p^{l-1-1/p}(B_1)$ into $W_q^{l-2-j}(B_1)$. Then

$$s_2^p \leq c \sup_{x\in B_\varepsilon, \rho\in(0,\varepsilon)} \frac{\langle\nabla\varphi; B_\rho(x)\rangle^p_{p,l-1-1/p}}{\rho^{n-1-p(l-1-1/p)}} + c\,\|\nabla\varphi; B_\varepsilon\|^p_{L_\infty}. \tag{4}$$

Let now $p(l-1)=n$. By Hölder's inequality and (2) we have

$$s_2^p \leq c\varepsilon^{1-p} \sup_{e\subset B_\varepsilon} (\operatorname{mes}_{n-1} e)^{1-p/q} \frac{(\int_e |\nabla_{l-1}(\eta_\varepsilon\varphi)|^q \, dx)^{p/q}}{\operatorname{cap}(e, W_p^{l-1-1/p}(R^{n-1}))} \leq c\varepsilon^{1-p+(n-1)(1-p/q)} \sup_{x\in B_\varepsilon, \rho\in(0,\varepsilon)} |\log\rho|^{p-1}\left(\int_{B_\rho(x)} |\nabla_{l-1}(\eta_\varepsilon\varphi)|^q \, dy\right)^{p/q}.$$

Hence

$$s_2^p \leq c \sup_{x\in B_\varepsilon, \rho\in(0\ \varepsilon)} \Bigg(\rho^{1-p+(n-1)(1-p/q)} |\log \rho|^{-1}$$
$$\times \sum_{j=1}^{l-2} \bigg(\int_{B_\rho(x)} |y|^{-jq} |\nabla_{l-1-j}\varphi(y)|^q \, dy\bigg)^{p/q}$$
$$+\rho^{1-p+(n-1)(1-p/q)} |\log \rho|^{-1} \bigg(\int_{B_\rho(x)} |y|^{-(l-1)q} |\varphi(y)|^q \, dy\bigg)^{p/q}$$
$$+\varepsilon^{1-p+(n-1)(1-p/q)} |\log \rho|^{p-1} \bigg(\int_{B_\rho(x)} |\eta_\varepsilon \nabla_{l-1}\varphi|^q \, dy\bigg)^{p/q}\Bigg). \tag{5}$$

We apply the following variant of Hardy's inequality:

$$\bigg(\int_{B_\rho} |y|^{-jq} |\nabla_{l-2-j} v|^q \, dy\bigg)^{1/q}$$
$$\leq c\rho^{1-1/p+(n-1)(1/q-1/p)} \langle v; B_\rho\rangle_{p,l-1-1/p}$$
$$+c\rho^{2-l+(n-1)(1/q-1/p)} \bigg(\int_{B_\rho} |v|^p \, dy\bigg)^{1/p}, \qquad j=1,\ldots,l-2,$$

with $\nabla\varphi$ instead of v. Then the first summand on the right-hand side of (5) is dominated by

$$c\, |\log \rho|^{-1} \left(\langle \varphi; B_\rho(x)\rangle_{p,l-1/p}^p + \|\nabla\varphi; B_\rho(x)\|_{L_\infty}^p\right)$$
$$\leq c\left(\langle \varphi; B_{2\varepsilon}\rangle_{p,l-1/p}^p + \|\nabla\varphi; B_{2\varepsilon}\|_{L_\infty}^p\right).$$

Obviously, the second summand in the right-hand side of (5) does not exceed

$$c\rho^{1-p+(n-1)(1-p/q)} |\log \rho|^{-1} \bigg(\int_{B_\rho(x)} |y|^{-(l-2)q} \, dy\bigg)^{p/q} \|\nabla\varphi; B_\rho(x)\|_{L_\infty}^p$$
$$\leq c\, |\log \rho|^{-1} \|\nabla\varphi; B_\rho(x)\|_{L_\infty}^p.$$

We turn to the estimate of the third summand. Let s be a number sufficiently close to q, $s>q$. By Hölder's inequality the third summand is majorized by

$$\varepsilon^{1-p+(n-1)(1-p/q)} \rho^{(n-1)p(1/q-1/s)} |\log \rho|^{p-1} \bigg(\int_{B_\rho(x)} |\eta_\varepsilon \nabla_{l-1}\varphi|^s \, dy\bigg)^{p/s}$$
$$\leq \varepsilon^{1-p+(n-1)(1-p/s)} |\log \varepsilon|^{p-1} \bigg(\int_{B_\varepsilon} |\eta_\varepsilon \nabla_{l-1}\varphi|^s \, dy\bigg)^{p/s}. \tag{6}$$

The inequality $\|w; B_\varepsilon\|_{L_s} \leq c\varepsilon^{(n-1)(1/s-1/p)+1-1/p} \langle w; B_\varepsilon\rangle_{p,1-1/p}$ holds for all w, defined on B_ε and vanishing outside $B_{(1-c)\varepsilon}$, $c\in(0,1)$. Hence the right-hand side of (6) is not more than $c\, |\log \varepsilon|^{p-1} \langle \eta_\varepsilon \nabla_{l-1}\varphi; B_\varepsilon\rangle_{p,1-1/p}^p$. By

(7.8.2/12) the last summand does not exceed

$$c\,|\log \varepsilon|^{p-1}\,(\langle \varphi; B_\varepsilon\rangle^p_{p,1-1/p}+|\log \varepsilon|^{-p}\,\|\nabla\varphi; B_\varepsilon\|^p_{L_\infty}).$$

Using the estimates which were obtained for the three summands in (5), we arrive at

$$s_2^p\leqslant c\Big(\sup_{x\in B_\varepsilon,\rho\in(0,\varepsilon)} |\log \rho|^{p-1}\,\langle \varphi; B_\rho(x)\rangle^p_{p,l-1/p}+\|\nabla\varphi; B_{2\varepsilon}\|^p_{L_\infty}\Big).\quad\square$$

7.8.4 *Estimate for* s_3

Lemma. *If (7.7/2) holds, then* $s_3\leqslant c\delta$ *for sufficiently small* ε.

Proof. We have

$$\|D_{l-1/p}(\eta_\varepsilon\varphi; B_\varepsilon); e\|^p_{L_p}$$
$$\leqslant c\Big(\sum_{\substack{|\alpha|+|\beta|=l-1\\ |\alpha|>0}} \|D_{1-1/p}(D^\alpha\eta_\varepsilon D^\beta\varphi; B_\varepsilon); e\|^p_{L_p}$$
$$+\|D_{l-1/p}(\varphi; B_\varepsilon); e\|^p_{L_p}+\int_e |\nabla_{l-1}\varphi(x)|^p\,dx\int_{B_\varepsilon}\frac{|\eta_\varepsilon(x)-\eta_\varepsilon(y)|^p}{|x-y|^{n-2+p}}\,dy\Big). \tag{1}$$

Clearly,

$$\|D_{1-1/p}(D^\alpha\eta_\varepsilon D^\beta\varphi; B_\varepsilon); e\|^p_{L_p}$$
$$\leqslant c\Big(\int_e |D^\beta\varphi(x)|^p dx\int_{B_\varepsilon}\frac{|D^\alpha\eta_\varepsilon(x)-D^\alpha\eta_\varepsilon(y)|^p}{|x-y|^{n-2+p}}\,dy$$
$$+\int_{B_\varepsilon} |D^\alpha\eta_\varepsilon(y)|^p dy\int_e\frac{|D^\beta\varphi(x)-D^\beta\varphi(y)|^p}{|x-y|^{n-2+p}}\,dx\Big). \tag{2}$$

Let at first $p(l-1)<n$. Using the definition of η_ε and Lemma 7.8.1/1, we obtain

$$s_3^p\leqslant c\sup_{e\subset B_\varepsilon}\sum_{j=0}^{l-1}\varepsilon^{-(j+1)p+1}\frac{\int_e |\nabla_{l-1-j}\varphi(x)|^p\,dx}{\operatorname{cap}(e, W_p^{l-1-1/p}(R^{n-1}))}$$
$$+c\sup_{e\subset B_\varepsilon}\sum_{j=0}^{l-1}\varepsilon^{-jp}\frac{\int_e dx\int_{B_\varepsilon}(|\nabla_{l-1-j}\varphi(x)-\nabla_{l-1-j}\varphi(y)|^p/|x-y|^{n-2+p})\,dy}{\operatorname{cap}(e, W_p^{l-1-1/p}(R^{n-1}))}. \tag{3}$$

The first supremum is bounded by the right-hand side of (7.8.3/4) (see the beginning of the proof of Lemma 7.8.3).

By q we denote a number, sufficiently close to p, $q>p$. By Hölder's inequality and (7.8.3/2) the second supremum on the right in (3) does not exceed

$$c \sup_{\xi\in B_\varepsilon, \rho\in(0,\varepsilon)} \sum_{j=1}^{l-1} \varepsilon^{-jp+n(1-p/q)} \rho^{p(l-1)-n}$$
$$\times\left[\varepsilon^{(p-q)/p}\int_{B_\rho(\xi)}\left(\int_{B_\varepsilon}\frac{|\nabla_{l-1-j}\varphi(x)-\nabla_{l-1-j}\varphi(y)|^p}{|x-y|^{n-2+p}}dy\right)^{q/p}dx\right]^{p/q}$$
$$+c\sup_{e\subset B_\varepsilon}\frac{\|D_{l-1/p}(\varphi;B_\varepsilon);e\|^p_{L_p}}{\operatorname{cap}(e;W_p^{l-1-1/p}(R^{n-1}))}. \tag{4}$$

We note that

$$\varepsilon^{(p-q)/p}\int_{B_\rho(\xi)}dx\left(\int_{B_\varepsilon}\frac{|\nabla_{l-1-j}\varphi(x)-\nabla_{l-1-j}\varphi(y)|^p}{|x-y|^{n-2+p}}dy\right)^{q/p}$$
$$\leqslant \varepsilon^{(p-q)/p}\int_{B_\rho(\xi)}dx\int_{B_\varepsilon}\frac{|\nabla_{l-1-j}\varphi(x)-\nabla_{l-1-j}\varphi(y)|^q}{|x-y|^{n-2+q}}dy$$
$$\times\left(\int_{B_\varepsilon}\frac{dy}{|x-y|^{n-2}}\right)^{(q-p)/p}$$
$$\leqslant c\left(\int_{B_\rho(\xi)}dx\int_{B_{2\rho}(\xi)}\frac{|\nabla_{l-1-j}\varphi(x)-\nabla_{l-1-j}\varphi(y)|^q}{|x-y|^{n-2+q}}dy\right.$$
$$\left.+\rho^{1-q}\int_{B_\rho(\xi)}|\nabla_{l-1-j}\varphi(x)|^q\,dx+\int_{B_\varepsilon}\frac{|\nabla_{l-1-j}\varphi(y)|^q dy}{|y-\xi|^{q-1}}\right)$$
$$\leqslant c\left(\langle\nabla\varphi;B_{2\rho}(\xi)\rangle^q_{q,l-1-j-1/q}+\int_{B_\varepsilon}\frac{|\nabla_{l-1-j}\varphi(y)|^q dy}{|y-\xi|^{q-1}}\right) \tag{5}$$

with $j=1,\ldots,l-2$. The last expression does not exceed

$$c(\rho^{(1/q-1/p)n+j}\langle\nabla\varphi;B_{2\rho}(\xi)\rangle_{p,l-1-1/p}+\rho^{-l+1+j+n/q}\|\nabla\varphi;B_{2\rho}(\xi)\|_{L_\infty})^q.$$

The item with $j=l-1$ in (4) has the majorant

$$\varepsilon^{-(l-1)p+(n-1)(1-p/q)}\rho^{p(l-1)-n}\|\nabla\varphi;B_{2\varepsilon}\|^p_{L_\infty}\left[\int_{B_\rho(\xi)}dx\left(\int_{B_\varepsilon}\frac{dy}{|x-y|^{n-2}}\right)^{q/p}\right]^{p/q}$$
$$\leqslant c\varepsilon^{1-(l-1)p+(n-1)(1-p/q)}\rho^{p(l-1)-n+(n-1)p/q}\|\nabla\varphi;B_{2\varepsilon}\|^p_{L_\infty}$$
$$\leqslant c\|\nabla\varphi;B_{2\varepsilon}\|^p_{L_\infty}.$$

Hence, the sum over j in (4) is bounded by

$$c\sum_{j=1}^{l-2}\varepsilon^{-jp+n(1-p/q)}\rho^{p(l-1)-n+(p-q)n/q+pj}\langle\varphi;B_{2\rho}(\xi)\rangle^p_{p,l-1/p}$$
$$+c\left(1+\sum_{j=1}^{l-2}\varepsilon^{-jp+n(1-p/q)}\rho^{p(l-1)-n+np/q+p(1+j-l)}\right)\|\nabla\varphi;B_{2\rho}(\xi)\|^p_{L_\infty}.$$

Finally, the second supremum in (3) is dominated by

$$c \sup_{\xi\in B_\varepsilon, \rho\in(0,\varepsilon)} (\rho^{p(l-1)-n}\langle\varphi; B_{2\rho}(\xi)\rangle^p_{p,l-1/p} + \|\nabla\varphi; B_{2\rho}(\xi)\|^p_{L_\infty})$$

$$+ c \sup_{e\subset B_\varepsilon} \frac{\|D_{l-1/p}(\varphi; B_\varepsilon); e\|^p_{L_p}}{\operatorname{cap}(e, W_p^{l-1-1/p}(R^{n-1}))}.$$

Taking into account the estimate for the first supremum in (3), obtained before, we complete the proof for the case $p(l-1)<n$.

Let $p(l-1)=n$. We have

$$\|D_{l-1/p}(\eta_\varepsilon\varphi; B_\varepsilon); e\|^p_{L_p} \leqslant \sigma_1+\sigma_2+\sigma_3+c\,\|D_{l-1/p}(\varphi; B_\varepsilon); e\|^p_{L_p}. \tag{6}$$

where

$$\sigma_1 = c\sum_{j=0}^{l-2}\int_e |\nabla_{l-1-j}\varphi(x)|^p\,dx\int_{B_\varepsilon}\frac{|\nabla_j\eta_\varepsilon(x)-\nabla_j\eta_\varepsilon(y)|^p}{|x-y|^{n-2+p}}\,dy,$$

$$\sigma_2 = c\sum_{j=1}^{l-2}\int_{B_\varepsilon}|\nabla_j\eta_\varepsilon(y)|^p\,dy\int_e\frac{|\nabla_{l-1-j}\varphi(x)-\nabla_{l-1-j}\varphi(y)|^p}{|x-y|^{n-2+p}}\,dx,$$

$$\sigma_3 = c\int_e dx\int_{B_\varepsilon}\frac{|\varphi(x)\,\nabla_{l-1}\eta_\varepsilon(x)-\varphi(y)\,\nabla_{l-1}\eta_\varepsilon(y)|^p}{|x-y|^{n-2+p}}\,dy.$$

By Lemma 7.8.1/1,

$$\int_{B_\varepsilon}\frac{|\nabla_j\eta_\varepsilon(x)-\nabla_j\eta_\varepsilon(y)|^p}{|x-y|^{n-2+p}}\,dy \leqslant c\,|\log|x||^{-p}\,|x|^{1-p(j+1)}.$$

Therefore,

$$\sigma_1 \leqslant c\sum_{j=0}^{l-2}\int_e|\nabla_{l-1-j}\varphi(x)|^p\,|x|^{1-p(j+1)}\,|\log|x||^{-p}\,dx$$

$$\leqslant c\sum_{j=0}^{l-3}\left(\int_{B_\varepsilon}|\nabla_{l-1-j}\varphi(x)|^{p(n-1)/(n-p(j+1))}\,dx\right)^{(n-p(j+1))/(n-1)}$$

$$\times\left(\int_e\frac{dx}{|x|^{n-1}\,|\log|x||^{p(n-1)/(p(j+1)-1)}}\right)^{(p(j+1)-1)/(n-1)}$$

$$+c\,\|\nabla\varphi; B_\varepsilon\|^p_{L_\infty}\int_e|x|^{1-n}\,|\log|x||^{-p}\,dx.$$

The function $t^{n-1}\,|\log t|^\alpha$ increases near $t=0$. So, among all sets e of a fixed mes_{n-1}, the integral $\int_e|x|^{1-n}\,|\log|x||^{-\alpha}\,dx$ attains its maximum at a ball with the centre $x=0$. Consequently, for $\alpha>1$, we have

$$\int_e|x|^{1-n}\,|\log|x||^{-\alpha}\,dx \leqslant c\,|\log \operatorname{mes}_{n-1} e|^{1-\alpha} \tag{7}$$

and hence,

$$\sigma_1 \leqslant c \sum_{j=0}^{l-3} |\log \operatorname{mes}_{n-1} e|^{1-p-(n-p(j+1))/(n-1)} \|\nabla_{l-1-j}\varphi; B_\varepsilon\|^p_{L_{p(n-1)/(n-p(j+1))}}$$
$$+ c\,|\log \operatorname{mes}_{n-1} e|^{1-p} \|\nabla\varphi; B_\varepsilon\|^p_{L_\infty}. \tag{8}$$

This and (7.8.3/3) (with $q = p(n-1)/[n-p(j+1)]$, $\rho = \varepsilon$ and $\nabla\varphi$ as v) yield

$$\sigma_1 \leqslant c\,|\log \operatorname{mes}_{n-1} e|^{1-p}\,(\langle\varphi; B_\varepsilon\rangle^p_{p,l-1/p} + \|\nabla\varphi; B_\varepsilon\|^p_{L_\infty}).$$

We now turn to the estimate for the sum σ_2. Note that

$$\int_e \frac{|\nabla_{l-1-j}\varphi(x) - \nabla_{l-1-j}\varphi(y)|^p}{|x-y|^{n-2+p}}\,dx$$
$$\leqslant c\Big(\int_{\{x\in e:|x|>2|y|\}} |\nabla_{l-1-j}\varphi(x)|^p \frac{dx}{|x|^{n-2+p}}$$
$$+ |\nabla_{l-1-j}\varphi(y)|^p \int_{\{x\in e:|x|>2|y|\}} \frac{dx}{|x|^{n-2+p}}$$
$$+ \int_{\{x\in e:|x|\leqslant 2|y|\}} \frac{|\nabla_{l-1-j}\varphi(x) - \nabla_{l-1-j}\varphi(y)|^p}{|x-y|^{n-2+p}}\,dx\Big).$$

Therefore, $\sigma_2 \leqslant \sigma_2^{(1)} + \sigma_2^{(2)} + \sigma_2^{(3)}$, where

$$\sigma_2^{(1)} = c \sum_{j=1}^{l-2} \int_e |\nabla_{l-1-j}\varphi(x)|^p \frac{dx}{|x|^{n-2+p}} \int_{B_{|x|/2}} |\nabla_j \eta_\varepsilon(y)|^p\,dy,$$
$$\sigma_2^{(2)} = c \sum_{j=1}^{l-2} \int_e \frac{dx}{|x|^{n-2+p}} \int_{B_{|x|/2}} |\nabla_{l-1-j}\varphi(y)|^p\,|\nabla_j \eta_\varepsilon(y)|^p\,dy,$$
$$\sigma_2^{(3)} = c \sum_{j=1}^{l-2} \int_{B_\varepsilon} |\nabla_j \eta_\varepsilon(y)|^p\,dy \int_{\{x\in e:|x|\leqslant 2|y|\}} \frac{|\nabla_{l-1-j}\varphi(x) - \nabla_{l-1-j}\varphi(y)|^p}{|x-y|^{n-2+p}}\,dx.$$

Using the definition of the function η_ε, we obtain

$$\sigma_2^{(1)} \leqslant c \sum_{j=1}^{l-2} \int_e \frac{|\nabla_{l-1-j}\varphi(x)|^p}{|x|^{n-2+p}}\,dx \int_{B_{|x|/2}} |\log|y||^{-p}\,|y|^{-jp}\,dy$$
$$\leqslant c \sum_{j=1}^{l-2} \int_e \frac{|\nabla_{l-1-j}\varphi(x)|^p\,dx}{|x|^{(j+1)p-1}\,|\log|x||^p}.$$

The majorant for this sum was found when estimating σ_1. It is

$$c\,|\log \operatorname{mes}_{n-1} e|^{1-p}\,(\langle\varphi; B_{2\varepsilon}\rangle^p_{p,l-1/p} + \|\nabla\varphi; B_{2\varepsilon}\|^p_{L_\infty}). \tag{9}$$

Clearly,

$$\sigma_2^{(2)} \leq c \sum_{j=1}^{l-1} \int_e \frac{dx}{|x|^{n-2+p}} \int_{B_{|x|/2}} |\nabla_{l-1-j}\varphi(y)|^p \, |y|^{-jp} \, |\log|y||^{-p} \, dy$$

$$\leq c \sum_{j=1}^{l-2} \int_e \frac{dx}{|x|^{n-2+p} \, |\log|x||^p} \int_{B_{|x|/2}} |\nabla_{l-1-j}\varphi(y)|^p \, |y|^{-jp} \, dy.$$

By the inequality

$$\int_{B_{|x|/2}} |\nabla_{l-1-j}\varphi(y)|^p \, |y|^{-jp} \, dy \leq c \, |x|^{p-1} \, (\langle\varphi; B_{|x|/2}\rangle^p_{p,l-1/p} + \|\nabla\varphi; B_{|x|/2}\|^p_{L_\infty})$$

we have

$$\sigma_2^{(2)} \leq c \int_e \frac{dx}{|x|^{n-1} \, |\log|x||^p} \, (\langle\varphi; B_\varepsilon\rangle^p_{p,l-1/p} + \|\nabla\varphi; B_\varepsilon\|^p_{L_\infty}).$$

This, together with (7), yields (9) as the majorant of $\sigma_2^{(2)}$.

The value $\sigma_2^{(3)}$ does not exceed

$$c \sum_{j=1}^{l-2} \int_{B_\varepsilon} \frac{dy}{|y|^{jp} \, |\log|y||^p} \int_{\{x\in e \, : \, |x|\leq 2|y|\}} \frac{|\nabla_{l-1-j}\varphi(x) - \nabla_{l-1-j}\varphi(y)|^p}{|x-y|^{n-2+p}} \, dx.$$

Changing the order of integration and using the monotonicity of $t^{jp} \, |\log t|^p$ near $t=0$, we find

$$\sigma_2^{(3)} \leq c \sum_{j=1}^{l-2} \int_e \frac{dx}{|x|^{jp} \, |\log|x||^p} \int_{B_\varepsilon} \frac{|\nabla_{l-1-j}\varphi(x) - \nabla_{l-1-j}\varphi(y)|^p}{|x-y|^{n-2+p}} \, dy.$$

We apply Hölder's inequality and (7). Then, for $q_j = (n-1)/(n-1-jp)$,

$$\sigma_2^{(3)} \leq c \sum_{j=1}^{l-2} \left(\int_e |x|^{1-n} \, |\log|x||^{(1-n)/j} \, dx \right)^{jp/(n-1)}$$

$$\times \left[\int_{B_\varepsilon} \left(\int_{B_\varepsilon} \frac{|\nabla_{l-1-j}\varphi(x) - \nabla_{l-1-j}\varphi(y)|^p}{|x-y|^{n-2+p}} \, dy \right)^{q_j} dx \right]^{1/q_j}$$

$$\leq c \sum_{j=1}^{l-2} |\log \operatorname{mes}_{n-1} e|^{1-p-1/q_j}$$

$$\times \left[\int_{B_\varepsilon} \left(\int_{B_\varepsilon} |\nabla_{l-1-j}\varphi(x+h) - \nabla_{l-1-j}\varphi(x)|^p \frac{dh}{|h|^{n-2+p}} \right)^{q_j} dx \right]^{1/q_j}.$$

Hence, by Minkowski's inequality,

$$\sigma_2^{(3)} \leq c \, |\log \operatorname{mes}_{n-1} e|^{1-p}$$

$$\times \sum_{j=1}^{l-1} \int_{B_\varepsilon} \left(\int_{B_\varepsilon} |\nabla_{l-1-j}\varphi(x+h) - \nabla_{l-1-j}\varphi(x)|^{pq_j} \, dx \right)^{1/q_j} \frac{dh}{|h|^{n-2+p}}.$$

Since $W_p^{l-1-1/p}(R^{n-1}) \subset B_{pq_j,p}^{l-j-1-1/p}(R^{n-1})$ with $j=1,\ldots,l-2$ (see Besov [1] Triebel [4], 2.8.1), then

$$\int_{B_\varepsilon} \left(\int_{B_\varepsilon} |\nabla_{l-2-j} v(x+h) - \nabla_{l-2-j} v(x)|^{pq_j}\, dx \right)^{1/q_j} \frac{dh}{|h|^{n-2+p}}$$

$$\leqslant c(\langle v; B_{3\varepsilon} \rangle^p_{p,l-1-1/p} + \varepsilon^{1-n} \|v; B_{3\varepsilon}\|^p_{L_\infty}).$$

Therefore,

$$\sigma_2^{(3)} \leqslant c\, |\log \mathrm{mes}_{n-1}\, e|^{1-p} \,(\langle \varphi; B_{3\varepsilon}\rangle^p_{p,l-1/p} + \|\nabla\varphi; B_{3\varepsilon}\|^p_{L_\infty}).$$

Taking into account the estimates for $\sigma_2^{(1)}$ and $\sigma_2^{(2)}$ which were obtained before, we find that σ_2 is bounded by the right-hand side of the last inequality.

To obtain an estimate for σ_3 we note that

$$\int_e dx \int_{B_{2|x|}\setminus B_{|x|/2}} |\varphi(x)\, \nabla_{l-1}\eta_\varepsilon(x) - \varphi(y)\, \nabla_{l-1}\eta_\varepsilon(y)|^p \frac{dy}{|x-y|^{n-2+p}}$$

$$\leqslant c\, \|\nabla\varphi; B_\varepsilon\|^p_{L_\infty} \int_e \frac{dx}{|x|^{(l-1)p}\, |\log|x||^p} \int_{B_{2|x|}} \frac{dy}{|x-y|^{n-2}}$$

$$\leqslant c\, \|\nabla\varphi; B_\varepsilon\|^p_{L_\infty} \int_e \frac{dx}{|x|^{n-1}\, |\log|x||^p}. \tag{10}$$

Moreover,

$$\int_e dx \int_{B_{|x|/2}} |\varphi(x)\, \nabla_{l-1}\eta_\varepsilon(x) - \varphi(y)\, \nabla_{l-1}\eta_\varepsilon(y)|^p \frac{dy}{|x-y|^{n-2+p}}$$

$$\leqslant c\, \|\nabla\varphi; B_\varepsilon\|^p_{L_\infty} \int_e dx \int_{B_{|x|/2}} \frac{(|x|^{2-l}\, |\log|x||^{-1} + |y|^{2-l}\, |\log|y||^{-1})^p}{|x-y|^{n-2+p}}\, dy$$

$$\leqslant c\, \|\nabla\varphi; B_\varepsilon\|^p_{L_\infty} \int_e \frac{dx}{|x|^{n-2+p}} \int_{B_{|x|/2}} \frac{dy}{|y|^{n-p}\, |\log|y||^p}$$

$$\leqslant c\, \|\nabla\varphi; B_\varepsilon\|^p_{L_\infty} \int_e \frac{dx}{|x|^{n-1}\, |\log|x||^p}. \tag{11}$$

In the same way we obtain

$$\int_e dx \int_{R^{n-1}\setminus B_{2|x|}} |\varphi(x)\,\nabla_{l-1}\eta_\varepsilon(x) - \varphi(y)\,\nabla_{l-1}\eta_\varepsilon(y)|^p \frac{dy}{|x-y|^{n-2+p}}$$

$$\leqslant c\,\|\nabla\varphi; B_\varepsilon\|^p_{L_\infty} \int_e dx \int_{R^{n-1}\setminus B_{2|x|}} \frac{(|x|^{2-l}\,|\log|x||^{-1} + |y|^{2-l}\,|\log|y||^{-1})^p}{|x-y|^{n-2+p}}\,dy$$

$$\leqslant c\,\|\nabla\varphi; B_\varepsilon\|^p_{L_\infty} \int_e \frac{dx}{|x|^{(l-2)p}\,|\log|x||^p} \int_{R^{n-1}\setminus B_{2|x|}} \frac{dy}{|y|^{n-2+p}}$$

$$\leqslant c\,\|\nabla\varphi; B_\varepsilon\|^p_{L_\infty} \int_e \frac{dx}{|x|^{n-1}\,|\log|x||^p}. \tag{12}$$

Summing estimates (10)–(12) and applying (7), we come to

$$\sigma_3 \leqslant c\,|\log \text{mes}_{n-1}\, e|^{1-p}\,\|\nabla\varphi; B_\varepsilon\|^p_{L_\infty}.$$

Now from (6) and the estimates for σ_1, σ_2, σ_3 it follows that

$$\|D_{l-1/p}(\eta_\varepsilon\varphi; B_\varepsilon); e\|^p_{L_p} \leqslant c\,|\log \text{mes}_{n-1}\, e|^{1-p}\,(\langle\varphi; B_{3\varepsilon}\rangle^p_{p,l-1/p}$$
$$+\|\nabla\varphi; B_{3\varepsilon}\|^p_{L_\infty}) + c\,\|D_{l-1/p}(\varphi; B_\varepsilon); e\|^p_{L_p}.$$

This, together with (4), gives the estimate

$$s_3^p \leqslant c\Big(\langle\varphi; B_{3\varepsilon}\rangle^p_{p,l-1/p} + \|\nabla\varphi; B_{3\varepsilon}\|^p_{L_\infty}$$
$$+ \sup_{e\subset B_\varepsilon} \frac{\|D_{l-1/p}(\varphi; B_\varepsilon); e\|^p_L}{\text{cap}\,(e; W_p^{l-1-1/p}(R^{n-1}))}\Big). \quad \square$$

Lemmas 7.8.3–7.8.5, and the inequality (7.8.2/2) imply that (7.8.2/1) holds for sufficiently small ε. As was pointed out at the beginning of Subsection 7.8.2, the latter estimate is equivalent to $N_p^{l-1/p}$. Thus Theorem 7.7.1 is proved.

References

Adams D. R.

[1] A trace inequality for generalized potentials. *Studia Math.*, 1973, **48,** N1, 99–105.

[2] A note on Riesz potentials. *Duke Math. J.*, 1975, **42,** N4, 765–778.

[3] On the existence of capacitary strong type estimates in R^n. *Ark. mat.*, 1976, **14,** 125–140.

[4] Lectures on L^p-potential theory. Univ. of Umeå, Dpt. of Math., Preprint N2, 1981, p. 1–74.

Adams D. R., Hedberg L. I.

[1] Inclusion relations among fine topologies in non-linear potential theory. *Reports Dpt. of Math., Univ. of Stockholm*, 1982, N10, p. 1–15.

Adams D. R., Meyers N. G.

[1] Bessel potentials. Inclusion relations among classes of exceptional sets. *Indiana Univ. Math. J.*, 1973, **22,** N9, 873–905.

Adams D. R., Polking J. C.

[1] The equivalence of two definitions of capacity. *Proc. Amer. Math. Soc.*, 1973, **37,** 529–534.

Agmon S., Douglis A., Nirenberg L.

[1] Estimates near the boundary for the solutions of elliptic equations satisfying general boundary values, I. *Comm. Pure Appl. Math.* 1959, **12,** 623–727.

Ahlfors L., Beurling A.

[1] Conformal invariants and function-theoretic null-sets. *Acta Math.*, 1950, **83,** 101–129.

Aronszajn, N., Mulla F., Szeptycki P.

[1] On spaces of potentials connected with L^p-spaces. *Ann. Inst. Fourier*, 1963, **13,** 211–306.

Bennet C., Gilbert J. E.

[1] Homogeneous algebras on the circle: II. Multipliers, Ditkin conditions. *Ann. Inst. Fourier*, 1972, **22,** N3, 21–50.

Besov O. V.

[1] Investigation of a family of function spaces in connection with imbedding and extension theorems. *Trudy Mat. Inst. Steklov*, 1961, **60,** 42–81.

Besov O. V., Il'in V. P. Nikol'skii S. M.

[1] Integral representation of functions and imbedding theorems. Nauka, Moscow, 1975.

Beurling A.

[1] Analysis in some convolution algebras. Symposium on harmonic analysis and related integral transforms. Cornell, 1956.

Bliev N. K.

[1] On products of functions in Nikol'skii–Besov spaces. *Izv. AN. Kazach. SSR, ser. phis.-mat.*, 1979, N5, 69–71.

[2] Homeomorphisms of Beltrami equation in fractional spaces. "Differential and integral equations. Boundary value problems", Tbilisi, 1979, 33–43.

Burago Yu. D., Maz'ya V. G.

[1] Certain problems in potential and function theory for domains with irregular boundaries. *Zap. Naučn. Sem. LOMI*, 1967, **3,** 1–152.

Calderon A. P.

[1] Lebesgue spaces of differentiable functions and distributions. *Proc. Sympos. Pure Math.*, 1961, **4,** 33–49.

[2] Commutators of singular integral operators. *Proc. Nat. Acad. Sci. USA*, 1965, **53,** 1092–1099.

Campanato S.

[1] Proprietà di hölderianità di alcune classi di funzioni. *Ann. Scuola Norm. Sup. Pisa*, 1963, **17,** 175–188.

Carleson L.

[1] Interpolation of bounded analytic functions and the corona problem. *Ann. Math.*, 1962, **76,** 547–559.

[2] Selected problems on exceptional sets. Van Nostrand Company, 1967.

Dahlberg B.

[1] Regularity properties of Riesz potentials. *Indiana Univ. Math. J.*, 1979, **28,** N2, 257–268.

Devinatz A., Hirschman I. I.

[1] Multiplier transformations on $l^{2,\alpha}$. *Annals of Math.*, 1959, **69,** N3, 575–587.

Eskin G. I.

[1] Boundary value problems for elliptic pseudo-differential equations. Nauka, Moscow, 1973.

Evgrafov M. A.

[1] Asymptotic estimates and entire functions. Nauka, Moscow, 1979.

Federer H.

[1] Curvature measures. *Trans. Amer. Math. Soc.*, 1959, **93,** N3, 418–491.

[2] The area of nonparametric surface. *Proc. Amer. Math. Soc.*, 1960, **11,** N3, 436–439.

Fefferman C. L.

[1] Characterizations of bounded mean oscillation. *Bull. Amer. Math. Soc.*, 1971, **77,** 587–588.

Fraenkel L. E.

[1] Formulae for high derivatives of composite functions. *Math. Proc. Camb. Phil. Soc.*, 1978, **83,** 159–165.

Gagliardo E.

[1] Ulteriori proprietà di alcune classi di funzioni in più variabili. *Ric. Mat.*, 1959, **8,** 24–51.

Gehring F. W.

[1] Lipschitz mappings and p-capacity of rings in n-space. Adv. in the Theory of Riemann surfaces (Proc. Conf., Stony Brook, N.Y. 1969) Ann. Math Studies, Princeton-New York: Univ. Press, 1971, p. 175–193.

Gelfand I. M., Shilov G. E.

[1] Generalized functions. Vol. 1: Operations on them. Fizmatgiz, Moscow, 1958 (Russian). English transl. Academic Press, New York, 1964.

Gelman I. W., Mazja W. G.

[1] Abschätzungen für Differentialoperatoren im Halbraum. Akademie-Verlag, 1981.

Gustin W.

[1] Boxing inequalities. *J. Math. Mech.*, 1960, **9,** 229–239.

Guzman M. de

[1] Covering lemma with applications to differentiability of measures and singular integral operators. *Studia Math.*, 1970, **34,** N3, 299–317.

Hansson K.

[1] Imbedding theorems of Sobolev type in potential theory. *Math. Scand.*, 1979, **45,** 77–102.

Hedberg L. I.

[1] Non-linear potentials and approximation in the mean by analytic functions. *Math. Zeitschr.*, 1972, **129,** 299–319.

[2] Approximation in the mean by solutions of elliptic equations, *Duke Math. J.* 1973, **40,** N1, p. 9–16.

[3] On certain convolution inequalities. *Proc. Amer. Math. Soc.*, 1972, **36,** N2, 505–510.

Herz C. S.

[1] Lipschitz spaces and Bernstein's theorem on absolutely convergent Fourier transforms. *J. Math. Mech.*, 1968, **18,** N4, 283–323.

Hirschman I. I.

[1] On multiplier transformations. II. *Duke Math. J.*, 1961, **28,** 45–56.

[2] On multiplier transformations. III. *Proc. Amer. Math. Soc.*, 1962, **13,** 851–857.

Hörmander L.

[1] Linear partial differential operators. Springer Verlag, 1963.

Janson S.

[1] On functions with conditions on the mean oscillation. *Ark. mat.*, 1976, **14,** N2, 189–196.

[2] Mean oscillation and commutators of singular integral operators. *Ark. mat.*, 1978, **16,** N2, 263–270.

John F., Nirenberg L.

[1] On functions of bounded mean oscillation. *Comm. Pure Appl. Math.*, 1961, **14,** 415–426.

Johnson R.

[1] Multipliers in H^p-spaces. Technical report. Washington, 1977.

[2] Maximal subspaces of Besov spaces invariant under multiplication by characters. Technical report, College Park, 1975.

Kalyabin G. A.

[1] Conditions for multiplicative property of Besov and Lizorkin–Triebel function spaces. *Dokl. Akad. Nauk SSSR*, 1980, **251,** N1, 25–26 (Russian).

Kohn J. J., Nirenberg L.

[1] An algebra of pseudo-differential operators. *Comm. Pure Appl. Math.*, 1965, **18,** 269–305.

Kondrat'ev V. A.

[1] Boundary value problems for elliptic equations in domains with conical or angular points. *Trudy Mosk. Matem. Obšč.*, 1967, **16,** 209–292.

[2] On the smoothness of the solution to the Dirichlet problem for elliptic second order equations in a piecewise-smooth domain. *Diff. Uravnenija*, 1970, **6,** N10, 1831–1843.

Kondrat'ev V. A., Eidelman S. D.

[1] On conditions on the boundary surface in the theory of elliptic boundary value problems. *Dokl. Akad. Nauk SSSR*, 1979, **246,** N4, 812–815.

Krasnosel'skii M. A., Zabreyko P. P., Pustyl'nik E. I., Sobolevskii P. E.

[1] Integral operators in the space of summable functions. Nauka, Moscow, 1966.

Kronrod A. S.

[1] On functions of two variables. *Uspehi Matem. Nauk*, 1950, **5,** N1, 24–134.

Lewis, L. C.

[1] Quasiconformal mapping and Royden algebras in space. *Trans. Amer. Math. Soc.*, 1971, **158,** N2, 481–496.

Lizorkin P. I.

[1] Boundary properties of functions from "weighted" classes. *Dokl. Akad. Nauk SSSR*, 1960, **132,** N3, 514–517.

[2] On function characteristics of interpolation spaces ($L_p(\Omega)$, $W_p^1(\Omega))_{\theta,p}$. *Trudy Mosk. Matem. Inst.* 1975, **134,** 180–203.

Lions J.-L., Magenes E.

[1] Problèmes aux limits non homogènes, IV. *Ann. Scuola. Norm. Sup. Pisa*, 1961, **15,** 311–326.

Maz'ya V. G.

[1] On the negative spectrum of multidimensional Schrödinger operator. *Dokl. Akad. Nauk. SSSR*, 1962, **144,** N4, 721–722.

[2] On the theory of multidimensional Schrödinger operator. *Izvestia AN SSSR, ser. mat.*, 1964, **28,** N5, 1145–1172.

[3] On weak solutions of the Dirichlet and the Neumann problems. *Trudy Mosk. Matem. Obšč.*, 1969, **20,** 137–172.

[4] On some integral inequalities for functions of several variables. *Problemi Mat. Anal., Leningrad Univ.*, 1972, N3, 33–69.

[5] On (p, l)-capacity, imbedding theorems and the spectrum of a self-adjoint elliptic operator. *Izvestia AN SSSR, ser. mat.*, 1973, **37,** N2, 356–385.

[6] On the coercivity of the Dirichlet problem in a domain with irregular boundary. *Izv. Vysš. Uč. Zaved., Matem.*, 1973, N4, 64–76.

[7] On the local square summability of convolution. *Zap. Naučn. Sem. LOMI*, 1977, **73,** 211–216.

[8] On capacitary strong type estimates for fractional norms. *Zap. Naučn. Sem. LOMI*, 1977, **70,** 161–168.

[9] Multipliers in Sobolev spaces. Application of methods of function theory and functional analysis to problems of mathematical physics, Oct. 1976. Novosibirsk 1978, 181–189.

[10] On summability with respect to an arbitrary measure of functions in Sobolev–Slobodezkii spaces. *Zap. Naučn. Sem. LOMI*, 1979, **92,** 192–202.

[11] Einbettungssätze für Sobolewsche Räume, Leipzig, Teubner-Verlag, Teil 1, 1979; Teil 2, 1980.

[12] On an imbedding theorem and multipliers in pairs of Sobolev spaces. *Trudy Tbilis. Matem. Inst.*, 1980, **66,** 59–69.

[13] On connection of two kinds of capacity. *Vestnik Leningrad. Univ.*, 1974, N7, 33–40.

[14] Zur Theorie Sobolewscher Räume. Leipzig, Teubner-Verlag. 1981.

Maz'ya V. G., Havin V. P.

[1] Non-linear analog of Newton potential and metric properties of (p, l)-capacity. *Dokl. Akad. Nauk SSSR*, 1970, **194,** N4, 770–773.

[2] Non-linear potential theory. *Uspehi Matem. Nauk*, 1972, **27,** N6, 67–138.

Maz'ya V. G., Plamenevskii B. A.

[1] Elliptic boundary value problems on manifolds with singularities. *Problemi Mat. Anal. Leningrad Univ.* 1977, **6,** 85–142.

[2] Estimates in L_p and in Hölder classes and the Miranda–Agmon maximum principle for solutions of elliptic boundary value problems in domains with singular points at the boundary. *Math. Nachr.*, 1978, **81**, 25–82.

[3] L_p-estimates of solutions of elliptic boundary value problems in domains with edges. *Trudy Mosk. Matem. Obšč.*, 1978, **37,** 49–94.

Maz'ya V. G., Preobraženskii S. P.

[1] On estimates for (l, p)-capacities and on traces of potentials. *Wissenschaftliche Informationen, Technische Hochschule Karl-Marx-Stadt*, N28, 1981.

Maz'ya V. G., Shaposhnikova T. O.

[1] On multipliers in function spaces with fractional derivatives. *Dokl. Akad. Nauk SSSR*, 1979, **244,** N5, 1065–1067.

[2] On multipliers in Sobolev spaces. *Vestnik Leningrad. Univ.*, 1979, N7, 33–40.

[3] On traces and extensions of multipliers in the space W_p^l. *Uspehi Matem. Nauk*, 1979, **34,** N2, 205–206.

[4] Multipliers in spaces of differentiable functions. *Trudy Seminara Soboleva, Novosibirsk*, 1979, N1, 37–90.

[5] Requirements on the boundary in the L_p-theory of elliptic boundary value problems. *Dokl. Akad. Nauk SSSR*, 1980, **251,** N5, 1055–1059.

[6] Multipliers of Sobolev spaces in a domain. *Math. Nachr.*, 1980, **99,** 165–183.

[7] Coercive estimate for solutions of elliptic equations in spaces of multipliers. *Vestnik Leningrad. Univ.*, 1980, N1, 41–51.

[8] The theory of multipliers in spaces of differentiable functions and its applications. *Trudy Seminara Soboleva, Novosibirsk*, 1980, 225–233.

[9] Multipliers on the space $\mathring{W}_p^m$ and their applications. *Vestnik Leningrad. Univ.*, 1981, *N*1, 42–47.

[10] Multipliers in pairs of spaces of differentiable functions. *Trudy Mosk. Matem. Obšč.*, 1981, **43,** 37–80.

[11] Multipliers in pairs of potential spaces. *Math. Nachr.*, 1981, **99,** 363–379.

[12] On regularity of the boundary in L_p-theory of elliptic boundary value problems. Akad. Wissenschaften DDR, Zentralinst. Math. und Mech., Berlin. Preprint P-28/80.

[13] On sufficient conditions for belonging to classes of multipliers. *Math. Nachr.*, 1981, **100,** 151–162.

Meyers N. G.

[1] A theory of capacities for potentials of functions in Lebesgue classes. *Math. Scand.*, 1970, **26,** 255–292.

Mikhlin S. G.

[1] Multidimensional singular integrals and integral equations. Fizmatgiz, Moscow, 1962.

Michlin S. G., Prössdorf S.

[1] Singuläre Integraloperatoren. Akademie-Verlag, Berlin, 1980.

Muckenhoupt B.

[1] Weighted norm inequalities for the Hardy maximal function. *Trans. Amer. Math. Soc.*, 1972, **165,** 207–226.

Nečas J.

[1] Les méthodes directes en théorie des équations elliptiques. Academia, Prague, 1967.

Nikol'skii N. K.

[1] Lectures on the shift operator. Nauka, Moscow, 1980.

Nikol'skii S. M.

[1] Approximation of functions of several variables and imbedding theorems. Nauka, Moscow, 1977.

Nirenberg L.

[1] On elliptic partial differential equations (Lecture II). *Ann. Scuola Norm. Sup. Pisa*, 1959, s. **3, 13,** 115–162.

Nykodim O.

[1] Sur une classe de fonctions considérées dans le problème de Dirichlet. *Fundam. Mat.*, 1933, **21,** 129–150.

Peetre J.

[1] New thoughts on Besov spaces. Duke Univ. Math. Ser., Durham, 1976.

Pohožaev S. I.

[1] On eigenfunctions of the equation $\Delta u + \lambda f(u) = 0$, *Dokl. Akad. Nauk SSSR*, 1965, **165,** N1, 36–39.

Polking J. C.

[1] A Leibniz formula for some differentiation operators of fractional order. *Indiana Univ. Math. J.*, 1972, **27,** N11, 1019–1029.

[2] Approximation in L^p by solutions of elliptic partial differential equations, *Amer. Math. J.*, 1972, **94,** p. 1231–1244.

Rešetnjak Yu. G.

[1] Space mappings with bounded distortion. *Sib. Mat. Journ.* 1967, **8,** N3, 629–658.

de Rham G.

[1] Variétés différentiables. Hermann. Paris, 1960.

Schwartz L.

[1] Théorie des distributions. Paris, 1966.

Shamir E.

[1] Une propriété des espaces $H^{s,p}$. *C.R. Acad. Sci. Paris, Ser. A-B*, 1962, **255,** A448–A449.

Sjödin T.

[1] Capacities of compact sets in linear subspaces of R^n. *Pacif. J. of Math.*, 1978, **78,** N1, 261–266.

Sobolev S. L.

[1] Certain applications of functional analysis in mathematical physics. Leningrad Univ., 1955.

Stegenga D. A.

[1] Bounded Toeplitz operators on H^1 and applications of the duality between H^1 and the functions of bounded mean oscillation. *Amer. J. of Math.*, 1976, **98,** N3, 573–589.

[2] Multipliers of the Dirichlet space. *Illinois J. of Math.*, 1980, **24,** 113–139.

Stein E. M.

[1] Singular integrals and differentiability properties of functions. Princeton Univ. Press, Princeton, 1970.

[2] The characterization of functions arising as potentials. *Bull. Amer. Math. Soc.*, 1961, **67,** 102–104.

Strichartz R. S.

[1] Multipliers on fractional Sobolev spaces. *J. Math. and Mech.*, 1967, **16,** N9, 1031–1060.

Szeptycki P.

[1] Extensions by mollifiers in Besov spaces. *Studia Math.*, 1975, **54,** N1, 55–72.

Treves J. E.

[1] Lectures on linear partial differential equations with constant coefficients. Rio de Janeiro, 1961.

Triebel H.

[1] Multiplication properties of the spaces $B^s_{p,q}$ and $F^s_{p,q}$. Quasi-Banach algebras of functions. *Ann. Mat. Pura Appl.*, 1977, **113,** N4, 33–42.

[2] Multiplication properties of Besov spaces. *Ann. Mat. Pura Appl.*, 1977, **114,** N4, 87–102.

[3] Spaces of Besov–Hardy–Sobolev type. Teubner-Texte Math. Leipzig, 1978.

[4] Interpolation theory. Function spaces. Differential operators. VEB Deutscher Verlag der Wiss., Berlin, 1978.

Trudinger N. S.

[1] On imbeddings into Orlicz spaces and some applications. *J. Math. Mech.*, 1967, **17,** 473–483.

Uspenskii S. V.

[1] On imbedding theorems for weighted classes. *Trudy Matem. Inst. AN SSSR*, 1961, **60,** 282–303.

Verbitskii I. E.

[1] Inner functions as multipliers of the space D_α. *Funk. Anal. i Pril.*, 1982, **16,** N3, 47–48.

[2] Inner functions in the theory of multipliers of spaces with fractional norms. Preprint, Inst. of Geophysics and Geology. AN Mold., SSR, Kishinev 1983, p. 1–44.

Veržbinskii G. M., Maz'ya V. G.

[1] On the asymptotics of the solution of the Dirichlet problem near irregular boundary. *Dokl. Akad. Nauk. SSSR*, 1967, **176,** 498–501.

[2] On the closure in L_p of the operator of the Dirichlet problem in a domain with conic points. *Izv. Vysš. Uč. Zaved. Matem.*, 1974, **6,** 8–19.

Vinogradov C. A.

[1] Free interpolation in spaces of analytic functions. Thesis. Leningrad State Univ., 1982.

Vodopianov S. K., Goldstein V. M.

[1] Structural isomorphisms of the space W^1_n and quasi-conformal mappings. *Sib. Mat. Journ.*, 1975, **16,** N2, 224–246.

[2] Quasi-conformal mappings and spaces of functions with the first generalized derivatives. *Sib. Mat. Journ.*, 1976, **17,** N3, 515–531.

[3] Functional characteristics of quasi-isometric mappings. *Sib. Mat. Journ.*, 1976, **17,** N4, 768–773.

Vodopianov S. K., Goldstein V. U., Rešetnjak Yu. G.
[1] On geometric properties of functions with généralized first derivatives. *Uspehi. Matem. Nauk*, **34,** N1, 17–65.
Warschawski S. E.
[1] On conformal mapping of infinite strips. *Trans. Amer. Math. Soc.*, 1942, **51,** 280–335.
Yudovič V. I.
[1] On certain estimates connected with integral operators and solutions of elliptic equations. *Dokl. Akad. Nauk. SSSR*, 1961, **138,** N4, 805–808.
Zolesio J. L.
[1] Multiplication dans les espaces de Besov. *Proc. Roy. Soc. Edinburgh*, 1977/1978, s.A., **78,** N1–2, 113–117.
Zygmund A.
[1] Trigonometric series. Cambridge 1959.

List of Symbols

Function Spaces

Norms

Operators

Functions

Set Functions

Sets in R^n

Classes of Domains

Author Index

Subject Index